Michael Eckert

Arnold Sommerfeld. Science, Life and Turbulent Times

Michael Eckert
translated by Tom Artin

Arnold Sommerfeld

Science, Life and Turbulent Times 1868–1951

Michael Eckert
Deutsches Museum
Munich, Germany

Translation of *Arnold Sommerfeld: Atomphysiker und Kulturbote 1868–1951*, originally published in German by Wallstein Verlag, Göttingen

ISBN 978-1-4614-7460-9 ISBN 978-1-4614-7461-6 (eBook)
DOI 10.1007/978-1-4614-7461-6
Springer New York Heidelberg Dordrecht London

Library of Congress Control Number: 2013939588

Printed on acid-free paper

Springer is part of Springer Science+Business Media (www.springer.com)

Acknowledgment

In the case of such a source-dependent work, first thanks are due Sommerfeld's heirs for access to the private papers and for permission to make full use of their content. Sommerfeld's granddaughter, Monika Baier, deserves particular thanks for ordering her grandfather's private papers and for providing valuable information on the family background. The archivists of the various public institutions in which Sommerfeld correspondence resides, too, are gratefully acknowledged for their readiness to help. The bibliography reveals how much support the project has thereby received. Cited passages from letters and texts from other source materials have been adapted to modern notation for the sake of legibility. Numerous letters cited here only in excerpts are printed in full in editions of correspondence. In these cases, references are given indicating the respective volume of the published correspondence (e.g., ASWB I and ASWB II as abbreviations for the two volumes of *Arnold Sommerfeld – Wissenschaftlicher Briefwechsel, Band I und II*: since in those volumes the letters are printed in chronological order, there is no need to specify page or letter numbers). Thus the reader can study the respective letters in full and in the context of the other correspondence reproduced there. Since Sommerfeld's private letters are not publicly accessible, they are identified only by date, and no source is given.

It is conventional for the published work of a scientist to take precedence over unpublished papers at the beginning of a biographical endeavor. This approach reflects his life's work as a scientist as it has been perceived by his contemporaries. Usually this kind of biographical work begins with the mile-stone birthdays marking decades of a scientist's life.[1] Thus, festschrifts were assembled to celebrate Sommerfeld's 60th, 70th, and 80th birthdays. In 1968, on the occasion of his 100th birthday, the physicists of the University of Munich, organized an "Arnold Sommerfeld Centennial Memorial Meeting," and an "International Symposium on the Physics of the One- and Two-Electron Atoms." On this occasion, his successor at the Institute for Theoretical Physics (Fritz Bopp) was commissioned by the Bavarian Academy of Sciences to arrange for Sommerfeld's most important scientific papers to be published in the form of a four-volume edition.[2] The biographer is gratefully obliged to all those involved in this preliminary work. The editors of the scientific correspondence (the author and Karl Märker) had the support of the Munich Physics Department (especially Harald Fritzsch and Herbert Wagner), and the Bavarian Academy of Sciences (in the person of Past President Arnulf Schlüter). Special thanks are also due the Dean of the Faculty of Physics (Axel Schenzle) of Munich University for the financial support that has enabled the translation of the German original of this biography into English.

1 See chapter 14.
2 Sauter, *Sommerfeld*, 1969.

The biographer owes thanks and acknowledgment also to his colleagues in the history of science who have concerned themselves from a historical perspective with Sommerfeld and his fields of research. Atop the list are John L. Heilbron, and Paul Forman, who as participants in the *SHQP* project actually first set Sommerfeld research in motion. Subsequently, above all Armin Hermann, Ulrich Walter Benz, and Karl von Meyenn have helped cause this spark to jump over to Germany. In recent years, a new project on the history of quantum physics[3] at the Max Planck Institute for History of Science in Berlin brought further impetus to analyze Sommerfeld's work within the network of modern atomic and quantum theory. For this biography, it was a piece of good fortune to participate in this project. It would lead too far afield to list all the names of colleagues and friends who in this and in previous decades have researched the history of quantum physics—and thus also important aspects of the field of Sommerfeld biography.

Even if, in retrospect, quantum physics be deemed Sommerfeld's most significant area of research, the less spectacular mathematical, physical, and technical work to which he devoted himself in the course of his long scientific career deserves inclusion in his biography. The gratitude of the historian of science and biographer is therefore directed to all those who have devoted their scholarly attention to these aspects of Sommerfeld's work, and thus in diverse ways have advanced the work of this biography. Their names and contributions are to be found in the bibliography. Special thanks to our colleagues at the Research Institute of the Deutsches Museum, which constitutes a particularly sympathetic setting for the work of history of science. Last but not least, thanks to the German Research Foundation, without whose financial support this project would not have been possible.

3 http://quantum-history.mpiwg-berlin.mpg.de/main/(28 January 2013)

Contents

Prologue

Who was Arnold Sommerfeld? Along with Max Planck (1858–1947), Albert Einstein (1879–1955) and Niels Bohr (1885–1962), he belongs among the founders of theoretical physics, which developed into an independent discipline during his lifetime (1868–1951). Among his best known achievements is the elaboration of the Bohr atomic theory established a century ago. Even among physicists of the twenty-first century, the "Bohr-Sommerfeld-Atom" and the "Sommerfeld fine-structure constant," remain current concepts. Older physicists associate Sommerfeld's name with the first "school" of modern theoretical physics, and with the work known as the "Bible of atomic physics," *Atomic Structure and Spectral Lines*. This legendary textbook was spread throughout the world in many editions and translations, and initiated generations of physics students into the field of nuclear physics. Additionally, Sommerfeld's *Lectures on Theoretical Physics*, published in six volumes, and reissued long after his death in ever new editions, conveys a sense of the charismatic teacher's personality. At the University of Munich, where he taught and pursued research from 1906 for over three decades, the tradition of the Sommerfeld school continues at the "Arnold Sommerfeld Center for Theoretical Physics." Here, the latest findings in string theory and other areas of theoretical physics are discussed. A hundred years ago, the Munich "nursery of theoretical physics" (as Sommerfeld liked to describe his institute) was a haven for the new quantum physics. Sommerfeld's students included Nobel Prize winners Peter Debye (1884–1966), Max von Laue (1879–1960), Wolfgang Pauli (1900–1958), Werner Heisenberg (1901–1976), Linus Pauling (1901–1994) and Hans Bethe (1906–2005). With his 81 nominations, Sommerfeld himself holds the sad record of having been proposed for the Nobel Prize more often than any other physicists . . . without ever receiving the coveted distinction.[1]

It is not just his contributions to modern physics that suggest the need for a biography of Sommerfeld, however. In mathematics too, and in technology, evidence of his influence is apparent. For engineers who deal with the theory of hydrodynamic bearings, the "Sommerfeld number" is a term; in this technological discipline, Sommerfeld is counted as one of the "Men of Tribology."[2] Moreover, Sommerfeld was recognized beyond the boundaries of his own subject. In the 1920s, he traveled widely in the capacity of "Cultural Ambassador" campaigning for Germany's reputation abroad.[3] In the time of the "Third Reich," he became the target of attacks from a group of Nazi ideologues who sought to replace the "Theoretical Cartel" around Sommerfeld, denounced as Jewish, with a "German physics." Following Sommerfeld's retirement, the most fanatical member of this

1 See chapter 14.6

2 See chapter 5.3.

3 This was his son's designation for him; see Chapter 10.

group, a professor of aerodynamics, was appointed his successor. In 1939, the Institute, once renowned for modern physics, became for several years an arena of aggressive Nazi propaganda; later, it served the history of science as an example of the devastating effects of an ideologically driven science.[4]

Methodology

The biography of a physicist of such broad influence cannot content itself with a presentation of his scientific life's work. Needing to assess the significance of the historical events occurring during the life of the individual it depicts, it thus assumes its place in the domain of science history. Many biographies of great thinkers and learned scholars of earlier periods, however, evince little of the findings of history professionals, so that historians have long stood in an uneasy relation to biographies in general, and those of scientists in particular. The genre—so the allegations have run—suffers from weak theoretical underpinnings, and tends to be outmoded. Biography has been regarded as a relic of historicism, a "fossil of a long surpassed historiography of great events and personalities, when the historian still believed it possible through intuition and imagination to grasp and represent the inner logic of the historical evolution and actions of his biographical subjects." This is how, from the perspective of recent social history, this professional unease has aptly been described. That this characterization prefaced a biography of Fritz Haber (1868–1934) shows to what extent the genre of scientific biography today has distanced itself from historicist hero worship. For the modern discipline of history, biography constitutes simultaneously both "challenge and opportunity" to demonstrate "the interweaving of an individual life with its historical context." The socio-historically stamped biography traces "the historical scope of action of the individual," and is "as it were microscopically" focused on historical details to which history otherwise pays scant attention. "For this, Fritz Haber's transparently lived life presents itself almost ideally."[5]

This applies equally to the biography of Sommerfeld, the same age almost to the day as Haber, and whose path crossed Haber's on several occasions. But even if a modern science biography is committed to both social history and the history of science, the unique traits characterizing a life which, either through family background or in some other way lend a subject his quite individual personality, require of the biographer—more than of a "microscopically" oriented historian—considerable empathy to convey the personality of his subject to the reader. In an analytically oriented historical study, such an intimate view would rather be frowned upon since historical science values objectivity and critical distance. For biographers, however, empathy is nonetheless a prerequisite, on which theorists of this genre

4 See chapters 11, 12.

5 Szöllösi-Janze, *Fritz Haber*, 1998, p. 12; Szöllösi-Janze: *Lebens-Geschichte*, 2000; Daston/Sibum: *Scientific Personae*, 2003.

place great value. This proximity must also find narrative expression: the biographer needs to employ formulations that are as authentic as possible, and not obscure them through idiosyncratic terminology.[6] In the field of history of science, it has been and continues to be especially difficult for biography to establish itself as an independent genre. On the one hand, biographers of renowned scientists are liable to the stigma of outmoded hero-worship if they display empathy; on the other hand, biographies often serve as vehicles to illustrate scientific processes in a social context. The scientific biography is most convincing, however, precisely when it presents not only the scientific issues of the period, but also the ambitions, passions, and moral choices of a life in science. This at any rate was the judgment of one science historian on the incorporation of biography into her field.[7]

The scientific biography has yet to emancipate itself in this respect, and to reflect on the strengths of its own genre, claimed Thomas Söderqvist, who as a biographer and historian of science has made contributions in both spheres. Biography is a genre that deals with existential choices. Söderqvist coined the term "existential biography." The science biographer needs to convey the existential choices that life as a scientist entails.[8] This far exceeds the demand that history of science inspired by the social sciences be integrated into its social context. The genre of scientific biography adheres to its own set of rules and possibilities, and is not simply a tool in the service of history of science.[9]

The biography of a modern theoretical physicist like Sommerfeld, however, is a tight-rope balancing act in yet another way. Anyone who has not studied physics will hardly know where to begin with the details of Sommerfeld's work. It is the peculiar challenge for the biographer of mathematicians and physicists to be true to the scientific content and yet not pitch the demands on the reader so high that he can follow the biography only in the context of a technical course of study. If high priority is given to scientific content, the biography of a physicist or mathematician most often falls into chapters containing general accounts of their lives on the one hand, and chapters of physical or mathematical detail studies, larded with formulas and technical jargon, on the other. If broad accessibility is given priority, scientific content tends to fall by the wayside. The narrative essence of biography, however, requires a balanced presentation not divided up into chapters of divergent reading styles. Among scientific biographers, these problems have long been debated.[10] In mathematical biographies, a compromise between technical precision and broader accessibility seems impossible when very abstract topics are in question. In physics, though, even in highly complex areas, there is often an apparent relation to objects in the world of experience, so that at least a rough presentation of the essential

6 Frank, *Other*, 1985.
7 Jo-Nye, *Scientific Biography*, 2006.
8 Söderqvist, *Existential projects*, 1996.
9 Söderqvist, *History and Poetics*, 2007.
10 Hankins, *Defence*, 1979; Carson/Schweber, *Studies*, 1994.

concepts at issue seems possible. Such was, for example, demonstrated masterfully in a biography of Werner Heisenberg.[11] What was possible in the case of Sommerfeld's prize student ought to apply equally to his teacher. It serves at least as an incentive to achieve in a similar manner a successfully integrated presentation of life and work that satisfies the demands of modern scientific biography.

In addition to the methodological requirements of a scientific biography, the question of the premises of the history of science also presents itself. This concerns first of all the indispensable sources for any biography. In the history of physics, Sommerfeld has long attracted attention, so that efforts extend far back to gather source material relevant to his work from the perspective of the history of physics. It began in the 1960s, with the work of the project staff of *Sources for History of Quantum Physics (SHQP)*.[12] Together with other sources for the history of quantum physics, many of Sommerfeld's letters and manuscripts were archived as microfilm in the *Archive for the History of Quantum Physics (AHQP),* and made available to historians of science. Focus was primarily on the development of atomic and quantum physics, so that Sommerfeld's work was first assessed from this perspective.[13] But the *AHQP* archive also offered the first opportunity for a biographical approach.[14] In the wake of the *SHQP* project, finally Sommerfeld's students such as Alfred Landé (1888–1976), Wolfgang Pauli, and Werner Heisenberg became subjects of publications in the history of physics, so that the importance of the Sommerfeld school came to the fore.[15]

In the 1980s, the sources on Sommerfeld were augmented by material from the partial estate of the former Sommerfeld Institute for Theoretical Physics at the University of Munich, and from the holdings of the Sommerfeld family. In the next two decades, an exhibition at the Deutsches Museum, a monograph on the Sommerfeld school, and a two-volume edition of Sommerfeld's scientific correspondence, together with an Internet correspondence database, offered a wealth of insights into the Sommerfeld papers.[16] This opened up as well the prospect of an overall view of Sommerfeld's life and work. Specifically, the extensive correspondence in the possession of Sommerfeld's heirs now allows light to be shed on the more private aspects of his personality. This is the most important precondition to allowing Sommerfeld himself to speak regarding many aspects of his life, conveying narrative authenticity, and remaining true to the genre in the sense of an "existential biography."

11 Cassidy, *Uncertainty*, 1992.

12 Kuhn et. al., *Sources*, 1967.

13 Hermann, *Diskussion*, 1967; Hermann, *Frühgeschichte*, 1969; Nisio, *Formation*, 1973; Kragh, *Structure*, 1985.

14 Forman/Hermann, *Sommerfeld*, 1975; Benz, *Sommerfeld*, 1975.

15 Forman, *Alfred Landé*, 1970; Forman, *Environment*, 1967; Forman, *Doublet Riddle*, 1968; Heilbron, *Kossel-Sommerfeld Theory*, 1967; Cassidy, *Core Model*, 1979; Meyenn, *Paulis Weg*, 1980, 1981.

16 Eckert et al., *Geheimrat*, 1984; Eckert, *Atomphysiker*, 1993; Eckert/Märker, Arnold Sommerfeld. *Wissenschaftlicher Briefwechsel*, 2000, 2004 (abbreviated hereafter as ASWB I and ASWB II); http://sommerfeld.userweb.mwn.de/AS_WWW.html (28 January 2013).

1 Königsberg Roots

"You think you're pushing something aside; instead you get pushed into it."[1] Thus, in a letter to his parents, the 26-year-old Arnold Sommerfeld recalled the critical step on his path to a career as a university teacher when he gained qualification to teach at the University of Göttingen. That he now belonged to "Germany's houses of learning" was hardly worth making a big fuss over. "It's only circumstances that have forced me into it," he said. Nonetheless, the occasion seemed significant enough to remember his Königsberg roots. "Seeing our father sit each night at his books from the pure joy of work and the acquisition of knowledge; seeing our mother slave away day in and day out simply from her need to fulfill her duty—this had to have the effect of instilling earnest aspiration in us children too."[2]

The "circumstances," though, comprised not just his home environment but also the milieu of the city in which he was born and spent his childhood and youth. In 1701, Frederick I (1657–1713) had been crowned the first King of Prussia in Königsberg. The presence of Immanuel Kant (1742–1804), who spent his entire life in Königsberg, shed the luster of German Enlightenment on this city. Though Königsberg was actually the capital only of East Prussia, anyone whose family history like Sommerfeld's was deeply rooted here, would not have felt like a provincial, but like a proud resident of the intellectual and cultural center of Prussia.[3]

1.1 Childhood

The Königsberg tradition of the Sommerfeld family dates back to 1822, when Arnold's grandfather settled here in the position of "Court Postal Secretary" and established a career in Prussia's service. He had grown up in the Prussian hinterland, the son of a musician, but had then quickly felt at home in Königsberg, and established a family there. One of his eight children recorded the fortunes of this first Sommerfeld generation in the Prussian province in a family chronicle.[4] In miniature, this chronicle mirrors the larger tides of German history in those years.

1 "Man glaubt zu schieben und man wird geschoben!".

2 To his parents, March 12, 1895. Letters to and from Sommerfeld are cited by the name of the sender and the addressee only. If no archival source is given (like in this case), the letter is located in Sommerfeld's private papers in the possession of the family.

3 Gause, *Geschichte*, 1996.

4 Mütterchens Erinnerungen, Familiengeschichte der Sommerfelds und Rauschnings, erzählt von Emma Rauschning, geborene Sommerfeld, und zu Papier gebracht von deren Tochter, Anna Rauschning, February, 1926. DMA, NL 89, 016, folders 1,2.

M. Eckert, *Arnold Sommerfeld: Science, Life and Turbulent Times 1868-1951*,
DOI 10.1007/978-1-4614-7461-6_1,

As a student and fraternity member, the family's eldest son, Wilhelm Sommerfeld (1817–1866), joined the freedom movement that led to the revolution of 1848. He abandoned his law studies because he "did not wish to serve the king," as we read in the chronicle. Instead, he became editor of a newspaper and landed in prison for the first time for publishing an article in opposition to the military. His detention was brief, however. "The Freedom Year of 1848 found him in the front ranks of the advocates of freedom," the family record proudly states. Wilhelm authored pamphlets against the rule of the Prince and was consequently convicted to 3 years in prison on the charge of *lèse majesté*. He fled to Hamburg, planning to emigrate to America, but was arrested and extradited back to Königsberg. Exhausted nervously and physically after 3 years in prison, he was unable to readjust to regular middle-class life and shot himself on June 6, 1866, one day past his 49th birthday.

When Arnold Sommerfeld was born in 1868, his uncle's tragic death was still fresh in family memory. As his younger brother, Arnold's father, Franz Sommerfeld (1820–1906), had been especially close to the unfortunate Wilhelm. They had both attended the Altstädtisches Gymnasium and studied at the University of Königsberg. Franz's interests, however, lay not in law but in medicine and science. "Even then, flowers and minerals were his passion, and he was an avid collector," the family chronicle informs us. Following his medical studies, Franz Sommerfeld lived for several years as a bachelor, looked after by his sister, "Minchen," who remained single, and became a teacher. She too seems to have demonstrated "earnest aspiration" to learning. When Franz married Cäcilie Matthias, daughter of a Potsdam builder, in 1862, he was already 42 years old and had a respected medical practice in Königsberg. "Nonetheless, the couple was compelled to economize significantly, first because many of brother Wilhelm's debts remained to be paid, and second, because Franz's amateur passions consumed large sums."[5]

Occasional financial difficulties notwithstanding, Franz and Cäcilie Sommerfeld surely did not look towards the future with anxiety. Her first son, born in 1863, was christened Walter. So far as the housekeeping was concerned, Cäcilie had the help of her mother, Ottilie Matthias, who lived permanently with them. In addition, there was Amalie Prawitt, a "faithful housekeeper and governess to the children," who soon became a permanent member of the family. Despite occasional misfortune—a son baptized as Arnold died in infancy in 1864, and in 1866, came the news of Wilhelm's suicide—there was always reason to look to the future undaunted.

When Cäcilie gave birth to another son on December 5, 1868, as though defying fate the parents again named the child Arnold. Two years later, Arnold got a little sister, baptized Margarethe. "Gretchen" was the apple of the family's eye.

In photographs of the children from the 1870s, little Margarethe's expression shows she knew how to place herself confidently in the picture among her brothers. Even if their unfortunate uncle's debt and their father's passion for collecting allowed

5 Mütterchens Erinnerungen, DMA, NL 89, 016, folders 1,2.

Fig. 1: Arnold (*left*) with his brother, Walter, 3 years older than he, and his sister, Margarethe, 2 years younger (Courtesy: Deutsches Museum, Munich, Archive).

for no ruled out extravagances, the children lacked for nothing. At the piano as well as with brush and paint, Arnold showed artistic talent, as demonstrated by some of his autographed pictures. But the family was not spared additional blows. In 1880, Margarethe died of scarlet fever. Her parents and her two brothers had difficulty getting over this tragedy. The anniversary of Margarethe's death, the 2nd of March, was long observed in the Sommerfeld family as a day of remembrance.[6]

It may have been such family tragedies as the death of his beloved sister and his uncle's tragic end that led Arnold still as a child to a philosophy of life oriented towards defying adversity. Arnold had not yet been born when his uncle shot himself, but Wilhelm's death was perhaps one more reason for his father to prepare his children for "the serious business of life." In other ways, too, circumstances in the Königsberg of the 1870s and 1880s dictated what in a nutshell is known as "Prussian virtues." "Duty and self-discipline, application of one's entire effort to the place in life we are assigned, setting one's own convenience and pleasure aside, assimilation

6 To his mother, March 1, 1899.

into the greater scheme, disdain for incompetent sham, belief in the power of good—all this is what we understand by the good German word 'idealism,'" wrote the adult Arnold Sommerfeld of these virtues in a 1905 newspaper article.[7] This description was occasioned by the 90th Birthday of Otto von Bismarck (1815–1898), who as Prussian Prime Minister and German Chancellor had appeared to a younger Sommerfeld the incarnation of Prussian virtues.

1.2 School Years

But Arnold's childhood and youth consisted not only of duty and self-discipline. Many years later, a school friend recalled how he together with Arnold and three other friends enjoyed the daily life of the gymnasium student in Königsberg. The pranks of the "Five Inseparables," as they called themselves, reminded him of the *Lausbubengeschichten* of Ludwig Thomas, whose student years coincided closely with Sommerfeld's. And when it came to their teachers, his school friend came to a not very flattering judgment. "Such role models! Do you know?! Louis Schwiedop, despite his veneer of erudition, surely a semi-idiot, who always wore a brown scarf, and toyed incessantly with his long gold watch chain with its various pendants." The school friend recalled another of their teachers only as "the theme of inebriation." The director of the Altstädtisches Gymnasium, too, was in this reminiscence given a not very flattering mention. He "epitomized the true academic tyrant," for whom in questions of discipline, "the rod was the *ultima ratio*."[8]

Committed to paper more than 30 years after the fact, his high school friend's memoir is hardly a reliable historical source. But it does show that Prussian drill and training in the attitude of submissiveness had proved not especially fruitful in the case of the "Five Inseparables." "In tyrannos" ("Against the Tyrants"), the motto of Schiller's drama "*The Robbers*" was the device Arnold and his school friends inscribed on their banner too. At communal reading evenings they were enthralled "by Wallenstein, Joan of Arc, Tasso, etc."— that is, with the spirit of the *Sturm-und-Drang* era of German classicism.[9] "I want freedom in thought and in writing: in action, the world already hedges us about": in verses like this from Goethe's

7 Der 90. Geburtstag Otto von Bismarck. In: *Aachener Allgemeine Zeitung*, April 4, 1905. DMA, NL 89, 022.

8 From Ernst Ellendt, June 20, 1920. DMA, HS 1977-28/A,81. The school principal in question was Rudolf Möller, who directed the Altstädtische Gymnasium from 1863 to 1885. In the Königsberg City history, Möller and the principals of the other high schools of those years (Friedrichskolleg, Wilhelmsgymnasium, and Kneiphöfisches Gymnasium) were characterized as highly learned men. Gause, Geschichte, 1996, vol. 2, pp. 598–599.

9 *Wallenstein* and *Die Jungfrau von Orleans* by Friedrich Schiller and *Torquato Tasso* by Johann Wolfgang von Goethe are the works alluded to.

"*Torquato Tasso*, the "Five Inseparables" in the Königsberg of the 1880s would have recognized their own attitudes."[10]

The private "reading circle" inspired Arnold's enthusiasm for literature, especially the German classics, which is apparent in his academic achievements. The Altstädtische Gymnasium in Königsberg, which Arnold had attended since 1875, was one of the oldest humanistic high schools in Germany, and at these elite schools, German, Latin, and Greek were more than mere language courses. Here, an idealistic value system was instilled in students by way of models from classical antiquity and the German classics, as promoted by neo-humanism, and formulated in the "Königsberg School Plan" (1809) by Wilhelm von Humboldt as the guiding principle of Prussian educational policy. Arnold usually received a "good" in these subjects, which corresponded on the five-point scale of the Altstädtisches Gymnasium in those years to the top grade of "1."

When the 15-year-old high school student once had to write an essay on the subject "Still true today: the richest gifts are never to be had for money," he rejected the materialistic view of life so definitively that next to the grade of "satisfactory" (equivalent to "2"), his teacher commented: "Wealth is given far too short shrift here!"[11] Two years later, his graduation certificate had this to say of his performance: his competence in German, "both in reading and in writing were quite good," "His overall performance may therefore—although his final essay received the grade only of 'satisfactory'—be judged 'good.'"[12] In his other subjects, too, his performance was marked on the graduation certificate with the top grade of "good." Only in physical education was he rated "adequate." Ten years earlier, his grammar school teachers had awarded the 8-year-old Arnold in almost all subjects the grade only of "sufficient"—in German and arithmetic even just "minimally sufficient," the second-worst grade.[13]

Many years later, when Sommerfeld recalled his school days at the Altstädtisches Gymnasium, he felt it important to emphasize that he was interested not only in mathematics and physics. He had been "equally good in all subjects, including ancient languages," and was "interested more in literature and history perhaps than in the exact sciences."[14] He certainly retained a lifelong appreciation for the German classics.

10 Goethe, *Torquato Tasso*, 2005, Act 4, Scene 2.

11 Schulheft mit Aufsätzen 1884/85. Essay of May 12, 1884. Sommerfeld had written: "But how all these amenities pale beside those loftier goods, health, happiness, contentment, talent, that nature distributes according to its own discretion among rich and poor, and without which life even amidst the greatest material abundance is empty and sad . . . In spiritual life, money is merely the mechanism by which the inner treasures of the inner man are brought to light."

12 Reifezeugnis, issued September 15, 1886. DMA, NL 89, 016.

13 Zeugnis, Vorschule 1. Klasse, report covering July 31 to September 30, 1876.

14 Autobiographische Skizze, ASGS IV, pp. 673–679.

Fig. 2: Paintings and drawings like this portrait of a girl display Arnold Sommerfeld's artistic talent (Courtesy: Deutsches Museum, Munich, Archive).

1.3 University Study

In the wake of his successful completion of the Abitur, Arnold Sommerfeld seems to have been undecided about his future career path. His graduation certificate states that he wished to devote "himself to the study of the construction industry." It may be that "Ochen," as his maternal grandmother Ottilie was called, had depicted this profession as desirable to her grandson since her husband, who had died at an early age, had been a royal government building contractor. Or was the graduate, after his life as a high school student, now in the mood for a more practical field of study? In the Königsberg of his youth, construction was going on everywhere, so that the profession of building contractor must have appeared to be a meaningful activity. Already in the 1860s, the last of the city's gates were demolished because they were regarded as no more than obstacles to traffic. Not even historic buildings such as the Königsberg Powder Tower, demolished in 1888, were spared from the building mania. "You've no doubt heard of 'The Battering-Ram Club of Königstrasse;' we broke off a little piece of the old casino wall every day because it was an obstruction to traffic," recalled his classmate of those years.[15] With the increasing traffic, the historic wooden bridges over the Pregel River had also become outmoded. Beginning in 1879, they were replaced by modern iron structures. Demand for drinking water also increased with the growth of the city's population, which the construction of a dam in 1887 addressed. Some years earlier, sewage and toilet waste disposal had already begun to be fundamentally modernized through construction of a metropolitan drainage system. The introduction of electricity to Königsberg, too, occurred during the time of Sommerfeld's university studies. In 1888, the city built a power plant that by 1890 was supplying electricity to the first households and businesses.[16]

With his good grades in math and science, Sommerfeld would have brought good qualifications to a course of engineering or architectural studies. But he did not persevere in the aspiration noted on his graduation certificate. In order to pursue civil engineering, he would have had to leave Königsberg and choose a place of study with a technical university. At the Albertus University of Königsberg, he was not able to study engineering, but since he obviously was unsure of his future career choice, moving his location did not seem urgent. In addition, the "Albertina" (as it was known colloquially) had an excellent reputation. Founded in 1544 and with Immanuel Kant in the eighteenth century its most famous professor, it had achieved international fame. In the nineteenth century, Carl Gustav Jacob Jacobi (1804–1851), Friedrich Wilhelm Bessel (1784–1846), and Franz Ernst Neumann (1798–1895) brought it great respect in the fields of mathematics, astronomy, and theoretical physics. Other fields, too, were represented by renowned professors. The Königsberg town historian summarized the development of the University up to the mid-nineteenth century: "All in all, the Biedermeier decades were a good time for the Albertina." In addition, the professors felt committed to maintaining the "civic life" of the Albertina "by keeping the local educational community abreast of the results of their scholarship." The professors met regularly with the citizens of Königsberg in their town houses or at other appropriate locales. "Instead of glittering salons, they met in more intimate circles and gatherings. These took place in a variety of civic groups, among merchants, bankers, as well as academics." Thus, the Königsberg historian characterized the influence of the University on the intellectual life of the city on the eve of the 1848 Revolution. "Evenings were spent with readings, discussions, and the presentation of original poetry. Readings occurred even at parties and within the family circle."[17]

In the second half of the nineteenth century, too, the Albertina boasted celebrated professors. Before making a reputation in Berlin as "Imperial Chancellor of Physics," Hermann von Helmholtz (1821–1894) was active here in a department of physiology. In the field of mathematics, the University of Königsberg was stellar. Ferdinand Lindemann (1852–1939), who as a professor in 1882 worked out the proof that Pi (π)

15 From Ernst Ellendt, June 20, 1920. DMA, HS 1977-28/A,81.

16 Gause, *Geschichte*, 1996, vol. 2, pp. 570–577, 639–661.

17 Gause, *Geschichte*, 1996, vol. 2, pp. 450–461.

is a transcendental number, had garnered fame outside the field of mathematics as well. The classic problem of geometry, "squaring the circle," was herewith laid to rest; Lindemann proved definitively the impossibility of constructing from a circle with just ruler and compass a square of equal area. In 1884, Adolf Hurwitz (1859–1919), associate professor, and 2 years later, David Hilbert (1862–1943), lecturer, added their luster to the reputation of the Albertina in the field of mathematics.

In September, 1886, Sommerfeld matriculated at the University of Königsberg. He spent the first semester finding his bearings amidst the wealth of course offerings. The entries in his notebook clearly show that in this winter semester of 1886/1887, he had come to no clear decisions. He registered for lectures on political economy, economics, ethnography, political parties, German civil law, calculus, foundations of ethics, Kant's Critique of Pure Reason, comparative anatomy, and vertebrate taxonomy.[18] Some of these lectures remained in his memory all his life. He had attended a lecture "about ancient Germanic legal documents," by the lawyer Felix Dahn (1834–1912), who had made a reputation in Königsberg also as a writer and local historian, Sommerfeld wrote the Dean of Natural Sciences of the Albertina many years later, who had sent him a diploma honoring the 50th anniversary of his doctorate. He also mentioned lectures not listed in the matriculation notebook, like those of archaeologist Gustav Hirschfeld (1847–1895) on "Ancient Greek archeology." In retrospect, he felt that his course of studies at the Albertina had profited him "not only in technical matters but also in general terms."[19]

From the distance of half a century, Sommerfeld's recollection of his years of study, however, had been significantly transfigured. Contemporaneous sources tell a different story. He had been anything but enthusiastic about his physics professors. He had not reckoned Carl Pape (1836–1906), whose lectures in experimental physics he attended in the third semester, or Paul Volkmann (1856–1938), with whom he studied from the fourth semester to the end of his studies in theoretical physics, among the professors who had made this subject appealing to him. On one occasion he spoke of "useless Pape, Volkmann, etc.," in recalling the physicists at the Albertina. At the time, his studies were only 2 years behind him, and memory was still fresh.[20] Of course, we are no more dealing in this instance with an objective assessment than in Sommerfeld's retrospective view after 50 years. But even among historians of science, the time around 1890 is viewed as a period of decline of the once brilliant Königsberg physics establishment, which had inspired students like Kirchhoff in Neumann's seminar with enthusiasm for theoretical physics.[21]

After his first semester, when Sommerfeld began to organize his curriculum towards more technical subjects, his focus was nonetheless not primarily on physics.

18 Matriculation Notebook, entries, winter semester 1886/1887.

19 An den Dekan der naturwissenschaftlichen Fakultät der Universität Königsberg, November 10, 1941. DMA, NL 89, 017, folder 2,4.

20 To his mother, January 5, 1894.

21 Olesko, *Physics*, 1991, pp. 442–443.

Entries in his matriculation notebook show that his main interest was primarily in mathematics. From his second semester on, he registered for nearly all of Lindemann's lectures (Analytic Plane Geometry, Solid Geometry, Arbitrary Functions and Definite Integrals, Calculus, Non-Euclidean Geometry, Invariant Theory, Theory of Partial Differential Equations, Analytic Mechanics, Theory of Functions, On Euclid, Theory of Abelian Functions, Foundations of Geometry, Calculus of Variations, and The Use of Abelian Functions) and regularly took part in his seminar, in which the students had to solve mathematical problems through their own work. From the third semester on, he added lectures by Hurwitz (on Elliptical Functions, Conformal Mapping, and Algebraic Equations), and from 1889 on, lectures by Hilbert (on Number Theory and other more advanced mathematical subjects). As he later wrote in an obituary, he felt a bond of friendship with Hilbert, who was only 6 years older than he, "based on youthful memories and common nationality."[22]

Actually, at that time moving from one university to another was a matter of course in student life. Contrary to the trend of academic "roaming dynamics,"[23] spending one's entire training in only one place was rather the exception. Sommerfeld later based his devotion to the Albertina on the excellent personnel of the Königsberg mathematics departments.[24] But there was another reason he did not want to leave Königsberg: he was a fanatical devotee of the ethos of student fraternities. "A boy is rooted in his parental home; a man, in the family he establishes. A youth is rooted in the community of like-minded young men," a fraternity companion said by way of clarifying this emotional connection.[25]

Sommerfeld joined the "Germania." The other student organizations at the Albertina were "Gothia," "Alemannia," and "Teutonia." These names evoked the nationalistic fervor to which the anti-Napoleonic wars of liberation had given birth, in which the student fraternities—though they often fought violently among each other—saw themselves reflected. The Germania was founded in 1843 and was, at the time of Sommerfeld's student years, the most important of the Königsberg fraternities. Membership in Germania was a family tradition for many Königsberg families. "Walter was a good scholar and a stylish student (Germania)," the Sommerfeld family chronicle noted of Arnold's brother.[26] In a list of "the Old Guard," Arnold was registered as "Sommerfeld III, Arnold," following his brother Walter, and another relative, Fritz Sommerfeld, who had in 1882 and 1885, respectively, become members of Germania in Königsberg. By 1893, the Rauschnings, related to the Sommerfelds, brought the number of family who had become members of Germania to ten.[27] To be a "stylish student" meant cutting a dashing figure

22 Sommerfeld, *David Hilbert*, 1943.
23 Pyenson/Skopp, *Physicists*, 1977.
24 Autobiographische Skizze, ASGS IV, p. 673.
25 Popp, *Geschichte*, 1955, p. 43. On the history of the German student fraternities generally, see Jarausch, *Studenten*, 1984.
26 Mütterchens Erinnerungen, DMA, NL 89, 016, folder 1,2.
27 Germania, *Festschrift*, 1933.

Fig. 3 Arnold as "stylish student" with friends from his fraternity, Germania, ca. 1890 on a rowing excursion (Courtesy: Deutsches Museum, Munich, Archive)

at fencing on the dueling floor. At social events, too, a fraternity man had to hold his own, which included such diverse talents as holding his liquor and taking pleasure in song.

In addition to exemplifying the virtues of a Germania student, however, Sommerfeld also recognized his social responsibility as a citizen. "We students, during our university years and in our future occupations, assume a privileged position in civic life," he exhorted his fellow Königsberg students in 1890 in the fraternity newsletter *Burschenschaftliche Blätter*. "Thus, we more than others have reason to take our obligations to the state seriously." The impetus for this came from the "Association of Volunteer Wartime Medical Orderlies," an organization of the Red Cross founded a few years earlier. Since initially this association had made almost no impact among the students, Sommerfeld appealed to the fraternities to become involved. In peacetime, too, it was possible to "accomplish much good through knowledge of the simplest medical procedures," he argued in urging participation in the Red Cross medical courses. One could "treat people rationally who would either be totally neglected or forced into the hands of ignorant quacks."[28]

Aside from this appeal to the social conscience of his fellow students, other bits of memory give evidence that Sommerfeld was an enthusiastic fraternity member. All his life, he kept the "Complete German Book of Drinking Songs," decorated with metal fittings ("beer nails") containing the songs which, after an adequate

28 Sommerfeld, *Genossenschaft*, 1890; Riesenberger, *Das Rote Kreuz*, 2002, pp. 111–112.

consumption of beer, were sung in the student hangouts. His "stylish" student days were visibly recognizable too: a scar on his forehead was evidence that he had "held his own" on the dueling floor of the Germania in Königsberg. Later on, this aspect of his student days seems to have been something of an embarrassment. His student and friend Paul Ewald (1888–1985) wrote, "I remember only one conversation with him on the subject of his Königsberg days, about which, pointing to his scar, he said of himself: 'when I was still young and bloodthirsty.'"[29]

1.4 A Competition

The imminent end of his student years, however, admonished the "stylish student" ineluctably that he had yet other challenges to master than just those of the dueling floor of the Germania. "When I heard Hilbert lecture on Ideal Theory, I thought my interest was primarily in the direction of the most abstract mathematics," he recalled about his final university semester.[30] "Ideal Theory" is division of abstract algebra. It was just in those years that Hilbert was elaborating this area in critical ways.[31] He must also have conveyed to the small group of students attending his lectures a feel for research on the highest levels of pure mathematics. For Sommerfeld, however, the path to his own scientific work led rather along the lower-lying plains of applied mathematics. The occasion was a competition sponsored by the Association of Physics and Economics of Königsberg: "The Association desires as comprehensive a theoretical utilization as possible of the observations of ground temperature at Königsberg," as the problem was formulated, "for the understanding of thermal variation in the earth and its causes." The winner was to receive a prize of 300 Marks.[32]

The Association of Physics and Economics had been founded in 1790 by a district administrator who hoped thereby in the first instance to promote East Prussian agriculture. His biographer characterized him as a man who had set himself the goal of "improvement of the fatherland's agriculture." This explains the designation "economic" in the name of the organization, for in the eighteenth century the word had a primarily agricultural connotation. Consequently, the first patrons were East Prussian land owners. Its statutes proclaim, "The purpose of the organization is dissemination of information about and improvement of the various branches of the natural world." "It strives to promote agriculture and the economy and all the sciences connected with them, above all the natural sciences."[33] Over the course of

29 Ewald, *Sommerfeld als Mensch*, 1969, p. 9.
30 Autobiographische Skizze, ASGS IV, p. 674.
31 Reid, *Hilbert*, 1996, Chap. 5; Frei, *Briefwechsel*, 1985, pp. 89–91.
32 *Schriften der physikalisch-ökonomischen Gesellschaft zu Königsberg in Pr.* (abbreviated hereafter SPGK) 31, 1890, pp. 4–6.
33 Stieda, *Geschichte*, 1890, p. 40.

its history, the Association of Physics and Economics went through numerous changes. Its historian delineated four periods. The first two and a half decades had been dedicated entirely to agriculture. The years from 1814 to 1829, he designated the "literary" period. This was followed by a "popular science" period that lasted until 1858. Only thereafter did the natural sciences come more strongly to the fore as the organization's focus. With the construction of a station to measure ground temperatures on the grounds of the Königsberg Botanical Garden in 1872, "the promotion of scientific works, in particular those that concern the Province of Prussia" was put into actual practice.[34]

These initiatives of the Association of Physics and Economics intersected with the efforts of Neumann from the University of Königsberg, who had long before made the measurement of temperature a principal area of his pedagogical and research operations. Neumann saw this as an effective means of demonstrating the principles of physics research to prospective high school teachers. Using temperature measurement as examples, problems of working with instruments, measurement methodology, estimation of error, and theoretical analysis of measurement data could be dealt with. From the late 1830s onwards, Neumann had set the students in his seminars the assignment of measuring ground temperatures at various depths and of reaching conclusions from the measurement data regarding diffusion of heat in the earth. This led to more wide-ranging mathematical, meteorological, and geophysical questions. Since Neumann had often made such ground temperature measurements the subject of doctoral theses he supervised, it was only logical to put the scientific measurement station at the Botanical Garden into the hands of Neumann and his students.[35]

When new construction on the grounds of the Botanical Garden in 1890 threatened the continuation of temperature measurements, the Association of Physics and Economics saw itself duty-bound to make an intelligent assessment of the flood of data already amassed. In sponsoring a competition, the Association was following the tried and true precedent of the great academies of the eighteenth century, which, with similar competitions, had offered ambitious researchers the opportunity to make a name for themselves in the world of science. A seven-member commission was charged with the conception and execution of the competition. This commission first met on December 9, 1889, at the Institute of Experimental Physics of the University of Königsberg, and with a survey of comparable measurement stations both within and outside Germany, emphasized initially the global dimension of the task. At two subsequent meetings on December 11 and 19, 1889, agreement was reached on the wording of the competition and the amount of the prize. February 1, 1891, was set as the deadline for submission.[36]

34 Stieda, *Gedächtnisrede*, 1889.
35 Olesko, *Physics*, 1991, pp. 348–360.
36 Meeting of January 2, 1890. In: SPGK 31 (1890), pp. 4–6.

At the next meeting, Volkmann, as a student of Neumann and Director of the theoretical physics seminar at the University of Königsberg, undertook to acknowledge the scientific significance of ground temperature measurements in the spirit of his teacher. As an example of what the commission hoped to achieve with the competition, he cited the elucidation of the question how the sun heats the earth and what proportion of its radiant energy remains in the atmosphere. In 1868, another of Neumann's students had in a theoretical study reached the conclusion that in the case of vertical radiation of sunlight, only about a quarter of its heat is absorbed by the earth. On behalf of the Prize Commission, Volkmann expressed the hope that with measurement data gathered since 1872, more precise results could be achieved. Beyond this, there remained a whole list of further questions: "how does average temperature rise with increasing depth, what is the temperature of the earth's interior, what is the secular cooling of the earth, how long have earth's current temperature conditions, which appear necessary for organic life, prevailed: in other words, what is the age of the organic world?"[37]

Big questions, then, to which it was hoped answers would emerge from analysis of the enormous bank of temperature measurements gathered over more than two decades. If this could be realized, the competition would be addressing the great questions of the nineteenth century. The "secular cooling of the earth" stood at the center of the geological debate over the origins of the earth. Neptunists argued with Plutonists as to whether the earth's solid crust was composed of sedimentary rock deposited gradually from a primal ocean or of igneous, volcanic material that had cooled and solidified over the course of eons. By the middle of the nineteenth century, the Plutonist view was achieving ascendency. But if the earth at its creation had been a glowing hot liquid fireball that had gradually cooled and solidified, how could the theory of heat conduction supply information about the age of the earth? Further assumptions were necessary: Was the earth's core still composed of molten rock, or had it already solidified? And if the earth's interior was liquid, where was the boundary of the earth's solid crust?[38] William Thomson (1824–1907), later Lord Kelvin, compared the earth to a hard-boiled egg that gradually cools once it has been removed from the pot. For its initial state, he assumed a temperature just at the point at which molten rock solidifies. With the help of Joseph Fourier's (1768–1830) theory of heat conduction, formulated in 1822, he concluded the earth could not be older than about 400 Ma. Physics thus injected itself into the feuds raging among geologists, arguments over the questions of rock formation and other geo-historical processes relating to the geological periods at issue. Moreover, Thomson's estimation of the age of the earth conflicted with Darwinism; according to the theory of evolution, 400 Ma seemed too short a time

37 Meeting of January 2, 1890. In: SPGK 31 (1890), pp. 3–4.
38 Brush, *Debates*, 1979.

span for the appearance of higher species.[39] Volkmann alluded to this when he spoke of "the age of the organic world" as one of the great problems to be resolved by the competition. Thomson's theory rested on data from the Scottish station in Edinburgh, which had served as a model for the researchers at Königsberg. But so far as the scope of the series of observations was concerned, Volkmann asserted, "our Königsberg station is one of the most important, and certainly beside Edinburgh—where observations were made only from 1837 to 1854—the most important."[40]

With such high expectations, the competition occasioned lively discussions at the University of Königsberg. Pape, Volkmann, and Lindemann, professors of experimental physics, theoretical physics, and mathematics, respectively, whose lectures Sommerfeld had attended and whose seminar assignments he had completed, were members of the Prize Commission. The competition must have been a topic of conversation in the Sommerfeld household, for Arnold's father had long been a member of the Association of Physics and Economics. As an enthusiastic nature lover, the associated geological and biological ramifications must surely have interested him. From wherever the first impetus may have come, Arnold Sommerfeld warmed to this subject. Together with other students, he approached the assignment initially "by night and day carrying out continuous observations" of the thermometer.[41] He had already "for some time been occupied with the observations of the local thermometer station," he wrote in a comprehensive manuscript that was presumably intended as the draft of a competition entry. He had come to the conclusion "that a thorough treatment of the data is to be achieved only if the development of the temperature function can actually be carried out in a Fourier series, as the theory demands."[42]

In accordance with the formulation based on the Fourier series, every periodic function however complicated, can be represented as the sum of sine and cosine functions. The fact that beneath the data gathered over the course of many years a periodic temperature function lay hidden was suggestive, because the rise and fall of temperature values should reflect daily cycles, annual periods, and possibly other periodicities, such as the regular recurrence of sunspots. Since the temperature function $f(t)$ was not given analytically but only graphically represented in the form of a curve on the basis of the recorded data, Sommerfeld decided to represent it as a Fourier series:

$$f(t) = \sum_{0}^{\infty} a_n \cos(nt) + b_n \sin(nt)$$

39 Burchfield, *Darwin*, 1974.
40 Meeting of January 2, 1890. In: SPGK 31 (1890), p. 4.
41 Bericht über das Jahr 1891. In: SPGK 32 (1891), p. 68.
42 Ms., undated [presumably from the year 1891], DMA, NL 89, 026.

whereby the Fourier coefficients can be represented as an integral over the temperature function multiplied by the respective sine or cosine function

$$a_n = \frac{1}{\pi}\int_0^{2\pi} f(t)\cos(nt)dt \text{ and } b_n = \frac{1}{\pi}\int_0^{2\pi} f(t)\sin(nt)dt.$$

Thus, the problem was directed back to the calculation of integrals. Because of the purely graphically represented function $f(t)$, these calculations could be carried out only by way of instruments. Devices for mechanical integration existed already in the nineteenth century in numerous forms. Even special apparatuses for the calculations of Fourier coefficients, so-called harmonic analyzers, were in use.[43] They were employed for example in the calculation of tides, which presented a similar problem. Acquisition of such an apparatus would have been costly, however, so that Sommerfeld hit on the idea of personally arranging for the construction of a harmonic analyzer. For this project, he turned to Volkmann's assistant Emil Wiechert (1861–1928), whom he had come to know as a skilled experimentalist in the course of exercises in the mathematical physics seminar, and Wiechert saw to it that the idea evolved into a concrete plan. Though the institute's technician was tasked with carrying out the project, in the end it took so long to complete that Sommerfeld was unable to put the apparatus to use before the deadline for submission, February 1, 1891. When then he also committed an error in theoretical execution, he withdrew the entry he had already submitted. However, he viewed this failure as significant enough that in his autobiographical sketch he referred to it: "The competition entry I submitted contained quite a bit that was original and, as it seemed to me at the time, new, but it was incorrect in one essential point of the conditions, and had therefore to be withdrawn. My work had not advanced to the point of a numerical treatment, but had gotten stuck, significantly, in mathematical generalities."[44]

Even if he did not achieve his personal goal of winning the competition, nonetheless his work with the Königsberg ground temperature measurements constituted an important overture to his further academic career. On May 14, 1891, "Mathematical Doctoral Candidate A. Sommerfeld" demonstrated the apparatus to the Association of Physics and Economics, as his report concerning the harmonic analyzer recorded in the annals of the Association. This was Sommerfeld's first scientific publication.[45]

43 Dyck, *Katalog*, 1892; Fischer, *Instrumente*, 1995; Fischer, *Instrumente II*, 2002.

44 Autobiographische Skizze, ASGS IV, pp. 673–679.

45 Sommerfeld, *Maschine*, 1891.

Fig. 4: The 22-year-old Arnold Sommerfeld at work with the "harmonic analyzer." Sommerfeld, together with Emil Wiechert, had constructed this apparatus at the physics institute of the University of Königsberg for analysis of the Königsberg ground temperature measurements. From the temporal course of temperature at various depths, he sought to establish periodically occurring regularities that could, for example, be attributed to the cycles of sun-spot activity or other causes (Courtesy: Deutsches Museum, Munich, Archive).

1.5 The Dissertation

At this time, Sommerfeld completed his last semester of study at the University of Königsberg. He demonstrated the harmonic analyzer also to the mathematical physics institute, as his transcript attests,[46] but then devoted himself primarily to the theoretical foundations on which the representation of arbitrary functions by Fourier series and Fourier integrals rests. He was so captivated by the subject that he made it the subject of his doctoral thesis.[47] "I conceived and wrote it in a few weeks," he recalled of this final phase of his studies.[48] On July 28, 1891, he passed the "Examen Rigorosum" and on October 24, 1891, was awarded the Ph.D. degree. To be sure, his grade left something to be desired: "*rite*," as it reads on the Latin doctoral diploma.[49] What led to his receiving this "satisfactory" is unknown. In any case, the newly minted Ph. D. had to exercise patience before he was ultimately rewarded with recognition for his doctoral work. Lindemann, his dissertation

46 Abgangszeugnis, November 14, 1891. DMA, NL 89 016, folder 1.7.

47 Sommerfeld, *Functionen*, 1891.

48 Autobiographische Skizze, ASGS IV, pp. 673–679.

49 Doktorurkunde, DMA, NL 89 016, folder 1.7.

advisor, seems to have tolerated rather than valued the work. Nor did the Association of Physics and Economics, whose competition had provided the impetus for it, evince any interest in Sommerfeld's extended mathematical conclusions. "No one in Königsberg read it," Sommerfeld wrote his mother 3 years later.[50]

Was the work so little regarded in Königsberg perhaps because it appeared as pure mathematical theory and did not, like other works in mathematical physics of the Neumann school, indicate the connections among practical measurement, theoretical analysis, and physical conclusions? If by way of comparison one considers the entries awarded prizes by the commission for the Königsberg Ground Thermometer Competition, this becomes clear. The winner, a high school teacher from Gotha, was not himself a member of the Neumann school to be sure, but his winning entry did correspond broadly to the normal expectations of this tradition. He had "conformed in every respect to the relevant intentions of the assignment," the Prize Commission said in his praise. "Highly successful, in part exemplary," it commended the entry of a geophysicist from a "magnetic-meteorological" observatory in Russia. In this work, in light of the complicated circumstances in the ground, a mathematical analysis was declared impossible from the start.[51] All in all, in the case of these two entries, the Association of Physics and Economics awarded prizes for what they wanted to hear. It is true that, in his concluding "Evaluation of the Königsberg Ground Thermometer Station, 1872–1892," Volkmann referred in passing to the harmonic analyzer Sommerfeld and Wiechert had constructed; it was possible with this apparatus "to track the course of temperatures of each year with all its daily occurring irregularities." Sommerfeld's doctoral thesis, however, went unmentioned.[52]

Even if in the broader arc of his research it represents only a passing episode, a scientist's first work usually occupies a special place in his life. More than 50 years later, when Sommerfeld published his lectures on theoretical physics in book form, the problem of the Königsberg ground temperatures came belatedly back into favor. The task was for a given temporal course of temperature variation at the earth's surface which in the course of the year periodically rose and fell, to determine the course of temperature variation at increasing depths as a function of time. If the solution was represented as a Fourier series, there appeared already with the first members of the series at increasing depths a delayed temperature drop with respect to the passage of time at the surface. In the example Sommerfeld chose, the temperature dropped at a depth of 4 m by 1/16th the value of at the surface; to be sure, this had an effect at the depth with a phase displacement of a half year. Because of this phase displacement, in a deep basement it is "warmer in the winter than in the summer (or would be if all air supply could be excluded)," Sommerfeld wrote by way of illustrating this result.[53]

50 To his mother, June 9, 1894.

51 Meeting, June 4, 1891. In: SPGK 32 (1891), pp. 33–37. Olesko, *Physics*, 1991, S. 356–360.

52 Volkmann, *Beiträge*, 1893, p. 61.

53 Sommerfeld, *Vorlesungen*, vol. VI, 1948, pp. 68–71.

As a pedagogical example, then, the problem of heat conduction presented by the Königsberg competition proved useful even half a century later. For the time being, it played no role in Sommerfeld's career. That all the effort he put into it ought to have garnered him far greater recognition than his Königsberg professors had conceded became clear to him only when he encountered the circle of the mathematician Felix Klein in Göttingen. For now though, there were still other hurdles to overcome on his way to an academic career.

1.6 A Mechanical Basis of Electrodynamics

That Sommerfeld regarded the Königsberg heat conduction problem primarily as a mathematical challenge makes it clear why Lindemann became his dissertation advisor, and not Volkmann, in whose institute he had, together with Wiechert, conceived the harmonic analyzer. Lindemann's interest in physics was no less marked than that of other mathematicians who wished to demonstrate their competence in the field of physical differential equations. He lectured to the Association of Physics and Economics in Königsberg, for example, "On Molecular Physics," a subject actually more the province of his colleagues in physics. Lindemann explained this excursion into the neighboring field of physics by saying that his interest had been wakened by William Thomson. He thought he might be able to solve some of the problems Thomson had left unsolved "through more precise discussion of the relevant formulas by means of series development."[54] Thomson's and Lindemann's molecules were mechanical entities, which like solid bodies could display oscillations. At this time it was thought electricity, magnetism, and optics could also be explained mechanically in terms of the ether, which governed the exchange of forces among the molecules embedded in it according to specific mathematical laws.

The British physicist James Clerk Maxwell (1831–1879) had already determined these laws 20 years earlier and had formulated them in four equations. "Maxwell's equations" also led to the conclusion that light waves were transverse ether oscillations. Heinrich Hertz (1857–1894) went a step further. He showed with his celebrated experiments of 1888 that the ether oscillations postulated by Maxwell were applicable not only to light waves but also to invisible electromagnetic waves of far greater wavelength. Wiechert reported on the Hertz experiments to the Association of Physics and Economics in Königsberg.[55] Many years later, Sommerfeld recalled that "university lecturers and students were all at pains to assimilate the results of the Hertz experiments, which were emerging at that time piece by piece, and to explain them in terms of the equally abstruse presentation of Maxwell's original treatise."

54 Lindemann, *Molekularphysik*, 1888.

55 Meeting of June 6, 1889. In: SPGK 30 (1889), pp. 33–34.

Their significance was clear to him in a flash, however, when he saw the paper Hertz published in 1890 in the *Annalen der Physik*: "On the Basic Equations of the Electrodynamics of Bodies in Motion." There, without an attempt at mechanical explanation, Hertz placed Maxwell's equations "axiomatically at the top." The essential matter is expressed in the equations themselves, not in any mechanical conceptions, on which Maxwell himself based his theory.[56]

However, Sommerfeld had not actually arrived at this understanding quite as instantaneously as he describes. In his second scientific paper, which he took up shortly after receiving his degree, he was still far from setting Maxwell's equations "axiomatically at the top." He was rather undertaking the effort to establish them mechanically. The impetus thereto was the third volume of Thomson's *Mathematical and Physical Papers*, which had just been published, and immediately stirred considerable interest in Königsberg. Aside from an analysis of measurements from the Edinburgh ground thermometer station, the volume also contained a paper on the mechanical representation of Maxwell's equations.[57] Thomson ascribed these equations to the functioning of the ether, which he conceived variously, now as a kind of elastic body, now as a frictionless liquid, as it suited his purpose. Sommerfeld latched on to this idea and worked it up into a paper for the *Annalen der Physik*. He modified Thomson's hypothesis of a "quasi-elastic" ether such that he was able to derive a system of equations consistent with Hertz's equations for the electromagnetic phenomena in nonconductors. By attributing to the ether a "quasi-viscous" property (i.e., treating it like a greasy fluid instead of a rigid elastic body), he obtained the Hertz equations for electrically conductive bodies. He concluded that, "accordingly, the difference between conductors and non-conductors is that in the case of a conductor, the ether behaves like a liquid with friction, in that of a non-conductor, like a solid body."[58]

In contrast to his dissertation, this paper was not committed unread *ad acta*. No less distinguished a person than Ludwig Boltzmann (1844–1906), an authority in the field of theoretical physics, must have studied it with great interest, because he took the trouble to write a letter to Sommerfeld, still completely unknown in the world of physics. "For this purpose, I have added a few notes to my latest paper, soon to be published in Wiedemanns Annalen," Sommerfeld read in Boltzmann's letter, "among them, a note containing my objections to your paper—in the sense (it should go without saying) that I acknowledge the importance of your approach."[59] Boltzmann saw in Sommerfeld's work evidence "that all mechanical representations, if they only satisfy certain general conditions, have to lead to the equations of electromagnetism."[60] In his lectures on Maxwellian theory, published in book

56 Sommerfeld, *Vorlesungen*, vol. III, 1949, pp. 2–3.

57 Thomson, *Motion*, 1890, p. 462.

58 Sommerfeld, *Darstellung*, 1892, p. 139.

59 From Ludwig Boltzmann, November 17, 1892. DMA, HS 1977-28/A,31. Also in ASWB I, p. 49.

60 Boltzmann, *Medium*, 1893, p. 96.

form in 1891 and 1893, Boltzmann also referred to Sommerfeld's work when he surveyed the attempts to interpret Maxwell's equations mechanically. In order to explain certain electrodynamic phenomena, he said, Maxwell had hypothesized small particles in the ether that functioned like "ball-bearings." Sommerfeld rejected this assumption, "and so falls into new difficulties," Boltzmann criticized. But with criticism like that, Sommerfeld found himself in distinguished company, for in this context Boltzmann found Thomson's "quasi-rigid ether" vulnerable to criticism too.[61] Boltzmann had himself attempted several mechanical explanations and had even had demonstration models built for display at his lectures.[62] To be mentioned in the same breath—even if critically—with Maxwell and Thomson must have given the 23-year-old Sommerfeld a heady sense of success. "For me the most valuable thing about this work was that it attracted Boltzmann's interest," he recalled later. So far as its actual content was concerned, this first publication in the *Annalen der Physik* was in retrospect almost painful for him: "that not much else came out of such efforts at mechanical explanations soon became clear to me."[63]

Sommerfeld next registered for the state examination to qualify to teach at a high school. He needed to submit papers in the fields of mathematics, physics, and philosophy. In chemistry and mineralogy, he was tested only orally. In the subjects of religion and German, too, he had to stand for oral examination to demonstrate a broad general education. In June, 1892, this phase of his studies was also completed successfully. The "Royal Scientific Testing Commission" declared Sommerfeld certified to teach the subjects of mathematics, physics, chemistry, and mineralogy in the upper grades of a high school. His dissertation was accepted as the paper required in mathematics. As the paper in physics, his publication in the *Annalen der Physik* under a slightly altered title ("On Demonstration By Mechanical Representations in Physics According to Recent Work of Sir W. Thomson") was accepted. "It satisfies the requirements," read the certificate; "especially praiseworthy is that through his own ideas the author widened the field of view." In the subject of philosophy, honoring Königsberg's Kant tradition, Sommerfeld wrote a paper with the title "An Investigation of the Relationship Between Number and Time in Kant's Conception." It had been—so the testing commission wrote in its praise—"carried out with independent reflection and thoroughness."

In the oral examination, Sommerfeld demonstrated in the subject of mathematics "in all areas good, in some, very good knowledge." In the physics examination, he demonstrated "quite good knowledge in electrical science, and the same in mechanical theory of heat and theory of elasticity. In the area of optics and knowledge of instruments, however, there are several gaps." In chemistry, he demonstrated "quite comprehensive and confident knowledge;" in mineralogy he was attested to have been "well versed in the physical and crystallographical properties

61 Boltzmann, *Vorlesungen*, Part II., 1982, First Lecture, p. 6.

62 Dyck, *Katalog*, 1892, pp. 405–408.

63 Autobiographische Skizze, ASGS IV, p. 675.

generally, and consequently was also able to correctly identify the minerals presented to him. On closer examination, here and there gaps were found, although the candidate quickly oriented himself in every case. He was also familiar with the general theories of geology." In the philosophy examination, too, he showed no weaknesses. Sommerfeld had demonstrated "a most welcome familiarity with the important systems and disciplines of philosophy."[64]

64 Oberlehrer-Zeugnis, certified June 25, 1892. DMA, NL 89, 016.

2 Setting the Course

After 6 years of studies, his achievements—a Ph.D. and the teacher's certificate—let Sommerfeld look confidently to the future. Graduates of the mathematical physics seminar at the University of Königsberg, among whom he now could count himself, normally went on to careers as high school teachers.[1] The teaching certificate qualified him only in the subject matter for this profession, however. Before final qualification, he had yet to complete a probationary year, to demonstrate his practical abilities for a teaching career. But Sommerfeld was in no hurry to move ahead with this. For now, he still had his military obligation to fulfill. As the graduate of a humanistic high school, he belonged to the privileged class who, as "one-year volunteers," could opt for an abbreviated term of military service instead of the normal 3-year term.

2.1 Missed Opportunities

First, though, Sommerfeld treated himself to a trip to the south. For an East Prussian from Königsberg, the mountain world of the Alps held a magical attraction. "To the Alps! You know the magic in that word, of course," he wrote home. The letter bespeaks the impetuous lust for life of a 23-year old, hungry for experience, turning to all the things life offers up for which during his years of study there was too little time. Among these were excursions into the world of nature, as well as the pleasures of music and art. "Went hiking for the first time today, and also for the first time got soaked to the skin," he wrote on his arrival at Garmisch-Partenkirchen. The Zugspitze showed him "graciously its somewhat dour face, but then, smack! A curtain of fog like the theater at Bayreuth." That facing the cloud-hung mountains recalled Bayreuth was no coincidence; he linked this trip with a visit to the Wagner Festival. No less than with the Alps and Wagnerian operas, he was fascinated by the Bavarian lifestyle. In "a variety of beer pubs," he obtained a "heavenly impression" of Munich. Here, everything was "much more relaxed, reasonable; in Berlin, more businesslike, ostentatious." Thus, he compared Munich and Berlin, where he had spent several days at the start of his journey. Amid all the congenial atmosphere of the Bavarian beer pubs, he did not neglect art and culture. He went to see the ancient sculpture at the Glyptothek and enthused over the paintings of Arnold Böcklin (1827–1901), Anselm Feuerbach (1829–1880), and other nineteenth century masters in the Schack Gallery. "Then I went to one of the Munich attractions you won't know about, the Würm Baths. Scattered among flower beds is the greatest variety of basins, grottos, and springs of different

1 Volkmann, *Franz Neumann*, 1896, pp. 59–67; Olesko, *Physics*, 1991.

M. Eckert, *Arnold Sommerfeld: Science, Life and Turbulent Times 1868-1951*,
DOI 10.1007/978-1-4614-7461-6_2,

temperature, fed from the clear water of the Würm. You get sprayed by the giant frogs, swim through the blue grotto, then through a red one, and take showers of all sorts. The whole thing is as lush as a painting by Böcklin."[2]

Why not mix business and pleasure, he must have said to himself, for the visit to the Bayreuth Festival, the Alps, and the attractions of Munich were not the only reason for this trip. The annual congress of the German Mathematics Association was to take place in Nürnberg in September 1892. Walther Dyck (1856–1934), professor of higher mathematics and analytical mechanics at the Munich Technical University and organizer of this congress, was preparing an exhibition of mathematical models and instruments that was to include harmonic analyzers. Sommerfeld wished to use his stay in Munich to introduce himself to Dyck to discuss the demonstration of the harmonic analyzer he and Wiechert had constructed.[3] "Prof. Dyck not at home," he wrote his parents, and requested in the same breath that Wiechert be instructed "in the niceties of formal attire."[4] Apparently, observing social conventions on such occasions was not a matter of indifference to him.

But the timing of a visit to Dyck at the end of August 1892 was conceivably ill chosen, for shortly before, cholera had broken out in Hamburg, and Dyck was feverishly occupied with averting the threat of cancellation of the Nürnberg Congress. It was feared that large gatherings of people would spread the epidemic uncontrollably. On September 1, Dyck decided regretfully to cancel after all, since Nürnberg "was afraid of 50 mathematicians," as he wrote a colleague.[5] For Sommerfeld, this came as a bitter disappointment. He heard about the cancellation when he was in Meran, where each day, "hungry for Cholera news," he pored over the newspapers. "Not even the lovely setting of Bolzano mitigates the aggravation," he complained, bemoaning the fact that he would now not have the chance in Nürnberg "to get closer socially" to the luminaries of mathematics.[6] By chance, he met one of these luminaries, the Tübingen mathematician Alexander von Brill (1842–1935), at his holiday venue. Brill was an enthusiastic mountain trekker and was to preserve a "happy memory of our first meeting in the mountains," as he wrote Sommerfeld many years later.[7] At the Nürnberg Congress, Sommerfeld would have been able to deepen his "cordial acquaintance" with Brill and broaden his acquaintance with other leading mathematicians as well. He could have reaped the fruits of his Königsberg efforts with a demonstration of the harmonic analyzer.

"It's damned bad luck!" he said airing once more his irritation over the cancellation of the Nürnberg Congress; "I'd gladly forego a couple of my 4 Oetztal peaks." But he didn't let it spoil his vacation. "Otherwise, everything has gone as I wished, in the most daring sense of the word," he wrote proudly. "My energetic mountain climbing

2 To his parents, August, 25, 1892.
3 Hashagen, *Walther von Dyck*, 2003, pp. 419–424.
4 To his parents, August 25, 1892.
5 Hashagen, *Walther von Dyck*, 2003, p. 424.
6 To his parents, undated [early September, 1892].
7 From Alexander von Brill, December 5, 1928. DMA, HS 1977-28/A,41.

has won general admiration," he reports a mountain guide having told him; he could "climb on any of the peaks." To be sure, his feet reminded him of the rigors endured crossing snow fields wearing inadequate shoes; he was unsure "whether it was frostbite, or merely general overexertion." On the way, he befriended an American, with whom he hobbled about, competing over who was the greatest "martyr for this sport." At the same time, he dispelled his parents' anxieties that he was indulging in overly daredevil climbing adventures. He had undertaken "no dangerous, only strenuous" mountain treks, and with the fine weather and his "total immunity to vertigo," even "the so-called dangerous routes" had been "perfectly safe." He felt such a total absence of vertigo that he considered "vertigo either humbug or simply fear."[8] Ultimately, the tone of his letter reveals also how close his ties to his parents were. "Your loving, anxious letters, dear Mother, have filled me with the wish to be with you. I have to confess that earlier, I was focused almost entirely on the here-and-now, and thought of Königsberg only when I heard something about the cholera."[9]

In the nineteenth century, cholera recurred constantly and was accepted almost fatalistically as an unavoidable epidemic. But the cholera epidemic that broke out in Hamburg in mid-August 1892 went down in history.[10] It broke out with unparalleled ferocity and occasioned fear of the epidemic far beyond the borders of Hamburg, as Sommerfeld's reaction from the South Tyrol attests. The public followed its course as scarcely any previous epidemic. Up to August 20, 115 cases and 36 deaths had been recorded. Two days later, the number of cases had tripled, and 200 had died. Thousands of Hamburg citizens fled the city in panic. At the same time, measures were now finally taken to control the epidemic. Robert Koch (1843–1910) traveled to Hamburg on commission from the Imperial government. Almost a decade earlier, he had discovered the pathogenic cholera agent and now headed a newly established Institute for Infectious Diseases. Koch identified the cause of the epidemic as Hamburg's drinking water. The cholera bacterium, resident in the patient's intestine, found its way through sewage into the Elbe and from there back into the drinking water. On August 26, 995 cases and 317 deaths were recorded. The same day, the Hamburg city authorities responded to Koch's assessment and through a police decree warned against the use of water that had not first been boiled. Already in the first days of September, a decline in recorded infections and deaths was recorded. Purification of Hamburg's drinking water took a while longer to be realized; nonetheless, further outbreaks of cholera occurred neither in Hamburg nor elsewhere in Germany. The preventative measures proved effective, even if here and there (such as in the case of the cancellation of the Nürnberg Congress of the German Mathematics Association, in light of the small number of participants) they might have appeared exaggerated. The conference, along with its exhibition of

8 Here, Sommerfeld employs a pun impossible to render as a pun in English: "Schwindel" means both "dizziness," "vertigo," and "swindle," "humbug."

9 To his parents, undated [early September, 1892].

10 Evans, *Tod*, 1990.

mathematical models and instruments, was rescheduled and held a year later in Munich.[11] This was no consolation for Sommerfeld; in 1893, as a 1-year volunteer, he was fulfilling his military service and therefore could not take part in it.

On his return to Königsberg from his vacation in the Alps, he learned that Adolf Hurwitz, whose lectures on function theory and elliptical functions he had attended, had invited him to be his assistant at the Federal Technical University (ETH) in Zürich, where Hurwitz had been appointed successor to Ferdinand Georg Frobenius (1849–1917). Sommerfeld's mother had informed Hurwitz, however, that following his return from his trip, her son needed to begin military service. As soon as he learned this, Sommerfeld wrote Hurwitz, "When she wrote her letter, my mother assumed my year of service could not be postponed." As he had learned from the military authorities, a postponement would certainly have been possible. "I telephoned you at once, but to my great disappointment, it was too late." He would have "snapped up the offer," for his "love of scientific work was too great for him to pass over unconsidered any connection with scholarship." So nothing was left for him "but to curse those days I had spent too long on my trip, and to ask you to keep me in mind in the event of a future vacancy. I would of course be especially grateful for a position immediately upon completion of my year of service, that is, October, '93. I hope it not indiscrete of me to trouble you for an indication whether you might be able to offer me any prospects for that time-frame?"[12]

2.2 Military Service

The notification of a renewed offer for which he had hoped did not materialize, but the fact that he was even considered as a candidate for the position of assistant strengthened Sommerfeld's resolve to pursue an academic career. Following service as a 1-year volunteer at Königsberg, he would have been able to complete his probationary teaching year and thereby complete the final qualifying stage for a secure career as a high school teacher. But now his heart was set on something loftier. He wanted to complete the habilitation in order to become a university professor. Should this prove impossible with Hurwitz at the ETH in Zürich, then at some other university. Perhaps he was already now dreaming of pursuing mathematics under the wing of Felix Klein, who was in these years transforming Göttingen into a world center of mathematics. Klein must already have become an idol for Sommerfeld when he attended Hilbert's lectures, for since 1886 Hilbert and Klein had been engaged in an intense correspondence. It would be surprising had Hilbert not raved to his Königsberg students about Klein, whom he soon followed to Göttingen.[13]

For the time being as a soldier, though, Sommerfeld was in no position to realize his dreams of a career. On October 1, 1892, he reported for duty with the 43rd

11 Hashagen, *Walther von Dyck*, 2003, pp. 419–436; Dyck, *Katalog*, 1892, Foreword.

12 To Adolf Hurwitz, September, 1892. SUB, Mathematiker-Archiv 79, 260. Also in ASWB I.

13 Frei, *Briefwechsel*, 1985.

Fig. 5: After completing his education, Sommerfeld fulfilled his military obligation as a "one-year volunteer" in an infantry regiment at Königsberg. He took little pleasure in the often tedious daily round of the soldier, however. He had "never felt [himself] to be strong militarily," he wrote a colleague at the outbreak of World War I (see Chap. 7) (Courtesy: Deutsches Museum, Munich, Archive).

Infantry Regiment.[14] This regiment was stationed at Königsberg, and in fact the barracks were located at Steindamm very close to Sommerfeld's parents' home. One-year volunteers were allowed to choose their unit and the location where they would fulfill their year's service. They were also not required to live at the barracks, so that Sommerfeld could fulfill his service while living at home. The privileges granted 1-year volunteers required a military-friendly orientation within the educational community.[15] With his essay on the "Fellowship among Volunteer Medics in War," Sommerfeld had already clearly reflected this attitude. "And just at this time, when the people are making the greatest efforts at keeping the Fatherland defense-ready, we will place special value on remaining faithful in this regard," he had written in *Burschenschaftliche Blätter*.[16]

14 Bescheinigung, DMA, NL 89, 016, Mappe 1.7.

15 Mertens, *Bildungsprivileg*, 1990.

16 Sommerfeld, *Genossenschaft*, 1890, p. 220.

Physically, too, military service was no very great challenge for Sommerfeld, who was an Alpine-tested mountain trekker. "Today, just the brief note that despite the heat I have not expired, and despite the dust, have not silted up," he wrote in midsummer of 1893 reporting a maneuver from Masuria home to Königsberg. "One day, though, it was a little crazy; I can take it though." There was no enthusiasm for military routine, however. "Listening to the soldiers' songs being belted out makes my eardrums burst," he wrote on one occasion. Another time, he poked fun at a General who "was displaying his brilliance to the officers," through whose speech he "had slept on a pile of straw." The letters from the maneuver show that he could hold his own even under nasty circumstances and knew how to make the best of any situation. "Beautiful countryside, good fellowship, good food," he wrote on a postcard. "The hard work is doing me much good." Nor did he allow the stresses and strains to detract from his appreciation of the beauty of the Masurian lakes. "My present quarters are fabulous, our situation on Lake Kosno, fabulous" he wrote at the end of a strenuous day's march. "Imagine a large lake, 5 kilometers long, 2 wide. Its shoreline, now wooded, now pasture, winding right and left until it disappears behind a green backdrop. Water, of a beautiful clarity, whipped by the wind to white waves, and beautifully colored by the blue of the sky."[17]

2.3 Mineralogical Interlude

Even before he had concluded his military service, Sommerfeld heard through a friend of the family that the mineralogist Theodor Liebisch (1852–1922), appointed some years earlier at Göttingen from Königsberg, was seeking an assistant. "I spoke of you to Adelheid," the friend confided, "and suggested that you would likely go to Göttingen." Adelheid was the wife of the mineralogist in question. Another professor's wife, who was to hear nothing about the situation however, was Lisbeth Lindemann (1861–1936), the wife of Sommerfeld's doctoral advisor, who was appointed at Munich the same year. The professors' wives presumably did not wish to be liable to the charge of indiscretion; thus, the situation was treated as a matter of secrecy. "At present, Lisbeth knows nothing," the family friend added by way of explanation. She described the crux of the matter thus: "Adelheid asked me whether you might wish to accept the position of assistant to her husband; I said I didn't want to mention it to you before I knew for sure it would still be available in October, and what Professor Liebisch thought of the matter. Just today, I received a letter from the Professor saying he would be very happy if you would be in touch with him, since he had heard only good things about you. First of all, then, we need to know whether you are at all inclined to accept this position. Adelheid thinks you could learn much from it."[18]

17 Letters to his mother, end of July to beginning of September, 1893.

18 From Margarete Erdmann, July 7, 1893.

The observation "that you would likely go to Göttingen" indicates that even without the offer of a position as assistant to Liebisch, Sommerfeld would not have continued his career as a high school teacher in the Prussian educational system. Since 1886, Klein had been working to turn Göttingen into a Mecca of mathematics. In collaboration with the equally legendary Friedrich Althoff (1839–1908), who was setting the course for a new high school policy at the Prussian Ministry of Culture, Klein's significant influence extended far beyond Göttingen, where appointments to mathematical teaching positions and other consequential matters in his field were concerned. Even in the area of science policy, Prussia played a major role in the German Empire, and in Klein, Althoff had a fellow warrior and confidant, who shared his aspirations, and spared no efforts to realize them. Faced with limited resources, Althoff wanted to erect beacons of German science by concentrating scientific talent in universities where local tradition had raised one field of study or another to particular prominence. The archaeological sciences, for instance, would flourish at the University of Berlin. At Göttingen, where such important mathematicians as Carl Friedrich Gauß (1777–1855), Gustav Lejeune Dirichlet (1805–1859), and Bernhard Riemann (1826–1866) had worked, mathematics would shine above all.[19] The natural sciences, in which Göttingen was also traditionally strong, would of course benefit from this glory too. One has only to think of the physicist Georg Christoph Lichtenberg (1742–1799) or the chemist Friedrich Wöhler (1800–1882). In the "Althoff System,"[20] even well-established professors became chessmen in a university policy calculus, which—depending on the particular chess move—aroused embitterment or admiration. Klein's position was at all events substantially enhanced by his association with Althoff.

To his reputation as a mathematician could be added the aura of organizer of science. This role came into play, for instance, in the 1892 Göttingen appointment of Heinrich Weber (1842–1913), who had taught at the Albertina in Königsberg from 1875 to 1883, and numbered Hilbert among his students there.[21] Together with Weber, Klein founded the Göttingen Mathematical Society and reorganized the Göttingen Society of Sciences. Weber must have enhanced the attraction of Göttingen for Sommerfeld still further, even if his principal interest lay in "Felix the great," as Klein had admiringly been dubbed in mathematical circles. In America during the summer of 1893, Klein represented German mathematics at the World's Fair in Chicago.[22] With its universities exhibition, the Empire of Kaiser Wilhelm used this occasion to place itself in the limelight as a leading cultural nation among world powers. In this regard, Mathematics was a particular asset. With portraits of Dirichlet, Riemann, and other great mathematicians, a bust of Gauß, along with a

19 Rowe, *Felix Klein*, 1989; Tobies, *Development*, 2002.

20 Brocke, *Hochschul- und Wissenschaftspolitik*, 1980.

21 Peter Roquette: Heinrich Weber, David Hilbert, and Königsberg, 1992. http://www.rzuser.uni-heidelberg.de/~ci3/weber.pdf (29 January 2013).

22 Parshall/Rowe, *Emergence*, 1994, ch. 7.

display of textbooks, technical journals, and hundreds of dissertations and habilitation theses, Klein demonstrated to the world that in this field, Germany was world class.[23] Even the harmonic analyzer Sommerfeld had so dearly wished to have demonstrated in Nürnberg in 1892 was exhibited in the form of large format photographs (the apparatus itself was exhibited around the same time in Munich, where the exhibition planned for Nürnberg, but cancelled due to the cholera outbreak, was realized). For several weeks following the World's Fair, Klein remained in Chicago to highlight in guest lectures mathematical accomplishments of German provenance. He wrote later to Hilbert that "In Chicago in a series of lectures," he had had "the opportunity to present a kind of particular program."[24]

The offer of a position as assistant at Göttingen must have come at just the right time for Sommerfeld, then, even if he was to work only at a mineralogical institute. If one pursued this field as a theoretician, there were many bridges from mineralogy to mathematics. Perhaps the offer also awakened memories of the time when Arnold and his father had roamed through the East Prussian countryside searching for mineral specimens. The Association of Physics and Economics of Königsberg, of which Franz Sommerfeld and, since December 1891, Arnold too were members, set a high value on the science of mineralogy; its archive contains a number of reminiscences of his father's passion for minerals and that of other members as well.[25]

Why not mineralogy, then? Sommerfeld may well have asked himself when he hung up his soldier's uniform in September of 1893 to devote himself once more to his career. Liebisch had lectured on mineralogy as professor at the Albertina from 1884 to 1887 and had pursued his mineralogical research "in a strict mathematical spirit," which put him into a close collegial relationship above all with Lindemann and Volkmann.[26] Thus, even at the Göttingen mineralogical institute, Sommerfeld did not have to feel himself entirely cut off from his Königsberg roots. As soon became evident, Adelheid Liebisch was to treat her husband's assistant with almost maternal solicitude. Liebisch himself proved "exponentially more amiable," Sommerfeld wrote his parents shortly after his arrival in Göttingen; he had even "gone with him to pubs" and had helped him search for an apartment. Liebisch showed his new assistant his best side. "He says we'll write a book this winter, we have papers to prepare, we have to go to Vienna to the Natural Scientists's Assembly." Sommerfeld probably sensed that working for this boss, there would be precious little time left over for his continuing interests in mathematics. "He apparently wants total collaboration. Am I going to be able to manage that? I'm very fearful."[27]

In addition to these initial qualms whether with the job as assistant at the mineralogical institute he had set the right course for his academic career, there was also

23 Hashagen, *Walther von Dyck*, 2003, p. 434.

24 Klein to Hilbert, October 27, 1893, printed in Frei, *Briefwechsel*, 1985, p. 99.

25 Franz Sommerfeld, *Familie der Quarze*, 1900.

26 Schulz, *Theodor Liebisch*, 1922, p. 419.

27 To his parents, October 9, 1893.

a certain amount of homesickness. Through his correspondence at least, he assured his mother, he was going to maintain his contact with home. "In any case, you will hear everything from me that might interest you, both happy and sad. That I'm committed to. For what is the good of scribbling, if only to jot down humbug?" He reassured himself with respect to his work with Liebisch. "Mornings from 9:00 to 1:00 I'm at the institute; afternoons, from 4:00 to 8:00. During this time, though, I've been busy with institute-related work only minimally up to now. The only point of concern is whether I will end up in the right field of work." He certainly did not want to devote himself entirely to mineralogy; he was too fond of mathematics for that. On this point, from the very beginning, he left Liebisch in no doubt. Liebisch tried to make the field of work palatable to him through a compromise proposal: "Liebisch thinks I ought to do work in crystallography, which requires mathematical experience; he expects that after one year, I will have completed my habilitation thesis in this branch." As attractive as this proposal was, it couldn't dissipate Sommerfeld's fundamental doubts. "Work like that," he explained to his mother, "would, on the other hand, require significant experimental skill. It would be physics, not mathematics."[28]

A glance at Liebisch's publications shows that Sommerfeld's doubts were entirely justified. In February 1893, the Göttingen Academy published a report by Liebisch "On the Spectral Analysis of the Interference Colors of Optically Biaxial Crystals, I," which dealt with microscopic observations of crystal-optic phenomena, not with mathematical analysis. The addition "I" indicated that Liebisch intended to deepen this kind of crystal research with subsequent papers.[29] Future work at the mineralogical institute, Sommerfeld feared, would leave him little time for his true interests. "Once involved in the activity of observation, not much would come of any thorough, purely mathematical work. Professor Wallach advised me rather against a career such as Professor Liebisch has in mind; Liebisch-style mineralogy in Germany is not, in his opinion, an item much in demand."[30] In the person of the Göttingen chemist Otto Wallach (1847–1931), Sommerfeld found an advisor he quickly came to trust. Wallach was also from Königsberg and was even distantly related to Sommerfeld. His father was the vice president of the East Prussian administration. Wallach family roots on his mother's side are traceable to the time of Frederick the Great, when an ancestor in Königsberg was assigned the reorganization of the Prussian administration. Because they both came from Königsberg, Sommerfeld regarded the nearly 20-year older Wallach as a fatherly friend, to whom he could open his heart. Occasionally, too, he was invited over by Wallach, who led the life of a bachelor. "He has an awful lot to do, and is very diligent," Sommerfeld reported after an evening at Wallach's, at which he was served a "rather dangerous Spanish wine" and no doubt a few internal tidbits from the world of the

28 To his mother, October 13, 1893.
29 Schulz, *Theodor Liebisch*, 1922.
30 To his mother, October 13, 1893.

Göttingen professors as well.[31] The better he came to know the chemist, the more he was impressed by him. "Always authentic, hard-working, utterly incisive, engaging, helpful," was Sommerfeld's characterization of Wallach. To his assistants he was "the kindest boss," who in the case of every assignment at the institute explained "how it should be begun, carried out, and completed. Everything works together according to a great plan in order to scour an area of chemistry (the terpenes), and he is the integrating brain. He does not show off his importance. I visit him every two weeks or so for about an hour, and sometimes stay for supper."[32]

When Sommerfeld penned these lines, he had already been assistant at the mineralogical institute for 4 months and may very well have been regretting that he had not followed Wallach's advice to resign his post. But at the start, Liebisch had shown his assistant only his best side. Two weeks after his assistant's arrival in Göttingen, Liebisch still seemed "a magnificent man, a whole man" to him. "He stands like a mighty oak in the forest of the other professors, who (according to what I have heard) bend like small willows at the touch of an influential colleague (Klein, for instance) or a ministerial big shot (Althoff)." Particularly "enchanting" were his noon walks with Liebisch, from whom he heard details about academic life in Göttingen that he "would rather not commit to paper." Liebisch was at pains "to educate him diplomatically in weathering the tricky local complexities of university life and gossip." These included the ritual of house calls in which prospective lecturers[33] had to introduce themselves to university professors. Next Sunday, he informed his parents, he planned to begin his "visitation tour," which "here has to be done dressed formally. Any other attire is a cardinal sin, and would mean an end to all hopes of habilitation." His mother's admonition to guard against the "temptations" of a strange town he found superfluous. "From the description of my way of life you can see there is no room for pursuit of extravagant pleasures. Besides, the temptations you probably have in mind are out of the question here in any case. Everybody keeps far too close tabs on his neighbor for that." He found Göttingen "awfully small-townish," and had "so far not gone out one single evening. Life here proceeds like clockwork. Besides, not many mothers can have sent their sons off to foreign places so confident in their respectability as you have sent me. So, no worries!"[34]

31 To his mother, October 29, 1893.

32 To his mother, February 17, 1894.

33 Here and throughout this work, "lecturer" translates "Privatdozent," a German academic rank, pre-professorial, but granted permission to lecture at the university. Privatdozenten are essentially free-lancers: with or without a stipend, they are occasionally employed as assistant professors (as Sommerfeld was for two years). When this term of employment is over, they return to the status of Privatdozent (lecturer). This status requires the "habilitation"—another qualifying ritual peculiar to the German university system. A successful "habilitation" carries with it permission to lecture. See chap. 3.4 for greater detail.

34 To his parents, October 13, 1893.

Two weeks later in the context of this "visitation tour," sporting a top hat purchased expressly for the purpose, he had already paid his respects to several Göttingen professors, among them the theoretical physicist Woldemar Voigt (1850–1919), vice chancellor of the university, who soon thereafter would offer him an assistantship.[35] In other respects as well he gave his mother no grounds for disquiet. He was leading "the healthiest lifestyle in the world," he assured his parents in far off Königsberg. Munich, with its beer halls, would have offered greater temptations. Here, by contrast, he was acquainted with "the drink 'beer,' supposedly so beloved of Germans, in name only." The noontime stroll with Liebisch had in the meantime become a fixed feature of his daily routine. If it was not raining, these often turned into hours-long promenades. Additionally, heeding his father's advice, he joined a gymnastics group, which included "several lecturers, some candidates for civil service, and a lieutenant," meeting for an hour one evening a week at a gym. "I have a lot of fun with that."[36]

For Christmas, his parents wished to send him a piano so he could develop his musical talents, too. But the apartment Sommerfeld had rented did not allow for this. "I'd be thrown out at the first notes. The whole house is filled with studiousness; absolute quiet is sacred. So, much as I would like to, I wouldn't be able to accept the Christmas gift in question." He did not have to renounce the piano altogether however. He had opportunities to play at the house of his boss, albeit only at times of his absence, for Liebisch had no understanding of "such nonsense." Sommerfeld also described his lodging in great detail for his parents. "A very large writing desk, a huge floor-to-ceiling bookcase, a small sofa, along with a table, mirror, clothes cabinet, and an iron stove. I suffer with the stove here. Either the room is cold, or the iron rascal cooks one's brain dry." All in all, though, he was content with his apartment. For a modest supplement to his rent, the landlady gave him breakfast, with coffee and rolls, and in the evening, bread, butter, sausage, and cheese. "In addition, she makes me—this is the main thing—a nice pot of tea. I'm confident she isn't overcharging me. No extra charge for polishing my boots, service, etc.; it's all covered in the 120 Mark [annual] rent." At lunchtime, he ate at "the Hotel Royal for 1 Mark—with beer and tip, 1.15." Liebisch had advised him to do so; eating at cheaper places would damage his reputation. This genteel dining style gave him little pleasure, though. "For one sits at a long table chatting, or more often, not chatting."[37]

At 120 Marks yearly rent, Sommerfeld had found a relatively economical apartment. His assistant's salary was 1,200 Marks per annum.[38] The normal rent for a small two- to three-room apartment in Göttingen in the 1890s was about 180 Marks. Any prospective lecturer like Sommerfeld who had just climbed the first rung of the academic career ladder and had no other income was far from being able to count

35 To his mother, October 31, 1893.
36 To his parents, October 29, 1893.
37 To his parents, October 29, 1893.
38 To his parents, November 27, 1893.

Fig. 6: Sommerfeld felt his duties as assistant at the mineralogical institute of the University of Göttingen were "killing time mineralogically." He rested all his hopes on Felix Klein, who offered the prospect of a career as lecturer in mathematics (Courtesy: Deutsches Museum, Munich, Archive).

himself among the middle or upper class. In 1890, a Göttingen police official at the lowest income level, for instance, earned between 1,125 and 1,275 Marks per annum plus a housing supplement of 240 Marks. A bookkeeper might earn between 2,400 and 4,000 Marks.[39] So keeping an eye on his daily expenditures was, for Sommerfeld, not exactly a peripheral issue. When conforming to academic etiquette was an added consideration, Sommerfeld's report of his daily expenses becomes a kind of metric of how much he was prepared to lay out for the sake of his career.

As far as his duties at the mineralogical institute were concerned, Sommerfeld soon experienced them as not so "fearsome" as at first, but rather increasingly with a sense of tedium and a waste of time. He had to assemble "a very tedious catalogue" for a yearbook of mineralogy. In the mineralogical institute, which more resembled a museum, he saw students face to face at a practicum only once a week. "The collections are kept covered and in obscurity, but are otherwise in exemplary order." Among the more pleasant of his daily tasks was occasionally helping his boss "with more mathematical matters . . . about which he doesn't know much." Thus, he had recently been able to show, to Liebisch's "infinite delight," that in a publication Voigt "had dug deep into Maxwell," without acknowledging this source. "Rather often Liebisch has had occasion to ask as we sat cozily by ourselves in the museum: Don't we have marvelous fun?"[40]

39 Saldern, *Göttingen*, 1999, pp. 16–17.
40 To his parents, October 29, 1893.

But it was not long before Sommerfeld came to know a somewhat less agreeable side of his boss. "Liebisch was very surly today," he wrote in early November. "I think his wife came home too late yesterday. So we have not gone for a walk today."[41] Liebisch's moods and the frequently monotonous work at the mineralogical institute soured his days. "As nice as he generally is, he can be equally grouchy on occasion," he complained again a week later about his boss. "He has now had one, or rather four critical days. I have been able to speak with only about the strictest necessities during this time. What has been annoying him? Has he been angry with his wife (she wanted to go walking with us on a weekday on which he hadn't invited her to do so), or with me (the preparation of a catalogue had dragged on somewhat; the work was also tedious, though; I had to sit all day long arranging small slips alphabetically, and shifting them about so much my arm was sore), or was it the onset of a cold?"[42] His boss's moods occasioned real mental anguish for Sommerfeld. "It's so against my nature to relate to another person other than openly, and yet with him I now have to be cautious," he wrote home again a week later. "Mrs. Liebisch certainly doesn't have it easy. I would very much like to get together with her more often, but it can't happen—precisely this can't happen. It must be my nature that I need a person with whom I can relate completely frankly. Mrs. Liebisch is quite so inclined. Our relationship is as friendly as could be, but we're not permitted to see each other."[43]

The greater the distance between him and his moody boss grew, the stronger his empathy for his boss's wife. "She really is a good person," he wrote his mother, who had long been a friend of Adelheid Liebisch. "She is very fond of you, and since she lives an absolutely cloistered life here, speaks only to her 4 children, and sees only her husband, she cherishes fondest memories of her distant friends."[44] The situation evolved more and more grotesquely: "Alright, here comes a brief and completely frank report of my relationship with Liebisch," he wrote to Königsberg a few weeks later. "Othello is nothing compared to Liebisch. Isn't it ridiculous? Liebisch is jealous, insanely jealous. He locks his wife in, and won't countenance her speaking with any male individual. I have this from the most reliable source. What do you say to the following exchange, which Mrs. Liebisch reported to me? She: 'Dr. Sommerfeld could come to see me on Monday evening, when you have Eskimo (his men's club).' He: 'Then I'll resign from Eskimo.' So there it is! Liebisch is pathologically jealous. There you see the crazy sources of the difficulties one can face in life! I'm sure I don't need to reassure you that he has no grounds for his jealousy; the whole thing is really too ridiculous, too stupid, too crazy."[45] On the same day, Sommerfeld committed this "Othello" letter to paper, Adelheid Liebisch,

41 To his mother, November 7, 1893.
42 To his mother, November 14, 1893.
43 To his mother, November 19, 1893.
44 To his mother, November 27, 1893.
45 To his mother, December 20, 1893.

the "Desdemona" of this story, also wrote back to Königsberg. She would look out for Arnold's best interests, she assured his mother.[46]

From the perspective of the raging "Othello," his wife's maternal inclinations towards his assistant were not exactly apt to assuage his anger. But besides the jealousy, there were other reasons Liebisch was "thoroughly distrustful and moody to excess," as Sommerfeld wrote home in a character study of his boss. "He hates Göttingen and all his colleagues. At every opportunity, he disparages the situation at Göttingen, its formality and ceremoniousness. Mind you, there is no one so set on form as he, and no one more easily offended by trivial lapses in form." Additionally, Liebisch now resented him for showing greater interest in mathematics than in mineralogy. "On this point, I don't fault him too much. Perhaps I should not have accepted this post merely as a means to the end of attending some lectures in mathematics. That annoys him. But I never made a secret of this." Now, Liebisch's initial friendliness had turned to its opposite. "The role I've played in this manner at the institute for weeks now has been truly unenviable. I have been miserably annoyed."[47]

2.4 Stick It Out, or Resign?

After 3 months, there was no longer any talk of strolling with his boss, which before had given Sommerfeld such pleasure. Liebisch avoided direct contact with his assistant as much as possible. "At first, we always walked home together," Sommerfeld wrote about such details of his daily life. "But recently, the ass has several times gone off alone on trivial pretexts just to avoid having to walk with me. Whenever possible now I avoid subjecting myself to this. I respect myself too much for that. I've told him that if he thinks he can find a better assistant, I would gladly step aside, and that it would be against my principles to pocket the 100 M per month if there is nothing more for me to do than just put in my time. Then he gets a blank look, and says absolutely nothing."[48]

His readiness to resign his assistantship reveals Sommerfeld's determination no longer to purchase an academic career at the cost of compromises that ran counter to his inner motivation. "Now you will say that I could safely go on working at the institute, all the more so as I have nothing to do for him," he wrote anticipating the objection his parents might make. "Yes, but scientific work is different from a trade; one has to be enthusiastic and comfortable in it. For me in the institute now, ideas are largely precluded." He reviewed once more what in the beginning had made daily life at the mineralogical institute bearable for him. "Saturday mornings were always my favorites; students were there to whom I had to explain crystals.

46 Adelheid Liebisch to Sommerfeld's mother, December 20, 1893.

47 To his mother, December 20, 1893.

48 To his mother, December 20, 1893.

I had fun with that, and learned a great deal in the process. I would like to have had practicum every day." At the conclusion of this densely written, eight-page letter, he tried to come to a sober evaluation of his situation and his boss. "I've berated myself thoroughly in the foregoing. Not without reason; with his moods, he has really offended me. But I also want to be fair. He has his virtues. He is very energetic, and has an iron discipline, as well as a keen intellect and wit. He can also do much for others, so long as they don't touch any of his many crazy spots." But Sommerfeld got a certain satisfaction from learning that Liebisch's previous assistants had not stuck it out for long with him. They had "stayed on average 1 year at most. Apparently no one can get along with him." That in spite of everything he was invited to the Liebisch home for the coming Christmas Day, he attributed to the good offices of his mother's friend. "I'd rather spend it with Wallach, but he will be in Berlin. So I'll probably go to the Liebisch's after all. I have now become quite inured to his sourpuss face. I doubt I'll make it up with him, but I don't care about it either. The other circumstances of my life are very comfortable, after all; basically, I can content myself with having landed on a path not smoothed in advance for me by maternal solicitude and well-meaning friends. So, no worries on my account!"[49]

His parents did not take the report as badly as might have been supposed in light of his threatened resignation of the assistantship. "You're quite right to emphasize the humorous aspect of the situation; that is how I see it too," Sommerfeld wrote to Königsberg on Christmas Eve. He did not spend Christmas Day at the Liebisch home after all, though; instead, he used his day off to visit his brother, who had become a doctor, working at a neurological clinic near Hattenheim am Rhein. Regarding his own career problems, he wrote only that he wished to convey New Year's greetings to Lindemann, "discretely hinting the question whether he might be able to use me." To preclude any gossip between Königsberg and Göttingen concerning "the aforementioned tragedy," he swore his mother to secrecy. "Likewise if you must write to Mrs. Liebisch. Here too, of course, not a word about jealousy," he asked urgently because he had promised Adelheid Liebisch not to mention it. "But I owed it even more to you to break my promise, than to her to keep it."[50]

Actually, it was clear to Sommerfeld just several weeks after his arrival in Göttingen that he was not in a good situation at the mineralogical institute. But that was not his only worry. During these autumn weeks of 1893, he went through a whirl of shifting emotions. His grandmother was dying.[51] "Ochen" had been a presence his entire childhood and youth in his parents' Königsberg home. "So, I'll see one loved person less at home, and one grave more!" he wrote to Königsberg in response to the news of her death. This sad occasion brought back the memory of his sister's death. "How bleak and hopeless we all felt then! Please be so kind, dear Mother, as to lay a wreath for me on Ochen's grave. Flowers are the only expression

49 Ibid.
50 To his parents, December 24, 1893.
51 To his parents, October 29, 1893.

of love one can show the dead." In addition, he was concerned about his brother who had, already as a medical student, become addicted to morphine and in his work as a neurologist continuously faced the threat of relapse. The ups and downs of Walter's condition were the subject of sorrowful remarks in countless letters. He had just "had a very nice letter" from Walter, Arnold wrote his parents in an effort to lighten their woes in this respect at least. "He seems quite happy." He did not however want to listen to the gloomy ruminations about growing old his mother had associated with the sad news. "I think you will remain my young mother for a long time yet, and we will still often be young and happy together."[52]

Homesickness, the tensions of his relations with Liebisch, doubts about the career path he had entered into, and worries about his brother—such were Sommerfeld's thoughts as the year 1893 reached its conclusion. When he returned from his trip to the Rhein at the end of the Christmas holiday, the "Othello" affair retreated temporarily into the background. "Walter is physically run down, looks rather miserable, has very little appetite," he reported to his parents in far off Königsberg about his brother's condition, which, however, he attributed to a recent bout of the flu. He thought there were no grounds "to suspect morphine." Because of differences with his boss, Walter's situation was not exactly simple; he would "prefer to be with a different director, which I understand completely." About Walter, he had "absolutely never had the impression of someone neurologically ill; we spent the days in brotherly affection, discussing all sorts of things perfectly rationally; this would have been impossible for me had he shown any significant signs of illness." So the Christmas holidays spent together "had been a very pleasant, relaxing time for us both."[53]

It was only a few days, however, before Göttingen's "Othello" made clear to him how precarious his daily life as an assistant really was. Apparently, Adelheid Liebisch had read her husband passages from her correspondence with Sommerfeld's mother in order to demonstrate how baseless his jealousy was. This only served to confirm Sommerfeld's judgment that his boss was "a little crazy." Around this time, he had already firmly decided to put an end to this grotesquerie as soon as possible. In a New Year's letter to Lindemann, he had reported "all sorts of Göttingen circumstances" and had asked him "whether he can use me." In response—he reported to his mother—not Lindemann himself, but his wife had sent him a "Munich beer post-card." "She writes on it that she has spoken with Dyck about me, and will not give up, etc., etc. God save me from my friends, is all I can say."[54]

Among the faculty wives, Lisbeth Lindemann's reputation was hardly such that discretion numbered among her greatest virtues. For this reason, Sommerfeld's parents feared that gossip about Arnold's difficulties with Liebisch might now cause a stir also between Göttingen and Munich. "To set your minds at rest, I've also sent

52 To his parents, November 4, 1893.

53 To his mother, January 2, 1894.

54 To his mother, January 5, 1894.

off something quite stern," Sommerfeld wrote his mother in an attempt to relieve her of this fear. "I don't see any danger here. First of all, she can neither help my problems nor cause me any harm. Second, she talks more than she acts." So far as the "Othello" affair was concerned, he said he had resolved this for himself. "The jealousy story is by now no longer current. But it had planted a grudge in him, not to be uprooted from his thick skull."[55] From her side, Adelheid Liebisch confirmed in a letter to Sommerfeld's mother that she would not abandon Arnold in his troubles. "But I am truly sorry for him; he is as dear to me as a son, and I have always treated him like my good, honest friend. And so it will be going forward."[56]

This was no solution for the damaged relationship between boss and assistant, rather the contrary. "For my stern tyrant demands that I put in my full office hours," Sommerfeld wrote home. "The other professors (Wallach, Voigt, etc.) require their assistants to come in only when there is something for them to do. This Liebisch rule is truly childish." That he was allowed to attend Klein's lectures during his work hours was the only concession he was able to wring from Liebisch. "We're amazed that you have trained 'Theodore' so well that he permits you to go to the university daily from 11:00 to 1:00," Sommerfeld described a conversation with the assistant at Voigt's institute. "He is otherwise so strict with the work hours. And on top of that to allow you to go to Klein! The other assistants have always had problems with him." That the other assistants regarded this as a sign of his good relationship with Liebisch, he found highly amusing: "Oh simple innocence!" He found it especially annoying that he was compelled to put aside his further mathematical ambitions. "He seems to keep me away from the society of mathematicians by conveniently just at such moments giving me something to do. This is bitter."[57]

Meanwhile, Sommerfeld had long sought out contact with others who were, like him, just at the start of their academic careers. His housemate, Paul Drude (1863–1906), had brought to his attention the "lecturers' table" at a pub in the Göttingen town park, which he happily joined. This was "an extremely comfortable, jovial, and stimulating group," and there was much "shop talk and joking," he wrote admiringly of this circle.[58] Working out at the gym once a week came as a welcome break in his routine. Here, for a short while at least, he could forget logging the tedious work hours under the watchful gaze of his surly boss. When he picked up his pen again, his optimism once more gained the upper hand. "I've just come home from the gym, or more precisely from the beer pub," he wrote home on one such evening. "As for the rest, I live a thoroughly merry life." He had "picked up a new piano at a tablemate's," so that now in this respect he was no longer dependent on the infrequent opportunities at the Liebisch house. Now, all that was painfully

55 To his mother, January 20, 1894.
56 Adelheid Liebisch to Sommerfeld's mother, January 16, 1893.
57 To his mother, January 20, 1894.
58 To his mother, November 19, 1893. Also in ASWB I.

missing was the time to devote himself to mathematics. "When I come home in the evening after a day of peering at crystals, I am usually exhausted; then nothing comes of work."[59]

On "The Kaiser's Birthday," January 27, a holiday in Wilhelmine Germany, Liebisch once more gave him occasion for a comprehensive report home. In the spirit of the holiday, Sommerfeld delivered this latest turn in his relation to his boss in the form of a political news dispatch: "Have you heard about the great national event? The ruler of the mineral empire has reconciled with his most honorable First Minister. He invited him to partake of a bottle of wine on Sunday, the 28th, and promised him the best seat at the highest table in the ground-floor rooms of his villa. The minister thanked him for this supreme manifestation of his most supreme favor, and there was great joy among the people. This supreme grace was manifested just as capriciously as his supreme wrath." He had no idea how to explain this sudden change of attitude. "For the fact that Bismarck has measured a few crystals cannot possibly be reckoned such a special service. He has always upheld the interests of the Empire, and really does not know what new policy his services may have enabled."[60]

But he sent the letter only after his "reconciliation dinner" and had to report yet one more incident that showed Liebisch's surly side. "The Kaiser's Birthday" fell at carnival season, during which it was customary among the Göttingen professors to enact the less serious side of scholarly life. One of the favorite amusements at such carnival parties was acting out "living pictures." Scenes depicted in famous paintings were represented by the guests, which often compelled dignified Göttingen academics to strike quite unaccustomed poses, much to the delight of all participants. At the home of the art historian Robert Vischer (1847–1933), these parties were especially carefully prepared; they were considered local events. "While I was there, the wife of Professor Vischer comes over to Liebisch," Sommerfeld reports, "to ask him to represent Holofernes in one of the pictures, and let his head be cut off. Mrs. Vischer is a lady whose every aspect and gesture typifies formality, from a well-to-do Viennese family, a handsome woman. She made her request to Liebisch graciously, but he declined in a correspondingly witty and pleasant way. Scarcely had she walked away, though, than he showed his boundless annoyance, uttered not a word at the table, left the party without saying goodbye. He considered it an insulting impertinence, and was furious with his wife for not having kept Mrs. Vischer away from him." By way of showing Liebisch as even more of an oddball, he contrasted his behavior with the exuberance generally reigning in Göttingen. At a "Kaiser's Birthday" party at the home of the physical chemist Walter Nernst (1864–1941), he and the other guests had gotten themselves "thoroughly inebriated." "It was divine," Sommerfeld raved of this evening; "unrestrained merriment"

59 To his mother, January 25, 1894.
60 To his mother, January 25, 1894.

had reigned. "We behaved as uninhibitedly as if we were all at home. It was wonderful. Our host, who finally lighted our way out, needed a bit of help to get back up the stairs! And these were all the scientific stars—except for me, who was blooming in secret."[61]

"Carnival magic" at the Vischers' gave Liebisch renewed occasion for Othello-like emotions and grouchiness. "Mrs. Liebisch made a bit too merry for her Lord and Master," Sommerfeld reported to his mother after the party. It had been "an evening of the most unrestrained merriment; the renowned Göttingen stiffness had totally vanished." The hostess had known how to arrange the various personalities for the "living pictures" so expertly "that one thought one was seeing the originals." Voigt, for example, acted the part of Albrecht Dürer in his self-portrait; the wife of a Göttingen professor appeared as a praying nun—and "for all these distinguished (and in part very musical) guests, I had to supply appropriate music." Whoever he had not presented himself to at one of his introductory visits as Liebisch's assistant now came to know Sommerfeld as a pianist. "I played the following: Meistersinger (Albrecht Dürer), Mendelssohn, Gondolier's Song (Venetian Lady), Pathétique Sonata (Angel), Magic Flute, Don Giovanni, Military March, etc. It all went very well." The host then "toasted the Director, Professor Voigt, and the Orchestral Conductor, Dr. Sommerfeld." "I have apparently entered into a similar relationship with Professor Vischer's wife as before with Mrs. Lindemann—although the former is much more refined. I'll probably be invited over for coffee sometime."[62]

That the professors' wives were showing such inclinations towards her son, Sommerfeld's mother appears to have viewed with some distrust. Adelheid Liebisch even confided her marital problems to him. "You're judging her somewhat incorrectly," Sommerfeld wrote to dispel his mother's suspicion. "She's not to be reproached for speaking to me about her husband. First of all, we are good friends; second, she does it in part in my interests. The first time at any rate, when she broke her silence on the subject, I saw it would be difficult for her to speak of it. Now that I'm aware of the wound, it's only natural that she pours her heart out to me." On the occasion of the "carnival magic" at the Vischer home, he had seen the gregarious side of the "scientific stars." The rapidity with which his reputation as a piano virtuoso spread through the small town is clear from the fact that Klein and his wife, who hadn't even attended the party, had been well informed of his musical services at the Vischers. Next, he was pleased by an "alimentary visit" to the Vischers, an invitation from Eduard Riecke (1845–1915), professor of experimental physics at the University of Göttingen, and a concert at which Voigt conducted.[63] Liebisch's assistant was well on his way to making a name for himself in Göttingen—if initially only as a sociable pianist and favorite of the professors' wives.

61 To his mother, January 27, 1894.
62 To his parents, February 10, 1894.
63 To his mother, February 17, 1894.

2.5 Approach to Felix Klein

The best thing about his position at the mineralogical institute was that he was able to pursue his mathematical interests. Sommerfeld had written this to his parents already 4 weeks after his arrival in Göttingen. He attended Klein's lecture, which in this winter semester of 1893/1894 dealt with the difficult area of hyper-geometric functions, and whenever he had the opportunity, the "mathematical reading room," where he could study Klein's previous lectures as well. Although Klein had been characterized for him by Liebisch as an unapproachable authority, it quickly became clear to Sommerfeld that this picture did not match reality. From his first meeting, he found Klein "quite amiable" and not at all unapproachable. In evident high spirits, he wrote his mother, "Isn't fame a terrible thing? When I meet Klein, he says, 'I've known your name for some time: you're the man with the harmonic analyzer.' I'll soon have to travel incognito."[64] At his first meeting, too, Klein had invited him to participate in the mathematical colloquium.[65] His growing familiarity with Klein went hand in hand with his alienation from Liebisch. "Among all his colleagues, he hates Klein especially," wrote Sommerfeld about his boss's relationship to Klein. "The expressions, 'mean,' 'false,' 'power-greedy,' are the mildest. He himself told me to attend Klein's lectures; but now he's furious about it. Every day, when I attend the lectures, I believe, he stews. On top of that, I've often praised Klein's lectures, which he's taken quite badly."[66]

The worse his relation to Liebisch became, the more Sommerfeld placed his hopes on advancing his career under Klein's wing. Unfortunately, Klein already had an assistant, he wrote home on his 25th birthday, and he was much further along. "If you feed him strychnine, it's possible that Klein will take me on. Actually, that's not even clearly the case."[67] But he didn't abandon the goal of an assistantship under Klein and determined "to impress him as soon as possible, and not hide my light under a bushel." He seems to have been successful in this, for in March 1894, Klein did actually offer the prospect of an assistantship. The offer would become actual only in a few months, but Sommerfeld told Liebisch about it immediately, before he might hear it from some third party. "Very nice," Liebisch responded with sarcastic friendliness … and served him 1 month's notice on the position at his institute. "Of course I said 'Exactly as you wish, Professor,' and felt myself suddenly discarded and driven away! So the joy of the assistantship is over, just like that! Let's hope this doesn't prove a case of the bird in the hand being better than two in the bush!!"[68]

For his parents in far off Königsberg, this did not come as good news. Around the same time, there were increasing indications that Walter would not keep his

64 To his parents, October 29, 1893.
65 To his mother, October 31, 1893.
66 To his mother, December 20, 1893.
67 To his mother, December 5, 1893.
68 To his mother, March 4, 1894. Also in ASWB I.

position at the clinic near Hattenheim am Rhein and that he was once more consoling himself over his distaste for his professional life with morphine. Their worries over their sons were a heavy emotional burden for the parents. "I'll do one thing more," Arnold promised his mother. "I'll go tomorrow to see Professor Liebisch's wife, explain the situation with Walter to her, and tell her you are very concerned about the termination of my assistantship." To be sure, he thought their worry over his own future was unfounded, but "the similarity between my case and Walter's is too painful for me, and I want to leave no stone unturned in relieving you of your burden of worry." In the event that Walter did not retain his position, he wanted to have him come to Göttingen, where there was a "very well administered mental institution," in which Walter might perhaps work as a volunteer. A member of the "lecturers' table" was well-acquainted with the director of this facility, so that he hoped by this route to put in a good word for Walter. "In any case, I'd rather assume the worry myself, than think of you being burdened with it."[69]

At bottom, Sommerfeld was glad Liebisch, with his termination of the assistantship, was making his departure from the mineralogical institute so easy for him. In retrospect, he felt it was "fundamentally wrong" for him ever to have accepted this position. "So, the sooner I'm done with it, the better," he wrote his mother. Besides, Liebisch was going to clothe the termination "in terms of a mutual agreement." In light of the fact that none of Liebisch's earlier assistants had lasted long with him, the situation was no embarrassment to him. In Göttingen, it would be well known whom to blame. To preclude any gossip at home, he had already prepared a formula:"Where Königsberg is concerned, I say simply, I did not get along with my boss, or if you prefer, I say, I'm still an assistant, simply merging the former position with the future one."[70]

Sommerfeld's letters to Königsberg do not cast a particularly favorable light on the mineralogist Liebisch. In view of these descriptions, one is astonished to read in an obituary of Liebisch in the *Centralblatt für Mineralogie, Geologie und Paläontologie*, that in the course of the 21 years of his work at Göttingen, Liebisch had "the pleasure of drawing talented students under the spell of his areas of research. Gratefully and with pleasure, he recalled the loyal collaboration of all his Göttingen assistants." Liebisch had "experienced this period as the most fruitful, and thus the happiest of his industrious life."[71] Among the assistants cited, Sommerfeld is named as one of several, but the register might also be read as evidence that in fact no assistant lasted long with Liebisch.[72]

In the event, the decision about the assistantship with Klein had still not been made in March 1894, but for Sommerfeld, it was only a matter of time. Shortly

69 To his mother, March 15, 1894.
70 To his mother, March 16, 1894.
71 Schulz, *Theodor Liebisch*, 1922, p. 420.
72 The frequent change of assistants at the mineralogical insititute is documented in Kuratorialakten der Göttinger Universität, UAG, Kur 1522.

thereafter, Liebisch withdrew his termination, however, so that externally, everything appeared just as it had been. Liebisch explained to the Curator of the University of Göttingen that his assistant "had made the request that he be retained in his post until September 30 of this year, since there has been an unexpected accident in his family that prevents him from devoting himself exclusively, as had been his intention, to preparation for the habilitation in the subject field of mathematics."[73] By the "unexpected accident" Liebisch was no doubt alluding to Sommerfeld's brother's morphine addiction. Sommerfeld's parents in Königsberg were in any case relieved by the reversal of the termination. Sommerfeld himself had mixed feelings. "It is to my great satisfaction that remaining in my position is pleasing to you," he assured his parents, but deep down, it annoyed him that now the mind-numbing time serving in the mineralogical institute was to continue and that he would be unable to devote his whole energy to mathematics. Wallach too would "inveigh against this properly," he guessed. His hopes of installing his brother as a volunteer at the nearby Göttingen mental institution were also dashed. Instead, Walter was taken in at another clinic … as a patient. "It is heart-rending, this mess of a life," Sommerfeld wrote home. But the word "useless," which his mother had voiced over the failures of her sons in their first professional positions, he protested as utterly unjustified. "That's not the case at all! Klein hasn't found me useless at all, and neither has Liebisch. I've proven useless only in the battle against his whims."[74]

2.6 Physics or Mathematics?

At semester's end, Sommerfeld traveled to Königsberg to serve in a 2-month-long military exercise. As a 1-year volunteer, he could gradually qualify as a reserve officer through such exercises, served during his vacations. In this way, he worked his way up from junior officer to staff sergeant and finally to reserve lieutenant.[75] During the exercises in April and May of 1894, he lived in his parents' house and enjoyed the familiar environment of his childhood and student years. "As I departed," he wrote following his return to Göttingen, "it occurred to me that I really should have thanked you for your affectionate welcome to the parental home; I am no longer a child, accepting thoughtlessly everything good extended to him by his father and mother."[76]

Although on paper he was still assistant to Liebisch, Sommerfeld experienced his return to Göttingen in June 1894 as the beginning of a new chapter of his life. To begin with, he took a new apartment at the edge of town. It was pleasanter than the

73 Liebisch to the Curator, March 29, 1894.
74 To his mother, April 1, 1894. UAG, Kur 7522.
75 Bescheinigung des Aachener Bezirkskommandos, April 20, 1911. DMA, NL 89, 016, folder 1.7.
76 To his mother, June, 8 1894.

one he had previously occupied. "Very pretty view, and very quiet. It is splendid to work here at the open window with the view of the green trees and mountains."[77] In addition, he hoped to be able to better pursue his musical interests here. "My landlords have a piano," he gushed; "such a temptation!"[78]

That he nevertheless now had to spend his time at the mineralogical institute, he felt was a pure waste of time. "This drudgery with Liebisch is just too ridiculous. I'm really heart-sore about it. By the way, he is at pains to be pleasant. He has a real collector's mania about wall charts. I've already filled several dozen." When he reported on his "current work," he did not mean work at the mineralogical institute, but the mathematical studies he had undertaken for his habilitation under Klein. But just when he longed deep down to concentrate entirely on mathematics as his true area of work, Woldemar Voigt offered him an assistantship in the institute for theoretical physics. He would have "almost nothing to do," he wrote his mother following his conversation with Voigt, and he would be able to stand for his habilitation in mathematics just the same. "These people have an amazing trust in me. I understand nothing about experimental methodology, and I've told Voigt so. I'm afraid of making a fool of myself." Voigt is "a nice man," but Sommerfeld didn't feel himself scientifically drawn to him. Voigt's previous assistants could have confirmed for him that in the institute for theoretical physics, he would have a pretty free hand in his choice of research subjects. But weighing most heavily in this decision was that he still had received no firm offer from Klein. "What shall I do?"[79]

Two weeks later he informed his parents he had declined the offer because he did not want to be "in an untoward position again" occupying himself with things to which he did not feel drawn. He had discussed the matter with Klein as well. "He virtually advised me to accept the position with Voigt. He imagines me as somewhat more inclined towards physics than I am. Today, however, he seemed to be happy that I will be free next semester."[80]

Thereafter, the daily round at the mineralogical institute became even more unbearable. "Working with Liebisch is dreadful, enough to drive one crazy. The height of tedium, from 9:00 or 10:00 in the morning to 6:00 or 7:00 in the evening." Liebisch was indeed very friendly now, but this was small consolation. "I'd rather he were a little less so, and just didn't waste my time in such an irresponsible manner."[81]

Only the evening hours and weekends were left him in this summer of 1894 for his mathematical studies. "Next semester will be a wonderful time, and the mineralogical time killing will seem like a bad dream to me." In this way he expressed his yearning for the assistantship with Klein. Whenever he spoke of a conversation

77 To his mother, June 9, 1894.
78 To his mother, June 15, 1894.
79 To his mother, June 15, 1894.
80 To his parents, June 27, 1894.
81 To his mother, July 4, 1894.

with Klein, he gushed. He had "chatted away . . . 2 fabulous hours" with Klein; Klein was "brilliant, erudite, open, and honest." He wrote very differently about his boss at the mineralogical institute: "Recently, I almost addressed Liebisch as 'Sergeant.' Ha ha!"[82]

As semester's end approached, Sommerfeld got a surprise that put him into a state of euphoria and caused him late at night to write a long letter to his mother. "I've just come from the anniversary celebration of the Mathematical Society, and feel so upbeat that I want to stay up and write myself to exhaustion. Alright, shall I, or shall I not relate to you what I've heard?" Then it spilled out of him. A former assistant of Boltzmann who had just arrived from Munich was surprised that Sommerfeld was not pursuing his habilitation in Königsberg in theoretical physics. "Boltzmann, he says, had been quite taken by my work in electrodynamics. Now here's the thing! Make sure you're sitting down for this! Boltzmann had put my name down in 7th or 8th place on the list to replace him!!!" In the summer of 1894, Boltzmann resigned his teaching position at Munich and returned to Austria to assume the professorship in theoretical physics at the University of Vienna. Sommerfeld was so incredulous at the news that he was on the list of candidates to succeed Boltzmann at Munich that he took it at first for a joke. For that reason, he ordered his parents to remain silent. "It's too crazy. That poor Boltzmann is crazy is really too sad. Sadder, though, is that I'm now in a dilemma as to whether I should switch over to physics."

In any case, he was momentarily unsure his decision to pursue mathematics was the correct one. "Klein also wanted to see me perhaps habilitate in physics," he mused. "This is how it goes for anyone working along the borders of a field. Mathematicians think of me as a physicist, and were I to become a physicist, they would no doubt take me for a mathematician." He wished to convey to Boltzmann that here in Göttingen he was going to "become a mathematician and bid farewell to physics." If Boltzmann wanted him as an assistant at Vienna, however, "I would go along and become a physicist." Finally, he gave his imagination free reign once more: "Just think of it: I as successor to the greatest German physicist (after Hertz's death and Helmholtz's stroke). It is too silly. Forget it! And I will become a mathematician after all. However, I will let Liebisch know about this latest development via his wife. I'll play him this trick; it'll get his goat!"[83]

Still days later, all he had heard at the anniversary celebration of the Mathematical Society was still running through his head. Some of it also touched on the wife of his doctoral advisor, Lindemann. In Munich, she "was already considered a very loose woman," he wrote in great amusement; she had made "a conquest" of Boltzmann and persuaded him "to get himself a big mutt, and since then Boltzmann has been running outside every 15 minutes for the creature to do its business. A scream!"[84]

82 To his parents, June 27, 1894.

83 To his mother, July 29, 1894. Also in ASWB I. Heinrich Hertz (1857–1894) died on January 1 of this year; Hermann von Helmholtz (1821–1894) died on September 8.

84 To his parents, August 3, 1894.

When his parents wanted to know whether, with the news about filling Boltzmann's position at Munich, they hadn't been pulling his leg, he responded, "It is a fact that Boltzmann named me among the people to be considered as his successor. It's likewise a fact that of course I was never seriously in contention, and that it amounted to considerable deference on Boltzmann's part to place such value on purely mathematical speculation such as I published in Wiedemann's *Annalen* at that time." He responded with simulated nonchalance to his parents' objection that his position with Liebisch would end the next month in any case, and under those circumstances he would have done better to have accepted the assistantship with Voigt: "Well, choosing a career in teaching at a university is to renounce money from the start." The position with Klein was certain; the only question was, when. "It's a matter of indifference to me whether I start in October of '94, or of '95. One year with Voigt would just be a colossal waste of time. Voigt also expected that I would become a physicist. So the same conflicts would arise as with Liebisch (even if in not quite so stupid a form). So believe me, this is an unavoidable consequence of the direction my life's little ship has now taken."[85]

85 To his parents, August 24, 1894.

3 Klein's Assistant

"Klein's previous lectures interest me extremely. One can look at them in the reading room," Sommerfeld had written home shortly after his arrival in Göttingen in November, 1893. In a lecture on "Partial Differential Equations in Physics" from the winter semester of 1888/1889, Klein had also dealt with heat conduction, and Sommerfeld saw with satisfaction that Klein's process bore "a quite amusing similarity" to the theoretical approach he had pursued in his competition entry on ground temperature measurements at Königsberg. "This is by no means evidence that I am a Klein, rather that with this method it's possible to come as far as both he and I did. In fact, my attempts to go a step further have so far been unsuccessful."[1] In the case of other "physical differential equations" too, Klein's lecture notes revealed so much fascinating mathematics that the physical content almost paled in comparison. The deeper Sommerfeld delved into it, the stronger grew his desire to make this his future research subject.

3.1 Physical Mathematics

In his lectures, Klein repeatedly emphasized how closely mathematics and physics were related to one another. Nowhere was this more clearly demonstrated than in the partial differential equations used in everything from the theory of electricity to acoustics. "To treat all these differential equations together in a systematic way: that, we can say, is the task I set myself in the present lecture." Thus had Klein staked out the objective for this semester on the day of his first lecture in October 1888. And as in a previous lecture on potential theory, his credo now too was "that mathematics belongs inseparably together with its applications, especially with physics, and that it is rewarding also for the mathematician, from his perspective, to explore this connection—rewarding not only for the study of these applications, but also for pure mathematics itself, which draws its force from the applications."[2]

These approaches were nothing out of the ordinary for Klein. "Many explorations of what is called mathematical physics are purely mathematical explorations," he had already explained in 1872 in his "Erlangen Inaugural Speech," later to become famous. "We should rather speak of them in another category, the category of physical mathematics."[3] This sort of mathematics was directly at the center of his

1 To his mother, November 19, 1893; Klein, *Abhandlungen 3*, 1923, Appendix, p. 7.

2 Lecture Notes for "Partielle Differentialgleichungen I, Winter 88–89." SUB Cod. Ms. F. Klein 15H., p. 1.

3 Rowe, *Antrittsrede*, 1985, p. 133.

M. Eckert, *Arnold Sommerfeld: Science, Life and Turbulent Times 1868-1951*,
DOI 10.1007/978-1-4614-7461-6_3, © Springer Science+Business Media New York 2013

lectures on partial differential equations in physics in the winter semester of 1888/1889. Klein stressed that this was not a question of a new method. In this respect, Riemann was the great model for Klein. For Riemann, the proximity of mathematics to physics had been almost axiomatic. Klein advised his students to acquire a personal sense of Riemann's work from the primary source literature. "For this, the mathematical reading room offers you substantial help."[4]

When Sommerfeld studied these lectures of Klein's and Riemann's work in the reading room, he saw a great deal that was already familiar to him with fresh eyes. Despite the wide diversity of physical phenomena—electricity, magnetism, optics, heat conduction, elasticity, acoustics, etc.—the differential equations involved were often of the same type. The partial differential equations of potential theory, the Laplace and Poisson equations, were as important for mechanics as they were for electricity and magnetism. The same was true for oscillation and heat conduction equations. Underlying the multiplicity of physical phenomena, a "physical mathematician" recognized a comprehensive foundation of differential equations. A large portion of these relate to so-called boundary value problems, in which from a given distribution of function values for a given domain, the general space-time behavior is derived.

According to Klein, there were two methods for this: In the first, for the solution function, series expansion is applied; further development results in the determination of the coefficients. The other method proceeds from "principal solutions" corresponding to the respective differential equation; these must then in a second step be fitted to the proposed boundary values. Above all, "the unconditional association with the physical concept" shows how the solution functions are to be obtained. Here, we see once more what Klein meant with the concept of "physical mathematics." Physics supplies "the existence theorem" for the solution of a differential equation, even in cases where mathematically no proof has yet been furnished. "With respect to this association, we hope not to profit just physics, but first of all to profit ourselves."[5]

In Klein's "Method of the Principal Solutions," Sommerfeld immediately recognized his own approach to the solution of the heat conduction equation. He had his mother send him the manuscript of his "thermometer work of that time." Because—following the Chicago World's Fair and the Munich exhibition of mathematical instruments and models organized by Dyck—Klein was also interested in the harmonic analyzer, Sommerfeld recognized a promising opportunity to demonstrate his talents. "In 4 weeks, I'm to lecture on the analyzer and my doctoral thesis to the local mathematics colloquium," he wrote on November 7, 1893, to Königsberg. He wished to impress Klein with his lecture, but also felt rather insecure, since he had made no significant progress on what he had achieved in his

4 Lecture Notes for "Partielle Differentialgleichungen I, Winter 88–89." SUB Cod. Ms. F. Klein 15H. p. 3.

5 Lecture Notes for "Partielle Differentialgleichungen I, Winter 88–89." SUB Cod. Ms. F. Klein 15H. pp. 77–81.

doctoral work. "Not a lot is going to come of the whole business; perhaps a sense of edification for you; for me not even that."[6]

Klein left nothing to chance. He invited Sommerfeld to a conference several days in advance of the lecture. "I left very satisfied," Sommerfeld reported to his mother. He had "made a good impression on him, and hope[d] to reinforce it with [his] lecture."[7] It was clear to him already at this conference that his subject had met with great interest on Klein's part. "The conclusion of the conversation was very effective. I said to him: 3-fold periodic potentials can be constructed in space with the aid of an elaboration of the Mittag-Leffler theorem. He told me a paper by Appel[l], a very good Parisian mathematician, deals with that. He opened it and read out to me that one can construct 3-fold periodic potentials with the aid of an elaboration of the Mittag-Leffler theorem. This looked very good. I went away highly satisfied. I had the feeling I had impressed him."[8]

The theorem of the Swedish mathematician Magnus Gösta Mittag-Leffler (1846–1927) deals with the existence of certain functions with poles. In the 1880s, it was a much discussed object of mathematical research. With his reference to it, Sommerfeld showed he knew his way around the highest spheres of complex analysis.[9] In the reading room, Sommerfeld must accordingly have studied closely the Parisian mathematician Paul Appell (1855–1930), so highly valued by Klein. In 1892, Appell had published a study on differential equations of the type of the heat conduction equation. What Sommerfeld read there must have thrilled him, for Appell had cited his doctoral thesis as a generalization of a theory established by Karl Weierstrass (1815–1897).[10]

With the lecture he delivered on December 5, 1893, his birthday, Sommerfeld made his official entrance among Göttingen mathematicians. "The author speaks on the 'Method of Principal Solutions in Mathematical Physics'," he wrote afterwards in the record book of the Mathematical Society, in which speakers entered abstracts of their lectures. First, he explained the concept of the principal solution as that solution of a physical differential equation that corresponded to a point source in an infinite domain without boundary conditions. From this principal solution, it is possible by means of Green's function to represent the solution for a bounded domain of given boundary values as an integral. Making Green's function the starting point for the solution of boundary value problems means also that in this way one can arrive at integral representation of arbitrary functions, as he described in greater detail in his dissertation. Thereby, he moved his dissertation into line with the method of principal solutions so valued by Klein.[11]

6 To his mother, November 19, 1893.
7 To his mother, December 4, 1893.
8 To his mother, December 5, 1893.
9 Turner, *Mittag-Leffler theorem*, 2007. On the history of complex analysis, see Bottazzini, *Calculus*, 1986.
10 Appell, *Sur l'équation*, 1892, p. 209.
11 Protokollbuch Nr. 1 der Mathematischen Gesellschaft zu Göttingen (Easter 1893–February 1896). SUB Cod. Ms. Math. Archiv 49:1.

In the practical application of this method, it was thus first necessary to determine the Green function corresponding to the respective boundary conditions.[12] Here in particular, the method of images introduced by William Thomson into potential theory proved successful, whereby the boundary conditions governing given surfaces through mirroring of the sources (in the case of potential theory, point-source electrical charges) are produced behind the surface. Sommerfeld applied this procedure to heat conduction theory. To be sure, this was successful only when a complete volume ratio could be obtained by reflection of point-source heat sources. "The number of problems to be solved in this way is therefore not great," Sommerfeld conceded. But the method might also be applicable to "arbitrary heat conduction." Then, "heat conduction on a surface with winding points" needs to be examined. Hereby, he procured for the method of images an as yet unrecognized significance for complex analysis.[13]

With elaborations of this kind, Sommerfeld was treading mathematical–physical virgin soil. With respect to the method of principal solutions in general, they were perhaps not as new as Sommerfeld represented them. In potential theory, as Klein had elucidated in his lectures, it was virtually a standard procedure. With respect to the problem of heat conduction, too, the method had already been applied. "Just today at noon," Sommerfeld wrote on the evening following his lecture, he had "looked at 2 papers by a Sign. Betti" in the reading room. In the heat conduction theory of the Italian mathematician Enrico Betti (1823–1892), as in other papers not cited by Sommerfeld, the "method of source points," as it was also called, had been employed as early as the 1860s. It may well be that his ignorance of these earlier papers had evoked a few critical comments, for after his lecture he wrote home that he could have done better. Nevertheless, he was in a good mood: "The people will have seen I know a thing or two." The most important thing for him was that he had "used this opportunity to draw significantly closer" to Klein.[14] Just how seamlessly Sommerfeld's lecture fit into Klein's program of "physical mathematics" is evident in light of other Klein lectures of those years. When Sommerfeld spoke of "heat conduction on a surface with winding points," he gave the concept of Riemann surfaces a physical meaning. In a quite analogous way, in a lecture in the winter semester of 1891/1892, Klein had characterized Riemann surfaces as "substrates of the potentials."[15]

Shortly before, Klein had arranged for his student Friedrich Pockels (1865–1913) to make the partial differential equations of physics and new ideas contained in "Lamé functions" the basis of a book on the differential equation $\Delta u + k^2 u = 0$,

12 Tazzioli, *Green's Function*, 2001.

13 Protokollbuch Nr. 1 der Mathematischen Gesellschaft zu Göttingen (Easter 1893–February 1896). SUB Cod. Ms. Math. Archiv 49:1.

14 To his parents, December 5, 1893. Burkhardt, *Entwicklungen*, 1908, p. 1239.

15 Klein, *Riemannsche Flächen*, 1985, p. 8.

"not only because of their signal importance for innumerable physical problems, but also because they can be seen as the next generalization of the potential equation."[16] In Sommerfeld, Klein saw a talented and ambitious mathematician, who would make the physical approach to mathematical problems enthusiastically his own. "My relationship with Klein is excellent," Sommerfeld wrote his mother several weeks after his lecture. "Recently, I got a *billet doux* from him: I was to come to see him; he wished to discuss work with me. Soon, I'm to give another lecture, on recent French work. Klein organizes everything around himself. He has no time to read all these things, and wants to be briefed on them. He has very cleverly determined a distinct research area for me. I'm to write up a short paper about my previous lecture for the *Mathematische Annalen* as soon as possible."[17]

In working up his lecture for publication in this renowned mathematical journal, Sommerfeld gave his subject a very particular color: he attributed the alternative procedure for the solution of partial differential equations—the method of principal solutions vs. series expansion—to the two different fundamental physical conceptions, action at a distance vs. contiguous action. The method of series expansion corresponds to the standpoint of the contiguous action, whereby changes are propagated only by immediately neighboring domains throughout from one locality to another. According to the action at a distance, point sources can extend their effect to distant localities even without the interaction of neighboring spatial regions. In potential theory, for example, the function $f(r) = \frac{1}{r}$ is a "principal solution." It indicates how a force diminishes at a distance r from a unit source. The effect of spatially distributed sources is found by summing or integrating the source distribution. In heat conduction, proceeding from "temperature poles" with a "principal solution," one can similarly construct the solution appropriate to a given temperature distribution. The special charm of this method lay in its elaboration by means of complex analysis. Sommerfeld gathered the last of three sections under the rubric "The Principal Solution on a Riemann Surface." As an example, he applied the method of images to temperature poles inside an area bounded by two intersecting straight lines. Depending on the angle, different results were obtained in the reflection of the temperature poles on the straight lines. In the symmetrical repetition of the wedge-shaped area that contained the temperature poles, the entire plane could be covered multiple times. Mathematically, this meant a generalization of familiar solution functions of the "common surface" for those that can be obtained through integration on Riemann surfaces.[18]

16 Pockels, *Differentialgleichung*, 1891, p. 1–2.

17 To his mother, January 5, 1894.

18 Sommerfeld, *Theorie der Wärmeleitung*, 1894.

3.2 Then I'll Grow into the Lectureship

Following this acid test, Sommerfeld was repeatedly summoned "to audiences"[19] with Klein, to be consecrated into higher orders by way of further challenges—lectures to be given or papers to be published. At the end of his first Göttingen semester, he lectured on "Functions of Real Variables, Obtained Through Partial Differential Equations."[20] After his first lecture in December 1893, should Sommerfeld still have harbored doubts about his fitness for a habilitation under Klein's wing, he abolished them with this lecture and with his article for the *Mathematische Annalen.* "For the moment, the bluest skies are smiling over Göttingen," he wrote to Königsberg so his parents could share his happiness. Klein had given him the prospect of an assistantship for the fall. "There's only one tiny hitch. His current assistant plans to habilitate and requires a fellowship to do so. If he gets it, everything is in order, and he'll vacate his post. He'll almost certainly get it, for just now a fellowship here in Göttingen has opened up. So everything is going according to plan. Then I'll grow into the lectureship, and work out the ideas Klein doesn't have time for. Heiohei!" His salary would be the same 1,200 Marks annually as his unloved position at the mineralogical institute, and he would have the chance to talk with Klein every day. "Hurrah!"[21]

His studies in the mathematical reading room and the respect of other mathematicians also allowed Sommerfeld to endure the daily grind of the mineralogical institute. It became more and more clear to him now that his doctoral advisor in Königsberg had not understood the significance of his work. "Here, people are interested in what interests me," he wrote his mother in June 1894. "Prof. Burkhardt, for instance, who has just been to Paris, read my dissertation, and praised it."[22] Heinrich Burkhardt (1861–1914) had done his habilitation at Göttingen in 1889 and was recognized as an expert in the area of complex analysis and special functions. Sommerfeld, too, quickly found a topic for his habilitation thesis. Once the method of the "principal solutions" had proven its validity for heat conduction, new solutions would subsequently open up for other physical differential equations too.

In this regard, above all the wave equation, which had already often been the subject of mathematical dissertations, presented itself.[23] Since Hertz's day, it was known that electromagnetic processes—including the diffusion of light—could be described by Maxwell's equations. For the diffraction of light, too, it was an easy thing from these equations to derive wave equations of the type $\Delta u + k^2 u = 0$ for the various components of the electrical and magnetic field, where the field

19 To his mother, February 17, 1894.

20 Protokollbuch Nr. 1 der Mathematischen Gesellschaft zu Göttingen (Easter, 1893–February 1896). Entry of February 27, 1894. SUB Cod. Ms. Math. Archiv 49:1.

21 To his mother, March 1, 1894.

22 To his mother, June 9, 1894.

23 For example, Pockels, *Differentialgleichung*, 1891.

component u is a function of place and time and k as the "wave number" represents the reciprocal of the wavelengths. The solutions followed from the condition that the field components assume the values given by the arrangement of diffraction. In the language of mathematics, this is, then, a classical boundary value problem. For even the simplest cases, however, this proved to be mathematically a gargantuan assignment. Heretofore, no one had even succeeded in giving a solution for the diffraction on a wall with straight edges, let alone a slit or a grating. The appearance of light and dark diffraction stripes behind a slit, seemingly so simple to explain as the overlapping of light waves, was a mathematically unsolved problem. Physicists employed the Huygens Principle, according to which from every point in an opening illuminated from one side, light rays emanate in all directions that overlap each other, and on account of the wave character are dimmed at certain places and amplified at others. This could be described mathematically by assuming that from every point of the oncoming wave front in the diffraction opening, a spherical wave emanated, and examining the overlapping of these spherical waves at the position of observation. Kirchhoff had clothed this conception in a strict theory in terms of which many diffraction phenomena could be well described quantitatively also. But the Huygens Principle is valid only by approximation and breaks down completely when, for example, slit aperture and wavelength are of the same order of magnitude. Besides, the boundary conditions for the different field components are incompatible with it. In short: for a physical mathematician, the Huygens Principle was an unsuitable means of solving the diffraction problem.

But to criticize heroes of the history of physics such as Christiaan Huygens (1629–1695) was no small matter. "Besides, Mr. Kirchhoff is making things difficult for me," Sommerfeld wrote his mother in October 1894. "My well-grounded view is that what this most thoroughly mathematical figure has wrought among physicists in optics is all humbug and verbiage. But I can't simply say that in my paper. In any case I'll have to read him thoroughly."[24] Presumably, Sommerfeld had already selected the diffraction problem as his habilitation topic months earlier and had reckoned with speedy success. "If only things would go more quickly," he wrote home impatiently in mid-June. "It's all very beautiful and new. The mathematical methods are quite elegant and the physical result is very important. I expect considerable success from it. Three weeks from today I'm to lecture on it."[25] Klein would take this opportunity, he surmised, to decide whether he could take on this work as a habilitation thesis.

But the mathematical execution turned out to be more difficult in practice than he had anticipated. "Killing time mineralogically" achieved something more to darken his mood.[26] Shortly before the scheduled date of his lecture, he wrote home that "the monster, Liebisch," would not allow him to "work intensively" in preparation.[27]

24 To his mother, October 3, 1894.
25 To his parents, June 15, 1894.
26 To his parents, June 27, 1894.
27 To his parents, July 16, 1894.

"Tomorrow evening I'll be with Klein," he wrote home on July 20. "My paper gets ever longer and more beautiful."[28] Then he ran again into further difficulties. "The last few days, namely, my diffraction work has not progressed, on the contrary, on two points it has gone backward, and I have been quite despairing. The lecture has had to be moved back to Friday of next week because Klein has no time. So, the grind continues."[29] In the end, however, the lecture did go off "to general satisfaction," as Sommerfeld reported to his parents on August 3, 1894. "So, this was the diffraction of light, properly treated mathematically. I've really given the physicists, who have approached the subject incorrectly up to now, something to think about." Klein had paid him the compliment of telling him this "had made a nice conclusion to the semester."[30]

Nor in summarizing his lecture in the proceedings of the Mathematical Society did Sommerfeld hide his light under a bushel. "The author shows that the exact solution of certain special diffraction problems consists in finding solutions of the differential equation $\Delta u + k^2 u = 0$ on Riemann surfaces." Thus he thrust directly to the essential point of his approach. Similar to his approach to heat conduction, which he had treated from the different perspectives of the action-at-a-distance and the contiguous-action hypotheses, here he compared the two fundamental conceptions in terms of which physics treated the optical problem, the "Emission theory," and the "Undulatory theory." The former could not fully deal with the diffraction problem since it accounted only for rays spreading in straight lines. Only the wave theory was appropriate for the problem. In the case of diffraction on the edge of a wall, the solution of the wave equation led to integration in a "Riemann double-space," with the edge of the wall as a "branch cut." He thereby generalized the concept of the Riemann surface in three dimensions. For the radiation of light in a plane with a straight line spreading from the origin of the coordinate system as "diffraction screen," the Riemann double-space reduced to a two-leaved Riemann surface with the origin of the coordinate system as branch point. Thereby, Sommerfeld had returned to the familiar domain of ordinary complex analysis. "Here, very comfortable functions occur, which can be expressed as Bessel functions."[31]

This lecture, too, Sommerfeld made into a paper; on December 8, 1894, Klein presented it to the Göttingen Academy for publication. The solution of the diffraction problem was not yet revealed; Sommerfeld wished to save that for his habilitation thesis. But he left no doubt that his approach could be "carried out up to a comparison with the observed phenomena," and that it had found a quite encouraging agreement. "The formulas of ordinary diffraction theory appear as more or less good approximations of our exact ones."[32]

28 To his parents, July 20, 1894.
29 To his mother, July 29, 1894.
30 To his parents, August 3, 1894.
31 Protokollbuch Nr. 1 der Mathematischen Gesellschaft zu Göttingen (Easter, 1893–February 1896). Entry of August 3, 1894. SUB Cod. Ms. Math. Archiv 49:1.
32 Sommerfeld, Theorie der Beugungserscheinungen, 1894, p. 342.

3.3 Reading Room and Model Collection

That in August of 1894, Sommerfeld had presented his habilitation subject "to the general satisfaction" of the Göttingen Mathematical Society altered nothing for the moment in his unfortunate situation at the mineralogical institute. Only a few weeks before the start of the winter semester, his hopes for an assistantship with Klein also evaporated. Ernst Ritter (1867–1895), Klein's current assistant, was unable to vacate this position, since he was not awarded the fellowship on which he had been counting. Sommerfeld, however, had definitely to give up his position with Liebisch by October 1, 1894. His successor had already arrived, "a somewhat shy person. Poor scared rabbit!!"[33] Then Klein informed him that Ritter's fellowship had in the last minute been approved after all, and that thus "the prerequisite of our earlier arrangement is realized." "So I will in all likelihood play the assistant after all," Sommerfeld wrote triumphantly. "Who had been right about the position with Voigt now? I would now be annoyed were I nailed down with Voigt. Nothing ventured, nothing gained."[34]

A few days later, the situation was also officially settled. Sommerfeld was appointed Klein's assistant for 2 years, retroactively from October 1, 1894.[35] His annual salary of 1,200 Marks remained the same. But, liberated from the work he experienced as drudgery under Liebisch, Sommerfeld could finally focus fully and exclusively on what he felt was his vocation, mathematics. And in contrast to his assumption of duties at the mineralogical institute 1 year before, what now awaited him as Klein's assistant was clear to him. The reading room was among his main responsibilities. The collection of mathematical instruments and models, which was part of Klein's domain, and of which he now had stewardship, had long been familiar to him. Shortly after his arrival in Göttingen, Klein had acquired a harmonic analyzer from England for the model collection.[36] He himself was for Klein "the man with the harmonic analyzer,"[37] so it was self-evident that "a certain importance" in the matter of the model collection fell to him.[38]

Klein regarded the reading room and the model collection as essential equipment of his mathematical-pedagogical enterprise. Already at the time of his appointment at Göttingen, he had required that a "mathematical reading- and work room" such as he had established at the University of Leipzig in 1881 be available to students. At that time, the role that now fell to Sommerfeld had been taken by Klein's assistant Walther Dyck.[39] "What I will need at Göttingen to begin with,"

33 To his mother, September 19, 1894.
34 To his parents, October 5, 1894.
35 Curatorialrescripte 1891–1894. SUB Cod. Ms. F. Klein 2B.
36 To his mother, November 1893.
37 To his parents, October 29, 1893.
38 To his mother, March 4, 1894.
39 Hashagen, *Walther von Dyck*, ch. 9.2.

Klein had stipulated in the run-up to his appointment there, "are several rooms in which written versions of my lectures from previous semesters can be made available to advanced students, older mathematicians from outside the University, etc., and in which additionally certain other literary aids, any models that I acquire, etc., can be displayed."[40] The "reading room of the mathematical–physical seminar," as it was officially designated from 1886 onwards, was set up in a room adjoining the model collection, which had some years earlier been established by Klein's predecessors. The University assigned Klein a "personal assistant" for this, whose job was to look after the model collection and the reading room. To make use of this room, one had to register as it were in a club. In the years after 1894, when Sommerfeld assumed this position, the reading room enjoyed growing popularity. The number of users rose from about 30 a year in 1894 to over 300 in 1910. Even beyond his retirement in 1912, Klein held on to the reigns of the model collection and the reading room, for these institutions were, as he argued to the University administration, so peculiarly tailored to his personal guidelines that a successor could not be expected to administer them properly absent his cooperation.[41]

As regards furnishing the reading room, Klein had already in 1892 made it clear to the Curator of the University that, not only for "pure mathematics," but also for the mathematically related sciences, the latest literature needed to be available. "Applied mathematics, mathematical physics, all the way to mathematical astronomy should be considered in a manner appropriate to our audience." He was striving to "represent the whole sweep of mathematics," and desired that "solicitous attention [be given] the relations to its neighboring disciplines and the problems of practical life."[42] Following his return from the World's Fair in Chicago in 1893, it was especially clear to Klein that he would fundamentally reform mathematical instruction at the University of Göttingen. In his personal notes under the date December 10, 1893, we find the entry "New Göttingen Program," with subheadings "Women Students," "School Matters," and "Technology." Each of these items was tied to a growth in student enrollment and consequently with an increased use of the reading room and the model collection. In regard to women students, he encountered "strong resistance" from the Curator of the University. "This is worse than the Social Democrats, who only want to abolish distinctions of property," he countered. "You propose to abolish the distinction between the sexes!"[43] But in 1894, a new Curator assumed the office, Ernst Höpfner (1836–1915), who proved more open in this regard than his predecessor. So far as "School Matters" were concerned, Klein wished, for example, to awaken an interest in mathematics among high school teachers through vacation courses. When he touched on "Technology," Klein was addressing a particularly sensitive issue, because education in engineering

40 Quoted in Frewer, *Lesezimmer*, 1979, p. 29.

41 Frewer, *Lesezimmer*, 1979, pp. 30–48.

42 Curatorialrescripte 1891–1894. SUB Cod. Ms. F. Klein 2B.

43 Quoted from Jacobs, *Felix Klein*, 1977, p. 17.

was the domain of the technical universities, which just in these years were fighting a bitter struggle for equality with the universities.[44]

In 1894, Sommerfeld could not yet imagine what awaited him as spokesman for the Klein program. At first, the new daily life seemed entirely tranquil. He would not have to rush to finish his habilitation, but rather "pursue it *con amore*," he wrote to Königsberg at the beginning of October 1894.[45] But the leisure did not last long. In the spring of 1895, Klein had the reading room renovated. "On Saturday, I moved," Sommerfeld wrote his parents. He was referring not to a move to a new apartment, but to hauling books and rearranging furniture in the reading room. "In the process, I looked like a chimney-sweep."[46] An interior wall was removed so that the number of work stations could be increased from 20 to 35.[47] During this time, Klein was in Montreux in Switzerland, where he was recovering from the flu. But he was kept informed by his assistant how the renovation was proceeding. "The main work in the reading room is done," Sommerfeld wrote after 2 weeks of renovation. "All that remains to do is wall-papering and painting. Lighting is by Auer glow-light." Everything had been thought of. The conversion of the lighting from conventional gas lighting to gas-glow lighting (in which fine-mesh cotton or silk fabric coated with a special material was induced to bright illumination by being heated in a gas flame) invented only a few years earlier was as much a part of the modernization as the erection of cabinets to house technical journals and offprints. Functionality was foremost, but aesthetic considerations were also taken into account. Sommerfeld suggested, for instance, that a double cabinet be built in order that the Gauß bust in the reading room not be squeezed in between two tall, preexisting cabinets: "I think he would feel a little oppressed," he explained to Klein concerning this measure. In the process, he was thoroughly conscious of the financial constraints to which the renovation had to conform. Acquisition of more furniture "will have to be deferred until the Seminar's account has recovered from its present strains," he wrote, anticipating Klein's directives.[48]

After the renovation, the reading room was more than ever the social and intellectual center of Göttingen mathematics. It became something more, though. The Göttingen mathematical reading room was a template for the establishment of special libraries at which students by way of current technical literature and through interaction with assistants and professors could get a vivid impression of the doctrine and research in their disciplines. "I would especially like to direct your attention to our mathematical reading room," Klein stressed in 1895 in an address to high school teachers. "Open all day, including vacations, it offers students the comprehensive relevant literature in the most convenient form."[49]

44 Manegold, *Universität*, 1970.

45 To his parents, October 9, 1894.

46 To his parents, March 12, 1895.

47 Frewer, *Lesezimmer*, 1979, p. 49.

48 To Klein, March 25, 1895. SUB, Klein 11, 1065 C. Also in ASWB I.

49 Klein, *Unterricht*, 1895.

What made the mathematical reading room at Göttingen especially attractive to students from their first semester through to their doctoral and habilitation work were the comprehensive reports of lectures written out meticulously by hand by the respective assistants to Klein. Only a few weeks after his arrival in the fall of 1893, Sommerfeld had written enthusiastically about this, and he showed himself still impressed by it decades later recalling his assistantship with Klein. The reading room had been "at that time a modest room on the third floor of the auditorium building . . . no Rockefeller mansion, but furnished with comprehensive, above all foreign literature. My main task was writing up the four-hour long lecture, which would then be presented to him each week for painstaking revision preparatory to autographical duplication."[50]

3.4 Habilitation

In 1894, Sommerfeld could not yet imagine how much of his time precisely this writing up of the Klein lectures would occupy. He concentrated at first on his habilitation. It has "progressed nicely," he reported to Königsberg early in October 1894. Euphoria and impatience pressed him to speed up the work. "If the point that I now have in view is published, I will be very proud. This morning, I've already howled a long song of triumph in all keys and melodies. This was premature, of course."[51]

One can only imagine the soaring flights and the crash landings Sommerfeld lived through in working out his habilitation thesis in the winter semester of 1894/1895; he did not attempt in his letters home to explain the mathematical details connected to it. Before he had finally completed the work, though, he gave one more lecture to the Mathematical Society, so at least the final act of his labors can be reconstructed. So far as the fundamental approach was concerned, he referred to his preceding lecture. Now he was concerned primarily with bringing the physics of the diffraction process into agreement with the mathematics. It is a "curious fact" that in geometrical optics for every arrangement, no matter how complicated, the solution can be given immediately, whereas in light of the wave character of light, even the simplest case runs up against great difficulties. "The reason for this lies solely in the fact that in geometrical optics, the wave-length is assumed to be infinitely small. Under this assumption, the decomposition of an arbitrary state in rays of plane waves, etc., is made clear by a boundary crossing in the differential equation." Sommerfeld voiced "considerable reservations" concerning Kirchhoff's theory, since it was "subsumed under geometrical optics." He did not expand on his critique but rather proceeded directly to opposing the Kirchhoff procedure (based on the Huygens Principle) with his exact theory. In doing so, he

50 Sommerfeld, *Geburtstag*, 1949.

51 To his mother, October 3, 1894.

depended entirely on Klein's concept of physical mathematics, which he had adopted already in his heat conduction theory in describing the "heat conduction on surfaces with turning points." For various regions in front of and behind the diffraction screen, his theory supplied approximate equations, whose terms he could interpret as incident, reflected, and diffracted rays. He found correspondence with Kirchhoff's theory only in the case of small angles of diffraction; in the case of large angles of diffraction, his equation corresponded to a solution discovered in quite a different way by Henri Poincaré (1854–1912). At the conclusion, he cited additionally the recently published results of a Königsberg dissertation, which also disputed the conventional diffraction theory.[52]

With this lecture, Sommerfeld wished once more, prior to preparing the fair copy of his habilitation dissertation, to confirm a few things he felt needed clarification. It is notable that this related not to the mathematics but to the physics of diffraction phenomena. In a manuscript that served presumably as a template of the habilitation dissertation, Sommerfeld distinguished a mathematical from a physical part, the latter constituting nearly two thirds of the whole. Here too, he criticized the Kirchhoff diffraction theory fundamentally.[53]

Three weeks after his lecture, Sommerfeld submitted to the Philosophical Faculty of the University of Göttingen his formal application "to be granted the *venia legendi* [permission to lecture] in the subject of mathematics." Thereby, the official habilitation procedure was set in motion, a process prospective university professors at German universities had to go through before they were permitted to lecture as lecturers.[54] Klein, who in this semester was serving as Dean, forwarded the habilitation application to the faculty and called a habilitation commission together. As second referee of the habilitation dissertation, Voigt was to assess it from the viewpoint of physics. He soon added his own expert opinion to the proceedings, in which he stressed in particular that Sommerfeld had proceeded "in his mathematical investigations throughout from the basis of physical formulations." "We need such young people who maintain the connections among neighboring disciplines ever more urgently." Sommerfeld had derived "his equations for the accurate calculation" of the diffraction problem "in a very original manner." Some things were "still incomplete" and "not as well ordered as I could have wished." But these critical remarks could not diminish his positive judgment. Sommerfeld had addressed "a large problem" and had taken "the first definitive steps" to its solution. From the viewpoint of theoretical physics, too, Sommerfeld's habilitation dissertation was "a

52 Protokollbuch Nr. 1 der Mathematischen Gesellschaft zu Göttingen (Easter, 1893–February 1896). Entry of January 15, 1895. SUB Cod. Ms. Math. Archiv 49:1; Poincaré, *Polarisation*, 1892; Maey, *Beugung*, 1893.

53 Manuscript, Mathematische Theorie der Beugung, undated. AHQP, Microfilm 23A, Sects. 3.3 and 3.4.

54 Habilitationsgesuch, February 6, 1895. Dekanatsakten, F. Klein, 1894–1895. UAG Phil. Dek. 180a.

most welcome achievement, which in certain respects will prove epoch-making in the history of the development of diffraction theory." The other members of the habilitation commission concurred in the positive assessments of Klein and Voigt. Thereafter, the rest of the process was mere formality. At the next meeting of the commission on February 21, 1895, it was unanimously agreed that Sommerfeld be permitted to proceed to the "further habilitation requirements."[55]

The next step for the habilitation candidate was to submit in a colloquium to questions posed by the faculty. Following that, the candidate had to deliver a probationary lecture, for which he could propose three topics, from which the commission selected one. Sommerfeld's first topic read: "On the parallels between mathematical and physical conceptualization," which offered him the opportunity to expound on the essence of "physical mathematics." With the second topic, "On general theta-functions," he would have been able to present himself as a virtuoso in the area of partial differential equations. The third topic was "On graphic methods in mathematics."[56] In contrast to the first two, here the object was to show how mathematical investigations could also yield practical results. Klein had informed him ("swearing him to secrecy") that he should prepare himself for the third topic. "This is very painful for me," Sommerfeld wrote his parents, "because I have already worked out the first, and don't have much to say about the third."[57] As Klein had signaled, the habilitation commission decided to have Sommerfeld lecture on this last-named topic.

These were hectic weeks for Sommerfeld. At the end of the winter semester, he gave one more lecture to the Mathematical Society, in which he made several aspects of his habilitation dissertation, remarkable from a mathematical point of view, the central matter.[58] For the upcoming probationary lecture, Sommerfeld had his parents send him large format photographs of the Königsberg harmonic analyzer that had been prepared for the World's Fair in Chicago. In order to prepare himself fully, he renounced other pleasures. "I was invited to dances on March 2nd and 3rd. Today, I've declined both; dawdling has to stop," he wrote to Königsberg. But he did not renounce all socializing. As in the previous year, in this carnival season once again "Vischer's magical fest" took place, and he could not and would not absent himself from it, for aside from the entertainment, it would also be presenting himself as a prospective member of Göttingen's learned society: "Half Göttingen was in attendance. All the big shots: Pro-rector, Curator, etc.," he reported the next day to his parents in a 6-page letter, which he wanted to be understood as a kind of "entertainment gazette."

As he had the previous year, he shone with his musical talents. "There were 7 lieder, 2 arias, 3 duets, one trumpet song, one four-hand, and one two-hand pieces.

55 Dekanatsakten, F. Klein, 1894–1895. Göttingen, UAG Phil. Dek. 180a.

56 Dekanatsakten, F. Klein, 1894–1895. Göttingen, UAG Phil. Dek. 180a.

57 To his parents, undated [late February 1895].

58 Protokollbuch Nr. 1 der Mathematischen Gesellschaft zu Göttingen (Easter, 1893–February 1896). Entry of February 26, 1895. SUB Cod. Ms. Math. Archiv 49:1.

My tried and true Chopin came out admirably on the beautiful grand piano. I have played him better, to be sure, but he came off with such passion, without music of course, and without getting stuck. In the accompaniment of the songs too I acquitted myself respectably. My reputation as musician here is solid as a rock." But he landed his "major coup" in quite another way. "Around 10:00 o'clock, the guests were to eat. It was a buffet, and everyone sat around together casually and convivially. Then Mrs. Vischer clinked her glass and made an altogether lovely toast to her "artists." Now it was incumbent to respond. This fell apparently to me. Now I must confess: I had prepared for this ahead of time, I had an inkling that a toast would be made to the "artists," and had thought out a little speech in which I planned to use just this word as a springboard. So, wasting no time, I clink my glass as well, and give my well thought out speech. I declined the tribute paid us artists and redirected it onto the artistic spirit of the Vischer home, alluded to the marble bust of the poet Vischer which stood in front of me, paid Mrs. Vischer a number of handsome compliments, and toasted the House of Vischer. Since I was able to take off from the previous toast, the whole thing came to be easily improvised. In short, I made a big impression. The big wheels and dignitaries may have been annoyed that I as the youngest had snatched the speechifying from them. Liebisch, who was also present, was annoyed at everything, of course. May he be so! Mr. von Wilamowitz-Möllendorf—aside from Klein, the brightest luminary of the university—said to his neighbor (as I heard today) 'He played well, and his speech was also not bad; I think we can give him the colloquium.' You can well imagine that I spent the rest of this evening on which I had thus played the principal role in the highest spirits."[59]

The habilitation colloquium took place on March 7, 1895, in the absence of Klein, who was ill, and was represented by Wallach. It went, as did the probationary lecture scheduled for March 11, "with signal success," as the lapidary notation in the habilitation record reads.[60] For the Philosophical Faculty of the University of Göttingen, the habilitation was a procedure that always ran according to the same agenda. For Sommerfeld, however, this academic ritual meant entry to the scholarly world of Göttingen. Even if he could be quite confident of success in advance, he delivered a detailed report of it to his parents in Königsberg. He reported he had been successful at the habilitation colloquium, "but it could perhaps have gone more smoothly." He had no reservations with respect to the probationary lecture, however. "First, a splendid auditorium. 8 full professors, every mathematical student still in town, 4 young ladies, many of my local acquaintances, and also the Privy Counselor himself, Dr. Höpfner, Curator of the University. I spoke as fluently as a waterfall." He had given the lecture also "a philosophical touch in that I began

59 To his parents, undated [late February 1895]. The classical philologist Ulrich von Wilamowitz-Moellendorff (1848–1931) belonged, like Klein, to the circle of Althoff's intimates.

60 Dekanatsakten, F. Klein, 1894–1895. UAG Phil. Dek. 180a.

by citing Kant; there was a terrific train of thought in it, and many fine points. The thing was well hammered home, and was daringly and devoutly delivered. I had a lot of fun preaching to this elite company." Within the circle of his friends and colleagues, the event was heartily celebrated. "We've drunk, joked, and even danced!" Now he could count himself "quite officially a denizen of the learned houses of Germany." The next thing was to have visiting cards printed, in order to make this publicly known. Here too, this university town had a peculiarly prescribed ritual. Sommerfeld would "amuse [himself] for 2 days by going around in formal attire to visit all the professors. One is not simply admitted, though; first one hands the servant one, or where one wishes to be invited in, two cards to take into the house. Unfortunately, I have to do the latter in a lot of cases since I am already acquainted with many people." At the conclusion of this tightly written, eight-page letter to his parents, he signed it gratuitously "With love, your Arnold, Lecturer."[61]

3.5 Lecturer

In his letters to Königsberg, Sommerfeld repeatedly described the academic milieu of Göttingen, to which as a prospective university professor he would soon belong. The word "student" had a "condescending connotation," he wrote his parents once, and for some professors "even the word 'lecturer' had a disparaging note."[62]

The social status of a prospective academic was discernible at every occasion, at the "magical fests" at the Vischer home, as well as at the "lecturers' table."

To grasp the academic hierarchy, it sufficed to glance at the lecture register of the university. There, the professors of each department were listed, not according to their areas of specialty, but according to their status under the rubrics "full professors," "associate professors," "lecturers," and "readers."[63] The Philosophical Faculty reported the outcome of each habilitation proceeding also to Berlin, where dossiers were compiled on prospective university teachers by the Prussian Ministry of Culture.[64]

With the official status as lecturer, a university teacher acquired at first only the right to deliver lectures in his department. But the notification of the appropriate ministry itself shows that more was attached to this status. As lecturer, one could be appointed to a professorship at a university, insofar as one was qualified for the position by virtue of personal scholarship and teaching experience. So it was incumbent upon a new lecturer still wet behind the ears to make a name for himself

61 To his parents, March 12, 1895. Also in ASWB I.

62 To his parents, August 3, 1894.

63 As noted above, ch. 2, footnote 33, "lecturer" translates the German rank "Privatdozent." "Reader" here translates the German rank "Lektor."

64 Philosophische Fakultät an den Curator, March 11, 1895. Dekanatsakten, F. Klein, 1894–1895. UAG Phil. Dek. 180a.

Fig. 7: Lunch at the "lecturers' table" was among the social events at which Sommerfeld, as a prospective member of the academic world of scholarship in Göttingen, showed his talents (Courtesy: Deutsches Museum, Munich, Archive).

in research and teaching beyond the confines of his own university. So far as research was concerned, the habilitation dissertation normally served as a foundation for further publication. In this regard, Sommerfeld was unconcerned. About his work, even the experimental physicist of the University of Göttingen had declared himself "laudatory to a very high degree," Sommerfeld wrote his parents. "The habilitation dissertation won't be published as such; instead, over the vacation, I'll extract 3 separate papers from it. The rest will be incorporated into a later comprehensive summary."[65]

Nor with respect to his pedagogical talents was he worried. If already at his habilitation lecture it had given him pleasure "to preach" to the "elite company" of professors of his department, lectures delivered to students would be trivial. And in Klein, he had in this regard an extraordinary model: "Such lectures those were!" he raved many years later. "Meticulously prepared, forcefully delivered, each hour a small, even stylistically rounded masterpiece. Every 10 minutes a summary statement in concise form."[66]

In his initial lectures, Sommerfeld treated special areas of mathematics. This gave him the opportunity to acquire a broad spectrum of mathematical subjects, for each semester he presented a different lecture topic. In the 3 years of his

65 To his parents, March 12, 1895. Also in ASWB I.
66 Sommerfeld, *Geburtstag*, 1949, p. 289.

lectureship at Göttingen, he lectured on probability theory, projective geometry, calculus of variations, theory of surfaces, and second-order partial differential equations. From the winter semester of 1896/1897, he gave additionally introductory lectures on differential and integral calculus. "In Sommerfeld's lectures, a great deal of material was always surveyed and brought together, applications of the most diverse sort, that made them extraordinarily stimulating," recalled Otto Blumenthal (1876–1944) many years later of his mathematical studies at Göttingen. Among the lectures that affected him "most memorably" was Sommerfeld's lecture on the differential equations of physics and on the calculus of variations, which he wrote up for the mathematical reading room. "When I consider what probably most decisively impelled me to mathematics, it was probably the wealth of offerings at that time. The education was extraordinarily multi-faceted; to physics also, there was a continuous bridge; one was confronted with boundless material on which work was being done everywhere. What perhaps contributed most to this impression was the reading room and the universally comradely relationships prevailing there."[67]

That Blumenthal preserved his student years at Göttingen so positively in memory was due not only to Sommerfeld's lectures. With Felix Klein as the driving organizational force, the University of Göttingen rose in those years to become the global center of mathematics.[68] "Fall, 1894: beginning of the encyclopedia," Klein noted in his personal journal concerning a project that would present mathematics as like no other in its comprehensive significance also for neighboring fields such as mechanics, physics, and astronomy. For the year 1895, he noted, "Whitsuntide: Advancement Association of Göttingen. Festschrift, elementary geometry (in which I thus treat exact mathematics. In this connection, tactics: the gyroscope considered as a second festschrift). August: Engineering Association in Baden. 'Peace of Aachen.' Fall: insurance seminar founded."[69] With these concise notations, Klein indicated the most important projects by means of which he intended to realize his wide-ranging organizational interests. The "festschrifts" on elementary geometry and gyroscope theory were meant to demonstrate his commitment to high school instruction to the Association for the Advancement of Mathematical and Scientific Education.[70] The "insurance seminar," established on October 1, 1895, became the germ of the scientific insurance industry in Germany; Georg Bohlmann (1869–1928), who had completed his habilitation under Klein 1 year before Sommerfeld, taught this new field of actuarial mathematics.[71]

67 Quoted in Lorey, *Studium*, 1916, pp. 351–353. Blumenthal's hand-written draft of Sommerfeld's lecture on the calculus of variations can be examined at the reading room of the Mathematical Institute at the University of Göttingen.

68 Rowe, *Felix Klein*, 2001.

69 Jacobs, *Felix Klein*, 1977, p. 18.

70 Tobies, *Felix Klein*, 2000.

71 Bohlmann, *Versicherungsmathematik*, 1900; Koch, *Bedeutung*, 2005; see also http://www.stochastik.math.uni-goettingen.de/index.php?id=18 (29 January 2013).

As assistant to "Felix the Great," Sommerfeld shared intimately in these activities. "All kinds of curious things have happened in Göttingen," he wrote to Königsberg in June 1895. "First, around Whitsuntide, a congress of mathematics teachers. In the past, they have always inveighed against the universities. They were too ethereal, and ignored the needs of the schools. What does my Klein do? He grabs his opponents by the scruff of the neck, attends the previous meeting last fall and, on behalf of the combined Göttingen school and university groups (of whose existence, N.B., no one had ever heard) invites the teachers to move their next meeting to Göttingen. All of a sudden now, all is sweetness and light. During the 3 days the rigamarole lasted, mutual admiration was given voice in countless toasts and speeches. Reception evening, ceremonial dinner, scientific talks by Klein and the philosophers, tours of the institute (on one afternoon, between 3 and 8, the teachers were herded through 10 institutes). I had various things to attend to in connection with this, and had to be in attendance through the whole business."[72]

Klein engaged in similar initiatives with respect to engineers, whose training at technical universities seemed inadequate to him. Since the engineers in their emancipation campaign for equality with the universities felt like poor cousins, Klein, at a congress of the Association of German Engineers (VDI) at Aachen, conceded to the technical universities indeed the right to engineering education (what Althoff, alluding to the famous historical peace negotiations called "the Peace of Aachen"). However, he did not fully renounce the prospect of establishing in the universities curricula in the service of technology.[73]

Klein's activities addressed not only the application of mathematics in engineering and in school instruction, however, but concerned pure mathematics as well. Here too, the decisive course was set during Sommerfeld's years as lecturer at Göttingen. The kickoff was Hilbert's appointment at Göttingen, which Klein in collaboration with Althoff had been working to bring about since December 1894. "The nearer the vacation approaches, the happier I am about the forthcoming move," Hilbert wrote Klein on March 4 from Königsberg.[74] Hilbert's first semester as a Göttingen professor alongside Klein was also Sommerfeld's first semester as lecturer. In this capacity, it fell to him to keep the minutes of the Mathematical Society when extraordinary circumstances arose, and the lecturer himself did not undertake the entry. "Special session in honor of Poincaré's presence in Göttingen," Sommerfeld wrote in the minutes under "June 10, 1895," for example. "Prof. Klein reported on the success of the teachers' gathering at Whitsuntide. M. Poincaré speaks on the existence proof of the spatially regular potential, when the values of the potential on a surface S [are] specified." Hilbert spoke on "the fundamentals of the discriminants of Galois number fields." The session concluded with

72 To his mother, June 15, 1895.

73 Manegold, *Universität*, 1970, pp. 136–144.

74 Quoted in Frei, *Briefwechsel*, 1985, p. 121.

"Demonstration of a conic section circle and two apparatuses for the solution of cubic equations."[75]

It was this productive side-by-side interaction of applied and pure mathematics that made Göttingen in these years a world center of mathematics. Additionally, there was the unconstrained style radiated by the young lecturers, who no doubt mitigated the impression of Klein's overwhelming authority. In 1895, Hilbert had just turned 33, Sommerfeld, 26. "In his case as well as in the case of Hilbert, though, I cannot separate my impression of the lectures from the experience of personal interaction," Blumenthal wrote, "for I came into close contact with Sommerfeld especially with regard to the gyroscope, and was a regular participant in the Hilbert 'number field promenades.' Of course most of this personal interaction occurred during my student days."[76]

The terms "gyroscope" and "number field" might seem to epitomize the opposition between pure and applied mathematics, but for Sommerfeld, assistant to Klein, the two were closely allied. "To my great joy, next semester he is lecturing only 2 hours on number theory, and 2 hours on gyroscope motion," he wrote to Königsberg on Klein's lectures in the winter semester of 1895/1896.[77] The joy expressed over the "only" stemmed from the fact that he had to write up both lectures for study in the reading room—the gyroscope lecture on top of this as a festschrift for the "Advancement Association" of the high school teachers. "But Klein makes heavy demands on my vacation time," he groaned after this semester.[78] He addressed the theory of gyroscopes with such zeal that it would become an area that would occupy him for many years. The epistolary sigh should not, in other words, be misconstrued as an expression of distaste for the topic. The same is true for number theory. "You have no idea how beautiful it is!" he gushed. "Nothing done so far approaches this for mathematical elegance. Number theory is normally treated with ponderous and very abstract concepts. But Klein stands there, draws a few figures on the board, speaks of seemingly quite remote matters, and then accomplishes with a wave of the hand the same and more than the number theorists."[79]

It testifies to Sommerfeld's conscientiousness that in working up the lectures for Klein as well as in the preparation of his own lectures he did not give "physical mathematics" short shrift. Klein felt that with his habilitation dissertation, he had "broached an area in which there was a great deal to be done," and urged him to publish some articles about it in the *Mathematischen Annalen* as well as in the *Annalen der Physik*. Later, he could also, "in a separate book," publish "an integrative survey" of his method. "We'll see. For the most part, things go as Klein

75 Protokollbuch Nr. 1 der Mathematischen Gesellschaft zu Göttingen (Easter, 1893–February 1896). SUB Cod. Ms. Math. Archiv 49:1.
76 Quoted in Lorey, *Studium*, 1916, p. 352.
77 To his mother, October 12, 1895.
78 To his mother, February 17, 1896.
79 To his mother, November 30, 1894. Klein, *Zahlentheorie*, vols. 1 and 2, 1896.

decrees."[80] In this case, however, things didn't go quite according to Klein's wishes, for Sommerfeld wanted "to let the diffraction theory lie fallow a while longer," as he confessed to his mother in April 1895, because just then he was more occupied with another subject, about which, however, he said nothing.[81] But Poincaré's visit led him a few weeks later to return his attention to diffraction: "Since Poincaré has also worked on diffraction (the only one to have brought out a sensible piece of work), my things interested him very much," he wrote following this visit. "I attempted to speak a little broken French, he some broken German, and the business soon petered out in the sands of mutual incomprehension. Klein pushed me to complete my notes by today specifically for the conversation with Poincaré. And that has happened!"[82]

Publication, however, did not happen for quite a while. That gave him the opportunity to investigate the Poincaré theory of diffraction more thoroughly than he had originally intended. "I've finished my paper. It is very beautiful, 'very fine indeed,' [83] everything has turned out more beautifully than I had dared hope even recently. Poincaré will be impressed," he wrote to Königsberg in July 1895.[84] Indeed, in a subsequent paper on diffraction theory, Poincaré declared Sommerfeld's method "extremely brilliant."[85] The announcement of Sommerfeld's stroke of genius somewhat anticipated the detailed paper in the *Mathematischen Annalen.* "Klein thought I should give a lecture on diffraction," Sommerfeld wrote his mother several weeks before the fall, 1895 annual congress of the Association of German Natural Scientists and Physicians in Lübeck. "Of course it's always a good idea to promote oneself when one can and the subject is suited to a lecture."[86] At the same time as the Natural Scientists, the German Mathematical Society, which Sommerfeld had joined shortly before as member nr. 224, also had its meeting. In joint departmental meetings, mathematicians and physicists could exchange ideas about recent developments of interest to both disciplines. In condensed form, Sommerfeld presented the results of his diffraction theory to this forum. "My lecture was happily launched," he wrote on a postcard to Königsberg. "I spoke very coherently and had correctly planned my timing. I made a very good impression on the mathematicians. Whether on the physicists, remains to be seen."[87] Three days later, he reported a bit more thoroughly on the response to his lecture. "The

80 To his mother, February 15, 1895.

81 To his mother, April 17, 1895.

82 To his mother, June 15, 1895. On Poincaré's visit, see Protokollbuch Nr. 1 der Mathematischen Gesellschaft zu Göttingen (Easter, 1893–February 1896), entry of June 10, 1895. SUB Cod. Ms. Math. Archiv 49:1.

83 This phrase written in English in the original.

84 To his mother, July 19, 1895. Sommerfeld, Mathematische Theorie der Diffraction, 1896, pp. 371f.

85 Poincaré, Sur la Polarisation, 1897, p. 313.

86 To his mother, July 25, 1895.

87 To his mother, September 21, 1895. Sommerfeld, *Diffractionsprobleme*, 1895.

physicists haven't entirely understood my stuff. Even Boltzmann got it only the next day when Klein explained several things to him. But the mathematicians were all very edified by it. For example Prof. Brill, my friend from the Alps, with whom I enjoyed reliving our experiences from that time." Even among the physicists, be it said, there was at least one who valued his theory. Siegfried Czapski (1861–1907), the expert for physical optics and representative of Ernst Abbes (1840–1905) from the Zeiss works in Jena, had "come to Lübeck primarily to hear my lecture" and had "most urgently" invited him to Jena to meet Abbe and get to know his work. "This will occur at the next opportunity." Czapski had written "a splendid book on optics" and was "the great Abbe's right hand (director of the enormous glass factory at Jena which produces microscopes and devices). Abbe has put forward a quite new theory of the microscope, and is generally acknowledged the primary authority."[88]

The Lübeck Natural Sciences congress was a memorable experience for the brand new university lecturer in other ways, too. He met Boltzmann personally here and witnessed one of the great debates of contemporary physics, the debate over energetics: whether all physical phenomena could be derived solely from the mechanical statement of the conservation of energy. With their arguments at Lübeck, Boltzmann and Klein so definitively shattered the fundamental concepts of energetics put forth primarily by Georg Helm (1851–1923) and Wilhelm Ostwald (1853–1932) that even half a century later, Sommerfeld preserved a lively memory of the event: "Boltzmann's arguments were decisive. We younger mathematicians were all on the side of Boltzmann; it was clear to us at once that from the single energy equation, it was impossible to derive the equations of motion of even a single center of mass, let alone of a system of arbitrary degrees of freedom."[89] In the contemporaneous description, Sommerfeld gave his parents a few days after the event, he described this as a struggle between a mathematical and an unmathematical direction in physics: "Very interesting was the duel that lasted through two sessions between Ostwald (Leipzig) and Boltzmann, who represented two divergent directions in physics, the unmathematical and the mathematical. Boltzmann hit his opponent with all the novelty and force at his command. The other, a witty charlatan, defended himself with what he could, but was in the unanimous opinion of the mathematicians soundly disgraced."[90]

On his return from Lübeck, Sommerfeld made the problems associated with his habilitation dissertation once more the subject of a lecture delivered to the Göttingen Mathematical Society. He showed by way of an example that the method conventionally employed in optics yielded only approximate solutions that agreed with the results of his mathematical theory only in the limit of very short wave lengths. His example this time concerned not diffraction but the reflection and

88 To his mother, September 24, 1895. On Csapski, see Flitner/Wittig, *Optik*, 2000.

89 Sommerfeld, *Ludwig Boltzmann*, 1944, p. 25; Körber, *Briefwechsel*, 1961, pp. 118–120.

90 To his parents, October 5, 1895.

December 5, 1893	Methods of principal solutions in mathematical physics
February 27, 1894	On the work of French mathematicians (in particular, Picard)
August 3, 1894	Solution of the equation $\Delta u + k^2 u = 0$ on Riemann surfaces
January 15, 1895	Towards the mathematical theory of diffraction
February 26, 1895	On certain divergent developments
May 7, 1895	Production of branched solutions of the equation $\Delta u + k^2 u = 0$
June 10, 1895	Extraordinary session in honor of Poincaré
July 7, 1895	Probability calculus
November 12, 1895	Problems in reflection and refraction
February 18, 1896	Projective geometry
June 9, 1896	On Poincaré's theory of tides (1)
June 30, 1896	On Poincaré's theory of tides (2)
July 14, 1896	On certain theorems in the theory of surfaces
November 24, 1896	On new work of Peano and Volterra on the theory of gyroscopes
March 3, 1897	Differential equations in physics
May 11, 1897	On electric waves along wires
Mai 28, 1897	On some Riemann manuscripts
July 2, 1897	On numerical calculation with elliptical functions
July 30, 1897	On calculus of variations

Table 1: Sommerfeld's lectures at the Göttingen Mathematical Society[a]

[a] Protokollbuch Nr. 1 and Nr. 2 der Mathematischen Gesellschaft zu Göttingen, SUB Cod. Ms. Math. Archiv 49:1 and 49:2

refraction of light, which emanating from a point source onto the separating plane is refracted or reflected by two contiguous media.[91] The lecture had been "very excellent," he wrote his parents. "Hilbert too thought the result very beautiful." Even the physicists present had shown great interest. Presumably, he wished to elaborate the topic for an extensive publication, because he announced that he wanted to "introduce a quite new problem into the work." Considering his duties

91 Protokollbuch Nr. 1 der Mathematischen Gesellschaft zu Göttingen (Easter, 1893–February 1896). Entry of November 12, 1895. SUB Cod. Ms. Math. Archiv 49:1.

as assistant—in particular working up Klein's lectures on number theory and the "gyroscope"—that this publication never materialized is not surprising. For his own lectures in this winter semester 1895/1896, he had decided on projective geometry. "Each hour's lecture time requires 2 hours of preparation," he wrote, describing the expenditure of time.[92] "I've found all sorts of new things in the process, which will, I think, result in a quite respectable paper." He wished to exploit this topic still further, too. "Tomorrow, I'm to speak at the Mathematical Society about my lectures."[93] At the conclusion of the semester, any rate, he was quite pleased with himself: "I ended my lectures last Friday in grand style. I'd saved one special tidbit for the final session, and was applauded at the end with great stamping of feet. I have the very agreeable feeling that the people have learned a great deal from me, and that I maintained their interest in the subject throughout."[94]

3.6 The Engagement

Quite apart from his duties as assistant to Klein, there were other cogent reasons why Sommerfeld did not convert all his plans for publication into reality. "The company at my table intends to give a masked ball or a sleigh ride for the amusement of Göttingen society," he had reported to Königsberg a few weeks before his habilitation.[95] This letter contains an intriguing addition in someone else's hand: "Arnold as ancient professor, I as his daughter." The addition dates to a time at which Sommerfeld had long been married. The "I" referred to Sommerfeld's wife, Johanna Höpfner (1874–1955), at that time, the 20-year-old daughter of the Curator of the University of Göttingen. The addition was apparently intended to inform the children of the occasion on which their parents had met. "The sleigh party went off quite to everyone's satisfaction," Sommerfeld wrote in his next letter to Königsberg. "The young Göttingen ladies are really quite nice and easily amused. There were 10 musicians who blew into our ears on the journey, and then played at night for dancing. On the way there, I rode with the favorite daughter of the Curator here. Thank God, that's all over and done with."[96]

We can assume that Johanna and Arnold became a couple only much later. Still in October 1895, Sommerfeld sniped about a lecturer who became engaged to a Göttingen professor's daughter and got an appointment at once as a professor at another university without having published. "Should I, too, find myself such a nice papa-in-law?"[97] He would surely have suppressed such sarcasm if at the time

92 To his mother, November 17, 1895.
93 To his parents, February 17, 1896.
94 To his mother, March 10, 1896
95 To his parents, December 22, 1894.
96 To his parents, January 10, 1895.
97 To his mother, October 16, 1895.

he had had an inkling he would become Ernst Höpfner's son-in-law. The curators represented the highest authority within a university in all affairs of university administration from building construction to matters of appointment and were accountable accordingly to the respective government ministries. "Do you still remember? It was in the spring": with this line of a love poem, Arnold recalled to his bride the start of their amorous relationship in the spring of 1896, more than a year following the sleigh ride.[98]

In the small university town, it was difficult for lovers who were not yet officially engaged, to express their feelings without transgressing proper etiquette. When the couple in question comprised the daughter of the Curator and a lecturer who was a popular guest at professorial parties, a violation of etiquette would have provoked a scandal. Arnold and Johanna apparently used invitations from Adelheid Liebisch, whom both called "Mother-in-law," as occasions to meet. Apart from this, only letters were left as a medium of communication: "The mailman, about whom otherwise there is not an ounce of charm, has for some time been for me the most welcomed person in Göttingen," Sommerfeld once wrote Johanna early in the summer of 1896, though just a few blocks separated the two. "Tomorrow, I'll make my pilgrimage along the designated route, although in general, I'd like to avoid rendezvous on the street."[99] In another *billet doux*, Sommerfeld invoked the aid of "the god of love" to preclude "possible colleagues" from accompanying him so he would not to lose the opportunity of a fleeting rendezvous by eye contact. "Of course our Mama-in-law will soon be available again," he reassured himself and his "Hannchen." Each time his route brought him past the Curator's house, his heart beat faster, but he dared not "glance up," as he once confessed. "It is certainly a crazy situation we find ourselves in."[100] But their feelings couldn't so easily be made to conform with Göttingen etiquette, and so Arnold and Johanna became engaged, secretly at first, in the spring of 1896.[101]

Upon their official engagement, Arnold actually preferred to wait until he had fulfilled his next military maneuvers, which were to take place during the vacation following the summer semester. But the love relation between Arnold and Johanna could not be kept secret that long. Sommerfeld's mother met Johanna, her future daughter-in-law, already at the beginning of July, 1896 on a visit to Göttingen. Adelheid Liebisch used this visit to invite Johanna together with Arnold and his mother to the Liebisch home. "Tomorrow is an anniversary day for us. Do you remember? It was in a green room at the Liebisch's on a hot July day," Johanna wrote to Arnold's mother 1 year later, recalling this July 9, 1896, "and then we

98 To Johanna, undated.
99 To Johanna, June 19, 1896.
100 To Johanna, July 1, 1896.
101 Many years later, in a letter to his wife on March 24, 1914, Arnold alluded to the secret engagement on that day in the years 1896.

Fig. 8: After their "secret" engagement in Spring 1896, in July of the same year, Arnold officially asked for the hand of Johanna Höpfner, daughter of the Curator of the University of Göttingen (Courtesy: Deutsches Museum, Munich, Archive).

walked through the fields among nodding red poppies. I see that day so clear and vivid before me, and thinking of that day, I greet you 1000 times, dear Mother."[102]

So long as Johanna's parents had not been asked their approval, however, the relationship had to remain secret. "Dear, dearest mother," Johanna addressed her future mother-in-law, who along with their pseudo-mother-in-law Adelheid Liebisch now numbered among the conspirators. "Decide it with him; of course I'll do whatever seems best to you both," she wrote giving her assent in advance to the timing of their official engagement.[103] As secret bridegroom in this situation, Arnold hardly radiated his usual self-confidence. "My dear little mother! You'll receive a letter today of a quite private nature," he wrote her 2 weeks later. In the meantime, his mother had returned to Königsberg. In Göttingen, there was "a great deal of gossip again." In spite of all precautions, his relationship with Johanna had not remained as secret as he had hoped. The forthcoming engagement was already being discussed before he had even asked the Curator for his daughter's hand. He had inquired of one of the lecturers he was friendly with "whether we are so obvious about it." The friend answered that he and Johanna had, at a recent social event, indeed "behaved unmistakably like a bridal couple." As a result, he was "thrown into indecision" as to whether the timing of their engagement had not

102 Johanna to Arnold's mother, July 8, 1897.
103 Johanna to Arnold's mother, July 11, 1896.

better be brought forward. "It may well be that I'll surprise you even before October. Maybe I'll go to the Curator tomorrow, maybe not until Saturday, or maybe I'll cool down again. At the moment, I hate the whole Göttingen scene, and am annoyed with myself and the world."[104]

Five days later he decided, "in a state of considerable palpitation," not to postpone the official engagement ritual any longer. In the previous days, rumors concerning the secret bridal couple had already reached the point that people were congratulating Johanna on her engagement. "That gave me the final shove towards that which I'd actually already decided on," he wrote his parents. At the Höpfner's, he was greeted by Johanna and her sister Helene, since the Curator was not yet at home. "We sat for quite a while chatting, and Papa was nowhere to be seen. He generally worked continuously at his office. Finally, he walked into the room quite unexpectedly. I stammered through the beginning of my well-prepared speech. He was quite moved, and understood at once that basically he had nothing more to say in the matter." The Curator's wife, however, was so surprised by the news that she did not wish to give her blessing at once. "So Hannchen and I decided to besiege her directly. I was immensely polite, kissed her hand as often as possible. Mama said she really had to get to know me, and complained that Hannchen hadn't confided in her. I stayed for the evening. We said "Du" to each other, and held hands as much as we could. After dinner I was to play some Chopin for Mama to assuage her. She was actually quite assuaged, only she preferred not to run up the white flag so soon. Papa-in-law is apparently rather hen-pecked. After dinner, when we sat drinking a bottle of wine, and a wordless toast was made us. Finally, I requested and was granted permission to call again next day. Tomorrow there will probably be some further sparring, but I'm convinced that by the end of the day we'll be totally in the clear." Now all that was needed was agreement on the official date of the engagement. "We were engaged on Thursday the 23rd of July at the Curator's celebration, not earlier," Sommerfeld briefed his parents. "Otherwise Mama-in-law would be very angry. Even our good Papa-in-law knows no different."[105]

As expected, the next day "Mama-in-law" also bestowed her approval, so that nothing now stood in the way of making the engagement public, and the announcements prescribed by convention for such events could be printed. The news made the rounds in far-off Königsberg, too. When a childhood friend of Arnold's learned of the engagement "in your professorial village," he reminded Arnold of a promise he had given years before: "Whichever of us gets engaged first," (the "us" referred additionally to a third friend) "owes the others a basket of Champagne (or was it more? I think so!) . . . So you see, my fine friend, engagement has its drawbacks."[106]

104 To his mother, July 20, 1896.
105 To his parents, July 25, 1896
106 From Arthur Heygster, August 5, 1896

4 Clausthal

In the mathematical milieu of Göttingen, Sommerfeld was happy as a clam. Though he may have failed to realize this or that planned publication, or postponed it from lack of time, his letters to his parents show he was quite confident about his future career. As a student of Klein, he could look forward to an appointment sooner or later as a mathematician at some university or technical university. His assistantship expired in September 1896, but he hoped on its expiration to receive a lecturer's fellowship. His predecessor had bridged the waiting period for a professorship with such a fellowship. In light of the close relationship of Klein and Höpfner with Althoff at the Prussian Ministry of Culture, Education, and Church Affairs, which decided on these fellowships, he gauged his chances as very good, but notification from Berlin had not yet come. "Althoff has so far said nothing about the fellowship," Sommerfeld wrote his fiancée from military maneuvers in September 1896.[1] Now, he was not especially in the mood to play soldier. He waited longingly for the occasional visits to Göttingen to see his fiancée; in addition, he hoped his future father-in-law would lend timely assistance in the matter of the fellowship so he would not have to depend on his parents' financial aid during the coming winter semester in Göttingen.

4.1 An Offer from America

For the time being, nothing came of the fellowship. Nonetheless, Sommerfeld looked rather confidently to the future. Göttingen was the world center of mathematics; if there was a professorial opening anywhere, Göttingen was the most obvious place to look to fill it. The year before, Sommerfeld's predecessor in the Klein assistantship, Ernst Ritter (1867–1895), had received an appointment at Cornell University, in Ithaca, New York. He had accepted the offer, but had fallen ill and died on the crossing to America before he could take up the professorship.[2] Thereupon, the offer was extended to Sommerfeld. "They would like to fill the Ritter position," he reported to Königsberg. The position had a remuneration of 1,000 dollars per annum, which at the contemporaneous exchange rate corresponded to 4,000 Marks. "I would have many lectures to prepare, but would have good prospects for promotion."[3]

1 To Johanna, September 6, 1896.
2 Klein, *Ritter*, 1895.
3 To his parents, January 6, 1896.

M. Eckert, *Arnold Sommerfeld: Science, Life and Turbulent Times 1868-1951*,
DOI 10.1007/978-1-4614-7461-6_4,

In these years, Cornell University was building up its mathematics department, and in this respect, Klein was considered a model in America.[4] And "basically" Klein advised Sommerfeld to accept the offer. "His rationale is expressed in the saying, 'a person grows to meet his larger purposes.' In the process, he would love to shake up the American university system." Thus, Sommerfeld described Klein's motivation in the matter. Klein considered "the German situation too narrow," and thought he [Sommerfeld] could always "if he felt like it" return to "decrepit Germany." His other advisor was Wallach, who counseled him "pretty forcefully to decline." Wallach knew that Sommerfeld felt "thoroughly comfortable" at Göttingen and warned him of the quite different circumstances in America.[5] After "a walk with Hilbert, specifically dedicated to a discussion of this matter," Sommerfeld proved ultimately "pretty well determined to decline," which came as a great relief to his parents in Königsberg, too, for seeing their son whose prospects were so bright emigrate to America would have pained them immensely.[6]

However, Klein's advice and Cornell's offer were not entirely without their effect. Half a year later, Sommerfeld was still in doubt whether he shouldn't accept the offer after all. His parents staked everything on preventing that. "America may be the land of the future, and the land of dollars, and may outdo Germany in many respects; nonetheless in science, Germany is up to now superior in both its pure purposes and its unceasingly serious work." Arnold's mother wrote this to the Curator of the University of Göttingen shortly after the engagement of Arnold and Johanna on the assumption that the loss of his daughter would be equally painful for him. Her son's threatened emigration was to her and to her husband "so disagreeable that we hope he will give up the idea." Perhaps she was counting on the University Curator's influence to help procure a professorship in Germany when she added the rhetorical question: "Why should he take his young energy, which may be of use to the Fatherland, abroad?"[7] The Curator assured Arnold's mother by return mail that his feelings "with respect to America [did] not differ essentially from yours and your husband's."[8] His father, too, did what he could to avert the danger. "How on earth is America any concern of Arnold's!" he wrote his future daughter-in-law. "In my view, he is too good for America. If he achieves something admirable, let it be to the benefit of his Fatherland, but not America! So, my dear Hannchen, work hard in both our and your own interest to see that Arnold gives up this crazy idea."[9]

But 4 weeks later, the call beckoning him to America had not yet been quite stifled, although it was becoming ever clearer that Arnold would not succumb to it.

4 Cochell, *History*, 1998, pp. 144–146; Parshall/Rowe, *Emergence*, 1994, p. 213.
5 To his parents, January 12, 1896.
6 To his parents, January 16, 1896.
7 Cäcilie Sommerfeld to Ernst Höpfner, July 30, 1896.
8 Ernst Höpfner to Cäcilie Höpfner, August 2, 1896.
9 Franz Sommerfeld to Johanna, August 11, 1896.

"Will I still meet your father on the 8th and the 9th in Göttingen?" He wanted to know from Johanna, before taking advantage of a short furlough from maneuvers to visit the Curator's family. "Otherwise I'll ask you to sound him as thoroughly as possible on his opinion regarding America." To mollify his fiancée, for whom the move to America would also have been painful, he added that he had determined to decline the American offer. "Fear not!" In his mind he was running through the various possibilities of securing a professorship in Germany. "There is a mathematician who has died at Greifswald. Couldn't Schönflies be appointed there, and Burkhardt at Zürich or Kiel? Pull your father's coat about it with a nice recommendation from me."[10]

Arthur Schönflies (1853–1928) was full professor of applied mathematics at Göttingen. Heinrich Burkhardt (1861–1914) was an assistant to Hilbert at Göttingen. It was not clear to Johanna why Arnold was concerning himself with the career opportunities of third parties, so he clarified his thinking in a subsequent letter. If those two were drawn away from Göttingen, a place for him might open up: "of course the combination Schönflies/Burkhardt was intended to lead to the combination Sommerfeld/Göttingen." He was giving himself good odds on succeeding to the Schönflies professorship, should he be wooed elsewhere. To be sure, Burkhardt, who as a lecturer had been waiting for a professorship since 1892, would be given preference, provided he had not previously been offered a professorship elsewhere. Schönflies had waited as lecturer from the time of his habilitation in 1884 until 1892, when he was appointed to the professorship of applied mathematics that had been established through the efforts of Klein. The lecturers waiting in line of necessity became rivals. "Schönflies's area is entirely my own area," Sommerfeld explained to his fiancée. "I read Schönflies's principal lectures last year not unintentionally, and to great advantage. And Hilbert and Klein both know that."[11]

4.2 The Appointment at the Clausthal School of Mining

It would be nearly a year yet before Sommerfeld was released from uncertainty about his academic future. In the meantime, Johanna also became familiar with the subtle observations lecturers made about their career opportunities. "Recently, though, Saint Felix himself said everything is going to be very beautiful," she wrote on one occasion to Königsberg. "Saint Felix" was Klein, and the "very beautiful" referred to Sommerfeld's elaboration of Klein's lecture on the theory of the gyroscope. The satisfaction of the manipulative Klein once more offered reason to hope her fiancé could count on an appointment soon—perhaps even at Göttingen. Together with Schönflies, Burkhardt, Hilbert, and other mathematicians, they had been "at dinner at his house," she continued, in the same breath as it were.

10 To Johanna, September 3, 1896.
11 To Johanna, September 6, 1896.

Present also had been the mathematician Franz Meyer (1856–1934), one of Klein's oldest students, from the Clausthal School of Mining. The get-together had been for the purpose of "celebrating [Burkhardt's] departure."[12]

Burkhardt was in fact appointed at Zürich a short time later.[13] But no position opened up at Göttingen for Sommerfeld as a result. Instead, the presence of the mathematician from Clausthal indicated a different arrangement, for Meyer was expecting an appointment at Königsberg. In April 1897, the Upper Mining Authority of Clausthal reported to the Ministry of Trade and Industry, which oversaw the School of Mining, the search for a successor to the mathematics professorship at Clausthal. The matter landed on the desk of Althoff, who conferred with Klein about it, and subsequently proposed as candidates the names of Schönflies, Sommerfeld, and Georg Scheffers (1866–1945) to the Minister of Trade and Industry, all three, students of Klein. Scheffers had attended Klein's lectures already in the 1880s at Leipzig and was Professor of Mathematics at the Technical University of Darmstadt.[14] Johanna was happy that Schönflies, not Sommerfeld, was first on the list for Clausthal, because compared to Göttingen, this town in the Harz Mountains held no great attraction for her. "It's really a good thing that Arnold doesn't have to go to Clausthal," she wrote to Königsberg; "it makes me happy every day now when his lectures bring him such joy." Schönflies would have to "make his way there moaning" instead.[15] She expressed her and Arnold's wishes for the future in verse, as well. Alluding to Sommerfeld's write-up of Klein's lecture on the theory of the gyroscope, she wrote lyrically:

Now the gyroscope's finished, Felix, I ask you please,
Quickly make my darling a professor;
A full professor with a salary,
So big shots can't disparage us![16]

In the matter of publication, so important to an academic career, Sommerfeld still had little to show. He hoped to establish himself in this area solely on the basis of the elaboration of his theoretical method of complex analysis on potential theory tried and tested in the theories of heat conduction and diffraction: "The number of boundary value problems solvable by means of my elaborated Thomson's method of images is very great," he wrote in the spring of 1897 to Klein, who was just then formulating plans for a trip to England. "I hope you will also find some pleasure in it." He asked Klein to offer the paper for publication, "perhaps to the London Mathematical

12 Johanna to Cäcilie Sommerfeld, March 4, 1897.

13 Liebmann, *Erinnerung*, 1915.

14 Ministerium der geistlichen Unterrichts- und Medicinal-Angelegenheiten to das Ministerium für Handel und Gewerbe, May 7, 1897. GSA, folder I, HA Rep. 121 DII, Sect. 6, Nr. 102, vol. 4.

15 Johanna to Cäcilie Sommerfeld, May 4, 1897.

16 Undated, presumably May, 1897.

Society. Otherwise the English won't read it. For the corresponding reason, though, I would like to publish it in a German journal (Crelle?)."[17] Hopes for publication in the respected German "Crelle Journal," the *Journal für die Reine und Angewandte Mathematik*, were dashed, but in England, Klein fulfilled his request. The paper appeared several months later in the *Proceedings of the London Mathematical Society*.[18]

Meanwhile, there was a new twist to the Clausthal appointment. The Upper Mine Authority revised its list, placing a certain "Dr. phil. W. Grosse, teacher at the Main School in Bremen" in the top position. Sommerfeld remained in the second spot. Schönflies was moved to the third spot. Scheffers's name was crossed off the list. In his place, two new "scientific assistant teachers" were added. After "thorough examination of the circumstances," Sommerfeld received the offer. He had been personally recommended to the Ministry by Althoff. In addition, at the age of 29, he was 10 years younger than the candidate at the top of the list, which meant a younger service age in the calculation of his salary.[19]

This turn of events in the matter of appointments soon became known around Göttingen. "What do you say to the fact that the Clausthal opportunity has suddenly surfaced again," Sommerfeld wrote his mother. He knew also of reasons behind the scenes not mentioned in the ministerial documents: "My friend Schönflies" was not proposed "basically because of his Jewishness, under the pretext that his application was only half-hearted, and thus he would not approach the job with the necessary ardor."[20] Three days later he added that the matter was "still far from being decided." Should he receive the offer, however, "there will be a Christmas wedding." The security of a professorship would permit him and Johanna to start a family. He also confessed, though, that there was for him "still a residue of resistance against the relatively unscientific nature of the position, as well as the winter isolation."[21] One day later, Johanna added, "Yes, it's still up in the air, and since fortunately there is nothing we can do to affect the outcome at this point, we just have to await the decision patiently as to what happens to us, consider it for the best and be happy."[22] A month later it was still not clear how the Clausthal appointment would be resolved, and the bridal couple expressed relief "that we'll no more go there than will Schönflies."[23] But then Franz Meyer wrote Sommerfeld: "Your situation appears to be very promising." He needed only to be patient a while longer. Meyer intimated there was not unanimity in Clausthal concerning the appointment and that Sommerfeld would probably not be welcomed with open arms by all his colleagues.[24]

17 To Klein, March 18, 1897. SUB, Klein 11, 1042.
18 Sommerfeld, *Potentiale*, 1897.
19 File note, June 25, 1897. GSA. folder I, HA Rep. 121 DII, Sect. 6, Nr. 102, vol. 4.
20 To his mother, June 5, 1897.
21 To his mother, June 8, 1897.
22 Johanna to Cäcilie Sommerfeld, June 9, 1897.
23 Johanna to Cäcilie Sommerfeld, July 8, 1897.
24 Franz Meyer to Sommerfeld, July 12, 1897.

But the decision about the Clausthal appointment did come before the end of the summer semester of 1897. "It is as good as certain that I will go to Clausthal by October 1," Sommerfeld reported to Königsberg. He described also the background maneuvering that either Klein or his prospective father-in-law must have confided to him: "At first it appeared my prospects were slim, because Berlin (consistent with Klein's wish) preferred Schönflies, while Clausthal preferred a high school teacher from Bremen, Dr. Grosse. The latter must have connections in Clausthal, and seems to be either friends with or related to other lecturers. The people at Clausthal are now resisting the Schönflies candidacy, partly on anti-Semitic grounds, partly because Schönflies personally made a weak impression on them and showed little enthusiasm. The Trade Ministry, however, to which the School of Mining is subordinate, raised categorical objections against the high school teacher." So far as the material circumstances were concerned, he could be very satisfied: "Starting salary, 3800 Marks + 480 Marks housing supplement, additionally ¼ of the lecture fees and a portion of the test fees," which at least for his predecessor had amounted to an additional 1,000 Marks. "So monetarily, I think the position is better than a professorship of the same rank, compared nominally," he summed up the advantages of the Clausthal position. As the "drawbacks," he listed the "scientific isolation" and the lower level of the lectures that he would have to give. "Lectures such as those I've given here, in which last semester, for instance, I was always able to present the material of my own research, much to my own and my students' pleasure, are naturally out of the question. It is true, however, that lectures of that sort are possible hardly anywhere outside of Göttingen." The scientific isolation would be "not so bad," he consoled himself; with his bicycle and the train he could, "every Sunday, if necessary," get to Göttingen.[25]

On July 23, 1897, the Upper Mine Authority in Clausthal notified the Berlin Ministry of Trade that Sommerfeld had accepted the appointment.[26] Meanwhile, Johanna had on an excursion in the Harz Mountains taken a look at Clausthal. She returned "quite charmed," Sommerfeld reported to his parents. He had also already received the offer of an apartment at a rent of 600 Marks per annum, "comprising 4 large and 2 small rooms on the ground floor, as well as several rooms 2 floors above. The building is said to be exceptionally grand by Clausthal standards, and built entirely of stone. We'll certainly rent this one."[27] Of course the appointment took another few weeks to be officially confirmed. The letter of appointment came only a few days before the start of the winter semester. "His majesty King and Emperor"—in the bureaucratic terms of German officialese, the Berlin Minister of Trade informed the Upper Mine Authority in Clausthal on September 23, 1897—"has on my graciously accepted petition, by means of the attached letter of

25 To his mother, July 17, 1897.

26 Königliches Oberbergamt (Referent Bannizer) an den Minister für Handel und Gewerbe, July 20, 1897 ("sent July 23, 1897"). GSA. folder I, HA Rep. 121 DII, Sect. 6, Nr. 102, vol. 4.

27 To his parents, July 22, 1897.

appointment of the 13th of September this year of Our Lord named the lecturer Dr. phil. Arnold Sommerfeld at Göttingen to the post of permanent professor at the United Mining Academy and Clausthal School of Mining."[28]

In Göttingen professorial circles, as Sommerfeld once wrote home, "even the word 'lecturer' had a disparaging odor."[29] His appointment as professor meant the end of this status. Financially, too, Sommerfeld could now breathe easier. His assistantship with Klein, limited to 2 years, had ended already in the fall of 1896. Thereafter, his parents had sent him money from Königsberg until, in April 1897, he received the lecturer's fellowship after all. "What will I do with all that money," he wrote confidently, when he learned of the fellowship, which at 1,200 Marks per annum came to exactly the same as his assistant's salary at which 3 years earlier he had begun at the mineralogical institute with Liebisch. "300 Marks from you, plus next semester the many lecture fees, and the limitless gyroscope royalties in the offing. I'll have to get myself a savings account book."[30] He had to wait several months longer for the "gyroscope royalties," to be sure, for the first "fascicle" of his write-up of Klein's lecture on the theory of the gyroscope appeared only in July 1897.[31] With the professor's salary awaiting him at Clausthal that aside from the "lecture fees" amounted to more than three times his lecturer's fellowship, he was at any rate putting the "disparaging odor" of the lecturer's existence behind him.

There were some things the appointment at Clausthal did not change, though. Among these was the write-up of the theory of the gyroscope, which was not remotely completed with the appearance of the first "fascicle" (196 pages!). For Klein, this was a special topic which he wanted to present as an example of the uses to which mechanics, astronomy, and physics could put pure mathematics. "The Klein lecture on the gyroscope is to be brought out by Teubner as a special book, and is to be dedicated to a congress of high school teachers at Whitsuntide"—thus Sommerfeld described the gyroscope project to his parents when it was still in its early stages. "Klein expects enormous success with it. In this way, he hopes to win over the teachers and simultaneously the engineers to the side of the universities, so that from now on they will not rail against them as they have done in the past."[32] For Klein it was a question of a mathematics book. Readers could "not miss gaining a certain familiarity with the methods of complex analysis," reads the introduction.[33]

What Klein had touched on just briefly in his lecture, Sommerfeld was to elaborate in detail in the book. The "gyroscope" became a long-term preoccupation, with which Johanna, too, had to suffer. "You must never ask me when the gyroscope will be finished," Sommerfeld wrote in a letter to his fiancée, paraphrasing a passage

28 Sommerfeld's personnel file, Archive of the TU Clausthal.
29 To his parents, August 3, 1894.
30 To his parents, April 1, 1897.
31 Klein/Sommerfeld, *Theorie des Kreisels 1*, 1897.
32 To his mother, 22. Februar 1896.
33 Klein/Sommerfeld, *Theorie des Kreisels 1*, 1897, p. 6.

from Wagner's *Lohengrin* to fit his situation when he was writing a chapter for the second "gyroscope" fascicle.[34] "I really can't write to you every day," Johanna read another time. "Klein's whip is ever at my back to speed things up."[35] Sommerfeld let it be known shortly before his move to Clausthal that nothing in this situation would change even after the appearance of the first fascicle. "I'm moving the 8th of October, and am still up to my ears in the gyroscope," he wrote to Königsberg.[36] "Just now intensely immersed in the gyroscope," read his terse postcard 3 weeks later to Johanna, who was staying in Göttingen until the wedding.[37]

It took months before finally the second "fascicle" of the theory of the gyroscope appeared.[38] Even then no end was in sight, for Sommerfeld, notwithstanding the time-consuming nature of the work, enjoyed presenting the subject in the clearest way possible, and sounding the depths of various applications Klein could never have remotely imagined in his lecture. It would be several years before the third and fourth fascicles of this mammoth work totaling nearly a 1,000 pages appeared.[39] In the end, the work seemed even to Klein himself to be coming apart at the seams; he attributed its "peculiar arrangement" to the complicated process of its genesis rather than as the consequence of an inner logic.[40]

In December 1896—the position as assistant to Klein had just expired, and the lecturer's fellowship had not yet been approved—Sommerfeld took on a job with Dyck, who along with Klein functioned as editor of the *Mathematische Annalen*, to produce an index to the first 50 volumes. Klein wished to spare his former assistant this labor: "But where will Sommerfeld find the time?" he asked Dyck to consider. "First and foremost, now, he has to finish the lectures on the gyroscope on which I have been working for a year."[41] Nonetheless, Sommerfeld took on this rather mechanical work, on which Johanna also collaborated, leading to the *Annalen*-Index being known in Göttingen mathematician's jargon as the "Hannalen-Index."[42] Even if this work required no tedious calculations, it proved very time-consuming. "Among other things, the Annalen-Index has to be finished," work which had to be completed within that year. "Hannchen has already written on it until her fingers are bleeding," Sommerfeld reported on the status of the index in March 1897.[43] All in all, this work dragged on for another year. The "general index" to the first 50 volumes of the *Mathematischen Annalen* appeared only at the end of 1898.[44]

34 To Johanna, January 12, 1897.
35 To Johanna, January 16, 1897.
36 To his mother, October 2, 1897.
37 To Johanna, October 22, 1897.
38 Klein/Sommerfeld, *Theorie des Kreisels* 2, 1898.
39 Klein/Sommerfeld, *Theorie des Kreisels* 3, 1903; 4, 1910.
40 Klein, *Abhandlungen* 2, 1922, pp. 658–659.
41 Klein to Dyck, December 25, 1896. BSB, Dyckiana, box 5.
42 Klein to Dyck, April 6, 1897. BSB, Dyckiana, box 5.
43 To his mother, March 11, 1897.
44 The Foreword is dated "June, 1898"; Sommerfeld, *Generalregister*, 1898, p. VII.

4.3 The School of Mining

Before he assumed his position as Professor at the School of Mining at Clausthal on October 1, 1897, Sommerfeld dove once more into the world of science. At the end of September, the German Natural Scientists and Physicians together with the German Mathematical Society held their annual congress in Braunschweig. "The Society of Natural Scientists continues to be nicely promising," Sommerfeld wrote his fiancée in Göttingen from this meeting. "I've spoken with all sorts of people of whom I've long been scientifically afraid." He discussed diffraction theory with the physicist Wilhelm Wien (1864–1924) from Aachen; vector analysis and geometry with August Föppl (1854–1924) and Sebastian Finsterwalder (1862–1930) from the technical university in Munich; and calculus of variations with Adolf Kneser (1862–1930), professor of applied mathematics at the university of Dorpat (today Tartu in Estonia). He had also "risen to speak in the discussion that followed Boltzmann's talk," he indicated, without going into detail, for "that is all fairly tedious for you. All that will interest you is this: I am literally swimming in mathematics, and am extremely happy."[45]

Even on the topic of his own lecture, he wrote only that he had "explained it to a narrower circle of mathematicians," and had "been met with lively agreement."[46] After the Natural Scientists' congress, Sommerfeld did not return to Clausthal immediately, but went first to Göttingen. "My darling has returned from Braunschweig quite happy, full of mathematics," Johanna wrote to Königsberg; "I am heartily glad he has so enjoyed these days, and now thus refreshed goes off into the loneliness of separation from all his colleagues." Arnold added that he had "feasted in Braunschweig from 9 in the morning till 7 at night at sessions, and that afterwards till late at night over beer had held 'detention' hours," and had "significantly refreshed his friendship" with Boltzmann.[47]

Meanwhile, Johanna had done her best to prepare things for her fiancé for his start at Clausthal. For the 3 months before the wedding, during which he still would have to manage without her "up there" (Clausthal is situated on a high plain in the Upper Harz, 450 m above the level of Göttingen), she had arranged for a housekeeper to do his laundry, wash the dishes, and take care of other small details of daily life. The rest would be seen to by the wife of the Clausthal Professor Wilhelm Hampe (1841–1899) in whose house they had leased their apartment. Hampe, who had taught chemistry at the School of Mining since 1867, would also initiate Sommerfeld into the other customs of the mining town in the Harz Mountains. That these would differ substantially from what they were accustomed to in Göttingen must have long been clear to them from excursions into the Harz Mountains. Even including neighboring Zellerfeld, the local population of 14,000 constituted less than half that of Göttingen. Clausthal and Zellerfeld had for

45 To Johanna, September 22, 1897.

46 Sommerfeld, *Beweis*, 1897.

47 To his parents, September 26, 1897.

centuries been associated with mining. A "Mining School" founded in 1811 in Clausthal provided education of officials for the Upper Harz Mine and Foundry Authority. In 1864, this became the "Clausthal School of Mining."[48]

After the 1866 war, the kingdom of Hannover, of which Clausthal had been a part, was converted to a Prussian province. Since Prussia already had a School of Mining in Berlin, the Clausthal School of Mining was deemed superfluous. "Under these circumstances, it was at first thought," we read in a 1907 monograph on the School of Mining, "that the comprehensive academic education of mining and foundry people was to be eschewed, that training here should rather be limited to those students who had already devoted themselves to studies in the natural sciences, and who wished here to attend lectures on technical aspects, and acquire the practical concepts of mining and foundry processing operations." But the Clausthal School of Mining had not let itself be shunted to the side and—all economizing measures notwithstanding—had evolved into a full-fledged academic institution. "Since 1892, the regular teachers, who formerly only after longer or shorter terms of service had been granted the titles "Professor," or "Mining Counselor," were now from the start (like their counterparts at technical universities) named permanent professors with the rank of fourth class counselor by the king."[49]

When Sommerfeld began his service at Clausthal, his landlords immediately "commandeered" him to tea and brought him completely up-to-date on everything he had to know about his new workplace and its most recent history. "The Hampes" are "very friendly," Sommerfeld reported to his fiancée in Göttingen. "My only reservation is that Father Hampe has all sorts of hiking plans, and wants to organize joint excursions to Harzberg. So we'll have to be somewhat guarded in this respect." In general, Clausthal had shown him "its friendliest face in welcoming me. Grinning from ear to ear, so to speak. Not a cloud in the sky, only little puffs of smoke from the chimneys of the cottages curling aloft and dispersed by the blue morning airs. All is colorful: a friendly green."[50]

But already the next day, the report on the Clausthal situation was no longer quite so idyllic. "Collegial relations appear not to be very edifying," he wrote his mother in Königsberg.[51] To Johanna he went into greater detail: "Yesterday morning I went visiting and was received by: Schnabel and Brathuhn," he related his first meeting with the professors of principles of metal foundry and mine surveying, Carl Schnabel (1843–1914) and Otto Brathuhn (1837–1906). "None of the others are at home. Much to say about Schnabel. A dignified, lively man. Began at once to rail against Hampe and Köhler. Hampe, nervous and quarrelsome, Köhler intellectually insignificant. In general, mutual invective very much the order of the day in Clausthal. Each person rails against the other, and everyone rails against our good

48 Clausthal, *Bergakademie*, 1883, p. 2.
49 Clausthal, *Festschrift*, 1907, pp. 37–42.
50 To Johanna, October 9, 1897.
51 To his mother, October 10, 1897.

Papa Hampe."[52] Gustav Köhler (1839–1923) was professor of mining science and Acting Rector of the School of Mining, "a fine and pleasant person," as Sommerfeld found 2 days later, after he had met with him rather by chance "a little tipsy from a morning tipple." In the meantime, he had also met the professors of technical mechanics and physics, Oskar Hoppe (1838–1923) and Ernst Gerland (1838–1910). Hoppe was "a very worthy old gentleman" and had at once invited him to come for a visit. He described Gerland as "relatively young, rather lively, very nice, speaks continuously about music. We surely get along well with these two, and with Köhler too." Only his landlord was a concern: "How will it go with Hampe? He asks me to visit him, to go out walking with him, while I am universally warned against it."[53]

Already the next day Sommerfeld knew these warnings had been justified. "He has begun to be unbearable to me with his hunger for company," Sommerfeld groaned after returning from a walk with Hampe. Instead of Hampe, he wished it had been Johanna at his side. "How lovely it will be when you and I stroll down that route!!!" He enthused of the landscape of the lower Harz mountains around Clausthal. Apart from this, Köhler introduced him into a kind of men's club, where "he enjoyed himself very much." "A game room, a reading room, a parlor. I was together mostly with 2 Upper- and 2 Mining Counselors."[54] He was most disturbed by the "mutual invective" among his colleagues, but he consoled himself with the knowledge that Johanna would soon be with him. The beauty of the autumn landscape did its part in mitigating many a disappointment: "It was just grand," he gushed after a hike through the environs of Clausthal. "The valleys were steaming in part, throwing a veil over the view, which as a result appeared twice as interesting. The magnificent fir forest, the green blanket of moss, the incredibly pure air! I'm really a lucky guy, to get to be continuously in this summer freshness. I want to savor nature gratefully!"[55]

His teaching duties, however, set narrow limits to his enjoyment of nature. Sommerfeld had to deliver three major lectures on algebra and analysis, analytic plane geometry, and differential and integral calculus, each meeting 4 h a week. In addition, there was a 2-h lecture on trigonometry.[56] "I was very satisfied," he wrote after the first lectures. "Whether the students were too, I don't know."[57] On top of the 14 h a week of lectures, there were the tests required of prospective mine officers, and which at the beginning challenged the examiner almost more than the taker, for the university procedures he was familiar with at Göttingen were much

52 To Johanna, October 10, 1897.

53 To Johanna, October 12, 1897.

54 To Johanna, October 13, 1897.

55 To his parents, October 15, 1897.

56 Programm der Königlichen Bergakademie zu Clausthal, Academic year 1897–1898. Archive of the TU Clausthal.

57 To Johanna, October 17, 1897.

looser in this regard. "I think I've made a good impression," he assessed himself in his still unfamiliar role as examiner. "The test lasted from 8 – 12. Afterwards, a little morning nip."[58] But it didn't always go so satisfactorily: "Today, another test," he wrote a week later. "Unfortunately, one student failed, partly my fault. But he was also just too absurdly dumb."[59]

To spare himself and "the Eager Ones" (as prospective candidates for the mining industry were called) all too bitter disappointments in the tests, Sommerfeld invited them to his apartment to show them in what ways they were still not up to the upcoming challenges. "These guys knew so little that they were simply going to fail *en masse*. They're coming back in two weeks; then I'll see how much they've crammed into their heads, and will adjust my questions accordingly. None may fail, for "the Eager Ones" are generally coddled here, and the Mining Supervisor takes a dim view if any of us lets a student flunk. So the test is a farce, to be sure."[60]

Two weeks later, when the prospective test takers had returned, and knew scarcely more than before, he admitted to himself that this was "actually a painful business." "The Eager Ones" had also complained they found his lectures too difficult. "If I can't strike the proper tone here, the pleasure of the professorship will be quite diminished." The difference between the liberal academic teaching regime prevailing at the University of Göttingen and the schoolmasterly instruction at the School of Mining had once more been brought sharply into focus when shortly before, mathematical friends from Göttingen paid him a visit. "In the end, I am truly above this job at Clausthal. Through the scholarly visit, I relived my lovely Göttingen lectures: that was something different entirely from this torture over trifles that in the end are not understood."[61]

Nonetheless, he proved himself determined to make the best of his situation. In this, his gregarious disposition was a great asset, for in this tradition-bound town of the Harz mining industry, a professor needed to demonstrate his allegiance to the School of Mining not only in the lecture hall. Each year, on the 4th of December, the "Feast of St. Barbara" offered such an opportunity. "Barbara pub very nice; made a speech, and went home barbarically early," Sommerfeld afterwards wrote his fiancée, who had apparently feared that on this occasion he would also have to prove how well he could hold his liquor. "And not the slightest meow," by which he indicated that the Feast of St. Barbara had left no trace of a hangover.[62] Several days later he was pleased that quite a few people spoke to him of his "Barbara speech," which had made a favorable impression generally.[63]

58 To Johanna, October 20, 1897.
59 To Johanna, October 27, 1897.
60 To Johanna, November 4, 1897.
61 To Johanna, November 19, 1897.
62 To Johanna, December 4, 1897. An untranslatable pun: colloquially, Kater (tomcat) = hangover.
63 To Johanna, December 7, 1897.

Sommerfeld showed his professorial colleagues his gregarious side most of all when he was entertained at the homes of other professors. "Of course, I'd much rather have stayed at home and written at leisure to you, and on the gyroscope," Johanna read in a short letter written before one of these invitations. In this small mountain town, where everyone knew everyone else, professional and family matters were hardly separable, and the new colleague was put under the microscope not only by the professors of the School of Mining, but by their family members as well. "So long! Have to go be entertained. Brrr!" was Sommerfeld's epistolary sign off for this evening.[64] Yet these tests of his sociability were mastered with bravura too, not least thanks to his pianistic skills, for which he "gathered heaps of laurels."[65]

4.4 The Wedding

To be completely accepted, a professor of the School of Mining at Clausthal had to extend reciprocal invitations to his colleagues. Before the wedding with Johanna and the establishment of their marital household, appropriate invitations to his professorial colleagues and their families were out of the question. Thus socially considered too, the wedding, planned for the Christmas holidays, was an extraordinary event. Since the bride was the daughter of the Curator of the University of Göttingen, and the wedding had been scheduled so that Johanna's sister Helene could also be married the same day, the double wedding, which was naturally to take place in Göttingen and not in Clausthal, assumed a not insignificant importance.

Already in September, Sommerfeld had written to Königsberg that the great event was to take place during the Christmas holidays. "The wedding will be right after Christmas."[66] His parents were first to come visit him in Clausthal for 2 days to "inspect his living arrangements," then they would travel together to Göttingen. There, they were to celebrate Christmas together with the Curator's family.

In the matter of the number of wedding guests expected, meticulous planning was necessary. Lists were drawn up of the relatives, friends, and acquaintances who were to be invited and of who could be accommodated where and when. Ideas for wedding gifts were compared. At times, the planning assumed the character of a strategic military operation, prepared down to the minutest detail. Now "finally the definitive wedding battle" has been fought, Sommerfeld wrote once to Königsberg. December 27, 1897 was set as the wedding day. It was decided to forego the traditional pre-wedding party. Instead, "only the immediate family" was to gather at the Höpfners' on the eve of the wedding. The invitation list numbered 80 people, "of whom ca. 60 will come."[67]

64 To Johanna, November 30, 1897.
65 To Johanna, December 17, 1897.
66 To his parents, September 2, 1897.
67 To his mother, October 31, 1897.

Fig. 9: The double wedding of the daughters of the University Curator during the Christmas holidays of 1897 was a significant social event in Göttingen. In the photo at the *left*, Helene Höpfner with her bridegroom Ludwig Rhumbler; *middle*, the Curator's son Willy Höpfner; *right*, Johanna and Arnold (Courtesy: Deutsches Museum, Munich, Archive).

In a subsequent report on the "wedding battle," Johanna filled in details Arnold had forgotten to mention, such as the "tuxedo question." Arnold's tuxedo was 2 years old and had "hardly [been] used," so that Johanna, against her mother-in-law's counsel, did not want to purchase a new one for the wedding. She was "not interested in guarding 2 good tuxedos from the predations of moths and mice," she wrote in anticipation of the rustic ethos of remote Clausthal. But alluding to her sister's future husband, Ludwig Rhumbler (1864–1939), she deployed yet another argument: "Ludwig does not wish to and will not acquire a new tuxedo; he will certainly be married in his old one, that has 6 years on it." "Discretion" would thus dictate not embarrassing him with a new tuxedo [for Arnold], "when as it is he might already be sensitive about his status vis-à-vis the 'Professor.'" Johanna also found the right tone in matters of the heart. "See, you mustn't always say 'if our happiness lasts,' and so on," she criticized her mother-in-law. "Alright? You won't say it again. It makes me sad. Now there are just 7 ½ weeks till I depart with my Love. If only the world doesn't end before that!"[68]

68 Johanna to Arnold's mother, November 3, 1897.

The world didn't end. From Göttingen, Johanna went on making preparations for the wedding, while Arnold taught "the Eager Ones" mathematics and proceeded with his elaboration of the theory of the gyroscope. "The gyroscope roars, and will have its victim," he consoled Johanna when on one weekend he did not, as usual, travel to Göttingen. "I can't be away from him 3 Sundays in a row if the 2nd fascicle is to be ready by Christmas. And I've promised myself this so I can enter the state of holy matrimony with a clear conscience."[69] By then, despite her physical absence, Johanna had already been integrated into the Clausthal milieu: "On Saturday I was at the club," Sommerfeld reported to Göttingen. There, in the speeches of the club members, Johanna's health as future wife of the Professor was toasted "with enormous solemnity."[70]

The nearer the Christmas holidays approached, the more hectic preparations for the wedding grew. The bridal couple had its hands full preventing misunderstandings among parents and parents-in-law regarding the celebrations of Christmas and the wedding. "Mother writes that she has received no invitation or request to attend your Christmas festivities. Your father told me in person, of course, and I've written that home." His mother had threatened to spend Christmas with him in Clausthal in the event the invitation did not arrive in time. Sommerfeld requested that Johanna make the invitation official with a short letter to her mother-in-law. "A few words from you will suffice; it's not necessary for your father to write."[71]

The impending wedding was an event for his Clausthal colleagues, too. One morning, Köhler and Hampe showed up, each with a picture under his arm. "'Castle by the Sea' (not a villa), and 'Spring Day,' both by Böcklin in an etching by Klinger." The Swiss painter Arnold Böcklin (1827–1901) was among Sommerfeld's favorite artists. He had expressed enthusiasm for Böcklin's pictures at the time of his first visit to the Schack Gallery in Munich in 1892. All the greater was his pleasure over this wedding gift. "Simply exquisite. Room décor of the very highest order. Splendid execution. The contrasts among the colors are beautifully rendered in the strongest tones of the etching. Framed in beautiful black frames with gold beading. Overall impression highly painterly."[72]

The wedding itself was likewise preserved in family memory in loving detail. "The House of the Lord," a female friend of the Höpfner family reported of the double wedding in the Göttingen University chapel, "was thronged with onlookers." The ceremony began with the entrance of the bridal couple. "At the sound of the organ, Helene on Rhumbler's arm, behind them Hannchen with Sommerfeld entered the chapel. The sisters both looked nice and proper. Happiness shone from Hannchen's friendly eyes. . . First the Rhumblers, then the Sommerfelds were wed. All four were blessed simultaneously, which I found very moving. To the sounds of

69 To Johanna, November 19, 1897.
70 To Johanna, November 23, 1897.
71 To Johanna, December 15, 1897.
72 To Johanna, December 21, 1897.

'Praise the Lord,' guests departed the chapel." At the wedding banquet, Klein gave the toast to the newlyweds. Speaking of Sommerfeld, the chronicler lapsed into gushing. "He is such [a] friendly, true-hearted person, who has a kind word for everyone—I think Hannchen will be very happy with the brave little man." Thus was Sommerfeld—at just 1.65 m, shorter than Johanna—registered in this reminiscence.[73]

Following the Göttingen festivities, married life commenced for Johanna and Arnold with a honeymoon journey. "The Sommerfelds traveled the same evening to Kassel," we read in a report of the wedding. From Kassel, the journey went via Marburg to Frankfurt. "We knocked about a little in beer pubs," Sommerfeld wrote his parents. "Next morning to the Goethe House, which is really a singularly atmospheric little place." They finished the following day in Würzburg with a bottle of "proper, local Bocksbeutel, on which we both got a little tipsy. Our two hosts, who gave us their recommendation of this bottle, spotted us (as does everyone, actually) as honeymooners, much to Hannchen's annoyance." The journey proceeded over the romantic Rothenburg ob der Tauber ("so full of antiquity, so tiny, so medievally fortified, so charmingly small-townish!") to Munich. "Facing the central railway station and the Glass Palace (art exhibition). Situated on the trolley-line," is written on the stationary of the Grand Hotel, Grünwald, where Sommerfeld committed this report of the honeymoon journey to paper. Here they planned to enjoy the art in the Munich museums, excursions into the foothills of the Alps, and their long-deferred being together as a couple. Sommerfeld assured his parents "that it has never in our lives gone continuously so well. Such wonderful weather we are having, and such a splendidly thought-out itinerary. We will probably greet the year 1898 in the Café Luitpold over a glass of punch."[74]

Eight days later they were already on their return journey. "Let me tell you a little about it," Johanna wrote in a chatty mood to her mother-in-law. "We wandered around Munich till yesterday. It was too beautiful—the air each day grew gentler, the sky bluer, and the sun more intense." On New Year's Day, they had undertaken an excursion to Lake Starnberg and climbed the Hoher Peissenberg. In ensuing days in Munich, they had "seen the Pinakothek, the Glyptothek, the Schack Gallery, the Basilica, Our Lady and Theatine Churches, and Arnold had visited several colleagues in his field." Among the "colleagues in his field" first and foremost, naturally, was Sommerfeld's doctoral advisor, Ferdinand Lindemann, a "charmingly good person," in Johanna's estimation. Whether, in light of the mixed feelings Arnold must have felt for his old professor following his doctoral thesis, she diplomatically suppressed further comments or did not have anything to add on that subject is an open question. Lindemann's wife seems to have monopolized the

73 Anna Wendland (a friend of Helene Höpfner's) to one of Johanna's aunts ("Die Doppelhochzeitsfeier meiner Nichten Helene und Johanna aus einem Bericht von Frl. Anna Wendland aus Hannover von M. Hübler in Berlin"), undated.

74 To his parents, December 31, 1897.

honeymooners more than desired. She was not to be deflected from introducing them to the Tyrolean painter Franz Defregger (1835–1921), who had achieved a bit of recognition on the Munich art scene. But Johanna did not get much from Defregger's folk-art style that Lisbeth Lindemann wished to acquaint them with. "Tuesday afternoon we were allowed to, or rather had to visit Defregger's studio under her aegis, where work on another of his happy, innocuous Tyrolean peasant interiors and a somewhat flabby Madonna was in progress. Defregger himself makes a fine impression, simple, honest, true-hearted. Lisbeth fawned on him continuously, was made a present of this here, finagled something else there," Johanna wrote wryly. All in all, though, the pleasures of Munich art did give the bridal couple "for many a winter's evening something beautiful to enjoy in retrospect," as Arnold added at the conclusion of his seven-page letter. They were "heartily glad to have made [the trip]."[75]

4.5 Gyroscope Matters and Electrodynamic Problems

Back in the winter remoteness of Clausthal, the trip to art-conscious Munich must have seemed like an excursion in another world. But they had little time to mourn their beautiful experiences, for with the end of the vacation, the teaching enterprise resumed for Sommerfeld at the School of Mining, and Johanna had her hands full moving into the marital apartment. Many of the wedding gifts, household furnishings, and wall decorations had been sent by the couple to Clausthal and awaited their proper disposition. To this was added art works acquired on the honeymoon journey, such as the reproduction of a relief depicting the Madonna by the Renaissance Florentine sculptor Andrea della Robbia (1435–1525). "We'll be taken for Catholics here on account of our numerous Madonnas," Arnold wrote his mother. Then they had to get caught up with the visits of introduction, and the reciprocal invitations postponed until after the wedding. "We did the whole thing in one morning. 60 visits, of course by carriage."[76]

The entertaining of his colleagues could not be accomplished in one fell swoop. In addition, there were invitations every afternoon to "silly teas," which she nonetheless "with or without grounds consistently" declined, Johanna wrote to Königsberg. Instead, she preferred spoiling her husband who was deriving less and less out of the modest lectures and the frequent visits of colleagues. "Today, my little Dick got a nice local rutabaga soup again; he is thriving quite nicely, despite the fact that he really suffers scandalously. We are both doing extraordinarily well in spite of everything, and we get along splendidly."[77] Reading between the lines and judging from qualifications such as "actually" and "in spite of everything," it is

75 To his mother, January 7, 1898.

76 To his mother, January 28, 1898.

77 Johanna to her mother-in-law, February 11, 1898.

clear that "Prof. and Mrs. Sommerfeld" had to put a great deal of effort into feeling at home here. When the weather allowed for sleigh rides into the snow-bound woods of the upper Harz, it could be "very, very beautiful." They enjoyed just being together in their apartment furnished quite to their tastes. But when it was a question of a party at a colleague's, the tone changed: "It was quite heartily dull, and we rejoiced at being home again, although we both behaved very properly and left a good impression."[78]

But the life of the young couple did not play out only in Clausthal. On March 1, 1898, Sommerfeld began a 7-week military exercise with the 82nd Infantry Regiment, stationed at Göttingen.[79] During this time, he stayed with his parents-in-law together with Johanna, who did not want to spend these weeks alone in Clausthal. Even though he had been promoted to the rank of Lieutenant in the reserves, he took little pleasure in it. "Military life makes one dull-witted, as usual. I have little chance to do any work," he wrote 2 weeks after beginning his service. "Constantly have to get up early. Have also done a nighttime exercise. In considerable cold, I spent the night in a barn," he wrote afterwards. He was determined, though, to make the best of the situation. "Nevertheless, I think the atmosphere of Göttingen will have a good influence on my mathematics. Much together with Klein."[80] One week later he bemoaned the military waste of time: "Such lovely time piddled away. A few mathematical visits to Hilbert, Klein, Schönflies were pleasant refreshment." He wrote this letter directed to his father on his birthday, during a break, and ended with the sigh: "Right now I have to go back to work, that is, to idle standing around."[81] Towards the end of the military exercises, the reserve officers were granted somewhat more free time, but that meant only "several hours daily of Felix-duty," as Johanna wrote to Königsberg concerning her husband's "mathematical visits" to Klein.[82]

In this spring of 1898, "Felix-duty" meant discussions of the theory of the gyroscope. "Teubner is pressing the gyroscope with dogged determination, now on the 5th chapter, while Arnold writes the 6th," Johanna had reported to her mother-in-law shortly before their departure from Clausthal at the end of February.[83] This chapter concerned elliptical functions and certain "spin parameters" by which the motion of the gyroscope could be especially elegantly described in mathematical terms. The equations of motion could be expressed in the form of so-called Hermite-Lamé differential equations. Physicists and engineers might not have

78 Johanna to her mother-in-law, February 26, 1898.

79 Certificate of Military Service, April 20, 1911. DMA, NL 89, 016; Leave of Absence for Military Exercise, 8 and 28 February, 1898. Sommerfeld's personnel file, Archive of the TU Clausthal.

80 To his mother, March 15, 1898.

81 To his parents, March 22, 1898.

82 Johanna to her mother-in-law, April 11, 1898.

83 Johanna to her mother-in-law, February 26. 1898.

registered the charm of this chapter, but a mathematician's heart must surely have beaten faster: the solution of these differential equations could, namely, "be written directly in the form of elliptical functions without any significant intermediate calculations," as stated in the prepublication announcement of the second fascicle. Therewith, the "pure theory" was complete. A "third and final fascicle" would be devoted to demonstrating "to what extent this theory coincides with experience, and what modifications need to be made so it is applicable to various empirical data from physics and astronomy."[84]

In working up this portion of the Klein lectures on the theory of the gyroscope, too, Sommerfeld did not play the role merely of executive agent. That from these lectures something more was to be made than a "pamphlet on the gyroscope" for the Association of High School Teachers[85] was clear already following the first "fascicle." Klein's lectures set the direction, but "St. Felix" gave his former assistant a free hand in their elaboration. "In Fascicle III, Sommerfeld essentially worked in" the geophysical and astronomical applications, Klein explained with reference to the authorship of the third gyroscope fascicle, which, at the time of publication of Fascicle II, of course, was still 5 years in the future.[86]

What Johanna called "Felix-duty" was thus not merely a matter of receiving Klein's instructions and converting them into sentences suitable for publication, but an interplay between teacher and student, to which the student increasingly brought his own conceptions until finally he gave the entire enterprise his personal stamp. While he was still formulating the "pure theory of gyroscopic motion" for the second fascicle, with an eye towards further chapters, he was already talking shop with practitioners about its applications. "I still have to write a very long letter to a man in Stuttgart who is treating the motion of artillery shells in terms of gyroscope theory, unfortunately incorrectly on essential points, though, and has asked for my advice," he had written to Johanna while still living the bachelor's life in Clausthal.[87] The "man in Stuttgart" was Carl Cranz (1858–1945) from the Technical University of Stuttgart, who had written his doctoral thesis on ballistics in 1883 and 20 years later, as Director of the Ballistics Laboratory at the Military Academy of Berlin, became an authority on this subject in Germany. He had sent Cranz "10 folio pages with calculations," he wrote shortly thereafter.[88] Cranz thanked him and promised in return to help Sommerfeld with a future "section on the motion of projectiles."[89] When Sommerfeld returned to Clausthal from his military exercises at Göttingen and his "Felix-duty" in April 1898, a subsequent letter from the Stuttgart ballistics expert was already waiting for him with a

84 Verlagsanzeige zu Klein/Sommerfeld: *Theorie des Kreisels 2*, 1898.
85 Klein, *Abhandlungen 2*, 1922, p. 509.
86 Klein, *Abhandlungen 2*, 1922, p. 659.
87 To Johanna, November 19, 1897.
88 To Johanna, November 23, 1897.
89 From Cranz, November 30, 1897. DMA, HS 1977-28/A,56.

manuscript on the gyroscopic motion of projectiles which Cranz wished shortly to publish. "You had formerly looked for another example from military practice," he explained; "I herewith present you with a larger number."[90] Particularly, the view widespread among practitioners about the role of the spin of projectiles and the concepts of "nutation" and "precession" required clarification, Cranz wrote in another letter. After thorough research in the library, he was convinced "that the greatest confusion rules here" and "one person cribs from another without acknowledgement."[91]

His correspondence with the ballistics expert marked just the prelude to ever novel applications of gyroscope theory. This interchange between theory and practice was made public in printed form only with the 1910 publication of the fourth "fascicle." In his correspondence, it soon became part of Sommerfeld's daily experience—not just with respect to ballistics, but also to gyroscopic applications for torpedo guidance and ship stabilization. Meanwhile, in the summer of 1898, he continued to be "urgently pressed" on the second fascicle, as Sommerfeld reported to Königsberg at the beginning of June. Publication was slated for "the end of July."[92] With the first two "fascicles," the theory of the gyroscope had already reached the sizeable dimensions of 512 pages. In mid-August, in time for the annual congress of Natural Scientists, at Klein's request the publisher sent a number of freshly printed copies of the second gyroscope fascicle to a select group of colleagues. "Many appreciative assessments of the gyroscope," Sommerfeld wrote home from this natural sciences congress, which took place that year at Düsseldorf.[93] After many months in the remoteness of Clausthal, Sommerfeld finally had in Düsseldorf once more the welcome opportunity to meet with colleagues in his field. Arnold was "highly satisfied with the whole group of natural scientists," Johanna passed on the report to her mother-in-law in Königsberg. "I'm very happy that he is having such a good time; in his situation at the (mathematically) so lonely Clausthal, he needs this stimulation doubly."[94]

The positive reception of the second fascicle of the gyroscope was not the only reason Sommerfeld enjoyed himself so much at the natural sciences congress at Düsseldorf. Already the year before in Braunschweig, the German Mathematical Association, meeting jointly with the natural scientists, had—in contrast to earlier annual congresses, and to make clearer the "inner relationships" among lectures—focused the meeting program more sharply, so that in Braunschweig they had succeeded "in placing the whole of mechanics alongside number theory largely at the center of the proceedings," as the annual report records this determination of

90 From Cranz, April 3, 1898. DMA, HS 1977-28/A,56. Also in ASWB I. Cranz, *Untersuchungen*, 1898.

91 From Cranz, April 23, 1898. DMA, HS 1977-28/A,56.

92 To his parents, June 3, 1898 .

93 To Johanna, September 19, 1898.

94 Johanna to her mother-in-law, September 23, 1898.

emphasis.[95] At the Düsseldorf congress, "theory of manifolds, on the one hand, and the mathematical theory of modern electrodynamics on the other [were moved] to the center of the proceedings."[96] The subject of electrodynamics was tailor-made for Sommerfeld. He had published his first physical paper on this subject in 1892 in the *Annalen der Physik*, which garnered him Boltzmann's recognition. For the natural sciences congress in Düsseldorf, he announced a lecture "On Some Mathematical Problems in Electrodynamics," in which he treated the question of the propagation of electromagnetic waves along a wire. He had also often discussed this problem with colleagues at Göttingen and had lectured on it to the Mathematical Society.[97] "Today, my lecture took place," he wrote on the evening of September 21, 1898, to Clausthal. "There was great interest, even if few were in attendance."[98]

Electromagnetic waves along wires might not at first glance seem a particularly inspiring subject. From the viewpoint of the physicist, it was a known phenomenon, explored already a decade earlier by Heinrich Hertz, that electromagnetic waves propagate not only in open space but also along an electrically conductive wire. It was assumed that the speed of transmission in air was exactly the same as along a wire, namely equal to the speed of light. It is true that in his experiments Hertz had measured distinctly lower values than the speed of light, but that was the result of an error Poincaré had corrected in 1892. But that had not really solved the problem. Hertz had assumed an infinitely thin wire, for which the boundary conditions for the electrical and magnetic field on the surface of the wire could not even be formulated; Poincaré had indeed proceeded on the basis of a wire of finite thickness, but had assumed the field lines to be perpendicular to the wire's surface, which anticipated the result of the calculation already in the formulation: wave transmission at the speed of light. Sommerfeld now regarded this problem as a mathematical boundary value problem, analogous to the diffraction problem in optics in his habilitation dissertation. By correctly observing all boundary conditions for a wire of finite thickness, "a determinate, non-zero degree of local attenuation and a determinate value of the speed of propagation somewhat below the speed of light" are obtained, Sommerfeld explained in his lecture.[99] In the experiments conducted until then, deviation from the speed of light had been immeasurably small, but with his theory, Sommerfeld demonstrated conditions whereby the transmission speed along a wire would be distinctly less than the speed of light. "I've just puzzled out an example," he wrote a colleague after the Düsseldorf

95 Jahresbericht der Deutschen Mathematiker-Vereinigung, 6, 1898, p. 3.

96 Jahresbericht der Deutschen Mathematiker-Vereinigung, 7, 1899, p. 3.

97 Lecture by Sommerfeld, May 11, 1897 on "Waves in Wires." Protokollbuch der Mathematischen Gesellschaft zu Göttingen. SUB Cod. Ms. Math. Archiv 49:2. Des Coudres to Sommerfeld, October 24, 1897. DMA, HS 1977-28/A,62.

98 To Johanna, September 21, 1898.

99 Sommerfeld, *Drahtwellen*, 1898; *Aufgaben*, 1898.

congress, "where the wave propagates along a wire at half the speed of light. This result will no doubt give the physicists goose-flesh."[100]

Even if in his lecture at Düsseldorf Sommerfeld felt himself more the mathematician than the physicist, he registered with satisfaction that he was respected by the physicists too and was regarded as nearly one of their own. "Earlier, Planck-Berlin spoke," he wrote Johanna, describing the afternoon on which the departments of mathematics, astronomy, and physics held their joint session, which concluded with Sommerfeld's lecture. Max Planck (1858–1947) spoke on "Maxwell's Theory of Electricity, from the Mathematical Point of View," a subject on which Sommerfeld immediately felt obliged to comment. "I joined the debate, and made a number of observations in agreement," he wrote afterwards, describing this first meeting with Planck, who taught theoretical physics as Professor at the University of Berlin and was one of the few representatives of theoretical physics as an independent discipline. He had gotten from Planck "a whole bunch of compliments on past and future work" and had "also afterwards gotten along very well" with him "over a beer," Sommerfeld wrote to Clausthal. "In short, I am quite at peace with myself and the world."[101]

Sommerfeld published the comprehensive theory of the propagation of electromagnetic waves along wires in the *Annalen der Physik*.[102] But also from a mathematical point of view, it paid for Sommerfeld to have acknowledged the thematic focus of the Düsseldorf congress. In carrying out the boundary value problem of the electromagnetic waves in wires, namely, he confronted a transcendental equation that in and of itself presented a nice problem. Sommerfeld developed an approximation procedure with which one could solve this equation stepwise. He sent Hilbert a brief paper on this "with the request that you present it to the *Göttingen Nachrichten*, if it doesn't seem too dull to you."[103] But there was additional success arising from the Düsseldorf congress. His "old diffraction papers" were also getting recognition, he wrote Johanna. Even "the great Poincaré" once again had praised his diffraction theory.[104]

Thus, after a year in the remoteness of Clausthal, Sommerfeld could be altogether satisfied with his scientific productivity. The list of his scientific publications grew and contained papers that were increasingly attracting the attention of mathematicians and physicists. After this natural sciences congress, he was, not only for Planck but also for Boltzmann, Wilhelm Wien (1864–1928), and Hendrik Antoon Lorentz (1853–1928), who were in Düsseldorf too, an ambitious young colleague who really deserved better than the intellectually unstimulating environment of the

100 To Carl Runge, November 3, 1898. SBPK, Nachlass 141.

101 To Johanna, September 12, 1898.

102 Sommerfeld, *Fortpflanzung*, 1899.

103 To Hilbert, November 22, 1898. SUB, Cod. Ms. D. Hilbert 379A. Sommerfeld, *Auflösung*, 1898.

104 To Johanna, September 19, 1898; Poincaré, *Polarisation*, 1897.

School of Mining at Clausthal. But there was little chance for an opening of a professorship in mathematics. And theoretical physics was represented at most universities only by lecturers and associate professors. Even Wilhelm Wien, who 2 years earlier had formulated the eponymous Wien's displacement law and had read a foundational paper on electrodynamic theory at Düsseldorf, had to content himself with the rank of associate professor at the Technical University at Aachen. "Theoretical physics in Germany lies virtually totally fallow," Wien judged the situation of this field in June 1898. Even so important a chair as that at the University of Munich had "practically ceased to exist," after Boltzmann relinquished it. "There is currently no market for theoretical physics."[105]

4.6 Encyclopedia Travels

Wilhelm Wien saw himself compelled to make this assessment of the situation of theoretical physics because Sommerfeld wished to persuade him to accept a task Klein had actually originally planned for Sommerfeld himself: He was to take over the editorship of the fifth volume of the *Encyclopedia of Mathematical Sciences* that Felix Klein had been pushing intensely, and that was to have physics as its subject. For this, Sommerfeld wrote Wien, "Klein wants me as editor. I have long been and will long continue to be fully occupied with the publication on the gyroscope, and would thus like to escape my fate if possible. Among the alternate editors I suggested, you were the only one he accepted eagerly. Would you be available to take this on? Unquestionably, of all the mathematicians and physicists in Germany, you are the one most suitable for this task." The assignment would consist in identifying authors for the designated topics, collaborating with them in defining the scope of the material, and subjecting the submitted articles to thorough critiques. "Klein himself wants to participate in the editing process in the following way. He plans to travel around the world for several months in order to gather the experts who are spread all about (which is difficult and necessary particularly in the technical fields, but would not involve you directly), to set up a provisional program with them, and to persuade them into cooperating. He is thinking, namely, of Italy, Holland, and probably also England." Sommerfeld tried everything to make the project attractive to Wien: "The beautiful thing here is that you will have the opportunity to a significant extent to put the stamp of your personal convictions onto a presentation of mathematical physics that may prove definitive for decades."[106] Wien's reply throws a spotlight on the theoretical physics of those years. Though it existed as a field of study at most universities, and with respect to research thoroughly was recognized by physicists as a specific type of scientific work, it would have been extremely risky for a physicist to make theoretical physics his entire occupation.

105 From Wien, June 11, 1898. DMA, HS 1977-28/A,369. Also in ASWB I.
106 To Wien, June 2, 1898. DMA, NL 56, 010. Also in ASWB I.

For many physicists, working as a theoretician was only an occasional thing.[107] After his papers on heat radiation, Wien had a good reputation as a theoretician, but he declined the invitation to take over theoretical physics for the *Encyclopedia of Mathematical Sciences*, referencing the poor career possibilities in theoretical physics. He had to "take the trends of the times somewhat into consideration" and would therefore "occupy himself with purely experimental work" so long as he had to operate with his "outward position" in mind.[108] Given this reply, Sommerfeld prepared himself for further "Felix-duty." Johanna's joy over this was "very mixed," as she confided to her parents-in-law. "At first, I was very much against it; but now that Arnold has gradually warmed to the idea, I am getting used to it."[109] Arnold assured his parents in Königsberg that the situation could be thoroughly to his advantage. He "continued to get along well" with Klein. Work on the Encyclopedia along with that on the gyroscope served him in maintaining contact with Klein and getting to know the scientific experts who were collaborating on it, which could only work to the advantage of his further career.[110]

The encyclopedia project as such was not new to Sommerfeld. Preparations for it had been underway since 1894. Already during his years as assistant at Göttingen, Sommerfeld had had ample opportunity to observe the initiatives with which, in conjunction with Walther Dyck and Franz Meyer, Klein had brought this ambitious undertaking into being.[111] Sommerfeld himself was to author one of the encyclopedia articles in the mathematical section: Heinrich Burkhardt, who edited the encyclopedia volume dedicated to calculus, had proposed in 1896 that he write the article on "Boundary Value Problems in the Theory of Partial Differential Equations."[112] Sommerfeld addressed this task only after the Düsseldorf Natural Sciences congress. "I'm now energetically attacking my paper on partial differential equations for the encyclopedia—this will be a lot of fun for me," he wrote Hilbert in November 1898.[113] It was a year before he was finished with it, however, and then another few years before the volume containing his article saw publication.[114] Matters of the encyclopedia demanded patience from the authors as well as from the editors.

The planned Volume V on physics began to take shape likewise only after the Düsseldorf Natural Sciences congress. Klein himself got the ball rolling in

107 Jungnickel/McCormmach, *Mastery*, 1990, pp. 159–165.

108 From Wien, June 11, 1898. DMA, HS 1977-28/A,369. Also in ASWB I.

109 To her parents-in-law, June 12, 1898.

110 To his parents, July 8, 1898.

111 Hashagen, *Walther von Dyck*, 2003, ch.. 21; Tobies, *Mathematik*, 1994.

112 From Émile Picard, June 12, 1896. DMA, NL 89, 012; To Klein, March 18, 1897. DMA, SUB, Klein 11, 1042.

113 To Hilbert, November 22, 1898. SUB, Cod. Ms. D. Hilbert 379 A.

114 From Burkhardt, January 12, 1900. DMA, NL 89, 006; from Picard, April 25, 1900. DMA, NL 89, 012; from Wirtinger, April 26, 1900. DMA, HS 1977-28/A,373. Sommerfeld, *Randwertaufgaben*, 1904.

undertaking a journey with Sommerfeld to recruit several of the authors envisaged for the project. "My wish would be," Klein wrote Lorentz before the journey, "first, to discuss the whole mathematical–physical section thoroughly with you, and then through your good offices to become better acquainted with Dutch mathematical–physical circles. With your approval, Prof. Sommerfeld, who will handle the editing of the math. phys. section, would join in our discussions."[115] Even during the "natural sciences business," as Sommerfeld wrote Johanna 1 day after the Düsseldorf congress from a Dutch seaside hotel in Zandvoort, Klein arranged "constant encyclopedia sessions" with the editors he had assembled at Düsseldorf (aside from Sommerfeld, these were Schönflies, Burkhardt, and Franz Meyer). Characteristically adventurous as a traveler, Sommerfeld wanted this excursion to Holland to be understood as not exclusively a business trip. "I'm really looking forward to Amsterdam. According to the map, it is pure Venice. Countless canals run through the whole city. Already from the train it looked unique. In places, the railroad has water on both sides. The air here is delightful."[116] One day later, his euphoric report of the Rembrandt exhibit to be seen there: "This is truly exquisite. Ten galleries chock full, the majority highly impressive portraits. All the same, this collection contains only about 1/10 of Rembrandt's total output."[117]

Next day, Klein and Sommerfeld were visiting Lorentz at Leiden. "Lorentz is one of the cleverest and at yet most charming people I have ever met," Sommerfeld wrote, noticeably impressed by his personality. "He readily consented to Klein's wide-ranging scientific demands, while at the same time like a proper Dutchman laying stress on the worldly pleasures of life in presenting us a most elegant dinner, a wife, and 3 children. When we returned to our guest house around 10:30 and I told Klein I still intended to write to you, he said in that case I might as well write his wife as well. You can just imagine how overjoyed I was to undertake this assignment!"[118]

After further visits to the Dutch physicists Diederick Johannes Korteweg (1848–1941) and Johannes Diterik van der Waals (1837–1923), Klein continued the journey alone to meet with mathematicians of his acquaintance in Paris. For his part, Sommerfeld met up with his wife who traveled to Cologne to meet him halfway. The Rhine valley was associated for Johanna with childhood memories from the time when her father still functioned as school superintendent at Koblenz.[119] Following the journey with Johanna along the Rhine, Arnold traveled on separately to visit his brother, who was seeking a cure for his morphine addiction at a clinic near Dresden. "After our happy journey down the Rhine, for me there followed a melancholy trip to Pirna," he reported to his mother in the wake of the

115 Klein to Lorentz, September 5, 1898. AHQP/LTZ-1.
116 To Johanna, September 25, 1898.
117 To Johanna, September 26, 1898.
118 To Johanna, September 27, 1898.
119 To her mother-in-law, October 22, 1898.

visit. Walter looked "wretched, was terribly depressed, particularly when he thought of you, was in general, though, quite rational. Except for a little nervousness, there was not much unusual about him. He admitted to having used some opium during the past weeks. He denied injecting morphine." On his return trip to Clausthal, Arnold made a stop at Goslar in order to explore the possibility of treatment for his brother at a "local institution. Of course I would rather find an institution dedicated to morphine addiction, which this one isn't. I'll naturally investigate further. There is to be sure always the question: is one looking for a temporary remission—ultimately, any reasonable institution is suitable for that—or does one want a lasting cure? That is probably not to be had anywhere." He also wished to assume the costs of Walter's treatment. "I'm regularly setting money aside, because my income is greater than what I spend (Holland and Rhine notwithstanding)."[120]

Back in Clausthal, concern over his brother mixed once more with the tedium of teaching at the School of Mining. To be sure, the lectures were no trouble for Sommerfeld since he was now delivering them for the second year, but the lack of enthusiasm on the part of his students gave him little pleasure. The "Eager Ones" also showed no enthusiasm when Sommerfeld invited them to his home. "One evening last week we had 15 students," Johanna wrote to Königsberg. "They were terribly boring; Albert tortured himself till midnight trying to enliven and converse with them, but in vain." Clausthal society generally was equally little to their taste. Nonetheless they made every effort to profit in some way from the social life of the Upper Harz mining town. They had "insinuated" themselves into a little reading circle, Johanna reported. "We were very classical, and read Aeschylus. The Clytemnestra was crocheting the whole time, and missed her entrances. We two read the best, of course. 5 couples and 2 daughters belong to it, quite reasonable people." But there was also something to report that she could write without irony and that would soon give a different direction to their life in Clausthal. "We hope in the spring to be able to call a little child our own," was how she conveyed the news of her pregnancy to her mother-in-law. "As well as I am still able to calculate, it will stick its little nose into the world at the end of April."[121]

The weeks and months of this winter in Clausthal stretched out at great length for this couple in joyful expectation. Sommerfeld sought relief from the unedifying daily grind at the School of Mining in his encyclopedia article and in his work on the gyroscope. One day he visited the mathematician Carl Runge (1856–1927) at the Technical University of Hannover, with whom he had corresponded on his work on electromagnetic waves in wires. "I've seen much of interest at the Polytechnic, and once more properly dabbled in mathematics." To be sure, Johanna

120 To his mother, undated [October, 1898].
121 To her mother-in-law, November 13, 1898.

had been "quite lonely at home."[122] She spent Christmas in Göttingen, where Arnold once more expected "Felix-duty." He would be "sitting all day at the library," he wrote in expectation of a very work-intensive Christmas vacation. "At the moment, I'm being pumped for information by a torpedo officer, who is giving his torpedoes a spin."[123] For Johanna, this sojourn at Göttingen was not at all salutary, as Sommerfeld reported to Königsberg. This time, the fault lay not in his "Felix-duty," but in the "very unfortunate situation within the Höpfner household." Johanna's mother was ill, and her father was physically and mentally exhausted. In response, Sommerfeld wanted to hasten their return to Clausthal as much as possible because he feared his wife, if she were further exposed to the depressing domestic circumstances in Göttingen, would "in 14 days, completely lose the round red cheeks, on whose account she was universally admired."[124] Johanna, normally full of confidence, now radiated no joy in life: "Oh, it's all so sad!" she wrote to Königsberg. "It's also very worrisome here. Mama, worse than in a long time; Papa, dead tired from the uninterrupted household woe, and all sorts of physical complaints, so exhausted that he hardly gets any pleasure from our being here."[125]

Back in Clausthal, the remaining weeks of this winter semester passed without any special events. "We have dispatched our social obligations," Johanna reported to her mother-in-law, "they were limited to 7 students whom we had over for dinner last Sunday lunchtime and the Sunday before that. The rest can be put off until next winter; almost no one has invited us over. So we finish the winter with a clear social conscience."[126]

The semester did not ultimately end as peacefully as expected, however. An incident occurred at the farewell party for a colleague who had been appointed at the Technical University of Aachen. "Among the faculty, there is an odd-ball, Schnabel, an old lush and an undignified character," Sommerfeld wrote to Königsberg. He was "on this occasion severely compromised."[127] He did not reveal in which way Carl Schnabel had occasioned friction, but on a later occasion, this colleague's behavior was again recorded. Schnabel, "very drunk," had been first "taken to the toilet" before being transported further to the lecture hall. "The lecture cannot have lasted even half an hour," as the Academy attendant deposed as witness in this matter reported, "because when after this time I went back there, the gentlemen had already gone."[128] Sommerfeld and a colleague hoped vainly that Schnabel would be

122 To his mother, December 5, 1898.

123 To his mother, December 21, 1898. The "torpedo officer" in question was the "Chief Torpedo Engineer" Carl Diegel of the "Imperial Torpedo Workshop at Friedrichsort" in Kiel. See Sommerfeld's Correspondence in DMA, NL 89, 007 and Broelmann, *Kreiseltechnik*, 2002, pp. 136–138.

124 To his mother, December 30, 1898.

125 To her mother-in-law, December 30, 1898.

126 To her mother-in-law, March 1, 1899.

127 To his mother, March 14, 1899.

128 Personnel Files of Carl Schnabel, Proceedings of June 28, 1899. Archive of the TU Clausthal.

dismissed from his post. They "have worn themselves out," as Johanna described these events to her mother-in-law, "trying to make sure the other toadies don't let the Schnabel matter slide again."[129] At the School of Mining, though, these events were magnanimously winked at. Schnabel had transformed himself from a practical man to a "scientist and 'son of the muses,'" backed by a majority of his colleagues. Even a century later, his memory was still preserved and honored in this mining industry town.[130]

The "Schnabel matter" and those colleagues characterized as "toadies" poisoned the professorial experience in Clausthal for Sommerfeld. But the day to day at the School of Mining retreated somewhat into the background when on April 30 the ardently awaited offspring arrived. "Dear Grandpapa, dear Grandmama!" Sommerfeld began his report next day to Königsberg. "So, it has arrived, and it is a he." The birth had been for Johanna an "inhuman torture," and the newborn was nearly choked in the process. "The umbilical cord had wound itself around the baby's neck, and it came into the world quite black-and-blue. The doctor immediately induced extensive breathing movements, and swung it about in the air. Since then, it has lain very comfortably in its little cradle, crows from time to time out of high spirits, and seems very strong." His further description reveals equally the high spirits of the father: "Temperature, 37.4. Little cheeks, red. Heart, happy. We can hope that everything will proceed smoothly and well. He is to be named Ernst after his deceased uncle. Had it been a little she, we'd have named her Gretchen. The gender accords with our wishes."[131]

In the days and weeks that followed, the letters to Königsberg from time to time took on the character of medical reports, for Johanna recuperated only slowly from the effects of the difficult birth. "If you only knew how Arnold has cared for me and how firmly and loyally he has stood by me when I have screamed and wept," Johanna wrote her mother-in-law when after more than 6 weeks she felt herself finally recovered from the exertions of the birth and the resulting complications. "For the husband, too, a child's birth is no child's play."[132]

During this time, Klein took further steps in Göttingen to drive the encyclopedia project forward. "Of course I've stayed away from Klein," Sommerfeld assured his parents, while was caring for Johanna. Nevertheless he signaled Klein his willingness to collaborate on the further planning for the encyclopedia. "It benefits me both inwardly and outwardly. So, for example, the planned trip to England. But naturally, within bounds."[133] Klein had resolved to assume editorship himself of Volume IV of the Encyclopedia, which was to cover mechanics. With a trip to England, Wales, and Ireland, he hoped to recruit British mathematicians and

129 To her mother-in-law, undated [late June, 1899].
130 Müller, *Carl Schnabel*, 2000.
131 To his parents, May 1, 1899.
132 To her mother-in-law, June 19, 1899.
133 To his parents, May 25, 1899.

physicists as authors for this volume. The mathematical–physical conception, as it was taught first and foremost at Cambridge, and given voice in textbooks such as Horace Lamb's *Hydrodynamics*, impressed Klein.[134] One or another British textbook author might also come under consideration for the physics volume of the Encyclopedia; therefore, Klein wanted to have Sommerfeld along on this trip.[135]

Sommerfeld had never before been in a country in which he could not make himself understood in his mother tongue. Lorentz and other scientists with whom he had met on the Encyclopedia trip to Holland the previous year spoke excellent German. For Klein, who had spent considerable time in the USA in 1893 and 1896, the English language was not a problem, but Sommerfeld's command of English had greatly diminished since his school days in Königsberg, where he had studied the language only as an elective subject. So preparatory for the trip to the British Isles, he exploited every opportunity to refresh his English language skills. "The sun is bright, no rain falls," he wrote on a postcard to Johanna from a train station when he was traveling with Klein to a meeting in Göttingen. "I think very much to you and I am very fond of you and of our little sweet puttl."[136] To Klein, too, he demonstrated his command of English. "Regarding the road, you will take for England, I beg you to deliver, if it is not nearer and cheaper to start from Cassel than from Hannover. I myself shall make requirements in this direction and will give you then information," he wrote 3 weeks before the commencement of the English journey. "Don't I write English as my native language?"[137]

As on the trip to Holland the year before, Sommerfeld did not wish to neglect art and culture. In the very first letter he wrote Johanna from London, he raved about the Renaissance art he had seen at the National Gallery. At Westminster Abbey, he stood before the gravestone of "your friend Shakespeare" and "my friend Newton." At the Museum of Science in the South Kensington district, he marveled at the first steam engine of James Watt, the "Babbadge [sic] calculating engine," and a harmonic analyzer by Lord Kelvin. He had to forgo St. Paul's Cathedral and the British Museum because Klein also demanded his due and, as he expressed it, had brought Sommerfeld along to be "at his disposal." But the relationship between the two must have been quite relaxed, for he had "joked with him in English." Sommerfeld closed his report on the ninth page of his letter with "Godby [sic] my dearest.[138] I still have to write to several Englishmen."[139]

The next several days were spent with meetings in London and Cambridge. "I'm enjoying the most wonderful hospitality imaginable," he wrote during their stay at

134 Klein, *Abhandlungen 2*, 1922, p. 508; Warwick, Masters, 2003.

135 Klein to Dyck, July 2, 1899. BSB, Dyckiana, box 5.

136 To Johanna, June 16, 1899.

137 To Klein, July 10, 1899. SUB, Klein 11. Also in ASWB I.

138 This adieu is in English; the remainder quoted is in German in the letter, but translated here.

139 To Johanna, August 5, 1899. By the "Babbadge calculating machine," he was referring to Charles Babbage's "Difference Engine."

Fig. 10: The 30-year-old Sommerfeld with Felix Klein (between the two seated women) on a visit to the English mathematician George Bryan (next to Sommerfeld), who was being recruited as an author for the *Encyclopedia of Mathematical Sciences*. Bryan's article on thermodynamics would later cause Sommerfeld some problems. (Courtesy: Deutsches Museum, Munich, Archive).

Trinity College, Cambridge. "There is apparently greater interest in my work here than in Germany."[140] He met Joseph John Thomson (1856–1940), and Joseph Larmor (1857–1942) presented him with his Collected Papers as a gift. A few days later he left Cambridge "with the feeling that life is good here, and that both physically and intellectually I have enjoyed the most pleasant hospitality imaginable here." Though J. J. Thomson was his "competitor in waves along wires," nevertheless he had come further than Thomson. "He acknowledged that very graciously." He was especially impressed by an invitation to visit the mathematician Edward Routh (1831–1907), who had authored significant works on mechanics. Here they also met another authority on mathematical physics at Cambridge, Sir George Gabriel Stokes (1819–1903), "a handsome octogenarian with a most illustrious scientific resumé."[141]

After this, a visit to Lord Rayleigh (1842–1919) was on the agenda. On his estate at Terling Place, Sommerfeld "unexpectedly found himself again amid the loftiest

140 To Johanna, August 9, 1899.

141 To Johanna, August 11, 1899. On his rivalry with J. J. Thomson see Sommerfeld, *Fortpflanzung*, 1899, p. 234.

English aristocracy." "Lord Rayleigh, discoverer of argon and author of important books, his wife, a sister of Minister Balfour, his sister-in-law, the daughter of a Scottish Duke of Algir." The country manor was situated in a park "it takes half an hour to travel through." Servants saw to the needs of the guests and even took care—to Sommerfeld's astonishment—of the unpacking of his trunk. "With my command of English, my presence amid this society is truly scandalous. The only somewhat extended coherent expressions[142] I've managed have been a movement of Beethoven, and a movement of Chopin, after lunch,. Presumably I'll be asked to resume after dinner. Lord Rayleigh is very diligent, and disappears continually to monitor his experiments. He has a private laboratory here."[143]

Other stops on this trip were Dublin in Ireland and Bangor in Wales. In face of the many changes of address on these postcards and letters from England, Johanna, who had been staying with her parents in Göttingen for these weeks, lost the overview of the journey. Her darling had traveled to London, she wrote her parents-in-law in Königsberg. "Since then, he has been knocking about with mathematics and museum visits, there and in Cambridge, at colleges and at the homes of real Lords, and has written me highly satisfied [letters] from everywhere he has visited." Most recently she had written yet another card to Bangor, but many of her letters must have arrived too late "because the good gentlemen are constantly flitting about."[144]

4.7 Gyroscope + Encyclopedia = Aachen-Recommendation

After the two journeys on the matter of the encyclopedia, Klein and Sommerfeld could set to work outlining in detail the volumes for mechanics (IV) and physics (V). To begin with, Sommerfeld asked Lorentz to confirm his collaboration, promised verbally in Holland, since Klein had charged him to report on the status of the project to the upcoming natural sciences congress in Munich.[145] He wrote similar letters to Carl Runge, Wilhelm Wien, Max Planck, and others. By this time it was clear that this editorial work, as well as the elaboration of the theory of the gyroscope, was going to grow into a very time-consuming activity, for the prospective authors would first have to be won over to the concept that Sommerfeld and Klein and several "principal authors" had developed. That the project would by no means meet with unanimous approval was made clear, for instance, by Paul Volkmann, Sommerfeld's former teacher at Königsberg: "With respect to the encyclopedia, I am too much a theoretical physicist to be able to agree wholeheartedly with the

142 An untranslatable pun, based on the German noun "Satz," which means both "sentence" and "musical movement." In the original: "Die einzigen längeren zusammenhängenden Sätze, die ich zuwege gebracht habe, waren ein Satz Beethoven und ein Satz Chopin nach dem Lunch."

143 To Johanna, August 13, 1899.

144 To her parents-in-law, August 19, 1899.

145 To Lorentz, September 2, 1899. RANH, Lorentz, inv.nr. 74. Also in ASWB I.

mathematicalization of physics lying at the heart of this program. Theoretical physics is an independent discipline, that has infinitely much to thank mathematics for, but for all that won't submit to a mathematical leash."[146] In this situation Sommerfeld, as the responsible editor, it represented an initial victory to have won over Boltzmann, the widely respected authority on theoretical physics, for the enterprise. "It was particularly valuable to me for Boltzmann to have lent his approval to my previous editorial measures," he reported to Königsberg from the Munich encyclopedia discussions. "He will also collaborate himself."[147]

Back in Clausthal once again, his meetings with figures such as Stokes and Rayleigh in England, or Boltzmann in Munich, must have seemed liked experiences from another world. At the School of Mining, quite different matters were the order of the day. "Schnabel has gotten a serious reprimand for his drinking from Berlin, as well as a 100 Mark fine," he wrote about the latest development in the Schnabel affair. "We've won a victory, then, despite the efforts of the Upper Mining Authority and its partisan hearing."[148] But he took no pleasure in the "victory." "Today is the 4th examination day on which I've had to sit beside Schnabel without exchanging a word all morning," he wrote shortly afterward. "An edifying relationship!"[149]

It had long become clear to Sommerfeld during this winter semester of 1899/1900 that in the long run he would not remain at Clausthal. It was only a matter of time when an appropriate professorship at a university or a technical university would open up for which he could now hope to qualify thanks to the scientific recognition he had enjoyed in the interim, even in England. When his colleague Friedrich Klockmann (1858–1937), who had been appointed at Aachen, and at whose farewell party in the spring of 1899 there had been another notorious "Schnabel" performance, informed him in June 1899 that a professorship in mechanics would soon open up there, Sommerfeld already represented to himself and to his parents in Königsberg "of course in strictest confidence" that he might be recommended to fill the new opening.[150] One month later, he wrote in a letter in English to Klein: "I write you that, to begg you, if you have any occasion, to do anything for me in Berlin [sic]." He thought to build on Klein's recommendation "because, as German people says, 'eine Liebe' (Kreisel + Encyclop.) 'der anderen wert ist'[151] (Aachen-recommandation)." His "secret special correspondent at Aachen" had informed him that he stood third on the list of candidates.[152] But in October, Sommerfeld

146 From Volkmann, October 3, 1899. DMA, HS 1977-28/A,348. Also in ASWB I.
147 To his mother, September 25, 1899.
148 To his mother, October 10, 1899.
149 To his mother, October 20, 1899.
150 To his parents, June 8, 1899.
151 "One good turn (gyroscope + encyclopedia) deserves another."
152 To Klein, July 10, 1899. SUB, Klein 11. Also in ASWB I.

reported to Königsberg that his father-in-law had learned from Althoff that Aachen wanted "a technical person, not a mathematician."[153]

While this possibility appeared to evaporate, another materialized that seemed even more desirable than an appointment at Aachen. Arthur Schönflies was appointed as professor at Königsberg, so that his position opened up at Göttingen. Sommerfeld as professor of mathematics beside Klein and Hilbert at Göttingen—what an idea! But Schönflies soon disabused him of this fantasy when he revealed "the Master's" intentions. "Klein wants a younger man who will look up to him as his superior." It would be better to wait until Klein "goes on the search himself, and in the process sets his sights on Clausthal."[154] The professorship at Göttingen vacated by Schönflies was filled by Friedrich Schilling (1868–1945), who had done his doctoral work under Klein and had chosen descriptive geometry as his mathematical area of concentration. Apparently, this orientation suited Klein's expectations better than Sommerfeld's.[155]

In between his hopes and disappointments over his professorial career, the divergent demands of his gyroscope and encyclopedia duties that led to voluminous correspondence on ballistics, torpedo guidance, partial differential equations, and various topics of theoretical physics, Sommerfeld did always find time to pursue his own research. Currently, he wrote his parents following his return from England, he was occupied with a paper on the diffraction of X-rays, "that promises to be very beautiful."[156] The nature of this radiation, discovered in 1895, was still entirely unclear even 4 years later. It may be that the first impetus to turn his attention to this subject came from Johanna, for shortly after their engagement, Sommerfeld had written his parents that his fiancée was enthusiastic "about Röntgen": "I'm to write up and send her everything I know about it."[157] In Göttingen he might very well have discussed this on occasion with Wiechert, who regarded X-rays as ether impulses generated on the anti-cathode on impact with the cathode ray particles inside the X-ray tube and spread according to Maxwell's laws of electrodynamics. Given this conception, it was to be expected that X-rays would show the diffraction phenomena characteristic of waves. The Dutch physicists Hermanus Haga (1852–1936) and Cornelis Wind (1867–1911) believed they had demonstrated these diffraction phenomena as X-rays pass through an extremely narrow slit, and Sommerfeld hoped to provide these experiments with the necessary theoretical foundation with his mathematical diffraction theory.[158] "I actually meant to ask you a favor," he wrote Wiechert, in requesting that he look over his manuscript on the diffraction of X-rays. Above all he wanted to know "whether it is written clearly and

153 To his mother, October 10, 1899.
154 From Schönflies, September 20, 1899. DMA, HS 1977-28/A, 311.
155 Schilling, *Geomentrie*, 1900.
156 To his parents, August 28, 1899.
157 Undated fragment [presumably September, 1896].
158 Wheaton, *Tiger*, 1983, pp. 29–33.

understandably from a physical perspective."[159] One week later he announced "his preliminary notes on X-ray diffraction" to Lorentz, which he hoped would also interest "your distinguished colleagues in Groningen," viz., Haga and Wind.[160]

After his earlier papers on electrodynamics, heat conduction, and diffraction, this was a further excursion into the realm of physics, albeit from a very mathematical position. The exact solution of the diffraction of a light wave on a semi-plane reinforced his hope that he could employ the "method of branched solutions" he had used there also for the diffraction at a slit. But an impulse-like motion in the ether necessitated a different treatment from that of a wave motion, the subject of his earlier work. Sommerfeld's new theory described the alteration of a temporally suddenly initiated and shortly thereafter just as suddenly ended electromagnetic excitation of the ether as it strikes an edge. The comparison with the slit experiment of Haga and Wind demands more far-reaching physical considerations, however, which Sommerfeld intended for a later paper. He therefore characterized the "Notes," submitted in November 1899 to the recently established *Physikalische Zeitschrift*, as only a provisional report.[161]

But he did not get to the continuation of the theory of diffraction of X-rays at once. "Well, to spill the beans at once," Johanna wrote her mother-in-law in Königsberg, "Arnold has been summoned to Berlin." The professorship at the Technical University of Aachen, on which they had no longer been counting, was suddenly current. "This news is of course as big a bomb-shell for you as it has been for us, because for all the joking we did last summer, we had given up hoping for an escape, and had oriented both thoughts and housing toward staying here. Now then, my 'little hare' has steamed off today at noon, and I sit here loaded with 1000 thoughts filling my head and my heart that I can express only to my sweet Puttl, who keeps silence of course. It is now quite probable that something will come of it; nonetheless, we ask your silence until you have further notice from us."[162]

This came 4 days later: "They now want me at all costs in Aachen," Sommerfeld wrote to Königsberg. Since the professorship was in mechanics, he once more faced a realignment. He had told the official at the Berlin Ministry of Culture, Education, and Church Affairs responsible for Prussian technical universities that he was "a mathematician, and at present unprepared," but the official replied merely that he could "always return to mathematics later." Sommerfeld requested a week's time to consider, but let his mother know at once that he would accept the offer. He had already applied to the Berlin Ministry of Trade to prepare for his release from the Clausthal professorship. He was to begin the position at Aachen on April 1, 1900. "Moving at Easter! Oh my! And all next year, terribly intensive working into the new duties." But the "Oh my!" stood over against the huge relief of finally being

159 To Wiechert, October 28, 1899. SUB (Wiechert).
160 To Lorentz, November 6, 1899. RANH, Lorentz, inv.nr. 74.
161 Sommerfeld, *Beugung*, 1900.
162 To her mother-in-law, November 26, 1899.

able to turn his back on Clausthal. "I am myself particularly happy to escape all the problems with colleagues and students here."[163]

He was only sorry that for the foreseeable future he would have to forego "X-rays and such tomfoolery," since at Aachen he was going to have to concentrate "fully on technical applications."[164] The theoretical physicist at Göttingen, where the news had immediately made the rounds, regretted this too. "Your beautiful work," as Voigt praised Sommerfeld's just published theory of X-ray diffraction, "leads me to fear a great loss for theoretical physics, if in your new position you are forced out of the track so splendidly begun! This tempers my pleasure at your appointment, on which I heartily congratulate you! Unfortunately, we have few broadly educated mathematicians who have a sense for theoretical physics."[165]

163 To his mother, November 30, 1899.
164 To his mother, December 6, 1899.
165 From Voigt, December 3, 1899.

5 Aachen

Sommerfeld shed no tears over Clausthal. "Now Aachen is definite, and we head there with spirits revived," Johanna wrote her mother-in-law on December 8, 1899. The departure from Clausthal was still more than 3 months in the future, but the anticipatory joy in "the improved atmosphere and the easing of numerous social conditions" made her last winter in the Harz Mountain hinterland bearable.[1] From the perspective of this rustic mountain town, Aachen, the "Imperial City" in the Prussian Rhine Province, must have held a particular attraction. Not just for the sake of the hot springs here did Charlemagne choose this town in the foothills of the Ardennes and the Eifel as his seat of government. From the eighteenth century on, Aachen had been a desirable address for the well-to-do, who traveled there from afar to enjoy a few weeks of recuperation in this brilliant town. In the nineteenth century, the industrial revolution brought explosive population growth to Aachen. In 1815, when the town was annexed to the Kingdom of Prussia, Aachen had around 30,000 inhabitants; by century's end, there were already more than 130,000. Around the middle of the nineteenth century, greater Aachen was part of the most intensively industrialized regions of Germany. The "Royal Rhine-Westphalian Polytechnical School at Aachen," founded in 1870, had as its mission to furnish engineers to the growing industries and soon was itself bursting at the seams.[2] Though situated far from Berlin at the western corner of the kingdom, Aachen was secure in its exalted status. "And our West particularly can boast that its technology has found singular recognition in the eyes of his Imperial Majesty," Nikolaus Holz (1868–1949), Professor of Hydraulic Engineering gushed over frequent Imperial visits from Berlin.[3]

In contrast to Clausthal, the appointment at Aachen—this much was clear—would mean an appreciation in his circumstances, but also a commitment to engineering, and this gave Sommerfeld pause. To represent the subject of mechanics in its broadest sense at a technical university was quite a different proposition from drilling down on just the mathematically interesting aspects of theoretical mechanics, as he had done in the theory of the gyroscope. "The engineers are now very jealous of the theoreticians, and are going to keep a devilishly sharp eye on me," thus he assessed the atmosphere awaiting him at Aachen. He proved

1 To her mother-in-law, December 8, 1899.

2 Laurent, *Entwickelung*, 1920, p. 7; Düwell, *Gründung*, 1970; Ricking, *Geist*, 1995.

3 Nikolaus Holz: Festrede zur Begehung des zweihundertjährigen Bestehens des Königreiches Preussen und zur Vorfeier des zweiundvierzigsten Geburtstages Sr. Majestät des deutschen Kaisers und Königs von Preussen Wilhelms II.: delivered January 18, 1901. http://darwin.bth.rwth-aachen.de/opus3/volltexte/2009/2782/pdf/1901_Holz.pdf (30 January 2013).

M. Eckert, *Arnold Sommerfeld: Science, Life and Turbulent Times 1868-1951*,
DOI 10.1007/978-1-4614-7461-6_5,

determined, however, "to fill that position conscientiously, and to set the example of a mathematician equal also to the demands of technology."[4]

5.1 Backgrounds of an Appointment

In light of his close association with Klein, it was not difficult to predict that the professors of the engineering disciplines would greet Sommerfeld with some suspicion. "For years, he has sought to bring science and engineering closer together," Sommerfeld had written his parents in Königsberg 5 years earlier concerning Klein's efforts. "He wanted to move the Polytechnic from Hannover to Göttingen in order to establish both a scientific practice and a practical science."[5] Klein even became a member of the *Verein Deutscher Ingenieure*—VDI (Association of German Engineers)—to give visible expression to his technological commitment. Representatives of the engineering sciences, however, saw in this move instead an attempt to undermine the technical universities in their ongoing campaign for equal status with universities. So Klein was compelled to renounce publicly his plans to take over the advanced engineering curricula from the technical universities and transfer them to the academic universities. This confrontation occurred at the annual meeting of the VDI at the Technical University at Aachen, which was celebrating the 25th anniversary of its founding in 1895. All the same, some representatives of the technical universities expressed support for a rapprochement with the universities, and Klein was conceded the right to establish university institutes oriented towards the applied sciences at Göttingen, so long as the monopoly of the technical universities in the matter of engineering education was not compromised.[6]

This "Peace of Aachen"[7] did not, however, result in a permanent cessation of hostilities. Several representatives of the technical universities, led by mechanical engineering professor Alois Riedler (1850–1936) of the Technical University at Berlin-Charlottenburg, declared war on the university mathematicians. They wished to recognize mathematics at the technical universities only as an ancillary science, to be taught by professors with closer ties to engineering than the university professors. Not only Klein and the university mathematicians allied with him but also the mathematics professors at the technical universities protested this view. Thus, the issue became a struggle within the technical universities themselves. Mathematics, physics, chemistry, mechanics, and other fields were split off from the specifically engineering subjects of architecture, civil engineering, and mechanical engineering, and belonged to the division of general sciences. In 1896, all 33 mathematicians of the general divisions of the technical universities published a

4 To his mother, December 6, 1899.

5 To his mother, November 30, 1894.

6 Manegold, *Universität*, 1970.

7 Klein, quoted in Jacobs, *Felix Klein*, 1977, p. 7.

manifesto. For the mission of the technical universities, it declared, their subject was "a foundational science, not, as was often contended, an ancillary science."[8]

This rallied the engineering professors onto the field afresh. "Mathematics does not bear the significance of an essential foundation for the education of engineers, but rather that of an ancillary discipline," began the declaration, signed by 57 representatives of the engineering divisions. At the latest, with this declaration, published in 1897 in the *Zeitschrift für Architektur und Ingenieurwesen*, the protest of the engineers against the university mathematicians had taken on the character of an "anti-mathematical movement."[9] Point #7 of their declaration dealt with the subject to which Sommerfeld had been appointed at Aachen: "Instruction in all branches of mechanics must be the province of engineers exclusively."[10] Quite in the spirit of this declaration, in April 1899, the representative of hydraulic engineering proposed to the Rector of the Technical University at Aachen that in the matter of the replacement of the mechanics professorship, "a technologically trained candidate" be chosen.[11]

There was a difference of opinion between the three engineering departments on the one hand (I Architecture, II Civil Engineering, III Mechanical Engineering) and the division of general sciences (V) on the other. Ultimately, Johann Jacob von Weyrauch (1845–1917) from the Technical University at Stuttgart was placed first on the list of candidates. Fritz Kötter (1857–1912), a mathematician from the Berlin School of Mining, was placed second. Fritz Kötter was the brother of Ernst Kötter (1859–1922), who had shortly before been appointed Professor of Descriptive Geometry at the Technical University at Aachen. Sommerfeld ranked third on this list of candidates. The engineering departments protested that predictably Weyrauch, as the oldest of the candidates, would decline the appointment, so that whether it be Kötter or Sommerfeld, "a pure mathematician [would be] assuming the position in mechanics." In a petition to the Prussian Ministry of Culture, Education and Church Affairs, they explained that for "truly profitable coverage of mechanics at a technical university, the appropriate candidate by far is someone who is first of all an engineer."[12] The general division regarded this as an encroachment on its vested right of recommendation and moved in the senate of the Technical University at Aachen that formal censure of the engineering divisions be pronounced and that the vote in question with respect to the Ministry not be authorized. This motion was defeated by a narrow majority. Consequently, the general division addressed itself directly to the Ministry with a special vote to confirm

8 Hensel, *Auseinandersetzungen*, 1989, p. 73 and Appendix 11, p. 284.

9 Ibid., pp. 55ff.

10 Ibid., p. 76 and Appendix 12, pp. 286–287.

11 Holz to the Rector of the TH Aachen, April 13, 1899. folder 875, Archive of the RWTH Aachen.

12 The heads of the divisions I, II, III to Minister Bosse, July 15, 1899. Berlin, GSA, I.HA Rep.76 V b, Sec. 6, Tit. III, Nr. 6, vol. III, pp. 66–68.

its list. If "only engineers" were to be considered, it would be "quite contrary to the long established practice in appointments to technical universities of taking into consideration only a candidate's competence, knowledge, and achievements."[13]

In far-off Clausthal, Sommerfeld and his wife were remarkably well informed about these internal matters at Aachen. Weyrauch was "ancient" and would not accept the appointment, Sommerfeld surmised in a letter to Johanna, who was staying in Göttingen at the time. "Kötter has a brother at Aachen who is lobbying for him. We just have to wait and see."[14] Apparently, Klein was not yet actively engaged at this stage of the appointment process, for the Berlin Ministry acceded to the petition of the engineering departments and instructed the Technical University at Aachen to name two further candidates.[15] The candidates put forward by the engineering departments were a trade school teacher and two government architects. Against this, the general division immediately raised sharp protest: "No information whatsoever has been adduced in support of these latter gentlemen," complained a counter-opinion submitted to the Ministry; these candidates are "not known to any members of the division."[16]

Then the situation hung fire. "My father-in-law has recently discussed Aachen with Althoff," Sommerfeld reported to Königsberg on October 10. "Althoff thought nothing would come of Aachen because the engineering departments want an engineer, not a mathematician."[17] Two days later, the Rector of the Technical University at Aachen sent an inquiry to Berlin "whether the replacement of the vacant professorship in mechanics is imminent, or if steps should be taken to organize an appointment."[18] On November 2, the Rector proposed that two professors from the Department II of Civil Engineering be entrusted for the winter semester of 1899/1900 with the task of filling the vacancy in the professorship in mechanics.[19] So for this semester at least, the engineering departments had a mechanism in place that corresponded entirely with their needs. What ultimately tipped the balance in Sommerfeld's favor cannot be determined from the documents. Weyrauch was indeed asked whether he would accept an appointment at Aachen,[20] but, as

13 Bredt, Kötter, v. Mangoldt, Wüllner to Minister Bosse, July 24, 1899. GSA, I.HA Rep.76V b, Sec. 6, Tit. III, Nr. 6, vol. III, pp. 69–71.

14 Johanna to her father, undated [late June 1899].

15 Kultusminister Studt to the TH Aachen, July 19, 1899. folder 886, Archive of the RWTH Aachen.

16 Gutachten der Abteilung V bezüglich Ministererlass No. 22436 T, July 25, 1899. GSA, I. HA Rep.76 V b, Sec. 6, Tit. III, Nr. 6, vol. III, pp. 72–74.

17 To his mother, October 10, 1899.

18 Mangoldt to Studt, October 12, 1899. GSA, I. HA Rep.76 V b, Sec. 6, Tit. III, Nr. 6, vol. III, p. 102.

19 Mangoldt to Studt, November 2, 1899. GSA, I. HA Rep.76 V b, Sec. 6, Tit. III, Nr. 6, vol. III, p. 61.

20 Naumann to Weyrauch, August 31,1899. GSA, I. HA Rep.76 V b, Sec. 6, Tit. III, Nr. 6, vol. III, p. 91.

expected, he would not leave his professorship at the Technical University at Stuttgart. Presumably, Klein then intervened on Sommerfeld's behalf, for Fritz Kötter, second on the list, was also passed over, like the two candidates proposed by the engineers. Still in November 1899, the advisor at the Prussian Ministry of Culture, Education, and Church Affairs summoned Sommerfeld to Berlin "to discuss the conditions in greater detail."[21] Shortly thereupon, Sommerfeld accepted the appointment. He tied his acceptance only to a request to the Minister, on account of the higher cost of living at Aachen to boost his yearly salary from 5,500 to 6,000 Marks.[22] This request was granted, so by April 1, 1900, nothing more stood in the way of the appointment.[23]

Sommerfeld's mathematical colleagues considered this appointment a victory. Paul Stäckel (1862–1919) of the University of Kiel thought it "very gratifying under present circumstances" that "a mathematician from the universities" be given a chance at the Technical University at Aachen and sent Sommerfeld "heartfelt best wishes."[24] Heinrich Weber (1842–1913) of the University of Strasbourg coupled his congratulations with the hope that "by your appointment, the rigorous mathematical direction at the technical university [may be] reinforced."[25] The mathematicians at the technical universities, too, looked on Sommerfeld's appointment with particular satisfaction: "I congratulate you heartily, and even more those at Aachen," wrote Sebastian Finsterwalder (1862–1951) of the Technical University at Munich, where the "anti-mathematical movement" had likewise given rise to concern. "You will certainly contribute to raising the engineers' respect for theory, and thereby help resolve unfortunate antagonisms."[26]

5.2 Rapprochement with Engineering

The behind-the-scenes of the appointment and the reactions of the mathematicians show how justified Sommerfeld was in his suspicion that his engineering colleagues would "keep a devilishly sharp eye" on him. For them, the outcome of this appointment process represented a defeat. On March 30, 1900, 2 days before Sommerfeld took up his professorship at Aachen, Adolf Slaby (1849–1913), Professor of Theoretical Mechanical Engineering and Electrical Engineering at the Technical

21 Naumann to Sommerfeld, November 24, 1899. DMA, NL 89, 019.
22 To Studt, December 1, 1899. GSA, I. HA Rep.76 V b, Sec. 6, Tit. III, Nr. 6, vol. III, pp. 103–104.
23 Bestallungsurkunde, January 13, 1900. GSA, I. HA Rep.76 V b, Sec. 6, Tit. III, Nr. 6, Bd. III, pp. 112–118; to Althoff, January 23, 1900. GSA, I. HA. Rep.92 Althoff B, Nr.178/2.
24 From Stäckel, February 2, 1900. DMA, NL 89, 013.
25 From Weber, February 4, 1900. DMA, HS 1977-28/A,356.
26 From Finsterwalder, January 29, 1900. DMA, NL 89, 008. Also in ASWB I. On the "anti-mathematical movement" in Munich, see Hashagen, *Walther von Dyck*, 2003, pp. 207–225.

University at Berlin, inveighed against Klein's initiatives in a celebrated speech in the Prussian House of Lords. Even 5 years after the "Peace of Aachen," Riedler, Slaby, and other spokesmen of the engineering movement had not wearied of their battle for the liberation of the technical universities, which they saw as threatened by Klein. The founding in 1898 of the "Göttingen Association for the Advancement of Applied Physics and Mathematics," by which Klein had incorporated heavy industry also into his plans, proved sufficiently to Slaby that Klein was sticking to his original intentions. As a personal advisor to Wilhelm II, Slaby was for Klein a formidable opponent to be taken seriously. Klein intended—in Slaby's pointed statement of Klein's own formulations—to train the "general staff officers" of engineering at the universities, so that all that remained to the technical universities was training of "field officers."[27]

Slaby's speech was reprinted in many technical journals and was daily conversation at the Technical University at Aachen as Sommerfeld was taking up his professorship there. As a student of Klein's, he saw himself on the defensive at once. Klein was also very preoccupied with the "Slaby affair" in April of 1900.[28] Even 2 months later, "Slaby & Co." were still the subjects of correspondence between Aachen and Göttingen. Sommerfeld wrote Klein that the engineering professors considered him a Trojan horse. "Recently, I spoke forcefully against the suspicions of your initiatives at a general session at which the topic under discussion was your 'guiding principles' and Riedler's commentary on them. I can't go into greater detail since the matter was confidential."[29] For Klein, the "cultivation of friendship with the engineers" that Sommerfeld reported to him from Aachen was "naturally very important." Althoff had let him know, namely, "that he would shortly pay a visit here with Slaby and wished to look after things."[30] On the occasion of this visit, there was between Slaby and Klein an "agreement concerning the apportioning between technical universities and universities" (as Althoff recorded it under the date July 8, 1900), whereby engineering education was declared the sole domain of the technical universities, and "any thought of competition in this area is ruled out from the outset." Althoff named this agreement regarding the partition of roles of universities and technical universities the "Second Edition of the Peace of Aachen."[31]

Among professors of the engineering fields at the technical universities, however, grudges against the theoreticians at the universities had to be resolved with action rather than words. So it was to Sommerfeld's advantage that he was currently in correspondence with the Kiel torpedo engineer Carl Diegel (1854–1931). "I'm glad you continue to be occupied with torpedoes, and are making yourself of service to the Navy out there in the far west," Diegel wrote him at Aachen. For that purpose, he

27 Manegold, *Universität,* 1970, p. 207.

28 From Klein, April 25, 1900. DMA, HS 1977-28/A,170. Also in ASWB I.

29 To Klein, June 13, 1900. SUB, Klein. Also in ASWB I.

30 From Klein, June 21, 1900. DMA, HS 1977-28/A,170. Also in ASWB I.

31 Manegold, *Universität,* 1970, pp. 213–214.

sent him a gyroscope such as was employed in the guidance of torpedoes. Sommerfeld was welcome, he said, to keep it for the collection at the Technical University at Aachen. He offered also to provide him with a complete "straight-line trajectory apparatus," as employed in torpedoes.[32] Sommerfeld happily accepted this offer. Sommerfeld reported to Klein 3 weeks later that he had demonstrated the apparatus to his colleagues. "This was so to speak his probationary lecture to his engineering colleagues, and was greeted with great approval." The Rector assured him afterwards that this had "been a step towards understanding between me and my Aachen colleagues, and between the universities and the technical universities."[33]

In these weeks and months, Sommerfeld was also increasingly occupied with editorial work on the *Encyclopedia of Mathematical Sciences*, so that he must have had a quite lively sense of the opposition between academic science and practical technology. "I can commence my electrodynamic article only this winter," Max Abraham (1875–1922), who had taken his doctorate in 1897 under Planck and now as lecturer at Göttingen was taking the first steps along the path to an eventful career as a theoretician, wrote in apology. "I am also filling in for Klein in the seminar." Sommerfeld must have been reminded of his own time at Göttingen. Abraham's habilitation dissertation had as its subject "Electrical Oscillations in an Open-Ended Wire" and thus bore a close connection with the subject of Sommerfeld's work on electromagnetic waves along wires. Their correspondence quickly evolved into shoptalk over the mathematical difficulties in the treatment of such waves. "I could have wished that the series converged in a larger domain, for then there might have been an application to Marconi," Abraham wrote in reference to the most recent experiments of Guglielmo Marconi (1874–1937). Sommerfeld also brought to his attention Jonathan Zenneck (1871–1959) who, as assistant to Ferdinand Braun (1850–1918) at Strasbourg, was likewise performing experiments on the propagation of electromagnetic waves. "I will certainly contact Dr. Zenneck," Abraham wrote back. "Unfortunately, one hears such a variety of opinions on the subject from the gentlemen associated with the technology, as different patents are applied for."[34] The names of Marconi, Braun, and Zenneck themselves indicate how great a practical importance attached to the subjects Abraham and Sommerfeld were talking shop about. Around the turn of the century, wireless telegraphy was becoming literally a global technology.[35] In his work on waves along wires, Sommerfeld had obtained results "of fundamental importance," Abraham wrote him in congratulation after a thorough study of this work.[36] Thus, a fine opportunity to demonstrate to the engineers the value of theory in the area of wireless telegraphy—an area, moreover, close to Slaby's heart, as a pioneer of the early

32 From Diegel, May 23, 1900. DMA, NL 89, 007; Broelmann, *Intuition*, 2002, pp. 136–138.
33 To Klein, June 13, 1900. SUB, Klein. Also in ASWB I.
34 From Abraham, May 28, 1900. DMA. HS 1977-28/A,1.
35 Aitken, *Syntony*, 1976; Hong, *Wireless*, 2001.
36 To Sommerfeld, April 27, 1899. DMA, HS 1977-28/A,1.

radio technology. Thoughts such as these may well have been on Sommerfeld's mind in his correspondence with Abraham in the summer of 1900, and in the event, he soon put major emphasis on wireless telegraphy in his work and continued to pay tribute to it his whole life. For now, though, he held back from it, because it had been made clear to the Professor of Mechanics that an excursion into radio technology constituted a diversion into territory outside his field. In addition, the engineers' movement—aside from Slaby himself—was represented less by electrical engineers than by engineers in the mechanical fields (hydraulic engineering, civil engineering, and architecture).[37]

For a theoretician like Sommerfeld, however, the question was not primarily to which engineering discipline a problem belonged, but rather what mathematical means were required for its solution. Fundamental questions of "pure" physics and applications in engineering might lie quite close together if they could be expressed in the same mathematical language, for example, as boundary value problems. For the time being, he would not continue working on diffraction theory, Sommerfeld wrote Karl Schwarzschild (1873–1916), "since in my new duties, the hydrodynamics of lubricants, for example, is more relevant than the electrodynamics of the pure ether."[38]

Sommerfeld used the opportunity of the next Natural Scientists Congress, which took place at Aachen in September 1900, to make a declaration to his colleagues from the engineering departments of his institution, in the presence of witnesses as it were, of his allegiance to technology. "Recent Investigations in Hydraulics" was the title of his lecture. With this formulation of the subject itself, it was clear he was addressing engineers more than physicists or mathematicians. "The speaker first takes up the opposition that exists between the theory of fluid motion in a mathematical–physical treatment (hydrodynamics) and in a technological treatment (hydraulics)," reads the day's program. Sommerfeld contrasted the law of friction derived from the hydrodynamic differential equations in laminar flow with the empirical law of friction in turbulent flow, heretofore not derivable from the theory. He recalled the experiments carried out a few years earlier in England by Osborne Reynolds (1842–1912) on the transition of flow from a laminar to a turbulent condition, which could be treated theoretically as a stability problem. He was familiar with similar questions from the theory of the gyroscope. As an example, he adduced the rotation of an elongated body that spins stably around one axis, but wobbles back and forth in rotation around the axis perpendicular to it. But even restricting the question to the stable, laminar motion of flow, which heretofore had been the only one accessible to hydrodynamics, engineering could profit from theory, for it was thereby possible to calculate the "lubrication action in machines."[39]

37 Hensel, *Auseinandersetzungen*, 1989, p. 75.
38 To Schwarzschild, July 16, 1900. SUB, Schwarzschild.
39 Sommerfeld, *Hydraulik*, 1900; Jackson/Launder, *Osborne Reynolds*, 2007.

The lecture program, formulated on half a printed page, was at first the only evidence of Sommerfeld's involvement in the needs of engineering. Whether this declaration of intent was sufficient to satisfy his Aachen engineering colleagues is an open question. He had "the impression," Sommerfeld wrote Klein in November 1900, "that our engineers are increasingly reconciled to my existence." On the other hand, Sommerfeld was not exactly registering admiration on the part of his colleagues in the engineering departments. "Most engineers have as little an idea of physical research as of mathematical."[40] It was proving somewhat troublesome for him "to adjust technologically." This was true of his teaching as well. He wanted to deliver his lectures not just with chalk on the blackboard but to give the engineering students a substantive experience of his field. To this end, he applied for means to acquire a "mechanics collection," for which Carl Runge, as an experienced colleague from the Technical University at Hannover, was to help him compile a list of apparatuses suitable for lecture demonstrations. He was thinking, for example, of "hydraulic apparatuses" and "an apparatus for demonstrating the buckling formula." "The wish-list is long."[41] He was also dissatisfied with the textbooks in his new field, in particular with the representation of hydraulics, in which, because of his lecture to the Natural Sciences Congress, he had become profoundly engaged. August Föppl (1854–1924) of the Technical University at Munich was the leading authority in matters of technical mechanics and a well-known textbook author—bona fides which did not, however, prevent Sommerfeld from criticizing Föppl's representation of hydraulics as taking theory insufficiently into account.[42]

With an eye towards hydraulics, Sommerfeld also addressed basic questions concerning the substance of the fundamental equations of hydrodynamics. Here was "one of those existence proofs mathematicians love so much and that physicists rightly find rather uninteresting," he wrote Lorentz, who had just authored a treatise on the onset of turbulence. If one could demonstrate that beside the laminar equations also "nonlinear integrals of the hydrodynamic equations" exist, then "one would be on solid ground . . . Unfortunately, no mathematician will dare approach this existence proof in the foreseeable future."[43]

Was he already no longer reckoning himself a mathematician? With or without such a proof, Sommerfeld wished to convert into action the words set out in his lecture and calculate the hydraulically interesting case of turbulent flows. Not only Lorentz, however, but also the British authorities so highly esteemed by Sommerfeld and Klein in this area, Reynolds, Lord Kelvin, and Lord Rayleigh, had foundered on this problem. Sommerfeld hoped to come further than his predecessors and would have dearly liked to shine with his contribution to a festschrift in honor of

40 To Klein, November 8, 1900. SUB, Klein 11. Also in ASWB I.

41 To Runge, November 14, 1900. DMA, HS 1976-31. Also in ASWB I; see also folder 941 in the Archive of the RWTH Aachen.

42 From Föppl, October 7, 1900. DMA, HS 1977-28/A,97. Föppl, *Vorlesungen*, 1899, p. 421.

43 To Lorentz, October 8, 1900. RANH, Lorentz, inv.nr. 74. Also in ASWB I.

Lorentz's 25th doctoral anniversary. In the end, though, he had to confess to Lorentz that he too had suffered "pathetic shipwreck" in his efforts.[44]

After foundering in hydraulics, Sommerfeld wished to demonstrate the usefulness of a mathematical approach to strength of materials. He made contact with Ludwig Prandtl (1875–1953), a mechanical engineer from Munich who had completed his doctoral work under the supervision of August Föppl on "buckling phenomena." Prandtl forwarded his dissertation to Sommerfeld and announced that he would shortly publish a paper "On the Buckling of Traveling Crane Beams." He had also calculated the "buckling load" of the beams of the Elberfeld suspension railway, though the relevant officials had ignored it. "Should you be interested in examining this more closely, however, I would be happy to send you a reprint of the paper."[45] This was in fact material to Sommerfeld's taste and actualized his approach to engineering in a way he could not have improved upon. The Elberfeld suspension railway was officially inaugurated on March 1, 1901. It had already earlier drawn the Kaiser to the west of the Empire, as Sommerfeld's colleague from the civil engineering department had stressed in his commemorative speech on the two hundredth anniversary of the founding of the Kingdom of Prussia.

Sommerfeld took Prandtl's dissertation along as reading matter on a trip to Königsberg to visit his parents and was so impressed by it that he recommended Prandtl for a mechanics professorship at the Technical University at Hannover. "He has significantly enriched our knowledge of unstable elastic states of equilibrium," he wrote to Runge. In Hannover too, the usual tensions existed between the engineering departments and the general division, to which Runge as mathematician belonged. "I am thinking," Sommerfeld wrote in allusion to this notorious feud at the technical universities of that time, "it might be desirable for your division to best Department III, perhaps even with suggestions from engineers. At all events, in Prandtl, we have a scientifically, mathematically, and physically educated man."[46]

5.3 Technological Expert

A good year after assuming his appointment at Aachen, Sommerfeld had come so far in his "technological acclimatization" that he almost gleefully toyed with the idea of styling himself "Consulting Engineer," along the English model. Half a year earlier, he was still attributing lack of comprehension of physical and mathematical research to the engineers; now he thought "that we can get along with the engineers very nicely, so long as they smell no university arrogance on our side. After all, in

44 To Lorentz, December 10, 1900. RANH, Lorentz, inv.nr. 74; Darrigol, *Worlds*, 2005, pp. 208–218; Eckert, *Birth*, 2010.

45 From Prandtl, February 11, 1901. DMA, HS 1977-28/A,270. Also in ASWB I; Prandtl, *Kipperscheinungen*, 1900.

46 To Runge, March 27, 1901. DMA, HS 1976-31.

most cases they are correct about that scent, as well as in their annoyance over it." Wilhelm Wien, to whom Sommerfeld wrote these lines, had himself spent several years at the Technical University at Aachen as associate professor of physics and may well have been taken aback at this declaration of allegiance to engineering. Of the engineers, Sommerfeld continued, he was now esteemed in technological questions as scientific advisor, "even by the fierce Köchy, who up to now has seemed to me the most unapproachable."[47] Otto Köchy (1849–1914) was professor of mechanical engineering at the Technical University at Aachen.

In the years around the turn of the century, as Wilhelmine Germany geared up for ever accelerating industrial growth, extracurricular assignments came increasingly to the engineering professors. As the Ministry of Public Works in Berlin officially informed him, effective August 1, 1902, Sommerfeld was named a member of the "Royal Technical Testing Office" in Berlin, for a term of 3 years.[48]

From the technical challenges, Sommerfeld was set as expert, scientifically interesting problems also emerged. "Let us imagine a building which must be supported on beams because a railroad line is to be routed under it. Inside the building, a steam-engine must be installed." Thus, Sommerfeld described the case presented to him by the professor of structural engineering of the civil engineering department Hermann Boost (1864–1941). The steam engine was to be located 10 m above ground level. The back-and-forth motion of its piston would be transferred to the entire building. "How strong must the beams be made—this was the question we saw put to us—so that the motion of the building remains under some set limit, let us say for instance under ½ millimeter?"[49]

Sommerfeld chose this problem for a demonstration to the Aachen District Association of the VDI. For the professors of the Technical University, it was good form to appear before this circle of technology enthusiasts, even if they themselves were not members of an engineering department.[50] The Association served Sommerfeld as a welcome forum in which to convey his commitment to technology. A steam engine that at a height of 10 m would make a building sway must have seemed an exceptional challenge to any engineer. This also clearly presented the necessity of a mathematical treatment of the problem, for no theory existed from which the required strength of the beams could be calculated. Sommerfeld demonstrated what this problem entailed by means of a motor screwed fast to a tabletop. He attached an unbalanced weight to the flywheel of the motor which, with increasing rotational speed, made the table rock back and forth. By increasing the drive force of the motor, Sommerfeld was able to induce powerful resonance vibration in the table. Horizontal deflection of the table legs was far greater than would have been expected had the table been distorted statically by a centrifugal force

47 To Wien, May 29, 1901. DMA, NL 56, 010. Also in ASWB I.

48 From the Ministry of Public Works, 31. July 31, 1902. DMA, NL 89, 019, folder 5,2.

49 Sommerfeld, *Beiträge*, 1902.

50 See the self-description of the Association at http://www.vdi.de/1672.0.html (30 January 2013).

equal to the force of the unbalanced weight. This was a dynamic phenomenon, in other words, that the engineers could not address by the methods of structural statics they were familiar with. A proximate phenomenon now appeared: Further increasing the drive force of the motor resulted not in higher revolutions of the motor, but in an increase in the intensity of vibration of the table. The increased energy induced in the motor was transferred to the vibration of the table, not to the motion of rotation. "In burning his expensive coal," Sommerfeld related the situation of his demonstration to the practical sphere, "the factory owner gets no more out of his machine; he merely loosens and compromises his foundation."[51]

In the steam era, resonance phenomena of this sort were the daily experience of the engineer. It was not just a matter of unnecessary consumption of "expensive coal," but also of heretofore inexplicable catastrophes. "In my opinion, namely, the collapse of the bridge at Mönchenstein, which in its day cost so many people their lives, is to be explained primarily by a resonance effect of the same kind you discussed," August Föppl wrote to Sommerfeld.[52] He was referring to a catastrophic event of the year 1891 in which 73 people were killed and 131 people injured, some critically. The accident was baffling since it could not be explained by the expertise of bridge construction technology of those years. Adhering to the rules of architectural statics, the bridge had been designed to support multiples of the weight of the trains that traversed it. In the speculation over the causes of the bridge collapse, it was inferred from eye-witness reports of survivors that it had probably resulted from some dynamic phenomenon: According to this conjecture, it had collapsed "not suddenly, but following prolonged waves of up-and-down heaving of the bridge."[53] With his demonstration to the Aachen District Association of the VDI, Sommerfeld had thus touched a nerve of the engineers of the time. The "rocking table" subsequently found its way into numerous lectures on resonance phenomena. Following a demonstration of the "rocking table" experiment, a mathematical colleague from Innsbruck, Wilhelm Wirtinger (1865–1945), reported to Sommerfeld that he had learned "all sorts of interesting things" from the engineers present, "such as that analogous things have been observed in the case of the swaying of locomotives."[54] Sommerfeld had thus demonstrated a phenomenon that generated the basis of further research not only by engineers but scientists as well. The phenomenon demonstrated by the "rocking table" went down in the history of nonlinear dynamics as the "Sommerfeld effect."[55]

Then, Föppl brought yet another oscillation phenomenon to Sommerfeld's attention over which quite various opinions had for some time been registered in the *Elektrotechnische Zeitschrift*: the so-called pendulation of machines connected

51 Sommerfeld, *Beiträge*, 1902.

52 From Föppl, October 27, 1901. DMA, HS 1977-28/A,97.

53 *Schweizerische Bauzeitung*, June 20 & 27, July 18 & 25, 1891.

54 From Wirtinger, December 18, 1901. DMA, NL 89, 014. Also in ASWB I.

55 Eckert, *Sommerfeld-Effekt*, 1996.

in parallel in a circuit of alternating current, which leads to a "back-and-forth surging of electrical supply between the individual machines connected in parallel."[56] For this problem, too, Sommerfeld quickly found a solution: "My point of departure was the study of an apparatus, 'sympathetic pendulums,' which I had constructed for instructional purposes, and which illustrates the essential phenomena here very well." This is how he illustrated the underlying principle of electrical "pendulation" in a parallel circuit by means of a familiar mechanical phenomenon. The origin here was the same as there: the coupling of a forced oscillation and an independent oscillation of one of the machines on the circuit. He illustrated this with a pocket watch suspended from a nail. The motions of the balance spring set the watch into forced oscillation, but simultaneously, an ordinary pendulum swing was set in motion, whose period was determined by the distance of the watch from the point of attachment. Between the swing forced by the balance spring's motion and the free pendulum swing, there arose a coupled oscillation, well-known in physics as "beat."[57] Gustav Benischke (1867–1947), an engineer of the Berlin General Electrical Corporation (AEG), showed great interest in Sommerfeld's pendulum models and invited him to Berlin for a factory tour.[58] Ultimately, though, the "beat" theory proved too remote from practical application. Sommerfeld had not taken into consideration "the most recent literature on the subject," Benischke criticized. In actual practice, the observable oscillations were far more complicated than those Sommerfeld had calculated.[59]

This did not deter Sommerfeld, however, and undaunted, he continued making engineering problems the subject of his own research. He was encouraged in this not only by colleagues from the engineering departments at the Aachen Technical University but also by commissions for expert testimony. The Association of German Iron Workers, for instance, asked him for an expert opinion on the buckling limits of I-beams. Sommerfeld did not content himself with a merely theoretical analysis, as in the case of pendulation of alternating current machines. He had his results tested experimentally with a hydraulic press at the steelworks of the Aachen Iron Works Corporation, Rothe Erde. He could assure the commissioners that in his expert opinion, the danger of buckling was insignificant, since the buckling threshold far exceeded the actual weight loads in practice.[60]

As with other assignments from practical engineering, Sommerfeld exploited this problem for a demonstration of his mathematical–physical approach to technological questions. He gave a lecture to the Aachen District Association of the VDI in which he clamped one end of a steel plate in a vise, then first standing upright, and then hanging down let it swing. When weights were added to the

56 From Föppl, January 31, 1902. DMA, HS 1977-28/A,97; Föppl, *Pendeln*, 1902.
57 Sommerfeld, *Pendeln*, 1904, p. 273.
58 From Benischke, April 17, 1902. DMA, NL 89, 005.
59 From Benischke, April 20, 1904. DMA, NL 89, 005.
60 Sommerfeld, *Knicksicherheit*, 1906; Sommerfeld, *Nachtrag*, 1907.

plate, the swings of the upright plate were consistently slower than those of the plate positioned downwards, until from a certain limit the swinging ceased altogether because the weight of the deflected plate now exceeded its elastic restoring force. With the mathematical treatment of these swings, Sommerfeld showed what physical quantities play a role in this.[61] He showed also how it was possible to calculate the elastic constants of a wire in a spiral spring from the longitudinal and rotational oscillations. This method was so pleasing to him that he made it the subject of his contribution to a festschrift for the Aachen experimental physicist Adolph Wüllner (1835–1908). He was functioning here "virtually as a physicist and photographer in the area of elasticity," he wrote Willy Wien.[62]

As a result, many who like Sommerfeld had begun their careers as mathematicians or physicists at universities were astonished to learn how quickly Sommerfeld adapted his career at Aachen to the expectations of his engineering colleagues. "You're really quite the engineer," Max Abraham, for instance, wrote in astonishment about a paper that Sommerfeld published in a memorandum of the Technical University at Aachen on locomotive brakes, in which he investigated the relationship among brake pressure, brake time, and brake travel in order to derive criteria for optimal brake functioning.[63]

The work on locomotive brakes was the prelude to a fundamental theory of the process of friction, as it occurs in the axle bearings of railroad cars, for instance. Here, one might imagine either unmediated contact of the frictional surfaces or friction reduced by some lubricating medium. In the former case of "dry" friction, there existed a theory traceable to Charles Augustin de Coulomb (1736–1806). Friction mediated by lubrication, on the other hand, was a hydrodynamic problem and as such still largely unexamined. Sommerfeld wished to explore this aspect, more significant for railroad technology, from the ground up. "I am now delighted to report that empirical practice seems to confirm your theory," a former Aachen student wrote Sommerfeld, who at his request had investigated the abrasion of the bearing housings of locomotive axles at a railroad factory. According to Sommerfeld's theory, the places of greatest abrasion from hydrodynamic friction would lie at places other than from dry friction.[64] Following this confirmation, Sommerfeld published his hydrodynamic theory of bearing friction.[65] Pure mathematicians like Edmund Landau (1877–1938) regarded this as a descent into the swamps of applied mathematics, which would now be characterized as "axel grease."[66] Engineers, on

61 Sommerfeld, *Vorrichtung*, 1905.

62 To Wilhelm Wien, April 15, 1905. DMA, NL 56, 010; Sommerfeld, *Lissajous-Figuren*, 1905.

63 From Abraham, December 9, 1902. DMA, HS 1977-28/A,1; Sommerfeld, *Eisenbahnbremsen*, 1902.

64 From Ernst Becker, April 3, 1903. DMA, NL 89, 005. Also in ASWB I.

65 Sommerfeld, *Schmiermittelreibung*, 1904.

66 Richard Courant, Interviewed by Thomas S. Kuhn and M. Kac, May 9, 1962. AHQP. http://aip.org/history/catalog/icos/4562.html. (30 January 2013). Ostrowski, *Zur Entwicklung der numerischen Analysis*, 1966.

the other hand, were brimming with praise.[67] Sommerfeld's theory of friction in lubricating media became a classic. By means of a dimensionless quantity introduced into it ("the Sommerfeld number"), different bearings can be compared with respect to their frictional properties. Therefore, in the engineering science of friction, tribology, Sommerfeld is recognized as one of the pioneers in the field.[68]

Just how greatly Sommerfeld had come to be esteemed in engineering circles is clear from his correspondence with August von Borries (1852–1906), a professor of railroad technology at the Technical University of Berlin. The impetus was a prize competition sponsored by the Association of German Mechanical Engineers "whose purpose [was] the realization of a textbook on locomotive construction."[69] The Berlin railroad professor would gladly have coauthored such a textbook with Sommerfeld. He would have left to him the "mechanical-theoretical portion, perhaps also the portion devoted to heat theory."[70] Sommerfeld accepted. Organizational drafts flew back and forth between Berlin and Aachen; there were discussions of the latest technical literature on railroad technology, particularly on what related to "the destructive motions of locomotives," the subject of several recent doctoral dissertations.[71] But the project of the locomotive textbook did not go beyond preliminary sketches and expired finally on its own when August von Borries died in 1906.

Sommerfeld's turn towards technology was apparent also in the matter of academic appointments. In 1904 he was being considered as a candidate for a position in mathematics at the Technical University at Hannover. Unlike his appointment at Aachen 4 years earlier, when the engineering departments regarded him as a mathematical Trojan horse, the circumstances at Hannover were now quite the reverse. A mathematician from the general division of the Technical University had spoken against his appointment, but "the engineers seem to want me at all costs," as Sommerfeld described the internal discussions at Hannover to his wife. But the negotiation came to a halt at the Prussian Ministry of Culture, Education, and Church Affairs. "On no account will we remove Sommerfeld from engineering mechanics," Berlin pronounced, thwarting the wishes of the engineers at Hannover.[72] Meanwhile, at the Technical University at Aachen, Sommerfeld's commitment to engineering was rewarded with the assignment of an assistantship in 1905. He filled this position with Peter Debye (1884–1966), who had completed his studies that year with an engineering diploma.[73] In 1904 and 1905, the Aachen District Association of the VDI elected Sommerfeld its protocolist,[74] and the

67 From Frahm, July 4, 1904. DMA, HS 1977-28/A,99.

68 Dowson, *History*, 1998, pp. 653–656, numbers Sommerfeld among the "Men of Lubrication".

69 In *Glasers Annalen für Gewerbe und Bauwesen*, June 1, 1904, p. 205.

70 From Borries, June 12, 1904. DMA, NL 89, 006.

71 From Borries, October 29, and November 18, 1904, January 15, 1906. DMA, NL 89, 006.

72 To Johanna, August 10, 1904.

73 Acta betreffend Etat und Rechnungswesen, August 19, 1904 to November 6, 1906. Akte 844; Debyes Diplomurkunde, folder 36c, Archive of the RWTH Aachen.

74 VDI Aachener Bezirksverein, http://www.vdi.de/1672.0.html (30 January 2013).

Technical University at Delft offered him a professorship of applied mechanics, which he nonetheless declined.[75] Sommerfeld appeared to have set the course for his academic future. "My work and position here are very comfortable, and life in Aachen is most enjoyable," he wrote to Wilhelm Wien in the summer of 1905.[76]

5.4 Family Life

When Sommerfeld moved to Aachen in the spring of 1900 with wife and child, in the expectation of a growing family, he had already rented a house with seven rooms.[77] Half a year later, Johanna gave birth to a baby girl, whom they christened Margarethe. Two months following her birth, the proud father inscribed in an album to his 1 ½-year-old Ernst, nicknamed Puttl, "On the 5th of August he was given his little sister, Gretchen. From the start, he treats her as something tender and mysterious. The 'Gretchen theme' he intones whenever he approaches her crib, or she is merely referred to is, Ha, ha, ha. Recently (October) he has become the model of chivalry. He kisses her little sisterly hands and little cheeks good night."[78]

When during the summer vacation, Sommerfeld was called up for duty at military exercises in Saxony from July to September 1903, Johanna went to stay with her parents for these weeks in Göttingen. In September, the annual Natural Scientists Congress took place in nearby Kassel, so that "lots of scientific visitors, among them Boltzmann," came to Göttingen, as Sommerfeld recorded in Ernst's album. What he remembered especially was the observation of his 4-year-old son on the table manners of his little sister: "You eat just like Uncle Boltzmann."[79]

When it came to Sommerfeld's friendly relations with his colleagues, family life and career were inseparable. In October 1900, Lorentz was a guest at the Sommerfeld home in Aachen.[80] Along with Boltzmann and Wilhelm Wien, Lorentz was among the "principle authors" of the physics volume of the *Encyclopedia of Mathematical Sciences*.[81] "The children speak a great deal about Lorentz-cookies and Cookie-Lorentz. Their associations with your name are almost as pleasant and estimable as those of their parents," Sommerfeld wrote in the wake of a visit to Lorentz in

75 From Jacob Cardinaal, July 4, 1906. DMA, HS 1977-28/A,48; from de Haas and Cardinaal, July 7, 1906 DMA, NL 89, 019, folder 5,3.

76 To Wilhelm Wien, July 4, 1905. DMA, NL 56, 010.

77 Nachweis Mietverhältnis. Aachen, folder 842, Archive of the RWTH Aachen.

78 Album for Ernst.

79 Album für Ernst.

80 From Lorentz, October 6, 1900. DMA, HS 1977-28/A,208. Also in ASWB I.

81 To W. Wien, July 6, 1901. DMA, NL 56, 010.

Fig. 11: Aachen family idyll from the year 1901: Gretchen and Ernst made for lively days in the Sommerfeld home (Courtesy: Deutsches Museum, Munich, Archive).

Holland.[82] On February 5, 1904, Johanna gave birth to another boy, whom they named Arnold Lorenz. "We wanted to keep the name Lorenz (though we don't quite spell it correctly) a secret from you and everyone else," Sommerfeld later wrote after a visit to Lorentz in Holland. "But when you greeted my wife at the station, she couldn't resist telling you. As I told you at Leiden, my wife is very fond of you. You will, we hope, see this from her verses, though no doubt better poetry has been written in the German language. I subscribe heartily to the sentiment of these verses."[83]

82 To Lorentz, August 27, 1902. RANH, Lorentz, inv. nr. 74.
83 To Lorentz, December 12, 1906. RANH, Lorentz, inv. nr. 74. Also in ASWB I.

Fig. 12: Hendrik Antoon Lorentz, patriarch of theoretical physics in Holland, became an elder friend of the Sommerfelds. When in 1904 Johanna gave birth to another son, they christened him Arnold Lorenz (Courtesy: Deutsches Museum, Munich, Archive).

Sommerfeld also interacted with his students more broadly than in his capacity as professor in the lecture hall. "You may remember me," a former engineering student once wrote Sommerfeld; he had had the pleasure "of taking part in a cycling tour to Lake Gileppe under your leadership, on which occasion we had a lengthy discussion on the stability of the bicycle in operation, and the underlying reasons for it."[84] Sommerfeld valued especially a student's pursuit of his scientific education not just because it was part of the lesson plan, but out of personal interest. Peter Debye had "privately worked his way into the higher realms of mechanics and theoretical physics," Sommerfeld wrote the Rector of the Technical University at Aachen in support of Debye's appointment as his assistant.[85] He invited students like these to his house to deepen these interests in a relaxed atmosphere. "Together, they studied works of theoretical physics far beyond the curriculum of the Technical University,

84 From Ernst Becker, November 12, 1902. DMA, NL 89, 005.

85 To Borchers, December 12, 1904. folder 844, Archive of the RWTH Aachen.

for instance Drude's optics, and even Maxwell's Treatise," Sommerfeld recalled many years later of Walter Rogowski (1881–1947) and Peter Debye. "How often those two scientific twins sat at my house to ask questions and absorb new ideas!"[86]

Even when Sommerfeld wrote home from a congress, scientific matters were mixed with private matters. At times, he gave his attestations of love verbal expression that reflected what he was just then working on: "Farewell my sympathetic little pendulum, my harmonizing soul," he closed a letter to his wife, a few weeks after Föppl had brought to his attention the "pendulum swings" of machines connected in parallel. "We two have our equal pendulum lengths, and are attuned to each another. When I'm back, I'll demonstrate this to you with my apparatus."[87] Shortly thereafter, he traveled to Königsberg to visit his mother, fatally ill. The stay lasted longer than planned, "since my mother's illness has me very concerned," he wrote by way of apology to Lorentz, who had meanwhile sent him a manuscript for the *Encyclopedia*.[88] His mother's condition deteriorated rapidly. "I have been summoned via telegram to my mother's deathbed, and travel this evening to Königsberg," Sommerfeld informed the Rector of the Technical University at the end of May.[89] A few days later, in a moving, eight-page letter to Johanna, he described the death and burial of his mother and his own emotions in the parental home at Königsberg following her death: "My diligent little mother peers out at me from every corner of the apartment; every wall and table is covered with the tokens of her activity. And she fills our hearts."[90] Only a few years later, Sommerfeld had to cope with the death of his father.[91] Family happiness at Aachen was dimmed also by concern for his brother. "I would be lying," Sommerfeld wrote after a visit to Walter, "were I to say I know he is morphine-free. But I have neither the courage nor the grounds to say he is ill." The uncertainty brought him to the brink of despair: "Should I write to him: If you take morphine, you might as well shoot yourself in the head today instead of tomorrow!? If that is the case, then I really no longer know any solution."[92] Ultimately, Walter entered a psychiatric clinic at Leubus near Breslau. When Sommerfeld visited him there, he was devastated by the "torpid, pitiable figure" his brother had become.[93] Nor was there any longer hope of recovery. Walter lived another 13 years at this clinic until his death in 1917, at the age of 56.[94]

86 Sommerfeld, *Lehrjahren*, 1950.
87 To Johanna, March 27, 1902.
88 To Lorentz, April 29, 1902. RANH, Lorentz, inv. nr. 74.
89 To Bräuler, May 31, 1902. folder 910, Archive of the RWTH Aachen.
90 To Johanna, June, 3, 1902.
91 To Johanna, January 25, and February 4, 1906.
92 To Johanna, April 1, 1901.
93 To Johanna, January 3, 1904.
94 Death notice, *Königsberger Hartungsche Zeitung*, September 18, 1917.

5.5 Duties and Inclinations

His precocious son Ernst, the little Margarethe, and Johanna, his "harmonizing soul" and sympathetically tuned spouse, helped Sommerfeld overcome these blows of fate. And his robust East Prussian nature, the ingrained sense of duty and diligence impressed on him from the cradle, no doubt also contributed to his transcending these periods of depression. "Yesterday I returned from my melancholy journey to Königsberg," Sommerfeld wrote Klein shortly after his mother's death to inform him that he would immediately resume his duties as editor of the *Encyclopedia*.[95] Soon thereafter, he assured Klein that he would, in the coming summer vacation, also take up "the gyroscope" again.[96] These two projects—the *Encyclopedia* and the forthcoming third "fascicle" of the *Theory of the Gyroscope*—would have been sufficient to fill up Sommerfeld's workdays entirely, even without his other duties at the Technical University at Aachen.

By 1902, "the gyroscope," almost 7 years after Klein's original gyroscope lecture and 4 years after completion of the second "fascicle," was for Sommerfeld a task he was fulfilling more from a sense of duty than personal inclination. With the second fascicle, though the work had already come to 512 pages, Sommerfeld had advanced only to the representation of gyroscopic motion by elliptic functions. The third fascicle was intended to bring the work to a conclusion with the applications of gyroscope theory. But the material Sommerfeld gathered for it multiplied so rapidly that it could not be left at that. "Therefore," he wrote in the Preface, "only the applications of gyroscope theory to astronomy and geophysics are presented in this fascicle; engineering and physical applications have been left to a fourth (and final) fascicle."[97]

For some time, Klein had been brooding "in quiet distress about our gyroscope," as he had once confessed to Sommerfeld at his impending appointment at Aachen.[98] Even a year later, this concern had not been diminished: "I have been meaning to write you to apologize for not going through the manuscript," Sommerfeld wrote to Göttingen. "Since the 3 weeks that lectures last here, I have managed to get to the gyroscope only one weekend."[99] Sommerfeld was constantly placating Klein. "The gyroscope is making strides, albeit with interruptions," he wrote at the start of 1902, shortly before a conference he requested with Klein to discuss a chapter treating the effect of friction on the motion of the gyroscope.[100] Even his promise to apply himself seriously to the work during the summer vacation seems not to have relieved Klein of his quiet distress, for he called in Karl Schwarzschild hoping to speed up the chapter on astronomical applications. Sommerfeld gladly accepted this support,

95 To Klein, June 9, 1902. SUB, Klein 11,1062.

96 To Klein, June 27, 1902. SUB, Klein 11,1062. Also in ASWB I.

97 Klein/Sommerfeld, *Theorie des Kreisels*, Heft 3, 1903, Vorwort.

98 From Klein, November 15, 1899. DMA, HS 1977-28/A,170.

99 To Klein, November 8, 1900. SUB, Klein 11, 1061.

100 To Klein, February 15, 1902. Stuttgart, Teubner-Archive, Sommerfeld.

but made it unmistakably clear that he was not going to relinquish control of the concept. "I assume that your participation in the matter of the gyroscope will be more along the lines of critique than of actual writing yourself," he wrote, affirming his own role as exclusive author. "The manuscript, if it is to have consistency, must ultimately be produced by me. If, then, you agree to rewrite portions subject to my approval prior to publication, no one will be more delighted than I."[101]

Thereafter, work flowed more easily. He sent Schwarzschild portions of the manuscript for revision and requested "maximally rigorous critique," since he felt quite insecure in astronomy.[102] Through their conjoined forces, this chapter now took shape.[103] The collaboration seems to have been enjoyable for both, for the formal salutation "Esteemed Colleague" now turned into "Dear Schwarzschild" and "Dear Sommerfeld." The tone of their intercourse became quite friendly, and they kept each other up to date on their respective work on gyroscope theory. "As Klein says, next time you gyrate in Göttingen," Schwarzschild wrote Sommerfeld when the astronomical chapter was finished, and Sommerfeld requested that Wiechert undertake a critical review of his manuscript on the geophysical applications of the gyroscope.[104]

In truth, it would have been opportune to take up at once the last gyroscope fascicle, which was to be devoted to technical applications. "You're entirely correct to remind me of the gyroscope," Sommerfeld admitted when Klein urged him to press on.[105] In any case, there was no lack of illustrative material such as the "straight-line trajectory apparatus" for torpedo guidance which Sommerfeld had received from Kiel at the assumption of duties at Aachen. Sommerfeld was also not lacking impetus from other quarters. "Gyroscopic effects of wheel-axle assemblies" were, for instance, the subject of discussion with the Berlin professor of railroad engineering on the projected locomotive textbook.[106] If Sommerfeld put off completion of the work on gyroscopes until later, it was neither for lack of material nor from paralysis in reaction to an oppressive sense of duty, but rather because other matters seemed to him to take precedence.

Primary among these was the editing of the physics volume of the *Encyclopedia*. The manuscripts for it were submitted to him and after thorough revision were published in the form of "fascicles," gathered later into partial volumes. During his Aachen years, Sommerfeld brought four such fascicles comprising a total of ten articles to publication. The first fascicle, published in April 1903, contained three articles filling 160 pages; this was followed in April 1904 by a fascicle with three articles in 280 pages and in October 1905 and March 1906 by fascicles with two

101 To Schwarzschild, July 26, 1902. SUB, Schwarzschild 743. Also in ASWB I.
102 To Schwarzschild, August 12, 1902. SUB, Schwarzschild 743.
103 To Schwarzschild, January 26, 1903. SUB, Schwarzschild 743.
104 From Schwarzschild, March 29, 1903. DMA, HS 1977-28/A,318.
105 To Klein, November 8, 1904. SUB, Klein 11, 1064.
106 From Borries, November 18, 1904. DMA, NL 89, 006.

articles each of 159 and 171 pages, respectively. In other words, in those years Sommerfeld edited contributions totaling nearly 800 pages for a work surveying subjects having barely anything in common with his own academic field.[107]

Considering the prominent names of most of the authors, and the high expectations placed on the *Encyclopedia* enterprise, preparation of an article could become a significant challenge for the editor. Even when as a rule the authors were experts in their fields, the editorial work could not be limited simply to improving details of style. In the case of the article "General Foundations of Thermodynamics" from the pen of the British mathematician George Hartley Bryan (1864–1928), Sommerfeld had first to translate the submitted manuscript from English to German. With this article, the burden of work was not eased for him by his enthusiasm for its contents. Bryan was known to be very eccentric,[108] and this aspect of his character must have been reflected in the manuscript, for Sommerfeld felt compelled to explain to Bryan changes he had made in the course of translation on four densely written pages of a letter.[109] After this there were further problems. The Dutch expert on experimental thermodynamics, Heike Kamerlingh Onnes (1853–1926), who had also prepared a topic in thermodynamics ("The Equation of State") for the *Encyclopedia*, indicated in a letter to Sommerfeld that he had not succeeded in moving Bryan to a happier coordination between their two articles.[110] Sommerfeld sent Bryan's manuscript also to Voigt and Lorentz to elicit an assessment from these authorities. Voigt "disagreed with quite a lot" and regretted that Planck could not be recruited to write this article. "He would surely have submitted something better than Bryan."[111] Lorentz likewise raised a string of objections, so that Sommerfeld had to embark on a second revision.[112]

By contrast, Sommerfeld was presented a challenge of quite a different sort by the two articles Lorentz sent him on electrodynamics. Here, his sense of duty was counterbalanced by his own predilections. Unlike thermodynamics, he felt at home with electrodynamics. Even with an authority such as Lorentz, in this area he could exchange ideas as expert to expert, bringing his own conceptions into the discussion. Already with his first reaction to the organizational outline for the article on "Maxwell's Electrodynamic Theory," he made it clear that he intended to have his say in the matter. "I understand very well," he wrote confidently to Lorentz, "that it would abbreviate the presentation for you to start from the field equations, and

107 *Enzyklopädie*, vol. V. The partial volumes of the *Encyclopedia* are also available online via http://de.wikipedia.org/wiki/Enzyklopädie_der_mathematischen_Wissenschaften (30 January 2013).

108 Anonymous [L. B.], *Bryan*, 1933.

109 To Bryan, February 10, 1902. DMA, HS 1977-28/A,45.

110 From Kamerlingh Onnes, September 28, 1902. DMA, HS 1977-28/A,160.

111 From Voigt, October 18, 1902. DMA, HS 1977-28/A,347.

112 To Lorentz, January 6, 1903. RANH, Lorentz, inv. nr. 74. Also in ASWB I. Bryan, *Grundlegung*, 1903.

thus this suggests itself." He recommended, though, not to cut to the chase, but to introduce a historical section devoted to Maxwell's predecessors. So far as the relevance of his own 1892 contributions to the relevance of the ether models to electrodynamics was concerned, he recommended that "the reader be left in no doubt as to the minimal importance of those investigations . . . You do me the honor of naming me in the outline. If that is to occur in the article too, then certainly Lord Kelvin should be cited first, from whose work I proceeded, and Reiff, who elaborated my idea further and more effectively."[113]

Richard Reiff (1855–1908), a mathematician from Tübingen, proceeding from Sommerfeld's ether model of the year 1892, had attempted to describe the nature of electricity purely mechanically by means of the properties of an elastic medium.[114] He was considered an expert in the field of ether theories and had been identified as a prospective author of an encyclopedia article on the historical background of electrodynamics. When Sommerfeld edited Reiff's manuscript, though, he was anything but enthusiastic. On the theories of Carl Neumann (1832–1925), for instance, Reiff had "written a section that was entirely meaningless, and that I have deleted," Sommerfeld wrote Wiechert, to whom he forwarded Reiff's manuscript for review and who himself had several things to object to. "I hope your letter inspires me to study Neumann myself and to present the matter from my perspective."[115] From Lorentz too he received a string of comments on Reiff's manuscript, so that he felt increasingly compelled to undertake his own study of the historical works Reiff had cited. In the end, his interpolations and revisions were so extensive that the article appeared under both names.[116]

So much the more satisfying, then, was his editing of Lorentz's article. The manuscript on "Maxwell's Electromagnetic Theory" was "quite wonderful" and corresponded "to the full measure the purpose of our compilation," he wrote, thanking Lorentz. After their thorough discussion about the structure of the article, it was clear that he viewed this article with particular interest. He could have "wished a somewhat more comprehensive treatment and greater physical descriptiveness in places," he permitted himself to say by way of critique. Nor did he immediately accept the first draft of the article, but asked Lorentz "for the reader's convenience" for a number of elaborations and urged him to make a clear statement regarding the various notational systems in electrodynamics. "I wanted so much for something truly useful in this regard to be achieved, which future authors could fruitfully use as a point of departure. In writing your article you have surely come to a clear judgment whether the notation used is effective. For the sake of the subject itself, I would very much regret if you have acceded in various points merely to

113 To Lorentz, March 21, 1901. RANH, Lorentz, inv. nr. 74. Also in ASWB I.

114 Reiff, *Elasticität*, 1893.

115 To Wiechert, January 29, 1903. SUB, Wiechert.

116 From Lorentz, February 24, 1903. DMA, HS 1977-28/A,208; to Lorentz, February 24, 1903. RANH, Lorentz, inv. nr. 74, also in ASWB I. Reiff/Sommerfeld, *Standpunkt*, 1904.

oblige me."[117] He also permitted himself commentary on subjects that recalled the "physical mathematics" of his Göttingen years. "In the matter of the solvability of Maxwell's equations with given initial conditions, I would like to distinguish the question of unambiguity from the question of existence." With a rigorous existence proof, one could also prove unambiguity. But he assured Lorentz that this would be "an equally superfluous and thankless task" and did not belong in the article. Only "the true mathematician" would find such "questions of existence" worth a closer look.[118] On the other hand, it was very important to him to reach a solution in the question of notation, which could serve not only the other authors of the Encyclopedia but could as such become the standard for electrodynamics. "Inspired by our encyclopedia notation, the German physics community is currently busy deliberating notation," he wrote, underscoring the importance of this question.[119]

Aside from the notation of electromagnetic quantities, the notation of the vector operations used for the representation of Maxwell Theory also required standardization. Sommerfeld reported in the *Physikalische Zeitschrift* that, "after many consultations" with Lorentz, Wilhelm Wien, and Emil Cohn (1854–1944), his suggestions had been realized. He hoped the authority of his encyclopedia authors would procure wide acceptance of his suggestions.[120] In this connection, Sommerfeld took part in a "Vector Commission," established in 1903 on the initiative of the German Mathematical Association. There, however, he did not succeed as he had with the *Encyclopedia* in accomplishing the standardization he had hoped for. Vector algebra and vector analysis long remained the arena of disparate spellings and notation.[121]

Editing Lorentz's second encyclopedia article on the "Elaboration of Maxwell Theory—Electron Theory" was for Sommerfeld likewise more than a mere fulfillment of obligation. After the discovery of the electron at the end of the nineteenth century, the previously prevailing conception of a mechanical ether was replaced by an electrodynamic model of the universe. Ether became a medium that had nothing in common with the mechanical ether of the nineteenth century, but rather merely mediated the electrodynamic effects among the electrons embedded in it. If previously (like Sommerfeld in his youthful work of 1892) one had attempted to ground the Maxwell equations mechanically, now the task became to describe all physical phenomena by the motions of electrons in an electrodynamic ether that itself required no further explanation. Like the axioms of mathematics, Maxwell's equations formed the ultimate foundation on which every theory of physics had to be built. Carrying out the mathematics of this conception, however, proved exceedingly difficult. Virtuosic facility with the Maxwell equations was required to derive

117 To Lorentz, April 29, 1902. RANH, Lorentz, inv. nr. 74.
118 To Lorentz, July 5, 1902. RANH, Lorentz, inv. nr. 74.
119 To Lorentz, January 6, 1903. RANH, Lorentz, inv. nr. 74.
120 Sommerfeld, *Bezeichnung*, 1904.
121 Reich, *Vektorrechnung*, 1995.

experimentally demonstrable physical conclusions. Since the early 1890s, Lorentz had treated this complex of problems in countless papers.[122] Hardly anyone other than Lorentz would have been in a position to deal with this subject with all the unresolved questions it raised in such a way as seemed required for the *Encyclopedia*. In his article on Maxwell Theory,[123] Lorentz could put the finishing touches to a more or less established state of knowledge, but in the case of electron theory, he was treading virgin soil. Investigation of the electrical and magnetic fields of an electron forced a differentiation between frames of reference in which the electron was at rest and frames of reference where the electron was moving and where things are set in motion in the truest sense of the word. Even the form of an electron, and consequently also that of larger bodies, had to be conceived as variable and dependent on the frame of reference.[124]

5.6 The "Super-Mechanics" of Electrons

Sommerfeld accompanied Lorentz on the journey into the new territory of electron theory. He corresponded on the subject with other encyclopedia authors as well, primarily with Max Abraham, who in 1902, with a dissertation on the "Dynamics of the Electron" had published a kind of manifesto of the electrodynamic model of the universe.[125] Abraham was overjoyed at Sommerfeld's interest in this work and announced to him the "final summing up" of his investigations of electron theory, published shortly thereafter in the *Annalen der Physik*.[126] Sommerfeld exchanged views on the latest developments in this field also with Wilhelm Wien, Wiechert, and Schwarzschild and let Lorentz know whenever any of it might be useful for his encyclopedia article.[127] From Klein, Sommerfeld learned that electron theory had become the focus of intense research at Göttingen. This may have spurred him to make a contribution of his own in this area. In January 1904, he confided to Schwarzschild that he was now working "with a full head of steam" on electron theory.[128] In response, Schwarzschild sent Sommerfeld the galleys of a paper "On Electron Theory" just submitted to the Göttingen Academy by Gustav Herglotz (1881–1953),[129] who was writing his habilitation dissertation under the wings of Felix

122 McCormmach, *Lorentz*, 1970.
123 Lorentz, *Maxwells elektromagnetische Theorie*, 1904.
124 Lorentz, *Elektronentheorie*, 1904.
125 Abraham, *Dynamik*, 1902; Goldberg, *Abraham Theory*, 1970.
126 From Abraham, December 9, 1902. DMA, HS 1977-28/A,1. Abraham, *Prinzipien*, 1903.
127 To Wiechert, January 29, 1903. SUB, Wiechert; from Schwarzschild, March 29, 1903. DMA, HS 1977-28/A,318; to Schwarzschild, March 31, 1903. SUB, Schwarzschild 743; also in ASWB I; to Lorentz, April 25, 1903. RANH, Lorentz, inv. nr. 74. Schwarzschild, *Elektrodynamik I-III*, 1903.
128 To Schwarzschild, January 10, 1904. SUB, Schwarzschild 743.
129 Herglotz, *Elektronentheorie*, 1903.

Klein. As it was for other Göttingen theoreticians, for Herglotz too this subject was primarily a mathematical challenge. And as so often the case, here too the same physical subject matter could be approached mathematically in a number of different ways. The Herglotz electron theory was "fundamentally different" from his own, Sommerfeld determined with some measure of relief. "I have a magic formula whereby I determine the field with an arbitrary set of straight-line motions of an electron—rigorously, in fact, and very simply. From it I have derived all known outcomes with respect to electrons."[130]

As though to reinforce his membership in the Göttingen circle of theoreticians, Sommerfeld submitted his electron theory to the Göttingen Academy for publication as well.[131] In the first part he derived formulas for the electromagnetic field of an electron of arbitrary motion, whereby he assumed the electron to have a spherical shape. For the distribution of electrical charge, he distinguished two possibilities: It could either be distributed evenly throughout the volume of the sphere or spread only over the spherical surface of the electron. In the second part, as he wrote to Schwarzschild in June 1904, "it really gets serious insofar as there I reconfigure both Abraham's and Herglotz's results in an entirely new way, and generalize them."[132] Many years later Sommerfeld assessed this theory quite critically. The "lengthy and difficult studies on which at first I placed great value" had been "judged fruitless."[133] The Theory of Relativity of Albert Einstein had pulled the rug from under the electrodynamic conception of the universe. Of the electron theory conceived before 1905, only fragments survived (such as the system of equations for the transformation of fields and potentials between different frames of reference, the eponymous "Lorentz Transformation").[134]

In the year 1904, however, speculations such as those Sommerfeld presented in his electron theory were regarded as trail blazing. "This morning, I gave my lecture," Sommerfeld wrote his wife from Heidelberg at the Third International Congress of Mathematicians in August 1904, where he had lectured "On the Mechanics of Electrons." "So far, it has decidedly been the best in the section for applied mathematics," he crowed in the fervor of his enthusiasm for this subject, even though he had the impression that his exposition "had not been entirely understood. Supermechanics, however, has taken wing."[135] Electron theory was the "youngest and most hopeful offspring of mathematical physics," Sommerfeld began his lecture. Since the electron moves in its own electrical field, quite unusual phenomena were to be expected. In this "super-mechanics" of the electron, there were force-free oscillations and rotational movements that did not occur in "ordinary mechanics."

130 To Schwarzschild, January 30, 1904. SUB, Schwarzschild 743. Also in ASWB I.

131 Sommerfeld, *Elektronentheorie I-II*, 1904; Sommerfeld, *Elektronentheorie III*, 1905.

132 To Schwarzschild, June 12. 1904. SUB, Schwarzschild 743. Also in ASWB I.

133 Autobiographische Skizze, ASGS IV, pp. 673–679.

134 Darrigol, *Origins*, 1996; Darrigol, *Electrodynamics*, 2000; Janssen/Mecklenburg: *Mechanics*, 2007.

135 To Johanna, August 10, 1904.

For example, if the "latent energy of the rotational motion" is converted to "kinetic energy of the translational motion," the electron could be ejected from an atom. "Nothing stands in the way of giving rotational energy to the electrons that are part of the structure of the radium atom," he wrote, applying this conception to the puzzling phenomena of radioactivity. Then this transformation of rotational to translational energy is perhaps the explanation of "Becquerel beta-rays."[136]

The mysterious radioactive rays were in these years the subject of intense research.[137] In 1903, Antoine Henri Becquerel (1852–1908), Marie Curie (1867–1934), and Pierre Curie (1859–1906) were awarded the Nobel Prize for their discovery of radioactivity. The nature of these radioactive rays, however, was unknown; indisputable only was that there were three distinct forms, labeled with the first three letters of the Greek alphabet, and that they must originate in the interior of the atom. Since it was certain that the relatively well-researched cathode rays involved electrons, presumably electrons were at work in radioactivity as well. The Tübingen physicist Friedrich Paschen (1865–1947) represented the view that electrons were involved not only in the case of Becquerel β-rays, which can be deflected in electrical and magnetic fields, but also in the case of γ-rays that are apparently non-deflectable.[138] Since the sample container in which his γ-ray source was enclosed was consistently positively charged, it suggested itself that they consisted of negatively charged particles flying away from it. That they could not be deflected could, according to Paschen's conception, be ascribed to the fact that they were passing by the target field at an extremely high velocity—perhaps even faster than the speed of light. To be sure, in all the electron theories particular importance was given to the relationship between the speed of electrons and the speed of light, so that the speed of light assumed a special role. But in 1904 it was not yet clear that the speed of light represented an insurmountable limit to velocity. For Sommerfeld, "surpassing the speed of light in spatial charge distribution is entirely possible," if only under somewhat extraordinary circumstances.[139]

In the case of γ-rays, then, are we dealing with electrons with a velocity above the speed of light? "Lately, I have primarily been considering the Paschen γ-rays," Sommerfeld wrote Wilhelm Wien in February 1904.[140] Clearly, he saw therein a possible application of his electron theory. Following his lecture at the Heidelberg mathematical congress, he took a detour to Tübingen to observe Paschen's γ-ray experiments in person.[141] His "super-mechanics" provided for the case in which electrons in their own field self-accelerated. "It seems appropriate," Sommerfeld argued in connection with an equation for the force exerted on the electron, "to consider the possibility of a force-free, quasi-accelerated motion with a speed above

136 Sommerfeld, *Mechanik der Elektronen*, 1905.
137 Hughes, *Radioactivity*, 2003; Malley, *Radioactivity*, 2011.
138 Paschen, *Strahlen*, 1904; Paschen, *Kathodenstrahlen*, 1904; Wheaton, *Tiger*, 1983, pp. 61–65.
139 Sommerfeld, *Elektronentheorie II*, 1904, p. 384.
140 To W. Wien, February 18, 1904. DMA, NL 56, 010.
141 To Johanna, August 12, 1904.

the speed of light." The question was "urgent with respect to the theory of the γ-rays of radium," but he did not yet want to commit himself to an answer. Nonetheless, it seemed to him "highly probable that this motion is of a very high velocity and self-accelerating."[142] At first Paschen thought this view so ill conceived that he "would not believe [it] until convinced by unimpeachable experiments."[143] Needless to say, in publishing these experiments, he would properly acknowledge Sommerfeld's "prediction of self-acceleration," he wrote, thanking the theoretician of "super-mechanics" for the "interesting discussions of the possibility of the speed of light."[144] When the experiments did not contradict this interpretation, Sommerfeld too became more confident in the "absurd results" of his theory, as he wrote Lorentz. He placed such importance on the matter that he asked Lorentz to present a short dissertation on it to the Amsterdam Academy. His assistant, Debye, "who understands Dutch as well as he does electron theory," undertook the translation of the manuscript into Dutch.[145] Two days later, Sommerfeld apologized to Klein for having to postpone "the gyroscope," because "the most absurd implications of my super-mechanics of electrons have recently been confirmed in γ-rays. . . So for the moment I cannot tear myself away from the electrons."[146]

The euphoria did not last long, however. Paschen was no longer confident of his experimental results.[147] In the short run, he did indeed believe he had demonstrated experimentally the self-acceleration of the γ-electron, but this confidence did not last. The experimental results were obtained from photographic plates, blackened to varying degrees, placed at varying distances from the radiating source, and exposed for varying time periods to radiation shielded by a platinum plate.[148] Now Sommerfeld was no longer quite sure of his super-mechanics. Paschen had indeed reported that he had "actually" confirmed "velocity above the speed of light," but—as Sommerfeld wrote Lorentz one week after Paschen's report—in his "Dutch memorandum," he had "omitted this because it did not seem quite certain to me."[149] Paschen, too, delayed the planned publication. "There is yet another reason I would like to postpone further publication on γ-rays," he wrote in apology to Sommerfeld after the Christmas holidays. "A number of writers have expressed reservations about the cathode-ray nature of the γ-rays." The positive charge of the radium container could also be a secondary effect of the radiation. "The γ-rays themselves could be X-rays."[150]

142 Sommerfeld, *Elektronentheorie II*, 1904, pp. 408–409.
143 From Paschen, October 23, 1904. DMA, HS 1977- 28/A,253.
144 From Paschen, October 26, 1904. DMA, HS 1977- 28/A,253. Also in ASWB I.
145 To Lorentz, November 6, 1904. RANH, Lorentz inv.nr. 74. Sommerfeld, *Afleiding*, 1904.
146 To Klein, November 8, 1904. SUB, Klein 11. Also in ASWB I.
147 From Paschen, November 14, 1904. DMA, HS 1977- 28/A,253.
148 From Paschen, December 6, 1904. DMA, HS 1977-28/A,253.
149 To Lorentz, December 14, 1904. RANH, Lorentz inv.nr. 74.
150 From Paschen, January 11, 1905. DMA, HS 1977-28/A,253. Also in ASWB I.

In the months that followed, it emerged that the critics had been correct. The γ-rays proved to be radiation without its own electrical charge. The charge of the test container observed by Paschen came from the ionization caused when the γ-rays left the radium source. [151] In February 1905, when Sommerfeld completed the third part of his paper on electron theory, he revised his earlier conception of electrons with a velocity above the speed of light. What before had garnered his "super-mechanics" so much attention, he now called "an unjustified extrapolation." His original conception of the genesis of the γ-rays as a sudden production of electrons with a velocity above the speed of light was thus invalid: "The γ-rays cannot be charges moved with a velocity above the speed of light, since such motions in a force-free field are totally impossible. Even the assumption that the γ-rays were electrons with the speed of light is hardly tenable."[152] When Paschen was confronted with this turnabout, he wrote Sommerfeld soberly that from all this he had "once again learned the powerful lesson that one must not assert something one cannot prove indisputably."[153]

Although Sommerfeld had been unable to explain the nature of γ-rays with his "super-mechanics," he nevertheless—his position as a professor at a technical university notwithstanding—made a name for himself among physicists with his effort. In Göttingen, Hilbert organized a seminar in the summer semester of 1905 on the subject of electron theory, in which Hermann Minkowski (1864–1909), Wiechert, and Herglotz took part as co-organizers, and among whose participants were Max Born (1882–1909) and Max Laue (1879–1960), who had just completed their studies. For 2 weeks, Sommerfeld's treatise on electron theory was on the program of the seminar.[154] At the end of September 1905, Sommerfeld himself once more addressed the Natural Scientists Congress in Meran on matters of electron theory. Motions of a velocity above the speed of light, he now thought, are "not physically realizable at all," but could also not entirely be ruled out. The value of his theory lay in its indicating what strange assumptions were required to allow one way or another for the physical possibility of velocity above the speed of light.[155]

5.7 "In Truth I Am No Engineering Professor; I Am a Physicist"

The electron theory also motivated Sommerfeld to take up other pressing questions of the physics of his time. Prominent among these was the question of the nature of X-rays. According to the conception of Wiechert and others, X-rays were electromagnetic impulses generated by the impact of electrons on the anticathode of an

151 Wheaton, *Tiger*, 1983, p. 65.
152 Sommerfeld, *Elektronentheorie III*, 1905, pp. 202–204.
153 From Paschen, June 12, 1905. DMA, HS 1977-28/A,253. Also in ASWB I.
154 Pyenson, *Physics*, 1979.
155 Sommerfeld, *Bemerkungen zur Elektronentheorie*, 1906.

X-ray tube and spreading at the speed of light through the ether. Sommerfeld adopted this conception when in 1900 he had calculated the diffraction of such impulses.[156] At that time, though, he had not dealt with the impact of the electron thought to generate such an impulse, but only with its widening behind a slit. The generation of an X-ray impulse at impact fell into the category of electron theory. Actually, he had intended, Sommerfeld wrote Wilhelm Wien in May 1905, "to write up a detailed treatment of the energy of X-radiation from my equations." He abandoned this project, however.[157]

A year before, Wien had stirred Sommerfeld's interest in these questions with a paper "On the Energy of the Cathode Rays in Relation to the Energy of X-rays and Secondary Rays," intended for the Wüllner Festschrift, edited by Sommerfeld. In it, Wien reported experiments on the measurement of energy at the impact of an electron, from which he drew conclusions about the width of an X-ray impulse.[158] In the case of the Wüllner Festschrift, as with the encyclopedia article by Lorentz, Sommerfeld did not eschew a thorough-going discussion well beyond merely editorial considerations of the problems addressed by Wien concerning the generation of X-rays. "The preceding," he wrote apologetically at the end of a ten-page letter, "has not been written in my capacity as 'editor,' who would have printed your paper unexamined, but rather as an 'electronics specialist.'"[159]

Thereafter, he continued to discuss these questions with Wien. On a visit to Würzburg with Wien, Sommerfeld reported on one occasion to Lorentz, they had "chatted extensively about electrons. . . We would have been happy to have had you there to referee a number of questions."[160] Sommerfeld even made X-rays the subject of a special lecture, something rather unusual for a professor of mechanics at a technical university. "Pursuant to your request," he wrote in April 1905 to Wien, "I have carefully considered the energy equation of X-radiation, and even lectured on it under the rubric of 'electron theory.'"[161] He went so far as to carry out experiments with Debye on the spatial intensity distribution of X-rays, since according to his theory there should be a relation between the deceleration of the electron and the energy emitted in varying directions. "I have myself made such qualitative experiments with the help of my talented assistant," he confided to Wien. The results of these experiments may have been one reason he excluded the generation of X-rays from his publication on electron theory, for they had demonstrated "virtually no interdependency whatsoever between the intensity of the X-radiation on the angle between an X-ray and the incident cathode ray." He concluded that on impact the electron describes a zigzag course, so that its direction in the process of

156 See Chap. 4 and Wheaton, *Tiger*, 1983, pp. 33–40.
157 To W. Wien, May 13, 1905. DMA, NL 56, 010. Also in ASWB I.
158 Wien, *Kathodenstrahlen*, 1905; Wheaton, *Tiger*, 1983, pp. 110–113.
159 To W. Wien, February 18, 1904. DMA, NL 56, 010. Also in ASWB I.
160 To Lorentz, May 29, 1904. RANH, Lorentz, inv. nr. 74. Also in ASWB I.
161 To W. Wien, April 15, 1904. DMA, NL 56, 010. Also in ASWB I.

deceleration is constantly changing. "The dependency of the direction can be extrapolated from the mean of the zig-zag course." But it was clear to him that this was extremely speculative. "It is truly scandalous that 10 years after Röntgen's discovery we still do not know what is actually going on with X-rays."[162]

While Sommerfeld was explaining his ideas about X-rays in lengthy letters to Wien, the latter received an inquiry from Munich as to whether he could recommend Sommerfeld for an appointment to the chair in theoretical physics originally created for Boltzmann. It had been vacant since 1894, when Boltzmann after only 4 years left Munich. In response to a similar inquiry from Leipzig (where Boltzmann had vacated his position in 1902 after only a short time, as he had done earlier at Munich), Wien had written in his referee's report that Sommerfeld would be "entirely incapable . . . of directing an institute [for theoretical physics], and of advising students in work in physics."[163] The discussions of electron theory and X-rays must have changed Wien's assessment, because the Munich Appointments Commission set down in its minutes that Sommerfeld had been brought to their attention by "very eminent theoretical physicists such as Boltzmann, Lorentz, and Wien." He was "described as a personable colleague and an excellent teacher." Though Sommerfeld's teaching and research at the Technical University at Aachen had been more in the realm of engineering than of theoretical physics, most recently he had devoted himself "particularly to electron theory," demonstrating thereby his affinity for the "circle of interests of theoretical physics." Röntgen, who had been a professor of physics at the Ludwig-Maximilians University in Munich since 1900, had long but vainly been at pains to fill the vacant chair. Only after declining an appointment as president of the Imperial Institute of Physics and Technology (*Physikalisch Technische Reichsanstalt*) in Berlin did he receive approval from the Bavarian Minister of Culture to fill the vacancy. Röntgen's preferred candidate was his fellow countryman Lorentz, considered the leading authority among "electronics experts," but loyal to his university at Leiden, Lorentz declined the offer, so that Munich had to go in search of another candidate.[164]

Sommerfeld knew nothing of these doings behind the scenes. Only when he received a telegram from Röntgen requesting that he send his *curriculum vitae* and a list of publications[165] did it become clear to him that apparently "something was going on in Munich." Not without reason did he assume Wilhelm Wien to be "the prime cause of this phenomenon." The prospect, as Boltzmann's successor in Munich, of making the physics of X-rays a major focus of his research fascinated

162 To W. Wien, May 13, 1905. DMA, NL 56, 010. Also in ASWB I.

163 Draft letter to Otto Wiener, undated. DMA, NL 56. Nr. 5882. Also in ASWB I, pp. 156–157.

164 Report of the Appointment Commission, Philosophical Faculty, 2. Section, July 21, 1905. UAM, Personnel folder Sommerfeld, E-II-N. Eckert/Pricha, *Boltzmann*, 1984; Jungnickel/McCormmach, *Mastery*, 1990, pp. 274–278.

165 From Röntgen, June 29, 1905. DMA, HS 1977-28/A,288. Also in ASWB I.

him. Nor would he limit himself therein to the merely theoretical. "Would I have the possibility of occasional experimental work at the Röntgen institute, or would I have to be very cautious with such requests?" he wanted to know from Wien. "Would I have an assistant at Munich or be able to bring one along?" On the other hand, he had grown quite used to life at Aachen and was not uncomfortable in the engineering-dominated environment at the Technical University. "It would be harder for me than one would imagine to leave Aachen."[166]

But Röntgen's inquiry was far from having concluded the situation of the Munich appointment. That Sommerfeld was being recommended for the Munich position by such renowned physicists as Boltzmann, Lorentz, and Wien did not yet mean that he would land atop the list of candidates. Within the Appointment Commission, consisting of the mathematicians Ferdinand Lindemann and Aurel Voss (1845–1931), the astronomer Hugo von Seeliger (1849–1931), and Röntgen, Lindemann objected that some of Sommerfeld's papers were "from the mathematical point of view at least, not unimpeachable."[167] Lindemann criticized, for example, a derivation in Sommerfeld's electron theory that was in his view based on an unreliable passage to the limit. In Sommerfeld's paper on waves along wires from the year 1899, he also found things to object to.[168] Sommerfeld went to great lengths to counter Lindemann's objections. "I would find it very sad if my old teacher should think I misuse the mathematical talents he passed on to me in those days," he wrote his old doctoral advisor at the end of an eight-page letter.[169] Lindemann was not to be placated, however. Since the other members of the Appointment Commission spoke in favor of Sommerfeld, however, Lindemann stood alone in opposition. Sommerfeld remained on the short list; he was placed second (*secundo loco*) on the list of candidates, after Cohn and Wiechert, who were tied for first place (*primo loco* and *ex aequo*).[170]

This discussion of his person and his placement on the candidate list did not remain concealed from Sommerfeld. He thanked Wien for having put in a good word for him "with the contrarian Lindemann" and remained calm. Wiechert was, he said, "superior to me in rather many ways," and even were Wiechert to decline the appointment, he assessed his own chances as not particularly great.[171] Not until July 1906 was there clarity as to the outcome of the Munich appointment. The offer was, as expected, extended to Wiechert; he however preferred to remain at

166 To W. Wien, July 4, 1905. DMA, NL 56, 010. Also in ASWB I.

167 Report of the Appointment Commission, Philosophical Faculty, 2. Section, July 21, 1905. UAM, Personnel folder Sommerfeld, E-II-N.

168 From Lindemann, July 5, 1905. DMA, HS 1977-28/A,203. Also in ASWB I.

169 To Lindemann, July, 7, 1905. Munich, DMA.

170 Report of the Appointment Commission, Philosophical Faculty, 2. Section, July 21, 1905. UAM, Personnel folder Sommerfeld, E-II-N.

171 To W. Wien, November 5, 1905. DMA, NL 56. Also in ASWB I.

Göttingen. Sommerfeld heard about this from Wiechert's mother, as he relayed to Wien. "So if my friend Lindemann does not mount a vigorous counter-offensive, it is not unlikely under present circumstances that I will be made the offer, and will say yes."[172] Cohn was passed over presumably because of his Jewish origins and even later never received an appointment to a full professorship.[173] On July 17, Röntgen telegraphed Sommerfeld that his appointment was imminent.[174] On July 23, the Bavarian Ministry of Culture, Education, and Church Affairs sent the official document of appointment.[175] A few days later, Sommerfeld requested the Prussian Ministry of Culture, Education, and Church Affairs to release him from his Aachen position. His rationale was that he regarded theoretical physics as his "actual professional area. . . Financially, it seems, I will be in no better a position than at Aachen."[176]

Anyone who knew Sommerfeld knew he was here pursuing not just an appointment, but a calling: "There, you will be entering the house of physics, mother of all sciences, in whose lap you will doubtless be happy," Hilbert congratulated him.[177] Wien, too, experienced Sommerfeld in these days as "delighted with his appointment at Munich," as he wrote his mother after a meeting with him. "He is very much looking forward to his new job."[178] An Aachen colleague from the Department of Mechanical Engineering certainly felt a "poignant melancholy" at the thought that Sommerfeld should turn his back on technological mechanics, but he understood nonetheless that this move to theoretical physics "corresponds to your true motivations." Hadn't Sommerfeld himself once said to him: "In truth I am no engineering professor; I am a physicist!"[179]

172 To W. Wien, July 5, 1906. DMA, NL 56, 010. Also in ASWB I.
173 Jungnickel/McCormmach, *Mastery*, 1990, p. 278.
174 From Röntgen, July 17, 1906. DMA, HS 1977-28/A,288. Also in ASWB I.
175 From Anton von Wehner, July 23, 1906. DMA, NL 89, 019, folder 5,2.
176 To Naumann, July 29, 1906. GSA, I. HA. Rep. 121 D II, Sect. 6 Nr. 10.
177 From Hilbert, July 29, 1906. DMA, HS 1977-28/A,141. Also in ASWB I.
178 W. Wien to his mother, July 31, 1906. DMA, NL 56, Nr. 5088.
179 From Rummel, August 3, 1906. DMA, NL 89, 012. Also in ASWB I.

6 Munich

It was not just Boltzmann's aura surrounding the chair in theoretical physics at the University of Munich nor the challenge of collaborating as theoretician beside Röntgen in research on X-rays that made the position at Munich so attractive for Sommerfeld. Munich, "City of Museums," was known in these years as a Mecca of art and joie de vivre.[1] Sommerfeld had gotten his first "heavenly impression" at the conclusion of his studies when he discovered the amazing Munich painting and sculpture collections, as well as his taste for Bavarian beer in various pubs. It was not by chance that he chose Munich as his honeymoon spot, and Johanna gladly let herself be infected by her husband's enthusiasm for this city as well.

At the turn of the century, Munich was a city in transition, where the traditional and the modern were tightly interwoven. Literati of the bohemian world such as that terror of the bourgeois, Erich Mühsam (1878–1934), were equally at home in the Munich suburb of Schwabing as the "literary prince" of the bourgeois, Thomas Mann (1875–1955). The "prince of painters," Franz von Stuck (1849–1940), a follower of Sommerfeld's favorite artist, Arnold Böcklin, was the talk of the art world, as were his students Paul Klee (1849–1940) and Wassily Kandinsky (1866–1944) and their "Blue Rider" school. Along with art, the city's much admired attractions also comprised technology. In 1903, Oskar von Miller (1855–1934) had established a "German museum of masterworks of the natural sciences and technology," which became world famous shortly after its opening. The Würm baths in Schwabing, whose "basins, grottos, and springs" had so enraptured Sommerfeld on his visit to Munich in 1892, belonged to an engineer named August Ungerer (1860–1921), remembered as a pioneer of streetcar technology in Munich history. Electricity, too, had been introduced to Munich, first in the transition from horse-drawn to electrically powered streetcars, then in public illumination, and finally in private households. The authoritarian state saw itself made the target of scorn and derision in the Schwabing satirical journal "Simplizissimus" and in political cabarets. In Schwabing, one could see art, science, technology, old and new, bohemian and bourgeois living side by side.

These things may well have weighed in Sommerfeld's decision to take up residence in Schwabing in 1906. Or perhaps he was guided only by practical considerations. As a family man with three young children–in 1908 a fourth child, Eckart, was born–and as a university professor, he needed a spacious house or apartment, not too far from the university, where visits from colleagues and students could be combined with promenades in pleasant surroundings. "I have rented Leopoldstrasse #87, quite at the end, near the Ungererbad; a beautiful apartment, though up three

1 Bauer, *Prinzregentenzeit*, 1988; Prinz/Kraus, *München*, 1988; Bauer, *Geschichte*, 2008.

M. Eckert, *Arnold Sommerfeld: Science, Life and Turbulent Times 1868-1951*,
DOI 10.1007/978-1-4614-7461-6_6,

flights," Sommerfeld wrote describing his new home.[2] What today appears as a residential area of the city was at that time the edge of town.[3] It was only a few minutes' walk from the English Garden with its idyllic Kleinhesseloher Lake, an ideal destination for walks with professors and students, as formerly Hainberg had been in Göttingen. To the north stretched a more or less open tract that was already targeted for future development but, in 1906, still functioned as a playground for children. The apartment comprised several large rooms and allowed for conversion to a hospitable space in which to entertain students and colleagues. At "semester parties," recalled Paul Ewald (1888–1985), a student of Sommerfeld's of those years, "the double doors between the three front rooms were opened wide so that 15–20 guests did not feel cramped in the large space."[4] On foot, Sommerfeld could reach the university on the opposite side of Schwabing in 20 min. Scientific and family lives were thus in harmony here in Munich, too. Although his departure from Aachen was not exactly easy for him, Sommerfeld wanted to create in Munich for himself and his family a lasting home. The picturesque landscape of the foothills of the Alps with its Bavarian lakes provided recreational weekend excursions and was a further reason the transition seemed a positive turning point in his life.

6.1 Academic Traditions

As Boltzmann's successor Sommerfeld became not just full professor of theoretical physics at the University of Munich. "The professorship is also tied to the position of Curator (of the board) of the state mathematical-physical collection," stated the document of appointment from the Bavarian Minister of Culture.[5] There was a long tradition behind this position.[6] In 1827, after the transfer of the university from Landshut to Munich, a "general conservatory" for the care and preservation of the scientific collections of the Bavarian state was established. It comprised mineralogical, zoological, and ethnographic collections, a cabinet of physical and mathematical instruments, the observatory at Bogenhausen, and other objects that had been confiscated from monastic collections in the secularization and, thereupon, remained in the custody of the Bavarian Academy of Sciences, the university, or the royal court. For the care of these collections, governmentally supported curatorial positions were established which as a rule were filled by the occupant of the university chair in each respective field. The Munich physics professors and Academy members Carl August Steinheil (1801–1870), Georg Simon Ohm (1789–1854), the mathematician and astronomer Philipp Ludwig von Seidel (1821–1896), and most

2 To W. Wien, September 12, 1906. DMA, NL 56, 010.
3 München, *München - wie geplant*, 2004, p. 86.
4 Ewald, *Arnold Sommerfeld als Mensch*, 1969, p. 11.
5 From Wehner, July 23, 1906. DMA, NL 89, 019, folder 5,2.
6 Bachmann, *Attribute*, 1966.

recently Boltzmann had been responsible for the "mathematical-physical collection." The collection included valuable instruments that were housed in special rooms at the Bavarian Academy of Sciences but that towards the end of the nineteenth century were hardly used any longer for the purposes they were originally intended for. By the time of Boltzmann's tenure, the duties of Curator had come to consist in no more than compiling inventory lists. It furnished him a sizeable supplementary income, however, as well as the use of rooms at the Academy. In addition, the services of an assistant and a machinist were at his disposal.[7]

If the collection in Boltzmann's day was more a museum than an institution for the use of teaching and research, this was all the more true in 1906 when Sommerfeld was entrusted with the role of Curator. Even before he arrived in Munich, a number of historically important instruments from the collection had been handed over to Oskar von Miller for his new museum, without abolishing the institution as such. "The personnel of the mathematical-physical collection consists of an assistant, with a starting salary of around 1,200 Marks, and a machinist who is also an attendant, with a starting salary of 1,500 Marks, plus 255 Marks supplemental; the actual budget of the collection comes to 1,800 Marks annually," Sommerfeld was informed in his document of appointment. His service as Curator increased his own annual salary by 2,000 Marks, bringing it to 6,900 Marks.[8]

Through additional negotiations, Sommerfeld managed to obtain further increases. He was "granted yet another 500 Marks to my salary," he wrote Willy Wien.[9] Thus, he was in a considerably better financial position than at Aachen, where his annual salary had been just 6,000 Marks. At the physics institute of the University, led by Röntgen as Director, he had only one workroom, but it was made clear at his appointment that this was only temporary. "In the new building project for expansion of the University, which is to begin shortly, dedicated rooms (of 42, 42, 35, and 57 sq. meters, as well as an adjoining lecture hall of 50 sq. meters) have been planned for an institute of theoretical physics."[10]

It was another 3 years, however, before construction on the expansion of the University on the Amalienstrasse was ready for occupancy, and he could, as Director of his own institute, feel on a par with Röntgen. In the meantime, he and Debye, whom he had brought along from Aachen as assistant, had to be content with workplaces that left much to be desired. Aside from the workroom at Röntgen's institute, really only the rooms of the collection at the Academy stood at his disposal, and those were housed half an hour away in a former Jesuit school on Neuhauser Strasse in the old town. "One reached them up a wide, flat wooden staircase with its nicely turned banister posts, past the entrance to a collection of stuffed animals on the second floor, which partly spilled out onto the spacious

7 Koch, *Konservatorenamt*, 1967; Litten, *Trennung*, 1992.
8 From Wehner, July 23, 1906. DMA, NL 89, 019, folder 5,2.
9 To W. Wien, September 12, 1906. DMA, NL 56, 010.
10 From Wehner, July 23, 1906. DMA, NL 89, 019, folder 5,2.

landing," recalled Ewald. "One of the rooms was furnished as a lecture hall for 20–25 students, with benches, desks, and a large blackboard. Next to this sat Debye, and the larger room was the Professor's."[11]

Sommerfeld also faced other academic traditions, which demanded a certain degree of patience. Although the chair of theoretical physics had stood vacant since 1894, the lecture routine had been only nominally affected. The main lectures in theoretical physics were given by two associate professors. One of these, Leo Graetz (1856–1941), had climbed the first rungs of his academic career ladder as lecturer at the University of Munich in the 1880s together with Max Planck, but had not advanced beyond associate professor even though in the area of electricity ("Graetz Rectifier Circuit"), and as a textbook author, his field owed him much. Röntgen did not regard him as the theoretician to whom to entrust the Boltzmann chair, which embittered Graetz and led to increasing distance on Röntgen's part. Sommerfeld's appointment came as a further blow to Graetz. He was thereby "somewhat demoted," Röntgen conceded, "and whatever one might think of him otherwise, that is not a pleasant experience for someone his age." At the age of 50, Graetz could no longer hope for a full professorship at another university. "To sweeten the change in his status somewhat, it was proposed that the ministry confer on him the title and rank (though not the full rights) of full professor of physics."[12] To Graetz, this was no more than a sop. His strained relationship with Röntgen was projected onto Sommerfeld as well. Sommerfeld made it clear that nothing much had changed in this regard when years later he wrote to Debye, congratulating him on an appointment that would put him in close contact with an unloved colleague: "If he is your only Graetz, count yourself lucky."[13] In turn, to Graetz, Sommerfeld was an upstart who should have stayed in mathematics. "There is a class of theoretical physicists who are skillful calculators, who understand how to apply their self-created equations to special, isolated problems," he wrote in a 1926 newspaper article, "Physics of the Last Hundred Years—its Practice in Munich." In thinly disguised allusion to Sommerfeld, he wrote that "there are mathematicians in physicists' clothing; typically, they are thought good physicists by mathematicians, and good mathematicians by physicists."[14]

Together with Graetz, Arthur Korn (1870–1945) had carried the whole burden of lecturing in theoretical physics, since 1895 as lecturer and since 1903 as associate professor. He too felt passed over by Sommerfeld's appointment. Korn himself instigated his dismissal, Röntgen wrote a colleague, "ostensibly because most of his students were drawn away by Sommerfeld's appointment, in reality, though, to get a promotion." Korn was a Jew and financially independent, which triggered

11 Ewald, *Erinnerungen*, 1968, pp. 538–539.

12 Röntgen to Zehnder, December 27, 1906. In: Zehnder, *Röntgen*, 1935, p. 112; Jungnickel/McCormmach, *Mastery*, 1990, pp. 278–281.

13 To Debye, August 6, 1920. MPGA, Abt. III, Rep. 19 (Debye). Also in ASWB II.

14 *Münchener Neueste Nachrichten*, November 26, 27, 1926.

Röntgen's anti-Semitic prejudices: "the stinking rich can engage in such ploys, and do so if they have the necessary Semitic impudence."[15] Röntgen's statement indicates what hurdles were placed in the way of the careers of Jewish lecturers and associate professors. Unlike Graetz, Korn did not remain in Munich, but went instead to Berlin—not without reflecting bitterly on his Munich years in retrospect: "The fact that I was passed over left me dumbfounded," he told the *Berliner Tageblatt* in 1909. "Undoubtedly the faculty abandoned me because of Professor Röntgen. He simply tyrannized the faculty in the whole affair."[16]

Aside from Graetz and Korn, there was another lecturer, Wilhelm Donle (1862–1926), who lectured on theoretical physics. Donle had done his habilitation under Röntgen's predecessor and had become a teacher of mathematics and physics at the Bavarian Cadet Corps, the officers' school of the Bavarian army. His lectures at the university dealt with specialty areas and were considered supplementary to the main lectures of Graetz and Korn. In 1907 he became full professor at the Royal Bavarian Artillery and Engineering School, which did not, however, interfere with his special lectures at the university, with which he retained his association.[17]

The fact that the Boltzmann chair stood vacant so long did not mean, then, that there was no theoretical physics at the University of Munich before Sommerfeld's appointment on October 1, 1906. For the winter semester 1906/1907, the lecture catalogue announced lectures by Graetz and Korn on analytic mechanics (5 h per week) and potential theory and spherical harmonics (4 h per week), as well as a special lecture by Donle on the electromagnetic theory of light (2 h).[18] In this semester, Sommerfeld lectured on "Maxwellian Theory and Electron Theory."[19] Though it was not announced in the lecture catalogue, this did not materially detract from the superficial appearance of theoretical physics at Munich. Even in the following years, Sommerfeld's lectures did not particularly stand out, but rather appeared alongside those of Graetz and Donle as replacements for those of Korn who had turned his back on Munich in 1908.

The offerings in the neighboring field of mathematics were also quite respectable. In the lectures of the three full professors, Lindemann, Alfred Pringsheim (1850–1941), and Aurel Voss, as well as the two associate professors, Karl Doehlemann (1864–1926) and Eduard von Weber (1870–1934), nearly all branches of mathematics were covered; anyone seeking a stronger emphasis on applied mathematics could find the appropriate supplementary curriculum in the lectures of Dyck, Sebastian Finsterwalder (1862–1951), Anton von Braunmühl (1853–1908), and Wilhelm Kutta

15 Röntgen to Zehnder, December 27, 1906. In: Zehnder, *Röntgen*, 1935, p. 112.

16 Litten, *Korn-Röntgen-Affäre*, 1993, p. 46.

17 Vorlesungsverzeichnisse der Ludwig-Maximilians-Universität München, http://epub.ub.uni-muenchen.de/view/lmu/vlverz=5F04.html (30 January 2013); on Donle, see http://litten.de/fulltext/donle.htm (30 January 2013).

18 http://epub.ub.uni-muenchen.de/1124/1/vvz_lmu_1906-07_wise.pdf (30 January 2013).

19 Lecture draft in DMA, NL 089/028.

(1867–1944) at the neighboring Technical University.[20] For a student deciding where to study who compared Munich's lecture offerings with those of other universities, Sommerfeld's presence certainly did not yet present grounds for favoring Munich over other cities. To a physics student of those years, Sommerfeld's name meant no more than those of Graetz or Korn. Although there was not yet a separate institute for theoretical physics at Munich, this was not obviously a shortcoming in its instructional offerings. Most students pursuing the relevant offerings in mathematics and physics graduated with a teaching diploma. Separate institutes of theoretical physics existed in 1906 only at the Universities of Berlin, Göttingen, Königsberg, and Leipzig. But so long as theoretical physics was a field lecturers and associate professors pursued only tangentially on their academic career path, this was no reason for a prospective physicist to favor one city over another.[21]

6.2 Quarrel over Electron Theory

With Sommerfeld's appointment, however, Röntgen saw already in 1906 the dawn of a new era in Munich physics. "In Sommerfeld I believe I have found a good colleague and fellow worker," he wrote to a collegial friend. "I can once more discuss physical things in a stimulating way, and his audiences are very interested in his lectures on Maxwellian and electron theory. We don't always agree on the questions that arise, but that isn't what matters after all. On the contrary, that ought to advance both the subject and our understanding."[22] For his part, Sommerfeld was "very happy" over his Munich environment; Röntgen was "scientifically and in his official capacity as friendly and obliging as can be."[23] He did not mention the animosity with which Korn and Graetz had reacted to his appointment.

Greater problems were caused him by his former doctoral advisor. Lindemann publicized his aversion to Sommerfeld's electron theory (which already caused trouble inside the Appointment Commission) in the form of a hundred-page dissertation for the Bavarian Academy. "Now Lindemann has also deposited his electrons at the Academy," Sommerfeld wrote Wien in January 1907. "According to him, everything is wrong, not just Lorentz—you, I, Abraham, even the Maxwell equations are mathematically self-contradictory."[24] Even if it was obvious to those knowledgeable that in this attempt to expose Sommerfeld's alleged errors Lindemann himself had gone awry, Sommerfeld could not take the situation lightly. Once the quarrel had reached the forum of the Academy, on Röntgen's

20 Toepell, *Mathematiker*, 1996, Chaps. 6 and 7; Hashagen, *Walther von Dyk*, 2003, Chap. 14.3.

21 Jungnickel/McCormmach, *Mastery*, 1990, p. 287.

22 Röntgen to Zehnder, December 27, 1906. In: Zehnder, *Röntgen*, 1935, p. 112.

23 To W. Wien, November 23, 1906. DMA, NL 56, 010. Also in ASWB I.

24 To W. Wien, January 15, 1907. DMA, NL 56, 010. Also in ASWB I. Lindemann, *Bewegung der Elektronen. Erster Teil*, 1907; for a comprehensive exposition, see Eckert, *Mathematik*, 1997.

advice he responded with a rebuttal which he also submitted to the Academy for publication. "Yes, it's good that he's publicly refuted," Abraham thought as well. "For there's nothing so dumb it won't find its audience."[25] The quarrel was lastingly embodied in a back and forth of rebuttal and counter-rebuttal in the minutes of the Academy.[26] "I'm minded not to respond to his most recent electron pap," Sommerfeld wrote unnerved to Willy Wien after 2 years of fruitless discussion.[27]

From a scientific standpoint, the quarrel had no detrimental consequences for Sommerfeld. Lorentz assured him that "not for a moment did he fear" the electron theory was as erroneous as Lindemann maintained. For that reason he hadn't thought it worth the trouble to write a rebuttal from his standpoint. "Had I known that the quarrel was a sequel to your appointment, though, it would have occurred to me that you might have liked for me to intervene. Probably that has now become superfluous; you will stand your ground splendidly without any help."[28] Felix Klein, whom Lindemann esteemed as the great mathematician who in 1882 had proved the transcendence of π, wrote to Sommerfeld: "Lindemann grieves me exceedingly. The upshot is this, that Lindemann, lacking suitable physical experience, relies on calculation alone, and in consequence of cumulative errors of calculation, there goes astray! For someone so naturally gifted, a tragic end."[29]

There were certainly grounds on which to criticize the electron theories in their various manifestations. Those lay not in the area of mathematics, however; rather they concerned the fundamental physical assumptions made in each case and the conclusions drawn from them. In all the electron theories, the mass of an electron was a quantity dependent on velocity; there was, however, no agreement as to whether the electron appeared as a particle laden with mass only in motion through its self-generated field (such mass was designated "apparent") or whether at rest too it possessed a "ponderable" or nonzero mass. The theories also provided differing answers to the question whether an electron is rigid or changes its form with increasing velocity. According to Lorentz's electron theory, it was supposed to assume the shape of an ellipsoid as it neared the speed of light—this was called a "deformable" electron. In opposition to this conception, Abraham's theory conjectured a rigid electron that always retained its spherical form. The theories of Abraham and Lorentz also gave differing dependencies on velocity for the mass of an electron in motion.

Which of the theories was correct would, one hoped, be determined through experiments in which the mass of electrons would be derived from the deflection of

25 From Abraham, June 18, 1907. DMA, HS 1977-28/A,1.

26 Sommerfeld, *Bewegung der Elektronen*, 1907; Lindemann, *Bewegung der Elektronen, Zweiter Teil*, 1907; Lindemann, *Elektronentheorie*, 1907; Sommerfeld, *Diskussion*, 1907; Lindemann, *Elektronentheorie II*, 1907.

27 To W. Wien, June 20, 1908. DMA, NL 56, 010. Also in ASWB I.

28 From Lorentz, November 13, 1907. DMA, HS 1977-28/A,208. Also in ASWB I.

29 From Klein, November 20, 1907. DMA, HS 1977-28/A,170.

electron rays in a magnetic field. Walter Kaufmann (1871–1947), an experimental physicist at the University of Bonn, had set himself the goal of making a determination by means of such experiments. Sommerfeld's electron theory was a generalization of Abraham's, and like it was based on the conception of a rigid, spherical electron. "Do you know," he thus wrote enthusiastically to Wien after Kaufmann had reported a new series of measurements, "that the rigid electron has triumphed splendidly? Lorentz's equations for the deformable electron lie altogether outside the margin of error of the observations."[30] Kaufmann's measurements would also have undermined the relativity theory Albert Einstein had published shortly before. In it, the laws Lorentz had formulated for electrons in motion were derived from more general principles; since the same end results were involved, it was thought at first that the Einstein theory was merely another variant of the electron theories—which added urgency to the determination.

Kaufmann's measurements were central also to the discussions at the Natural Scientists Congress of 1906. Planck, too, thought they lay "very much closer to the spherical theory than to relativity theory." The differences between the respective values of the two theories, however, were smaller than those between the theoretical and the experimentally measured values.[31] A lively discussion followed the lecture in which rigorous objectivity was inconsistently observed. Planck was more "sympathetic" to the Lorentz-Einstein theory than to Abraham's, which rested on what he regarded as the far too audacious principles of the electrodynamic worldview, according to which all physical phenomena are ultimately reducible to electrodynamics. Sommerfeld did not want to adopt the "pessimistic viewpoint of Herr Planck" and declared his allegiance to the Abraham faction. "On the question of principles formulated by Herr Planck, I venture to say that the gentlemen under 40 will favor the electrodynamic hypothesis, whereas the gentlemen over 40 will favor the mechanistic-relativistic hypothesis. I favor the electrodynamic."[32] In the minutes, the "amusement" Sommerfeld evoked by this contribution to the discussion was duly noted. He was still 2 years away from his fortieth birthday. Lorentz and Planck were obviously over 40, but Einstein, whom Sommerfeld threw into the mix of older representatives of the "pessimistic viewpoint," was just 27.

So electron theory was a subject of debate also among physicists. But unlike Lindemann, the physicists saw the main problem as lying not in the mathematics but in the unresolved questions of principle and the experimental difficulties of the deflection experiments. Since Röntgen hoped that elucidation of matters of electron theory would also furnish answers to unsolved problems on X-rays, Kaufmann's experiments soon became the subject of discussions between Sommerfeld and Röntgen. "He does not believe that Kaufmann's experiments weighed decidedly against the principle of the relativity movement," Sommerfeld wrote Wien

30 To W. Wien, November 5, 1905. DMA, NL 56, 010.

31 Planck, *Messungen*, 1906.

32 Diskussionsbemerkung zu Planck, *Messungen*, 1906, p. 761.

concerning Röntgen's position. "His observations were not so very precise." These discussions also led him to taking a closer look at relativity theory. "I have now studied Einstein, with whom I am very impressed, and I will shortly lecture on the subject at the Sohncke Colloquium." The colloquium named for Leonhard Sohncke (1842–1897) had a long tradition. It was a joint venture of the Munich physicists of the University and the Technical University, at which Sohncke had worked as full professor of physics. To lecture to this forum on Einstein's Theory of Relativity—1 year after its publication—is evidence of the importance Sommerfeld gave this theory, even if he, like most physicists, saw in it at first just a variant of Lorentz's electron theory. One could establish the inertia "with Einstein as well as with Lorentz," he answered Wien, who found the term "electrodynamic mass" missing in Einstein, one of the fundamental elements of the electromagnetic worldview. But the theory had to be further elaborated "before one could use it to deal with any sort of motion of electrons."[33] It was another year before the physicists learned to understand relativity theory otherwise than through the lens of electron theory. It was hoped primarily that Lorentz as a widely recognized authority would offer an enlightening word in this matter. "Now, though, we are all eagerly awaiting," Sommerfeld wrote in December 1907 to Leiden, "your opinion of the whole batch of Einstein papers." Though he thought Einstein's theory "brilliant," it was also somewhat dogmatic and obscure. "An Englishman would scarcely have presented such a theory; perhaps this is a manifestation of the Semitic abstract conceptual mode, as in the case of Cohn. I hope you succeed in filling out this brilliant conceptual outline with actual physical life."[34]

This statement, like that of Röntgen about Korn, shows that even the representatives of the "exact sciences" were not immune from anti-Semitic prejudices. This prejudice on Sommerfeld's part in Einstein's case, however, went hand in hand with boundless admiration. At the Natural Sciences Congress of 1907, he defended Einstein's theory against erroneous interpretations,[35] and shortly thereafter the two entered into lengthy correspondence that leaves no doubt over their mutual esteem. At this time, Einstein was still working at the patent office in Bern and was without the academic credentials that would automatically have lent him the status of colleague among professors. All the more did he appreciate the esteem he inferred from Sommerfeld's words. "Your letter affords me a rare pleasure; no physicist has approached me at once so openly and so generously," he wrote back. Nevertheless, he rejected Sommerfeld's electromagnetic worldview. "A satisfying theory should in my view be so constituted that the electron appears as solution, in other words that no external fictions are needed in order not to have to assume that its electrical masses are dispersed."[36]

33 To W. Wien, November 23, 1906. DMA, NL 56, 010. Also in ASWB I.

34 To Lorentz, December 26, 1907. RANH, Lorentz, inv.nr. 74. Also in ASWB I.

35 Sommerfeld, *Einwand*, 1907.

36 From Einstein, January 14, 1908. DMA, NL 89, 007. Also in ASWB I.

His correspondence with Einstein led Sommerfeld more and more away from the "fictions" of the electron theory, to which only a few years before he had devoted so much energy. Besides, new deflection experiments by Alfred Bucherer (1863–1927), a physicist at the University of Bonn, demonstrated that the experimental values did after all correspond more closely with the "Lorentz-Einstein" theory, than with the "rigid electron" theory of Abraham and Sommerfeld. "May I congratulate you on the victory of the relativity theory achieved by Bucherer?" Sommerfeld wrote thereafter to Lorentz.[37] "A great deal in Einstein is very clear to me, for instance that the *e* cannot remain alien to the theory," he confessed in a letter to Wien.[38]

Hereby he affirmed the declaration that 3 years earlier had provoked amusement. "I have now been converted to relativity theory," he wrote at the start of 1910 in a letter to Lorentz—now that he had turned 41 years old. "In particular, Minkowski's systematic form and interpretation has facilitated my understanding."[39] In 1907, Hermann Minkowski (1864–1909) had recognized that relativity theory could be especially elegantly presented if the coordinates of space and time were treated on an equal footing. "From now on, space as a separate entity, and time as a separate entity ought to be things entirely of the past, and only a kind of synthesis of the two should retain independence," Minkowski had begun his lecture titled "Space and Time" in 1908 at the Natural Scientists Congress at Cologne.[40] Sommerfeld had talked shop about electron theory with Minkowski earlier; personally, too, they were close.[41] He combined the meeting at Cologne with a hiking trip into the Siebengebirge range with Minkowski.[42] This was their last meeting, for Minkowski died shortly thereafter from complications of an inflamed appendix. "It is another one of those exquisite cruelties of fate to pick out the physically and mentally soundest among us at the height of his greatest achievements," Sommerfeld wrote to Runge at Göttingen, Minkowski's last place of work. "We too were very fond of Minkowski."[43] Thereafter, as it were to preserve his legacy, Sommerfeld made a personal project of elaborating Minkowski's ideas. They were the subject of a special lecture he gave in the winter semester of 1909/1910; later, with a two-part article in the *Annalen der Physik*, he paid tribute to "Minkowski's profound interpretation of space-time."[44]

Max von Laue, who came to Sommerfeld's institute in 1909, took the next step on the path towards an understanding of relativity theory freed from the "fictions" of electron theory. Laue had elaborated relativity theory "successfully, with several briefer papers," as Sommerfeld wrote in his report to the faculty, recommending the

37 To Lorentz, November 16, 1908. RANH, Lorentz, inv.nr. 74. Also in ASWB I.

38 To W. Wien, April 21, 1909. DMA, NL 56, 010. Also in ASWB I.

39 To Lorentz, January 9, 1910. RANH, Lorentz, inv.nr. 74. Also in ASWB I.

40 Minkowski, *Raum*, 1909; Walter, *Minkowski*, 1999; Walter, *World*, 2010.

41 To Johanna, September 22, 1908.

42 To Johanna, September 24, 1908.

43 To Runge, January 15, 1909. DMA, HS 1976-31. Also in ASWB I.

44 Sommerfeld, *Relativitätstheorie I, II*, 1910.

appointment of Laue as lecturer at his institute.[45] Two years later, pursuing the trail blazed by Minkowski and Sommerfeld, Laue authored the first textbook presentation of relativity theory, by which future generations of physicists would orient themselves without losing their way in the jungle of conflicting electron theories.[46]

6.3 The Origins of the Sommerfeld School

Liberation of relativity theory from the ballast of electron theory conceptions, as Sommerfeld experienced and contributed to it in his early years at Munich, was not the work of solitary researchers in the ivory tower of science, but rather the product of a lively exchange of ideas. It is not only the big names of theoretical physics who were involved in this but also physicists whose names never found their way into textbooks.[47] For the most part, these were doctoral students and lecturers on the first steps along the uncertain path of their careers who were unknown to each other. Several years after Sommerfeld's appointment at Munich, this began to change. Sommerfeld encouraged his students to debate electron theory and the principles of relativity passionately outside the classroom, too. He then got an earful about this when he invited students to his home. Alluding to utterances from his lectures on Maxwellian theory, they lampooned him in verses whose theme was the contradiction between relativistic mechanics and everyday experience[48]:

Mechanics is something pliable quite,
Velocities there pathetically slight;
When contradictory, I state pertly
The ether is pure, but matter is dirty.

Although relativity theory was a major topic of discussion among Munich physicists, it was by no means the exclusive one. Besides, the circle around Sommerfeld was no purely theoretical group but included experimental physicists from the Röntgen institute, the Technical University, and colleagues from other fields. At Aachen, Sommerfeld himself, along with Debye, had experimented with X-rays; at first, he had planned similar projects at Munich.[49] "To gather experience, he wanted to spend two hours a day at my laboratory," Abram Fedorovich Ioffe (1880–1960), a collaborator with Röntgen, recalled about Sommerfeld's first days at his new job. Ioffe suggested instead a visit to a café where "a sort of physics club" had established itself.[50]

45 To the Philosophical Faculty, Section, II, of the LMU, April 20, 1909. UA, OC I 36.
46 Laue, *Relativitätsprinzip*, 1911; Janssen/Mecklenburg, *Mechanics*, 2007.
47 Pyenson, *Collaboration*, 1978; Staley, *Generation*, 2008, Part III.
48 Ewald, *Sommerfeld als Mensch*, 1968, p. 12. "Die Mechanik, die ist ein gefügiges Ding,/Die Geschwindigkeiten sind so lumpig gering,/Und widerspricht was, behauptete keck ich/Der Aether ist rein, die Materie dreckig."
49 To W. Wien, July 4, 1905. DMA, NL 56, 010. Also in ASWB I.
50 Ioffe, *Begegnungen*, 1967, p. 39.

Fig. 13: Ski tours in the Bavarian mountains were for many Sommerfeld students unforgettable experiences of their student years at Munich (Courtesy: Deutsches Museum, Munich, Archive).

Over the years, this café in the Munich Hofgarten evolved into a particular institution. Here physicists gathered after dinner for coffee and cake, and spontaneously the conversation could morph into a seminar in which the marble tabletops served as blackboards. Here one also encountered out-of-town colleagues when they happened to be visiting Munich. For Sommerfeld's students, who were able to meet here so many of their professor's illustrious colleagues in a relaxed atmosphere, the Hofgarten café became an enduring memory.[51]

Excursions into the nearby mountains, hiking in the summer, and skiing in the winter provided further occasion for unconstrained physics talk. Sommerfeld belonged to the early and enthusiastic devotees of the sport of skiing and was an endurance cross-country skier, "if not displaying quite the most elegant form," as Ewald recalled. "He did belong to that first generation of ski enthusiasts in Germany who first had to shake off the style of the stave-riding Norwegian farmers of that time." At the conclusion of every winter semester, people gathered mostly in the area of Mittenwald, where Wien had a country house to which he regularly invited friends and colleagues. "In this way, many of us doctoral students could discuss our problems with our professor in total informality."[52]

51 Ewald, *Fifty Years*, 1962, pp. 33–34; Paul Epstein, Interview with Alice Epstein, November 22, 1965 and February 8, 1966. Archives of the California Institute of Technology, Pasadena, California. http://oralhistories.library.caltech.edu/73/ (30 January 2013).

52 Ewald, *Sommerfeld als Mensch*, 1968, p. 10.

Paul Ewald, who had actually come to Munich to study mathematics, was taken by a fellow student to a lecture of Sommerfeld's and was so impressed by it that he no longer regarded mathematics but theoretical physics as his vocation. Sommerfeld took him under his wing personally, as well. This was "the start of an enduring and deep family friendship," Ewald recalled. "Once you climbed the four or five flights up to the Sommerfelds' door, you stepped into an apartment with spacious, high-ceilinged rooms, the last of which was the family's living room, Mrs. Sommerfeld's domain, which with its lovingly tended flowers at once expressed its own special personality." These "somewhat bohemian times" in Munich remained unforgettable for Ewald: "Sommerfeld in his peak years, his forties; the children, still young; the tender and supremely sympathetic mother and spouse; the smallish, close-knit circle of students."[53] In the spring of 1908, the Sommerfeld family had increased once more. "Pet the Sparrow, and kiss the three big ones," Sommerfeld wrote home from a trip shortly after Johanna had given birth.[54] The "Sparrow" was Eckart (1908–1977), who together with "the big ones," Ernst (1899–1976), Margarethe (1900–1977), and Arnold Lorenz (1904–1919) now made for livelier family life.

Ewald's recollections may have undergone some degree of transfiguration in retrospect. But even in contemporaneous correspondence many examples attest that for Sommerfeld, the personal and the scientific went side by side. This was especially true of the ski trips at the conclusion of every winter semester. "We're here," Sommerfeld wrote home in March 1907 from Mittenwald. "In addition to Wien, there's a man from Greifswald, a Berliner, and the 2 Schwarzschilds."[55] The "2 Schwarzschilds" were Karl Schwarzschild and his brother Alfred Schwarzschild (1874–1948), a well-known painter on the Munich art scene. The "Berliner" was Max Laue, who 2 years later would become a lecturer under Sommerfeld. The "man from Greifswald" was Julius Herweg (1879–1936), who was making a name for himself as an experimental physicist with experiments on spark discharges and cathode and X-rays. Often joining the Mittenwald circle in subsequent years were the associate professor of theoretical physics from Greifswald Gustav Mie (1868–1957) and Willy Wien's cousin, Max Wien (1866–1938), who had devoted himself to the new field of wireless telegraphy. Within this circle, Sommerfeld and his doctoral students found tremendous stimulation, which went on to make for continued discussion among the Munich "physics club" at the Hofgarten café, and even provided the impetus for one or another doctoral dissertation at Sommerfeld's institute.

Debye became the first of Sommerfeld's doctoral students. "I have a quite brilliant assistant who wants to do his doctorate with you," Sommerfeld had written Willy Wien earlier from Aachen, calling his attention to Debye.[56] At that point he himself did not know that he would ultimately direct Debye's pursuit of a degree

53 Ewald, *Sommerfeld als Mensch*, 1968, S. 11.

54 To Johanna, April 7, 1908.

55 To Johanna, March 10, 1907.

56 To W. Wien, June 20, 1905. DMA, NL 56, 010.

with a topic in physics. At the Hofgarten café too, Debye attracted notice as a rising star in the firmament of Munich physics. "At times, it was not easy for Sommerfeld to follow our conversation," Ioffe recalled, "that's why his assistant Debye soon surpassed us all."[57] The problem chosen for Debye's doctoral thesis may have emerged from discussions with Karl Schwarzschild on the diffraction of light by small spheres. Schwarzschild was interested in the effect of sunlight on the tails of comets and to this end wished to investigate the scattering of light by small spheres. In this context, the eventuality of "sphere radius large relative to wavelength" came up, which was the case, for instance, with freely falling waterdrops.[58] "The rainbow was of course our actual goal," Sommerfeld wrote concerning the physical interest in this subject for Debye's doctoral work.[59] The differential equations of the problem could be solved only by approximation in the limit of very small and very large spherical radiuses relative to the length of the light wave. In the second case, Debye had "by a very elegant method of complex analysis" succeeded in deriving approximation formulas of the Bessel functions, as Sommerfeld put it in his report to the faculty.[60] Debye had thereby not only brilliantly mastered the diffraction of light by a sphere (rainbow), and a cylindrical wire, but also—quite in the sense of "physical mathematics"—made an important contribution in the mathematical area of special functions.

The second doctoral dissertation Sommerfeld supervised in 1908 had also grown out of a suggestion from one of his circle of colleagues. Max Wien had invented a technical measurement device with which to determine the smallest differences in the environment of electrical circuits. The "Wien Induction Meter," as it was called, was based on self-induction in solenoids. Although the physical principle had long been known, more exhaustive experimental and theoretical investigations into the conductivity of metals were necessary before it could be actualized. This seemed to Sommerfeld a suitable problem for Frederick W. Grover (1876–1973), an American doctoral student who had been working at the National Bureau of Standards in Washington, D.C., and was on a 1-year leave of absence to pursue his doctorate with him. The theory of the induction meter also had to be tested experimentally, which, in the spaces of the old Academy soon ran into difficulties. Grover complained about the inadequate experimental facilities. He had come to Munich primarily to attend Sommerfeld's lectures and to do theoretical work. So far as experiments were concerned, he had assumed erroneously that Sommerfeld's "laboratory" was connected to Röntgen's in the University building. Under the circumstances, he was hopeful that Sommerfeld would accept his dissertation even if the theory could not be confirmed by supporting experiments.[61] Sommerfeld promised

57 Ioffe, *Begegnungen*, 1967, p. 39.

58 To Schwarzschild, May 2, 1908. SUB, Schwarzschild. Also in ASWB I.

59 To Schwarzschild, May 9, 1908. SUB, Schwarzschild. Also in ASWB I.

60 To the Philosophical Faculty, II. Section, July 23, 1908. UA, OC-I-34p. Also in ASWB I; Debye, *Feld*, 1908; Debye, *Lichtdruck*, 1909.

61 From Grover, April 8, 1908. DMA, NL 89, 008.

Fig. 14: As an engineering student at Aachen, Peter Debye had already attracted Sommerfeld's attention. In 1905, Sommerfeld engaged him as his assistant and then offered him an assistantship following his move to theoretical physics at Munich. After his doctorate in 1908 and habilitation in 1910, Debye was appointed successor to Einstein at the University of Zürich (Courtesy: Deutsches Museum, Munich, Archive).

at first to acquire the necessary equipment[62] but ultimately accepted the theoretical portion. He extenuated this with the Faculty on the grounds of lack of time. Grover had to return to the USA to the Bureau of Standards, "where he hoped to continue the experimental aspects of his work."[63]

The case of the "induction meter" was not the only one in which a doctoral student had to carry out experiments too.[64] After the electrical equipment for Grover's experiments at Sommerfeld's "laboratory" had been dismantled, facilities for hydrodynamic experiments were procured to investigate the transition from laminar to turbulent flow. Sommerfeld wrote about Ludwig Hopf (1884–1939) on

62 Draft of reply to Grover, after April 8, 1908. DMA, NL 89, 008.

63 To the Philosophical Faculty, II. Section, 30. June 30, 1908. UA, OC-I-34p. On Grover's work at the Bureau of Standards, see Cochrane, *Measures*, 1966, pp. 74–109.

64 Eckert, *Mathematics*, 1999.

the completion of his dissertation that the candidate was "actually not experimentally skillful." Hopf had attended Sommerfeld's lectures from the winter semester 1906/1907 on and in July 1909 had submitted his dissertation with an experimental part "On Turbulence Phenomena in a River (Channel)" and theoretical part "On Ships' Waves." But Sommerfeld also attested to Hopf's "great love for his subject and his unusual perseverance." The topic Hopf had chosen lay close to Sommerfeld's own sphere of interest. In ship's waves, he saw a phenomenon that recalled his "super-mechanics" of electrons. Here, one could investigate what is encountered "otherwise in velocities above the speeds of sound or light." More important to him though was the experimental part of Hopf's work on the generation of turbulence, since it picked up from Reynolds' classic work on the transition from laminar flow through a pipe ("Poiseuille flow") to a state of turbulence, with which he had been occupied earlier, at the time of his appointment at Aachen. In his theoretical treatment, he had arrived only at the formulation of an approach ("the Orr-Sommerfeld equation") that did not yet enable any prediction of the details of turbulence transition.[65] Hopf was to investigate the turbulence transition in a flow with an open surface.[66] Like flow through a pipe, a predominantly laminar flow at a slow velocity should become unstable at some critical higher velocity and induce turbulence. To this end, Hopf constructed a rectangular trough, through which water flowed at varying angles of inclination. By adding sugar, he was able to adjust the viscosity of the water. The goal of the experiment was to determine the critical Reynolds number $R = Uh\rho / \mu$ (U = mean velocity of flow, h = width of the trough, ρ = density, μ = viscosity), at which the flow of water in the trough was no longer laminar but turbulent. "I still recall bringing him a whole sack of sugar so he could make sugar solutions," Debye said amusedly many years after these turbulence experiments.[67] The results, however, were not very conclusive, for in an open trough, the surface tension of the water came into play as an additional physical variable and made the already complex process of the turbulence transition even more intransparent.[68]

Another doctoral student was Demetrius Hondros (1882–1962), a Greek exchange student who, like Hopf, had attended Sommerfeld's lectures from the beginning and had completed his studies in 1909. In light of his mathematical inclinations, Sommerfeld suggested as the topic of his dissertation the elaboration of the theory of electromagnetic waves along wires. He himself had been occupied with this area 10 years earlier when he was still a professor of mathematics at Clausthal. Initially, he had laid out "a complicated (probably too complicated) topic," namely, the "emission of waves at a kink in the wire." With this problem, he had hoped to explain the principle behind the antennas employed by Marconi, whereby

65 Eckert, *Birth*, 2010.

66 To the Philosophical Faculty, II. Section, July 5, 1909. UA, OC-I-35p.

67 Peter Debye, Interview conducted by T. S. Kuhn and G. Uhlenbeck, May 3, 1962. AHQP. http://www.aip.org/history/ohilist/4568_1.html (30 January 2013).

68 Hopf, *Turbulenz*, 1910.

Fig. 15: Ludwig Hopf completed his studies under Sommerfeld in 1909 with a doctoral dissertation in hydrodynamics. The "turbulence problem," the calculation of the transition from laminar to turbulent flow, was a recurring challenge taken up by adherents of the Sommerfeld school (Courtesy: Deutsches Museum, Munich, Archive).

directivity of the propagated waves was achieved by means of a horizontally bent wire. Even though Hondros was unable to solve this problem, the results of his work were "informative, and in part surprising," Sommerfeld wrote in his report on the dissertation. The analysis had, namely, yielded two types of electromagnetic waves along wires that showed differing transmission behavior. Only the so-called main waves were actually observable. The other wave type, the so-called sub-wave, scarcely manifests itself because here the "skin effect," which under normal circumstances leads to a reduction in wave amplitude at the interior of the wire, effects a diminution towards the exterior![69] What initially had seemed exclusively of theoretical interest soon proved of practical significance too. One year after his dissertation, in a joint paper with Debye, Hondros showed that in "dielectric wires," the situation is reversed: Here, only sub-waves are emitted, not the main waves.[70] Later, these theoretical results proved important for telecommunications.[71]

69 To the Philosophical Faculty, II. Section, June 15, 1909. UA, OC I 35 p; Hondros, *Drahtwellen*, 1909.

70 Hondros/Debye, *Elektromagnetische Wellen*, 1910.

71 Zinke/Brunswig, *Hochfrequenztechnik*, Chap. 5.4.2.

Although Sommerfeld had to resign himself to the makeshift of an institute divided between university and academy during his first 5 years in Munich, the balance sheet of his teaching was impressive: By 1910, six doctoral students had completed their studies with the degree, Debye, Grover, Hondros, Hopf, Rudolf Seeliger (1886–1965), and Fritz Noether (1884–1941). Word was getting around that Munich was a good place for theoretical physics. As early as 1908, when he was still at the patent office in Bern, Einstein assured Sommerfeld "that, were I in Munich and had the time, I would attend your lectures to fill out my mathematical-physical knowledge."[72] Students who arrived with little preparation were quickly overtaxed by Sommerfeld's lectures, however. In connection with one lecture, Ewald recalled, he made the suggestion to Hondros that "the older students mount a discussion of the modern problems among themselves so we younger students could understand what it was all about." Sommerfeld was not to participate, only Debye, so the students could ask questions, even stupid ones, in a relaxed atmosphere, without being inhibited by the professor's presence. Hondros took the suggestion to Debye, who relayed it to Sommerfeld, who in turn, "wisely temperate," gave his blessing and contributed a box of cigars "to hone the faculty of thought."[73]

History-making innovations are often layered over in retrospect with the knowledge of their later significance. Thus, after the fact a foundational myth emerged about the "Sommerfeld Colloquium," as it soon came to be called in distinction to the larger "Sohncke Colloquium." Sommerfeld wrote in his "autobiographical sketch": "From the outset, I strived for, and let nothing deter me from establishing a nursery for theoretical physics at Munich through seminars and colloquia."[74] In the beginning, though, it was still not clear what was meant by the terms "seminar" and "colloquium." Debye did not distinguish between the two: "We wanted a seminar, but a seminar without professors, so we were free to be as dumb as we wished," he recalled many years later. At first, Sommerfeld had wanted to participate, "but we said no, he couldn't join in, though he was certainly free to provide the cigars."[75] According to another reminiscence, the initiative grew out of the desire to meet not just monthly at the Sohncke Colloquium but weekly and if possible in not so formal a setting.[76] Röntgen's assistant, Peter Paul Koch (1979–1945), claimed a share of this genesis story. He, together with Debye and Ernst Wagner (1876–1928), another Röntgen student, had established the "big-shot-free colloquium . . . out of which later the famous Sommerfeld Colloquium emerged."[77]

72 From Einstein, January 14, 1908. DMA, NL 89, 007. Also in ASWB I.

73 Ewald, *Erinnerungen*, 1968.

74 Sommerfeld, Autobiographische Skizze, ASGS 4, p. 677.

75 Debye, Interviewed by T. S. Kuhn and G. Uhlenbeck, May 3, 1962. http://www.aip.org/history/ohilist/4568_1.html (30 January 2013).

76 Epstein, Interviewed by Alice Epstein, November 22, 1965 and February 8, 1966. Archives of the California Institute of Technology, Pasadena, California. http://oralhistories.library.caltech.edu/73/ (30 January 2013).

77 Koch to Sommerfeld, August 6, 1944. Munich, UB, Sommerfeld estate.

The colloquium register, still preserved, confirms the origins of this institution in the winter semester of 1908/1909. The first entries show also that it was not a purely theoretical colloquium. The first lecture dealt with the Doppler effect in canal rays, and in the course of the colloquium, experimental physicists spoke frequently. Koch, for example, gave a lecture on January 27, 1909, on the "Zeeman effect on the sun." The subject of the last colloquium that semester was titled, "Momentum of a revolving stool with two beer-mugs."[78] Even the means "for honing the faculty of thought" were recorded. Since as a University lecturer, Debye had the right to hold instructional meetings, the Faculty had "no objection to there being held in the time period from 6 to 8 o'clock in lecture hall nr. 122 a colloquium under his direction for advanced students," the Dean determined. "As far as smoking at this colloquium is concerned, although in contravention of the general rule, it may temporarily be countenanced if all other colleagues using the lecture hall agree to it."[79]

Although the Sommerfeld Colloquium thus nearly achieved official status, it remained an informal institution. Over the course of years, it underwent numerous changes. The initial ambition to have a "big-shot-free" forum gave way in the interests of new research goals. While Sommerfeld participated only once in the winter semester 1908/1909, he subsequently used the colloquium regularly to present his latest research. Associated with it was a "seminar" for discussion of problems that had arisen in the context of the main lectures. "Seminar: exercise problems in mechanics, 2 h, Tuesdays from 5 to 7 o'clock," was the announcement for the winter semester 1910/1911, when Sommerfeld's main lecture series was on "analytic mechanics." After World War I, the seminar became a meeting for students to test their research skills. A lecture in the seminar became a sort of job interview for a doctoral thesis. What earlier had been designated a seminar, was now called simply "exercise." Neither sort of seminar should be confused with what had been understood by that term in the nineteenth century: the Neumann Seminar at the Albertina at Königsberg, the "mathematical-physical seminar" at the University of Munich, and other similarly organized seminars were institutions whose principal purpose was the education of high school teachers and which evolved later into the institutes of physics and mathematics at the various universities.[80] What was true of the colloquium and the seminar was true also of the lectures. It was a few years before Sommerfeld settled on a regular lecture cycle. From the winter semester of 1906/1907 the subjects were "Maxwellian theory/electron theory," "theory of radiation," "kinetic gas theory," "thermal conduction, diffusion, and electrical conduction," "electrodynamics, with a focus on electron theory," "optics," "vector calculus," and "partial differential equations in physics." Only thereafter did these topics evolve into the canonical, six-semester-long cycle that Sommerfeld later also made the basis of his textbooks: (1) mechanics, (2) mechanics of deformable media, (3) electrodynamics,

78 Münchener Physikalisches Mittwochskolloquium, DMA, 1997–5115.

79 Hertwig to Sommerfeld, November 3, 1910. DMA, NL 89, 030.

80 Olesko, *Physics*, 1991; Toepell, *Mathematiker*, 1996, Chap. 5.3.

(4) optics, (5) thermodynamics, and (6) partial differential equations in physics. Aside from these principal lectures, to which four weekly classes were attached, he gave 2-h-long special lectures on topics with which he was currently occupied in his own research and that were intended only for more advanced students.

Even though it took several years before Sommerfeld's pedagogical enterprise evolved to the level at which it later consistently garnered high praise, it was nonetheless soon clear that a remarkable school of theoretical physics was evolving at Munich. To be sure, there were institutes of theoretical physics at Berlin, Göttingen, Königsberg, and Leipzig, but none comparable to the Munich institute in terms of either physical plant or curriculum. As a rule, theoretical physics remained the domain of lecturers and associate professors. In 1911, when Jena instituted a search for a theoretician, its full professor of physics wished "to attract" preferably "a good lecturer from the Sommerfeld or Planck school. Of course he must be as non-Jewish as possible."[81] Just how laborious the metamorphosis of this field from lecturer's topic to independent discipline was is exemplified in the 1913 inquiry of an associate professor of mathematical physics at the University of Freiburg who asked Sommerfeld "how mathematical physicists at other universities are provided for." He had the use of only one room, "which [had to serve] simultaneously as work room, administrative office, and storeroom for equipment and an instructional collection."[82] Compared to this, Sommerfeld's situation at Munich must have seemed like paradise.

6.4 The Mathematical Attack

"It is unfortunate that mere opinions are expressed in this discussion, and no one tackles the subject mathematically," Sommerfeld in 1909 criticized the various approaches to a theory of thermal radiation.[83] But in this subject no breakthrough seemed achievable no matter how virtuosic the mathematics, so that in his own research he turned to other areas. Sommerfeld continued to regard the mathematical "attack" as an approach to new topics that seemed to promise success. Methods proven useful in one area could sometimes be successfully applied to other areas. So, for example, in connection with his work at Aachen on the buckling thresholds of rails in steel mills of certain profiles, he was hopeful "that an analogous calculation will lead to the theoretical calculation of the critical velocity in hydrodynamics and turbulence. Provisionally, I have a rather horrible transcendental equation, that still awaits discussion."[84] The mathematical method he alluded to and which he published 2 years later went down in the history of turbulence research as the

81 Max Wien to Wilhelm Wien, October 13, 1911. DMA, NL 56, Nr. 849.

82 From Johann Koenigsberger, January 8, 1913. DMA, NL 89, 010.

83 To W. Wien, April 21, 1909. DMA, NL 56, 010. Also in ASWB I. Kuhn, *Black-Body Theory*, 1978.

84 To Runge, June 9, 1906. DMA, HS 1976–31.

"Orr-Sommerfeld" approach. He was unable to solve the "horrible transcendental equation," but the related debates opened a fruitful new field of research.[85]

Similarly, he transferred the methods he had used to solve the problems of electromagnetic waves along wires, developed earlier in 1899 as professor at Clausthal, to a new application. It may have been his quarrel with Lindemann, which had cast doubt on Sommerfeld's mathematical procedures not only in electron theory but also with respect to electromagnetic waves along wires, that in 1906 he once again turned to his Clausthal work. In the process, it occurred to him "to apply what one usually does with electromagnetic waves along wires to electromagnetic surface waves that propagate on a smooth boundary, where everything is far simpler. I think that this type of wave is consistent with wireless telegraphy."[86] In other words, Sommerfeld assumed that the electromagnetic waves radiating from an antenna did not, like light waves, spread into surrounding space in straight lines, but rather remained tied to the earth's surface and so also followed the earth's curvature, quite along the pattern of the electromagnetic waves along wires, which as we know also follow the course of a bent wire. Jonathan Zenneck (1871–1959) pursued these ideas further. He thanked Sommerfeld for looking through his manuscript "On the Propagation of Smooth Electromagnetic Waves Along a Smooth Conductive Surface and its Relation to Wireless Telegraphy" and referred in it to Sommerfeld's work on electromagnetic waves along wires.[87] Zenneck had begun his career as assistant to the radio pioneer Ferdinand Braun (1850–1918) at Strasbourg and, like Sommerfeld, with whom as an author of an encyclopedia article he had been in regular contact since 1900, had kept technical applications in view. In 1908, he authored an *Introduction to Wireless Telegraphy*. In this book, Zenneck argued that only the spreading of waves "along the earth's surface" could explain why it was possible to transmit radio signals over distances great enough that the curvature of the earth precluded visual connection between transmitter and receiver. "For then the waves travel along the earth's surface and follow its curvature. In other words, the waves do not spread in straight lines as do those of light."[88]

Thus, Zenneck, and with him Sommerfeld, came out in opposition to the reigning conception that the electromagnetic waves of wireless telegraphy spread through space in the same way as light waves. Had not the celebrated experiments of Heinrich Hertz demonstrated that one can work with electromagnetic waves in the same way as with the optics of light beams? Hertz had reflected his waves on metal mirrors and refracted them through prisms of pitch. But electromagnetic waves along wires are surface waves. Even in the theory of elasticity, waves in space were distinguished from surface waves. In the case of earthquake waves, both types were demonstrable. Could this also be the case with wireless telegraphy? Ultimately, this

85 Sommerfeld, *Beitrag*, 1909; Eckert, *Birth*, 2010.
86 To Lorentz, 12. December 12, 1906. RANH, Lorentz inv.nr. 74. Also in ASWB I.
87 Zenneck, *Fortpflanzung*, 1907, pp. 849 and 865.
88 Zenneck, *Leitfaden*, 1909, pp. 221–222.

too was a matter of solving partial differential equations under certain boundary conditions, which however—despite limiting it to the case of a smooth plane surface—presented a nontrivial problem. Zenneck did not regard his analysis as a comprehensive theoretical treatment of the problem. His procedure had been carried out "primarily with respect to the circumstances of electromagnetic waves along wires," so he referred back to Sommerfeld's work as the source for the introduction of surface waves into wireless telegraphy.[89]

As he immersed himself in the problem, then, Sommerfeld may well have thought it was necessary here too for someone to "attack mathematically." In January 1909, he presented his results to the Bavarian Academy of Sciences; shortly thereafter, he published the comprehensive theory in the *Annalen der Physik*.[90] "To which type are the waves in wireless telegraphy to be ascribed? Are they comparable to the Hertz waves in the air, or to the electrodynamic waves along wires?" This was the fundamental question with which he began his analysis. It spanned 71 printed pages and offered a whole arsenal of methods of complex analysis with which to wrest physically interpretable expressions from the complex integrals he presented as solutions to the differential equations. In one case, he sought to demonstrate that space waves and surface waves are contained in the electromagnetic field spread from an antenna across flat ground. In another, he tried to make clear how the ground conditions influence the spread of waves. In earlier papers, propagation had always been calculated assuming an ideally conductive surface. In his theory, Sommerfeld distinguished among the various ground and space characteristics by means of complex materials constants and showed that the propagation at different ground and air properties could be expressed in a "kind of law of similarity of wireless telegraphy." In place of the actual distance from the antenna, he introduced a "numerical distance" incorporating the complex materials constants in order to compare the behavior of waves under varying ground and air conditions. The theory found immediate and enthusiastic acceptance among his colleagues in theoretical physics. "Your discovery of the 'surface waves,' through which the puzzle of the propagation across greater distances is solved, is very elegant," Lorentz congratulated him.[91] "How interesting and elegant this all is!" Voigt wrote from Göttingen.[92] "The solution appears splendidly, as though at the wave of a magic wand," enthused Schwarzschild.[93]

Given the currency of wireless telegraphy, it was expected that these results would meet with interest not just among theoreticians. So Sommerfeld published a version of his theory focused on the needs of technology in the *Jahrbuch der drahtlosen Telegraphie und Telephonie*, in which he laid major stress on the

89 Zenneck, *Fortpflanzung*, 1907, p. 856.

90 Sommerfeld, *Ausbreitung*, 1909.

91 From Lorentz, March 21, 1909. DMA, HS 1977- 28/A,208.

92 From Voigt, April 9, 1909. DMA, HS 1977-28/A,347.

93 From Schwarzschild, April 19, 1909. DMA, NL 56, 010. Also in ASWB I.

constitution of the ground.[94] He also made the diffusion of waves in wireless telegraphy a topic in his school. Paul Epstein (1883–1966), who had come to his institute following studies at Moscow, was tasked with calculating and graphically presenting the course of field lines in the environment of an antenna directly above ground level. The result had been intended as an illustration to Sommerfeld's publication in the *Jahrbuch der drahtlosen Telegraphie und Telephonie* and was printed there in connection with his paper as an original contribution of Epstein's.[95] He set another student to working out further technically important ramifications of the theory in the framework of a doctoral thesis. In it, "the operation of the bent Marconi transmitter in wireless telegraphy" was to be explained, "a rather puzzling problem up to now," as Sommerfeld explained in his report to the Faculty. He had been concerned with this problem already at the time of Hondros's doctoral work. The antenna with which Guglielmo Marconi (1874–1937) had carried out his famous transatlantic radio experiments displayed a horizontal as well as a vertical component, and it was puzzling how under such circumstances directionality could have come about. Sommerfeld conjectured that the "influence of the ground below the surface, which in the simpler arrangement of a symmetrical vertical antenna I have investigated in an earlier paper, must be considerable." In the dissertation, this conjecture was "completely confirmed." The doctoral candidate, Harald von Hörschelmann (1878–1941) showed that an effect was brought about through vertical ground currents in the vicinity of the transmitter that corresponded to those of two antennas set up in the chosen directionality with opposite phases. The directional effect is an interference effect arising from the poor electrical conductivity of the ground in the vicinity of the antennas. "The surrounding medium (sea water) with good conductivity then takes over the long-range transmission of the directional effect."[96]

One can infer from the Munich papers on wireless telegraphy around 1910 that "attacking mathematically" became a characterizing feature of the Sommerfeld "nursery," and that word quickly spread among the students of a theoretical bent. Epstein told his friend Paul Ehrenfest (1880–1933) enthusiastically that he was "in daily contact with Debye, Sommerfeld, and Laue" and encouraged him also to come to Munich.[97] Ehrenfest had taken his doctorate at Vienna under Boltzmann and would gladly have come to Munich to do his habilitation under Sommerfeld. However, to his inquiry to Debye whether he might get an appointment as lecturer at Munich, he received a negative reply. Sommerfeld preferred "to keep the position open for the 'progeny' of his school."[98] When Ehrenfest was turned down at other

94 Sommerfeld, *Ausbreitung*, 1910.

95 Epstein, *Kraftliniendiagramme*, 1910.

96 To the Philosophical Faculty, 2. Section, January 7, 1911. Munich, UA, OC I 37p. Hoerschelmann, *Wirkungsweise*, 1911.

97 Epstein to Ehrenfest, November 19, 1910. AHQP/EHR 19.

98 Debye to Ehrenfest, May 30, 1911. AHQP/EHR 19.

places too—no doubt on anti-Semitic grounds—he turned directly to Sommerfeld requesting at least to do doctoral work under his supervision. "For me to be able to do a habilitation at Leipzig, I need a German doctorate, because the Austrian doctorate is not recognized at Leipzig."[99] He hoped thereby "especially to learn this: how to carry through to completion a work demanding really a great deal of calculation."[100] Sommerfeld thought the refusal to recognize Ehrenfest's doctorate absurd and offered him, after a visit to Munich to get to know him, a position as lecturer—against Debye's advice, which reveals once more the anti-Semitism that so encumbered Ehrenfest's search for a position: "If you are now thinking of getting Ehrenfest, I can't refrain from expressing certain reservations. A Jew, such as he obviously is, of the 'high priest' variety with his insidious Talmudic logic can exert an extremely destructive influence. Many a fresh, not quite fully fledged idea, which normally one would express uninhibitedly, can all too easily be nipped in the bud by him. And in that regard I consider contact with him dangerous."[101] Nonetheless, Sommerfeld had gotten to know Ehrenfest as a "very sympathetic, sensitive human being," as he wrote Lorentz. "After learning through his visit that he is not—as I had thought based on his dissertations—an abstract dialectician, but on the contrary that he has a strong physical bent, I would be very happy to have him here."[102] That Ehrenfest notwithstanding did not become a lecturer at Munich had other grounds: viz., he was offered a professorship in Holland, even without the habilitation, and not just any professorship, in fact, but as successor to Lorentz's chair![103]

The evaluation of Ehrenfest's qualifications is instructive with respect not only to the anti-Semitism prevalent among academics of those years but also to what was expected of a theoretical physicist—at a time when this field was steadily gaining in prestige. Attacking a subject as a theoretician entailed, as the example of wireless telegraphy demonstrates, not only a mathematical analysis but also a physical interpretation. In contrast to the "physical mathematics" of his Göttingen years, physics was now Sommerfeld's actual goal. "Mathematics is for him properly not an end unto itself," Sommerfeld judged of Ehrenfest's suitability for theoretical physics. "He understands how to render the most difficult things clear and concrete. He translates mathematical considerations into pictures that can be grasped." This seemed most important to him with respect to teaching. In this regard, he gave Ehrenfest, after hearing him lecture at the Sohncke Colloquium, the highest marks: "His lectures are masterly. I have rarely heard a person speak so captivatingly and brilliantly. Concise verbal constructions, witty, dialectics at his disposal to an

99 To Sommerfeld, August 24, 1911. AHQP/Ehr 25. Also in ASWB I.

100 To Sommerfeld, September 30, 1911. DMA, HS 1977-28/A,76. Also in ASWB I.

101 From Debye, March 29, 1912. DMA, HS 1977-28/A,61.

102 To Lorentz, April 24, 1912. RANH, Lorentz, inv.nr. 74. Also in ASWB I.

103 Klein, *Ehrenfest*, 1970, Chap. 8.

unusual degree. Very characteristic is his manner of using the blackboard. The whole outline of his lecture is put on the blackboard for his audience in the clearest way."[104]

The qualities he praised in Ehrenfest represented his own ideals for his Munich "nursery." Teaching and research were closely associated for Sommerfeld. The talent for translating mathematical concepts into "graspable pictures" had already impressed him in his own teacher, Felix Klein. Now it was his turn to realize this pedagogical ideal for theoretical physics. He often ushered in new areas of research by trying out a "mathematical attack" in a lecture to an audience of his students. "My lecture gave me the opportunity to apply the approach normally taken with electromagnetic waves along wires to electromagnetic surface waves," he wrote Lorentz about the start of his work on wave diffusion in wireless telegraphy.[105] Likewise, advising students on the choice of a doctoral thesis, he strove to make the mathematical procedures required to solve the physical problems that would arise clear from the start. "Sommerfeld took a foolscap sheet of paper out of the drawer and I saw a list of some ten or twelve research problems written out in his large clear handwriting," recalled Ewald of the time he had asked Sommerfeld for a doctoral thesis topic. "He discussed and explained them to me one by one. Calculation of self-inductances of solenoids for alternating currents; propagation of radio waves over a surface of finite conductivity; an unsolved problem of gyroscopic theory; a new attempt at explaining the instability of Poiseuille flow, and further subjects. Each subject had its own merit and its own type of mathematical technique, and Sommerfeld pointed them out."[106]

6.5 The "*h*-Discovery"

Although Sommerfeld had succeeded within just a few years in transforming a makeshift into that "nursery of theoretical physics" he had envisioned on his appointment at Munich, he did have to accept several disappointments with respect to his own research. His "super-mechanics" of electrons proved not to be the key to understanding radioactive rays. Nor could the electron theory answer the question, "what is actually going on with X-rays?"[107] The research he pursued during his first Munich years, from hydrodynamics to wireless telegraphy, would certainly have brought distinction to a theoretician in a less prominent position, but did it fulfill the expectations placed on the successor to Boltzmann's chair?

104 To Lorentz, April 24, 1912. RANH, Lorentz, inv.nr. 74. Also in ASWB I; Ehrenfest had lectured at the Sohncke colloquium on February 5, 1912. Münchener Physikalisches Mittwochskolloquium, DMA, 1997–5115.

105 To Lorentz, December 12, 1906. RANH, Lorentz, inv.nr. 74. Also in ASWB I.

106 Ewald, The Setting for the Discovery of X-Ray Diffraction By Crystals, notes for a lecture delivered at the First Plenary Session of the International Union of Crystallography, August 2, 1948. pp. 21–22. DMA, NL 89, 027.

107 To W. Wien, May 13, 1905. DMA, NL 56, 010. Also in ASWB I.

One of the great problems theoretical physicists struggled with at the beginning of the twentieth century concerned thermal radiation.[108] In 1900, Planck had given a formula for the energy of thermal radiation as a function of wavelength and with it introduced a new constant, which Einstein in 1905 developed into a radical approach to the foundation of a new theory of radiation. Furthermore, Einstein used Planck's formula in order to describe the temperature dependence of the specific heat of solid bodies. It seemed as if the energy in heating or cooling changed only in amounts of $E = h\nu$, where ν is in this case the frequency of the lattice vibrations. In the former case, it is the frequency of the thermal radiation. The constant h, soon named for Planck, became the emblem of quantum theory, without there being a conclusive theoretical explanation of it. The quantum hypothesis became for theoretical physicists a particular challenge which Sommerfeld, too, wished to take up. "Who knows if I too won't soon play around a bit with radiation," he wrote in 1908 regarding his plans for future research.[109] As in the case of wireless telegraphy, he familiarized both himself and his students with this new area through a lecture on "Theory of Radiation" in the summer semester of 1907.[110] In the following winter semester of 1908/1909, he wanted to use his lecture on "Electrodynamics, Particularly Electron Theory" to "convert definitively to Planck's fundamental hypothesis," as he wrote Johannes Stark (1874–1957). Shortly before, Stark had discovered a conjectured quantum effect in a quite different physical phenomenon.[111] But here Sommerfeld had to confront the fact that a mathematical attack, be it ever so energetic, was of no further use. Unlike the case of wireless telegraphy, with radiation theory, he obtained no presentable result.

Withal, the Planck thermal radiation formula remained a stumbling block. Willy Wien believed it was impossible to resolve the puzzle of thermal radiation in the framework of Maxwellian electrodynamics; he hoped, as he wrote Sommerfeld, for "an extension of Maxwell's equations within the atom."[112] In 1908, Sommerfeld was still clinging, if somewhat halfheartedly, to the electron theory, whose potential he thought not quite exhausted. "There exists an endless spectrum of oscillation cycles there," he countered Wien, and so long as these are not accounted for, one ought not give up hope. "How to carry this out quantitatively is still unclear to me, however."[113]

His discussions with Einstein also revolved around quantum matters. Sommerfeld had met Einstein personally in September 1909 at the Natural Scientists Congress at Salzburg. Like most physicists, he regarded Einstein's light

108 Kuhn, *Black-Body Theory*, 1978. Darrigol, *Disagreement*, 2001; Seth, *Quantum Theory*, 2004.
109 To W. Wien, April 21, 1909. DMA, NL 56, 010. Also in ASWB I.
110 Lecture notes in DMA, NL 89, 028; Seth, *Quantum Theory*, 2004.
111 To Stark, October 10, 1908. SBPK, Stark. Also in ASWB I; Hermann, *Diskussion*, 1967; Hermann, *Frühgeschichte*, 1969.
112 From W. Wien, June 15, 1908. DMA, HS 1977-28/A,369. Also in ASWB I.
113 To W. Wien, June 20, 1908. DMA, NL 56, 010. Also in ASWB I

quantum hypothesis as a construction of quantum theory that went too far. If radiation itself consisted of energy quanta, how did this comport with its wave-like character, which, after all, was an established fact for radiation of all wavelengths from visible light to the invisible waves of wireless telegraphy? "On the other hand, I am old-fashioned enough," Sommerfeld wrote Lorentz, "to resist for now the light quanta in Einstein's conception. Stark's light quanta, against which I have recently spoken, are probably not to your taste either."[114]

This last remark referred to a publication by Stark on the spatial intensity distribution of an X-ray emitted from an anticathode. Stark had found that X-rays were not emitted in all directions at equal intensity and conjectured a quantum phenomenon behind it. But Sommerfeld demonstrated that this could be explained within the framework of classical electrodynamics.[115] He proceeded from the "well-known Wiechert-Stokes concept" that X-rays are electromagnetic impulses generated by the impact of electrons on the anticathode of an X-ray tube. Every variation in velocity of an electrically charged particle, including deceleration on impact, leads to the radiation of electromagnetic waves. Although this was actually just an application of Maxwellian theory, the consequences of this theory for X-rays had not heretofore been analyzed in detail. "It is long since anything in physics has made such an impression on me as that paper of yours on the distribution of the energy of X-rays in different directions," Einstein was pleased to write.[116] Sommerfeld's analysis kept quantum theory from going off on a wrong track. "You will, I hope, be convinced that the braking theory of X-rays itself accomplishes everything to which you apply the quite hypothetical and unsubstantiated light quantum theory," Sommerfeld wrote to Stark. "Not that I doubted the significance of the quantum of action. But the elaboration you have given it seems not only to me, but also to Planck, very dubious."[117]

Although in the debate with Stark that followed Sommerfeld maintained the upper hand, the many questions regarding the radiation of X-rays were by no means settled. It was not only the X-radiation referenced in braking theory that was emitted from the anticathode but also X-radiation characteristic of the anticathode material. While the *bremsstrahlung*[118] showed a uniform direction of oscillation (polarization) determined by the direction of the electrons, in the characteristic X-radiation, there was no predominant oscillatory direction. Sommerfeld had demonstrated only that *bremsstrahlung* could be explained by classical electrodynamics. He assumed that the characteristic X-radiation originated in the atoms of

114 To Lorentz, 9. Januar 1910. RANH, Lorentz, inv.nr. 74. Auch in ASWB I

115 Sommerfeld, *Verteilung*, 1909.

116 From Einstein, January 19, 1910. DMA, NL 89, 007. Also in ASWB I.

117 To Stark, December 4, 1909. SBPK, Stark. Also in ASWB I; Hermann, *Diskussion*, 1967

118 Literally, "braking radiation," or "deceleration radiation," conventionally left in its original German, "*Bremsstrahlung*".

the material of the anticathode. "It is very possible that the Planck quantum of action plays a role here."[119] The relation between characteristic radiation and *bremsstrahlung* in the total radiation from an X-ray tube, however, could be inferred only indirectly from polarization experiments. In his theory, Sommerfeld referred to a recently completed doctoral thesis at the Röntgen institute. It was also not clear for the "braking portion" how the electron is decelerated after impact on the anode. If it traced a zigzag course from atom to atom, as Sommerfeld had assumed following his experiments with Debye at Aachen, it changes its preferred direction and, thereby, also the degree of polarization. So the *bremsstrahlung* theory required further assumptions about the braking process itself.

Einstein brought an additional problem to Sommerfeld's attention, the wave-particle dualism in the photoelectric effect. In a thought experiment, he wrapped an X-ray tube with shielding that had a small opening at one spot. Behind it, he placed a metal plate. If an emitted X-ray impulse were now sent from the anticathode in the X-ray tube through the small opening as a wave in spherical mode into the whole space, how could an electron be ejected from the metal as if it had gathered the energy of the entire spherical wave and not just that portion of energy that arrived at the plate through the small opening? Would the metal plate be capable, he asked Sommerfeld, "of frugally storing up fragments of the spherical X-ray waves until capable of furnishing one of its electron offspring with sufficiently potent energy to make its journey through space with the vigor befitting its X-ray birth?"[120]

Shortly before, Einstein had adopted the wave-particle dualism for thermal radiation too.[121] In August 1910, Sommerfeld traveled to Zürich to discuss these questions with Einstein further, face-to-face. His student Ludwig Hopf also came to Zürich for several months to work with Einstein. In the morning, he had "played Bach," and now they were sitting together "in the quite atmospheric and low-ceilinged pub the Apfelkammer," Sommerfeld wrote home from the favorite bar of his favorite poet, Gottfried Keller (1819–1890). Einstein and Hopf added notes to the effect that they were taking refreshment "from a great many 'light quanta,' etc."[122] A few days later, Hopf and Einstein submitted two jointly authored articles on thermal radiation to the *Annalen der Physik*, in which they showed that Planck's formula could not be derived from the familiar laws of classical physics and that it necessitated a "more profound adjustment of the fundamental conceptions."[123]

What Einstein had discussed in his letter to Sommerfeld as a thought experiment on the photoelectric effect of X-rays, the Viennese physicist Egon von Schweidler (1873–1948) around the same time made a test case for the question whether γ-rays were waves or particles. The γ-rays, emitted from a point source of

119 Sommerfeld, *Verteilung*, 1909, p. 970.
120 From Einstein, January 19, 1910. DMA, NL 89, 007. Also in ASWB I.
121 Klein, *Einstein*, 1964.
122 To Johanna, August 25, 1910.
123 Renn, *Einstein's Annalen Papers*, 2005, p. 357.

radiation into space, were to strike a gas through an opening in an otherwise impenetrable shield, and there display their ionizing properties. Since the γ-rays originate in spontaneous radioactive decay processes, their intensity varies around a mean. Each decay event corresponds to a γ-impulse. Were we dealing with waves, at each decay a part of the emitted spherical wave should fall through the opening and ionize the gas; the degree of ionization would then depend on the angle of the opening. If on the other hand we were dealing with particles, for each particle that passed through the opening the same degree of ionization would occur. To be sure, in the latter case only a portion of the particles would pass through the opening. In the two eventualities, different degrees of dependence would be obtained. So Schweidler concluded that from a statistical analysis, one could determine which of these two interpretations was correct.[124] Certainly, even less was known about the generation of γ-rays than about X-rays. Since γ-rays always appeared together with β-rays (which because of their charge could be identified as electrons), it was assumed that in the radioactive decay, an electron was ejected from the atom as a β-ray, and that the γ-ray was a concomitant of the process. With the *bremsstrahlung* theory as background, the conjecture suggested itself that the γ-rays were generated by the same mechanism as X-rays. What is brought about on the anticathode of an X-ray tube by the deceleration of electrons could occur as a result of the ejection of electrons in radioactive decay. In both cases, Maxwell's equations required the radiation of electromagnetic waves for the accelerated motion of an electrical charge. Thus, Sommerfeld saw in the "Schweidlerian variations" an opportunity to develop a new field of application of the *bremsstrahlung* theory.

As plausible as the idea of looking for the origins of X-ray *bremsstrahlung* and γ-rays in the variation of velocity of electrons was, it was equally unclear what the step required from the idea to the actual formulation of a theory was, for there was no clear understanding of this variation in velocity. Sommerfeld reached back to earlier measurements carried out by Willy Wien to estimate the braking distance of the electrons caused by the X-ray *bremsstrahlung*. Then he calculated the "braking effect" as the product of impact energy and braking duration and obtained a numerical value that lay close to Planck's constant h. He applied this result to the situation of β-rays. The energies measured in various radioactive substances were consistent with the hypothesis that the "effect" at the ejection of an electron corresponded to Planck's constant. "As hypothetical as the previous observations may have been, we want to take them a step further experimentally," he explained by way of establishing his process. "We apply the fundamental hypothesis of Planck's theory of radiation, that is, to radioactive emission, and assume that at each such emission precisely the quantum of action h will be released."[125] In its details, realization of the theory in practice was very complicated. But in contrast to his earlier

124 Schweidler, *Entscheidung*, 1910; Wheaton, *Tiger*, 1983, pp. 147–150.
125 Sommerfeld, *Struktur*, 1911, pp. 24–25.

theories, in which the difficulty lay in the mathematical analysis, the challenge now consisted in assessing various experimental results about X-rays and radioactivity. Formulating new physical hypotheses from these with the aid of his own theoretical conceptions was a different sort of theoretical physics from the "mathematical attack" of his earlier topics.

At the Natural Scientists Congress in September 1910 at Königsberg, Sommerfeld demonstrated that mathematics had been and still was a central concern,[126] but when shortly thereafter he published his "*h*-hypothesis," he showed his other, more physically oriented side. He chose as his forum the Bavarian Academy of Sciences, which had just elected him to its ranks of sitting members.[127] His paper was titled "On the Structure of γ-Rays" and at once sparked lively debate.[128] "We have discussed my h-discovery a great deal," Sommerfeld wrote home in March 1911 from the Mittenwald Ski Club. "Einstein wrote to me about it at once, likewise someone from Leipzig (Wiener). When I get home, I really want to get to work on it."[129] He was looking forward most to Planck's reaction, who was currently working on a new formulation of his radiation theory. In Planck's "second theory," the absorption of radiation was supposed to occur according to the laws of classical physics, and only the emission process to occur in accordance with quantum theory. Planck expressed no opinion about the mechanism operative thereby in the individual atoms but rather derived his radiation formula from a statistical consideration of many "oscillators." Sommerfeld's *h*-hypothesis, however, rested on assumptions about atomic processes. "You specify a relation between them that in my hypothesis is left entirely open," Planck wrote in his initial reaction to Sommerfeld's theory.[130] After this opening, quantum theory moved to the center of the discussions. In July 1911, Sommerfeld proposed that the Bavarian Academy of Sciences name Planck, in recognition of his contributions to quantum theory, a corresponding member. Planck's quantum theory of thermal radiation was "of the most fundamental importance," even if not complete in every detail.[131] For his own quantum conception of the emission of γ-rays, he argued in a pointed statement, "A molecule always emits and absorbs an electron according to an action process *h*, which yields the associated electromagnetic radiation completely."[132]

126 Sommerfeld, *Greensche Funktion*, 1910.
127 To Karl Theodor von Heigel, December 3, 1910. Archive of the Bavarian Academy of, Proceedings, vol. 103.
128 Wheaton, *Tiger*, 1983, pp. 150–167.
129 To Johanna, March 20, 1911.
130 From Planck, April 6, 1911. DMA, HS 1977-28/A,263. Also in ASWB I. On Planck's "second theory," see Kuhn, *Black-Body Theory*, 1978, Chap. 10.
131 To the Bavarian Academy of Sciences, July 1, 1911. Archive of the Bavarian Academy of Sciences, personnel folder Planck.
132 From Planck, July 29, 1911. DMA, HS 1977-28/A,263. Also in ASWB I.

6.6 The First Solvay Congress

With his "*h*-discovery," Sommerfeld wished to demonstrate that the emission of electromagnetic radiation could be tightly focused without having to assign a quantum nature to the radiation itself. But with increasing distance from the radiation source, the focus would spread spherically. This was the decisive ground on which William Henry Bragg (1862–1942) rejected Sommerfeld's explanation. In all experiments with γ-rays and X-rays, energy appeared to be focused in the direction of the radiation, without the slightest indication of spreading out. "At least, that is how I read the facts," Bragg wrote Sommerfeld. "It seems to me that it is right to think of the X- or the γ-ray as a self-contained quantum which does not alter in form or any other way as it moves along."[133] Like Einstein, Bragg considered the photoelectric effect as the clearest proof of the quantum nature of radiation. "I do not see," he wrote in a subsequent letter, "how one can escape the conclusion that a single secondary cathode ray derives its energy from a single X-ray which up till then carried it along in an unvarying form."[134] In other words, the electron emitted from an X- or γ-ray behaves as though it had received its energy from the process of collision with a beam particle. For Sommerfeld, though, X- and γ-rays were not particles but impulses, which were indeed generated in a quantum process but which behaved otherwise like electromagnetic waves. Six years earlier, he had still regarded γ-rays as electrons of a velocity greater than the speed of light. In 1911, Sommerfeld must have asked himself whether the "h-discovery" wasn't also a mere chimera.

Like the electron theory of 1905, the quantum theory of 1911 was one—not to say *the*—area of research from which physicists hoped to obtain answers to their fundamental questions. Planck's constant *h* figured not only in thermal radiation or the photoelectric effect. All physical properties that are dependent on temperature at very low temperatures become in one way or another a case for the quantum theory. When the atoms of a solid body, imagined as arranged in lattice form, slowed their oscillations in response to cooling, the process appeared to obey a formula with Planck's *h*. For the electrical resistance of metal too, "Planck's law" (as this formula that was applicable to thermal radiation as well was called) furnished a good description for the approach to absolute zero: if the oscillations of the atoms near a state of rest represent the essential obstacle for the motion of electrons and thereby for an electric current, then the "freezing" of these oscillations according to Planck's formula with a corresponding increase in mobility, and thereby a lower electrical resistance, was responsible. Suddenly, Planck's *h* joined the group of familiar natural constants—the charge *e* and mass *m* of the electron, the speed of light *c*, and "Boltzmann's constant" *k*—as another fundamental constant that seemed always to appear whenever the electrical, magnetic, optical, or thermal

133 From W. H. Bragg, May 17, 1911. DMA, HS 1977-28/A,37. Also in ASWB I.
134 From W. H. Bragg, July 7, 1911. DMA, HS 1977-28/A,37. Also in ASWB I.

properties of matter were traced back to atomic processes. Unlike the idea of the γ-rays as electrons of a velocity above the speed of light, Sommerfeld's "*h*-discovery" was not the monstrous progeny of an overworked theory but rather a hypothesis which added an interesting variation to the bouquet of quantum conceptions around 1910 and was also regarded as such.

Although Sommerfeld initially applied the *h*-hypothesis only to γ-rays, he hoped in future to establish its significance also for the photoelectric effect and other elementary atomic processes. Debye, whose qualities as a researcher in physics he esteemed more highly than his own, was to assist him in this. In 1911, Debye was appointed successor to Einstein at the University of Zürich. When just a few years later he left to take up a professorship at Utrecht, Sommerfeld recommended his second lecturer, Laue for the Zürich professorship, whom he classified as "totally aligned with the fields of interest of Einstein and Debye," if not quite as distinguished as Debye. "I would not ascribe to him the same degree of special vision for physical reality, and extraordinarily quick comprehension that distinguishes Debye," Sommerfeld wrote the experimental physicist Alfred Kleiner of the University of Zürich as he was looking about for a successor to Debye.[135]

With such a keen sense of the qualities in his students indicative of a talent for physics, it must have concerned him that heretofore he himself had not published any truly outstanding research results that could be regarded as milestones of theoretical physics. The *h*-discovery was finally an achievement with which he could step to the fore as a physicist. The opportunity to do so arose in conjunction with a call from Brussels. The Belgian chemical industrialist Ernest Solvay (1838–1922), who for many years had been engaged in using his fortune for the furthering science, had informed his German colleague Walther Nernst in 1910 that he intended to promote research into the structure of matter. Nernst had in turn discussed this with Planck, and so the plan emerged to summon a number of prominent physicists the following year to Brussels to a "council" on the current foundational questions of physics. The congress was to meet under Lorentz's chairmanship and consist of around 20 internationally recognized physicists. We find ourselves at present—the letter of invitation stated—in the middle of an *évolution nouvelle* concerning the principles of a theory of matter. Eight topics were proposed for discussion, stretching from radiation theory to physical-chemical applications.[136] "As I have already informed my esteemed colleague Nernst, it will be a distinct pleasure and of great interest to me to take up the task assigned," Sommerfeld wrote in response to the invitation.[137]

135 To Kleiner, April 3, 1912.ETH, HS 412.Also in ASWB I.

136 Ernest Solvay to Henri Poincaré, June 15, 1911. http://www.univnancy2.fr/poincare/chp/text/solvay1.xml. (5 October 2012). Mehra, The Solvay Conferences, 1975, pp. 3–11; Barkan, *Witches' Sabbath*, 1993; Schirrmacher, *Konzil*, 2012.

137 To Solvay, June 28, 1911. Brüssel, Solvay-Archiv, Sommerfeld.

It was only at the end of October that the Solvay Congress took place. Sommerfeld used the annual Natural Scientists Congress held at Karlsruhe in September 1911 as a rehearsal for a comprehensive presentation of his *h*-hypothesis. The presenters had asked him for a lecture on relativity theory, but Sommerfeld saw fit to change the topic, since this theory had "already been incorporated into the secure canon of physics." In contrast, the "quantum theory of energy, or, as I prefer to call it, the theory of the quantum of action," is "distinctly current and problematic." It was no longer possible to think either of the theory of thermal radiation, or of conceptions "of the molecular structure of matter" apart from Planck's [Quantum] constant. The *h*-hypothesis had proven fundamental to all molecular processes. It would not make sense to explain Planck's constant by means of any atomic models; rather, one had to understand "the existence of molecules as a function and consequence of the existence of an elementary quantum of action . . . An electromagnetic or mechanical 'explanation' of *h* seems to me just as inappropriate and unpromising as a mechanical 'explanation' of Maxwell's Equations. Far more useful would be to pursue the *h*-hypothesis in its manifold consequences, and to trace other phenomena back to it. If, as is scarcely to be denied, our physics stands in need of a new fundamental hypothesis to be appended to the unfamiliar electromagnetic worldview, then the hypothesis of the quantum of action seems to me qualified above all others."[138]

It is not possible to infer from the discussion notes to Sommerfeld's lecture how the *h*-hypothesis was received.[139] He himself thought his lecture "very pretty," as he wrote home to his wife on a postcard. "Einstein here! Langevin sends warm greetings."[140] Paul Langevin (1872–1946) served later as Secretary of the Solvay Congress. Sommerfeld garnered special praise from Fritz Haber (1868–1934), who, as a physical chemist with a sense for practical applications, had up to now followed the discussions of the quantum questions with mixed feelings. "I've sniffed out quantum theory like a mouse round a sausage end," he confessed to Sommerfeld after the Karlsruhe meeting. "Your treatment of the matter has taught me to believe that one gets to the quantum [conception] even if one chooses a quite different starting point. My confidence that the house built on the quantum foundation will hold up has thereby been substantially increased."[141]

Acknowledgement from the mouth of a quantum skeptic such as Haber allowed Sommerfeld to travel to Brussels with renewed self-assurance and great expectations. "We meet daily for about 5 hours," he wrote home on the second day of the conference. "Last evening I had a Frenchman on my left and an Englishman on my right, and I spoke by turns." The luxurious ambience of the Hotel Metropole in which the Congress took place was "really stupendously swanky. We each have a

138 Sommerfeld, *Wirkungsquantum*, 1911a.
139 Sommerfeld, *Wirkungsquantum*, 1911b, pp. 1068–1069.
140 To Johanna, September 27, 1911.
141 From Haber, October 4, 1911. DMA, HS 1977-28/A,126.

private bath and W.C. in our rooms. I bathe each morning. We are the guests of Solvay even at all lunches, dinners, etc. No lunch of fewer than 5 courses! Crazy!" Etiquette demanded a wardrobe the equal of this splendor, in which he did not feel very comfortable. He was not the only one to whom such luxurious surroundings were unfamiliar. "Einstein of course was not wearing a tuxedo yesterday at the dinner with the Solvay family. He doesn't own one."[142]

Ostentation and etiquette, however, evaporated in the scientific debates. It is evident from the program itself how unconventional were the physical findings disseminated in the lectures and attendant discussions that went on throughout the 5 days at the Hotel Metropole. At present, the theories of the smallest particles of matter are in an unsatisfying state, Lorentz said in opening the conference. One feels oneself in a cul-de-sac. Emil Warburg (1846–1931), President of the Imperial Institute of Physics and Technology (*Physikalisch Technische Reichsanstalt*) in Berlin, and Heinrich Rubens (1865–1922), Professor of physics at the University of Berlin, lectured on the experimental testing of Planck's radiation formula. Nernst spoke on the application of quantum theory in physical chemistry. Heike Kamerlingh Onnes (1853–1926) from the University of Leiden presented new experimental findings on electrical resistance. All told, 12 of the 20 invited physicists gave lectures covering a broad spectrum of topics that all had one thing in common: They occasioned controversial discussion of the quantum question.[143]

Sommerfeld too thought of his Brussels lecture primarily as the occasion for discussion. His exposition, he conceded, was "hypothetical and incomplete." The quantum phenomena his theory encompassed were the following: (a) the ejection of electrons (β-rays) and the resulting emission of electromagnetic radiation (γ-rays) in radioactive decay, (b) the stripping of electrons from matter by electromagnetic radiation (the photoelectric effect), (c) the radiation of electromagnetic impulses at sudden braking (X-ray *bremsstrahlung*), and (d) the ejection of electrons through impact with other particles (ionization). The concept of Planck's "oscillators" was not applicable to these "nonperiodic" processes. The h-hypothesis ("In every purely molecular process, a specific universal quantum of action is either absorbed or emitted") described effects of a specific time duration τ (braking time in X-rays, accumulation time in the photoelectric effect, etc.), and in this interval was supposed either to store up or emit the energy E. In every such molecular process, the effect was supposed to run its course in quantum fashion according to the equation $E\tau = h$. To this, the photoelectric effect presented a particular challenge. Sommerfeld and Debye assumed that in a "photoelectric resonator," energy in the form of electromagnetic radiation is collected until sufficient to free an electron from its "molecular bond."[144]

142 To Johanna, October 31, 1911.

143 Langevin/de Broglie, *Théorie du Rayonnement*, 1912; Eucken, *Theorie der Strahlung*, 1914.

144 Sommerfeld, *Application*, 1912; Sommerfeld, *Bedeutung des Wirkungsquantums*, 1914.

The discussion that followed Sommerfeld's lecture was among the liveliest of the whole conference. It took up more than 20 pages in the conference report. One consequence of the *h*-hypothesis seemed absurd in light of everyday experience: In comparing two molecular processes in which different degrees of energy come into play, the one with the greater energy was supposed to be of shorter duration than the one with less energy, since the product of energy multiplied by time was in both cases supposed to equal Planck's constant. If this hypothesis were applied to the penetration of a projectile into a massive object, a projectile of high velocity would decelerate in less time and thereby achieve a lesser depth of penetration than a slow one, which manifestly contradicts the experience of ballistics. But the physics of the tiniest particles does not correspond to everyday experience. High velocity electrons appear to emit impulses on impact with the anticathode of an X-ray tube that correspond to a shorter braking distance than lower velocity electrons. (In the process, the extension of an X-ray impulse generated by the braking of an electron is to be understood as the area between the spherical wave fronts that spread from the starting and end points of the braking distance at the speed of light.) Though this result had been confirmed as quite certain by previous experiments with X-rays, it contradicted—as Sommerfeld stressed at the beginning of his Solvay lecture—"every analogy in the area of ballistic experience."[145]

Contradiction of everyday experience, then, was not a priori an argument against the *h*-hypothesis. In the case of γ-rays, quite to the contrary, some evidence actually suggested that with his theory Sommerfeld had hit the bull's eye. "A great success has been that my structure of the γ-rays appears to have been observed directly, and by Stark's assistant, to boot," Sommerfeld reported home following the Karlsruhe Natural Scientists Congress.[146] He was referring to experiments by Edgar Meyer (1879–1960), who had found the same spatial intensity distribution of γ-rays he had calculated according to his *h*-hypothesis.[147] Nonetheless, as the discussion demonstrated, there many grounds remained on which to regard the hypothesis with skepticism. Poincaré, for instance, derived a contradiction to the principle that for every action there is an equal and opposite reaction ("action = reaction"). If two molecules of different sizes collide, and according to Sommerfeld's quantum interaction fly apart again, the repulsion of the heavier would be of longer duration than the time necessary for the smaller molecule to rebound at the appropriate speed. "The reaction principle would accordingly have only statistical meaning," Poincaré said in criticism of the *h*-hypothesis—and not just of it, for "the same difficulty attaches to the conception of Professor Planck."[148]

145 Sommerfeld, *Bedeutung des Wirkungsquantums*, 1914, p. 253.
146 To Johanna, September 28, 1911.
147 Wheaton, *Tiger*, 1983, pp. 160–163.
148 Sommerfeld, *Bedeutung des Wirkungsquantums*, 1914, p. 301.

What portion of the quantum concepts discussed at Brussels would ultimately prove successful would have to await new experiments. Sommerfeld hoped for confirmation of his *h*-hypothesis in particular from new experiments with X-ray *bremsstrahlung* to determine the "impulse width" λ of an X-ray impulse emitted from a braking event dependent on the energy E of the electron. His theory gave a formula for this which involved no material constant of the anticathode whatsoever. Thereby, "the hardness of the polarized X-rays from the material of the anticathode [should be] independently and universally determined through the velocity of the impacting cathode rays." Ultimately, one could even use the formula to measure Planck's quantum of action exactly, provided, of course, that the formula itself could be confirmed experimentally. "Experiments to that effect are being prepared at my institute," he informed his Brussels colleagues.[149]

6.7 X-Rays and Crystals

The "Council" of Brussels did not lead to conclusions about one theory or another, nor was that the goal of this gathering. Quite the contrary. Following it, the quantum theory was wide open as never before. The various quantum conceptions had been so thoroughly discussed and illuminated from every imaginable angle that hardly a proposal escaped unscathed. Nonetheless, it had become clear that the future belonged to quantum physics. When in November 1911 Willy Wien was honored with the Nobel Prize for his discoveries about the laws of thermal radiation, Sommerfeld saw therein a further indication that the quantum theory was gaining ever increasing significance. "In the coming year, then, may another whole quantum be released for modern radiation theory," he wrote congratulating Wien, alluding to Planck as the next Nobel Prize candidate, for he added that Planck had just been admitted to the Bavarian Academy of Sciences for his contributions to quantum theory.[150]

Sommerfeld had to exercise patience with respect to the X-ray experiments at his institute from which he hoped for confirmation of his *h*-hypothesis. He offered Röntgen's doctoral student Walter Friedrich (1883–1968) his second assistant's position "to work on my X-ray problem."[151] Friedrich's doctoral thesis concerned the "Spatial Distribution of X-rays Emitted from a Platinum Anticathode," that is to say almost exactly what was to be measured in the experiments on the *h*-hypothesis. To be sure, these experiments were very expensive, for the *h*-hypothesis concerned only the braking portion. Distinguishing these from characteristic X-rays meant determining the polarization of the X-rays, which in turn necessitated additional experimental setups. When Lorentz inquired how the experiments Sommerfeld

149 Ibid., p. 266.
150 To W. Wien, November 12, 1911. DMA, NL 56, 010. Also in ASWB I.
151 To Johanna, July 22, 1911.

had referenced at Brussels were going more than 3 months after the Solvay Congress, Sommerfeld replied with a Goethe couplet:

Further efforts have no power;
If they are roses, they will flower.[152]

He could not yet make a pronouncement about "the probability of flowering," for the experiments were "still not complete."[153]

Even 2 months later, the experiments had still not been concluded. Robert Wichard Pohl (1884–1976), an assistant at the Physics Institute of the University of Berlin, who had just completed a habilitation dissertation on X-rays, brought to Sommerfeld's attention other experiments in which likewise a relation between the energy of the X-rays and the energy of the electrons impacting on the anticathode had been confirmed and which might occasion further investigation of the *h*-hypothesis.[154] Sommerfeld replied that he had already been made aware of this by Ernest Rutherford (1871–1937) at Brussels, but that these experiments were not applicable because they concerned the total energy of the X-rays, and not the braking portion separately, which was the issue here. "Dr. Friedrich has begun experiments here with me that have in view making the analogous examination of the polarized energy. These experiments will in fact—quite in the sense of your generous commentary—be decisive for my *h*-hypothesis."[155]

While awaiting confirmation by Friedrich's experiments, Sommerfeld pursued yet another trail. In 1899, he had calculated the diffraction of an X-ray impulse passing through a slit (see Chap. 4). He was trying to interpret the blackening that X-rays caused on photographic plates behind a wedge-shaped slit opening. The two Dutch physicists Hermanus Haga and Cornelis Wind had seen indications of the wave character in such photographic exposures, but this interpretation was controversial. Ten years later, Pohl and Bernhard Walter (1861–1950), another virtuoso in the area of experimental X-ray physics, repeated these slit experiments. By means of a photometer invented by Röntgen's assistant, Peter Paul Koch, visual interpretation of the photographic plates (which at the time had occasioned controversy) could now be carried out free from subjective evaluation. Sommerfeld hoped for evidence from this to support his conception of the generation of X-ray impulses, independent of Friedrich's polarization experiments.[156] The proof of diffraction at the slit would "form a kind of keystone of the theory, and definitely exclude any sort of corpuscular theory of X-radiation," Sommerfeld wrote in explanation of his rationale for the new analysis of such slit experiments. From the blackening of the

152 "Da hilft nun weiter kein Bemühn,/Sinds Rosen nun sie werden blühn." From Epigrams, "Kommt Zeit, Kommt Rat."

153 To Lorentz, February 25, 1912. RANH, Lorentz, inv.nr. 74. Also in ASWB I.

154 From Pohl, April 29, 1912. DMA, HS 1977-28/A,265. Pohl, *Physik*, 1912, p. 9.

155 To Pohl, May 1, 1912. Personal estate in the possession of Dr. Robert Pohl.

156 Pohl, *Physik*, 1912, pp. 23–37.

Fig. 16: Following the colloquium, participants gathered in a Munich beer cellar to bowl. Left front at table, Laue, to his left Epstein; opposite, leaning back, Ewald, to his right, Koch (Courtesy: Deutsches Museum, Munich, Archive).

available photographic plates, he established a wavelength (impulse width) of at most 4.10^{-9} cm. But because of the great experimental uncertainties, he did not want "to lay great quantitative stress" on this determination and recommended instead carrying out further diffraction experiments.[157]

After Sommerfeld had sent this paper to the *Annalen der Physik,* he went as he did at the start of every year to Mittenwald to relax after a strenuous winter semester by skiing. It would be an eventful semester break, for during these weeks further diffraction experiments successfully proved that X-rays were waves—albeit in quite a different way from what Sommerfeld had imagined. On March 9, 1912, several participants in the "big-shot-free colloquium" sent Sommerfeld at Mittenwald a postcard picturing them as a merry company in a Munich beer cellar, where they were concluding the last colloquium day with their customary bowling game.[158]

157 Sommerfeld, *Beugung*, 1912, pp. 474 & 506.
158 From Knipping, Koch, Laue, and Lenz, March 9, 1912. DMA, NL 89, 016, folder 1,4.

The bowling party included most of the doctoral students and lecturers from Sommerfeld's and Röntgen's institutes: Ewald, Laue, Koch, and Paul Knipping (1883–1935), who was just completing his doctoral work under Röntgen. Two months later Sommerfeld submitted a report to the Bavarian Academy of Sciences that had been drafted by Friedrich and signed also by Knipping and Laue. It stated that the three undersigned had been occupied at the Sommerfeld institute "since April 21, 1912 with interference experiments with X-rays passing through crystals… The leading idea was that interferences resulting from the space lattice structure of crystals occur because the lattice constants are ca. 10 x greater than the assumed wave-length of X-rays." The success of these experiments was illustrated with two photographs on which the interferences were visible in the form of regular arrangements of points.[159]

Aside from these documents, there are no contemporaneous sources that give information concerning the impetus and course of the interference experiments during these April days of 1912. In 1914, Laue was awarded the Nobel Prize for the discovery. In 1920, he described in his Nobel Prize speech what had brought him to the idea: "That the lattice constant in the crystals was of a size on the order of 10^{-8} cm, was sufficiently clear by analogy with other atomic distances in solid and liquid bodies; furthermore from the density, the molecular weight, and the mass of the hydrogen atom, which had just been extraordinarily well established, it was easy to verify the 10^{-9} cm order of magnitude Wien and Sommerfeld had estimated for the wave-length of X-rays. So the relation of wave-length and lattice constant was extraordinarily favorable if X-rays were passed through a crystal. I told Ewald immediately that I would expect to see interference phenomena in the X-rays."[160]

Ewald had just submitted his doctoral thesis on crystal optics. As Sommerfeld spelled out in his recommendation to the faculty, Ewald was to calculate "the dispersion and double refraction in an ideal rhomboid electron lattice."[161] As an expert in the area of optics, Laue (to whom Sommerfeld had entrusted the subject "wave optics" for the *Encyclopedia of Mathematical Sciences*)[162] must have been a frequent informational resource for Ewald. On January 8, Laue gave a lecture in the Sohncke Colloquium on "Light Interferences, Particle Diffraction."[163] Ewald's dissertation was also concerned with how light is scattered by many particles. To be sure, Ewald's scattering centers were arranged regularly into a spatial lattice. It is entirely plausible that in conversation with Ewald, Laue had hit on the idea of investigating the propagation of X-rays in such a lattice as well. But "the acknowledged masters of our science" would not have accepted his idea, as Laue explained in his Nobel Prize

159 Friedrich, Knipping, and Laue to the Bavarian Academy of Sciences, May 4, 1912. DMA, HS.
160 Laue, *Auffindung*, 1920.
161 To the Philosophical Faculty, Section 2, 16. February 16, 1912. UA, OC I 38p
162 Laue, *Wellenoptik*, 1915.
163 Münchener Physikalisches Mittwochskolloquium, DMA, 1997–5115.

speech, and it had required "a certain amount of diplomacy" to have the experiment carried out by the Sommerfeld institute.[164]

How Laue came to his idea in the discussion of Ewald's doctoral work, what this consisted of, and why "a certain amount of diplomacy was necessary" to help it to a breakthrough can only be reconstructed from sources of later years.[165] There is much to suggest that Laue was at first following a false trail and that the "acknowledged masters"—by whom he meant Röntgen and Sommerfeld—were absolutely correct in their doubts about the feasibility of the idea. In a letter to Sommerfeld, Debye expressed the suspicion that coincidence had played a role in "Laue's discovery" but left it at this innuendo.[166] To the outside world, the façade of a successful circle of physicists was maintained, but as in any group of independently minded individuals, they did not all get along with each other. In the Sommerfeld circle, Laue was isolated, which he recalled later "with a certain bitterness about the Munich years." He had "experienced [much] that was quite unpleasant," he wrote Sommerfeld in 1920, when fresh discord arose concerning the representation of the history of the discovery in Laue's Nobel Prize speech. "Let me touch on just one point—not the worst. Why did you exclude me when you celebrated the discovery of the X-ray interferences with Friedrich and Knipping and the other younger colleagues?" He conceded that he had "not always behaved correctly" with respect to Sommerfeld, but attributed that to his initially very delicate frame of mind, and would have hoped Sommerfeld might recognize "extenuating circumstances" in this case. On the underlying causes of the rupture, the letters reveal nothing, but it must have been deep, for with one of the annual gatherings at the Mittenwald country house of Willy Wien in the offing, Sommerfeld wrote that he hoped Laue would not be coming: "I'm afraid my enjoyment would suffer greatly from his presence."[167]

Given that one spoke publicly of "Laue's discovery," while privately the putative discoverer was excluded from the celebration of the discovery, no sober, objective representation of the story can be expected on the part of those involved. When all available clues, statements, and counterstatements are carefully weighed against one another, it is plausible to reconstruct the path from conception to discovery somewhat as follows: Laue had planned to excite a crystal to emission of X-rays by striking it with a primary X-ray. He anticipated the interference of the characteristic radiation of the crystal—not, as emerged only much later, the interference of the primary X-rays. With this idea, he probably elicited the demurral of the

164 Laue, *Auffindung*, 1920. (http://www.nobelprize.org/nobel_prizes/physics/laureates/1914/laue-lecture.pdf (30 January 2013): "the acknowledged masters of our science, to whom I had the opportunity of submitting it, entertained certain doubts about this viewpoint. A certain amount of diplomacy was necessary before Friedrich and Knipping were finally permitted to carry out the experiment according to my plan…").

165 Forman, *Discovery*, 1969; Ewald, *Myth*, 1969; Eckert, *Disputed Discovery*, 2012.

166 From Debye, May 13, 1912. DMA, HS 1977-28/A,61.

167 To W. Wien, February 10, 1916. DMA, NL 56, 010. Also in ASWB I.

"acknowledged masters," for the crystal atoms emit their characteristic X-rays in an uncoordinated manner. Among the emitted waves there is no phase relation as would be necessary for an interference. Why, then, should Sommerfeld assign his "experimental assistant" Friedrich, whom he had shortly before charged with X-ray experiments designed to confirm his *h*-hypothesis, to an experiment based on an erroneous conception? When Friedrich hesitated to carry out the experiment, Laue found in Knipping a companion in arms who, albeit extremely skilled as an experimental technician, was inexperienced insofar as the underlying theoretical conceptions were concerned. It may have been this Debye was thinking of when he attributed to coincidence a not insignificant role in the course of the discovery. He was not alone in this assessment. Ioffe expressed the same reservation. First, the photographic plate on which the interference phenomena were to be demonstrated was positioned sideways to the crystal, so that the characteristic radiation of the crystal could be photographed without the primary ray. The sought-after phenomena were so weak that an exposure of many hours was made. "And day by day, the X-ray tube crackled prodigiously, but the plate remained unblackened. The young physicist Knipping, working in the same room had to leave the laboratory in two to three weeks, but the continuously operating tube interfered with his experiments. In order to see at least something on the photographic plate, he repositioned it such that the X-rays fell on it—and the great discovery appeared . . ."[168]

This, or a very similar scenario, is how it must have gone that April of 1912 in the basement of the Sommerfeld institute where these experiments were carried out. The two experimenters, Friedrich and Knipping, were left mostly to their own devices, for Sommerfeld was traveling a great deal. At the beginning of April, shortly after returning to Munich from his ski vacation in March, he and Johanna traveled to Lake Garda. "We are treating ourselves to an Italian spring this year," he wrote Hilbert on the stationery of the Grand Hotel Torbole on April 10.[169] The day interference patterns first appeared on the photographic plate in Munich, Johanna was at home again, although Sommerfeld had traveled on to Vienna where he was to give a talk.[170] Back in Munich in time for the start of the summer semester, and confronted with the discovery, Sommerfeld's surprise must have been great. But however the interference phenomena may have come about, it was at once clear to him that this was a sensational discovery. Normally, scientific knowledge is announced to the profession through a publication, but in this case, Sommerfeld did not want so much time to elapse and risk that the discovery become known elsewhere, repeated by others, and published first. Disputes over rights of priority were already commonplace in science. Thus, he used the first meeting of the mathematical physics class at the Bavarian Academy of Sciences to submit the report, signed by Friedrich,

168 Ioffe, *Begegnungen*, 1967, p. 40.
169 To Hilbert, April 10, 1912. SUB, Cod. Ms. D. Hilbert 379 A
170 To Johanna, April 21, 1912.

Knipping, and Laue, "for the protection of priority of a scientific discovery" to the Secretary of the Academy.[171]

Thereafter, colleagues outside of Munich could also be made aware of the discovery, if initially only of the fact, not the substance. "On Laue's initiative, a very great practical discovery has been made in my laboratory," Sommerfeld wrote to Ehrenfest. "Not to spoil your surprise at the publication in question, I'll say no more about it."[172] A day later, he referred to "the wonderful interference photographs" also in a letter to Alfred Kleiner, who soon thereafter traveled to Munich to meet personally with Laue, then a candidate for the appointment to the chair of theoretical physics at the University of Zürich.[173] Although the interpretation of the interference pattern remained to be made, enthusiasm over the discovery was unanimous among all who heard of it. The X-ray diffraction by crystals was for many months the all-consuming topic in Munich, and soon elsewhere, too. A week of lectures was arranged at Göttingen for the end of the summer semester at which Sommerfeld was to present the current state of quantum theory. But Sommerfeld spoke, as he informed Schwarzschild, "not on quantum theory, but on our X-ray experiments."[174]

At the same time, the success in Sommerfeld's basement evoked feelings of rivalry in Röntgen. That the experiments of Friedrich and Knipping, his own doctoral students, were carried out next door at the Institute for Theoretical Physics made his own institute appear unproductive. "Röntgen wants to keep Friedrich to himself," Sommerfeld wrote his wife, who had traveled ahead of him to their summer lodgings in the Berchtesgaden Alps, where they were going to relax from the eventful summer semester. But Röntgen had been "so foolish" as to demand a decision from Friedrich on the spot, which he had declined. "I'm very happy about it, not only because I need F[riedrich], but also because one doesn't want to be bested in a test of wills. That R[öntgen] will feel the same way doesn't hurt."[175] Ten days later, he had put the Göttingen lectures behind him. "It all went very well, and I'm still alive, in fact fresher than 3 weeks ago," he wrote his wife. "The major event of the previous semester was the interference phenomena with X-Rays created in my institute," Sommerfeld summed up the efforts of the previous months in a letter to Langevin.[176]

At least one part of these efforts is to be attributed to having postponed a correct theoretical interpretation of the experimental discovery. Laue did offer a persuasive explanation of the interference pattern generated from a spatial lattice, but the

171 Transcript of the meeting of May 4, 1912, DMA, HS 1951-5; also in Forman, *Discovery*, 1969, p. 66.

172 To Ehrenfest, May12, 1912. AHQP/Ehr (25). Also in ASWB I.

173 To Kleiner, Ma13, 1912. ETH, HS 412. From Kleiner, June, 1912. DMA, HS 1977-28/A, 171. Also in ASWB I.

174 To Schwarzschild, undated [early July, 1912]. SUB, Schwarzschild 743.

175 To Johanna, July20, 1912

176 To Langevin, undated (presumably August/September, 1912). ESPC, Langevin.

origin of the overlapping waves that appeared in the process remained unclear. If this was a question of the characteristic radiation of crystals, the interference phenomena should have occurred only in the case of crystals, which are made up of heavy atoms, because only these display a marked characteristic X-radiation. In crystals such as diamonds, made up only of carbon, the phenomenon ought not to appear. Nonetheless, clear interference marks from radiation through diamond crystals appeared on the photographic plates.[177]

Such inconsistencies must have given rise to considerable discussion among the experimental and theoretical physicists at Munich. Elsewhere, too, confusion reigned. "The men who did the work entirely failed to understand what it meant, and give an explanation which was obviously wrong," the English physicist Henry Moseley (1887–1915) wrote in a letter to his mother when he himself began making a name for himself in this area of research.[178] William Henry Bragg and William Lawrence Bragg (1890–1971), father and son, first attempted to explain the phenomenon in the framework of a particle conception: Perhaps the dot pattern arose on the photographic plates because the X-ray particles could pass through the atomic lattice of the crystal only along "canals" in particular directions that were dependent on the lattice symmetry. They were quickly convinced, however, that this explanation could not be brought into line with the observed data. The younger Bragg was the first to present an explanation that appeared to be far more plausible than Laue's. The dot pattern on the photographic plates came about not as the result of an interference of the characteristic radiation of the crystal, but rather from the *bremsstrahlung* of the primary ray selected from the primary radiation of the crystal. One could imagine the shower of impulses of the primary radiation as an overlapping of many wave trains of differing wavelengths that are reflected on the lattice planes of the crystal. By a simple equation of the path lengths of two parallel rays reflected from neighboring lattice planes in the crystal, the younger Bragg formulated a condition determining which wavelengths at what angles overlapped positively and thus contributed to the interference pattern. This conception was brilliantly confirmed by an experiment, in which (unlike at Munich where an X-ray passed through a small perpendicular crystal plate vertically) the X-ray was reflected at a low angle of incidence on the surface of a crystal.[179]

Not long afterwards, the Russian crystallographer Georg Wulff (1863–1925) recognized that the interference pattern the younger Bragg had explained with his reflection theory was consistent with Laue's theory.[180] "If the crystal molecules send out the vibrations given them by the X-rays, this 'reflection' is the same phenomenon as the interference of the penetrating rays," he wrote Sommerfeld in January 1913.[181]

177 Friedrich/Knipping/Laue, *Interferenzerscheinungen*, 1912, 319–320.
178 Heilbron, *Moseley*, 1974, pp. 194–195.
179 Wheaton, *Tiger*, 1983, pp. 208–212; Jenkin, *Partnership*, 2001, p. 381.
180 Wulff, *Kristallröntgenogramme*, 1913.
181 From Wulff, January 16, 1913.DMA, HS 1977-28/A,377.

In March 1913, nearly a year after the discovery, Laue still regarded it as "a very strange and still unexplained fact that the extremely inhomogeneous impulses of the incoming X-radiation in the crystal [could] generate oscillations of quite precisely defined wave-length."[182] "All these things are still in flux," Sommerfeld consoled his audience in a lecture in May 1913. "It will take several years, still, and quite a few X-ray tubes will yet have to give up their lives before this area will have been systematically mined." The interference phenomena in crystals were for him also an example of how theoretical convictions can be corrected through experimentation. "I see an especially fine victory of the crystal photographs in that they have persuaded the worthiest and cleverest adherent of the opposing view, of the corpuscular theory of X-rays, namely Professor Bragg himself, and drawn him into the camp of the wave theory."[183]

One year after the discovery of the X-ray interferences in crystals, though essential aspects of this phenomenon remained unknown, this much was clear: With this discovery, two new branches of physics had been christened. For one thing, it was possible to use crystals to scan the shower of impulses from an X-ray tube to find the waves it contains. As it is possible to make visible the different constituent colors contained in a white beam of light with a glass prism, it was now possible to employ crystals for spectral analysis of X-rays, and this would prove to be the key to the elucidation of the processes inside atoms. For another thing, X-rays of suitable wavelengths could be used to determine the unknown structure of crystals. It was not by chance that several years later Sommerfeld characterized "Laue's discovery" as the "most important scientific event" in the history of his institute.[184]

In 1913, physicists could hardly guess what a boost X-ray spectroscopy and X-ray crystal structure analysis would get as new branches of physics. But it was already clear that the initial enthusiasm attending the discovery was not a flash in the pan. In September 1913, X-ray interferences were the focus of lectures at the annual meetings of both the British Association for the Advancement of Science in Birmingham and the Natural Scientists Congress in Vienna. At the second Solvay Congress, held in late October 1913 in Brussels, under the overarching theme "The Structure of Matter," the subject was also X-rays and crystals. At all these conferences, attendees were informed first hand of the latest advances: At Birmingham, the elder Bragg demonstrated a diamond lattice model he and his son had reconstructed from the X-ray interference patterns.[185] At Vienna, Laue and Friedrich gave lectures surveying the state of research as it had developed in the year and a half since the discovery.[186] At Brussels, Laue and the elder Bragg put their work up for discussion, followed by a lecture by Sommerfeld, who compared the two

182 Laue, *Prüfung*, 1913, p. 1000.
183 Sommerfeld, *Anschauungen*, 1913, p. 706.
184 Sommerfeld, *Institut*, 1926.
185 Ewald, *Bericht*, 1913.
186 Laue, *Röntgenstrahlinterferenzen*,1913; Friedrich, *Röntgenstrahlinterferenzen*, 1913.

procedures with a concrete example (zinc blende), and used the opportunity emphatically to praise the "brilliant experimental work" of the elder Bragg and the "magnificent theoretical research of his son."[187]

The second Solvay Congress was for Sommerfeld the highpoint of a period of extraordinary effort. Throughout 1913, he felt nervously exhausted. In April 1913, in the South Tyrolean Mountains, by "eating well, taking afternoon naps and very leisurely walks, not smoking, not talking physics," he hoped to regain the inner calm that had abandoned him during the turbulent months since the discovery of the X-ray interferences.[188] The plans he had had for experiments for which he had appointed Friedrich, and on which he had placed such great hopes, now seemed inconsequential. The questions arising from the diffraction of X-rays by crystals now assumed priority. After the first interference experiments, he had immediately arranged for Friedrich to receive better experimental equipment. He used his Brussels relationships to apply for support for further X-ray interference experiments from the Physics Foundation established by Solvay.[189] The X-ray interferences brought his institute the recognition he had been denied for his *h*-discovery. But the exertions this demanded also took their toll. Nearly every letter to his wife contained news of the current state of his health: "I'm doing quite well. The love is very good for me," he wrote home from Aachen, where he had made a stop on his trip to the Solvay Congress at Brussels and was cared for lovingly by old friends.[190] "I'm doing visibly better," he wrote reassuringly to Johanna a few days later, after his arrival in Brussels. "The drastic treatment seems really to be taking effect again, and shows that all my complaints are purely hysterical."[191] In truth, he was hardly doing better. "The last time in Brussels, I felt really quite miserable, and was almost not up to taking part in the discussions," he wrote in a letter to Langevin sometime later.[192]

His letters do not reveal the exact nature of his ailments. If they were actually "purely hysterical," the cause was perhaps the continuing strain of success to which he felt subject as Boltzmann's successor. The X-ray interferences went down in history, to be sure, but their discovery was only partly attributable to his own contribution. The great scientific accomplishment, by which he might feel the equal of a Lorentz or a Planck as a theoretical physicist, was still to be made. Or was it the "Age of Anxiety" that troubled so many Germans between 1880 and World War I? Nervous breakdowns seem to have been rather typical of the Wilhelmine Era.[193]

187 Sommerfeld, *Photogrammes*, 1921, p. 125. ("brillants travaux experimentaux de M. W. H. Bragg et les magnifiques recherches théoriques de son fils W. L. Bragg".); see also Ewald, *Intensität*, 1914.

188 An Johanna, April 4, 1913.

189 To the Solvay-Foundation, January 14, 1913. ESPC, Langevin.

190 To Johanna, October 25, 1913.

191 To Johanna, late October, 1913.

192 To Langevin, June 1, 1914. ESPC, Langevin, L 76/53.

193 Radkau, *Zeitalter*, 1998.

In this respect, Sommerfeld and his wife were not exceptional. Johanna especially appears to have been emotionally quite shattered shortly before the move to Munich. "Recently at Aachen, my wife has been quite unwell," Sommerfeld wrote Hilbert, just at the time he too had recovered from a nervous breakdown. In Munich, though, "at one stroke, the whole misery of fainting, and cardiac affection" disappeared.[194] Even then, though, worries over their emotional health persisted. "By being a little smarter with regard to our health, both our psychological conditions will in future be better, as they have been in years past," Sommerfeld wrote in March 1914, hopefully for himself and his wife.[195]

194 To Hilbert, April 28, 1908. SUB, Cod. Ms. D. Hilbert 379 A.
195 To Johanna, March 24, 1914.

7 Physics in War and Peace

"These days, all personal issues and problems are overshadowed by the question: when will the conflagration beginning tomorrow in Serbia leap across to Germany? Events are developing at a simply terrifying pace."[1] Sommerfeld wrote these lines to his wife at the vacation house at Berchtesgaden on July 26, 1914. Four weeks earlier, the Austro-Hungarian Crown Prince Ferdinand and his wife had been murdered in Sarajevo. Proceeding by a terrible logic, the assassination had brought in its train first an ultimatum from the Habsburg Monarchy to Serbia (July 23), then mobilization of the Serbian army (July 25), and finally the Austro-Hungarian declaration of war against Serbia (July 28). The politics of alliances among the European powers took care of the rest. Czarist Russia sided with Serbia, while the Wilhelmine Reich declared itself allied with Austria-Hungary. On August 1st, Germany declared war on Russia, which in turn triggered the mobilization of Russia's ally France. As German troops crossed the Belgian frontier, Great Britain, which had committed itself to the defense of Belgian neutrality, ordered the mobilization of its army. World War I had begun.[2]

7.1 "For Me, the Political Future Lies in Utter Darkness"

The "July Crisis" and the outbreak of war threw Europe into a delusional frenzy. "How will this all turn out? We live in mad anxiety. The automobiles seem to be tearing about the streets at unusually breakneck speed." To the end, from a naïve faith in the noble intentions of politicians in the Imperial capital, Sommerfeld hoped for a diplomatic resolution of the looming conflict: "It's clear to see: the Emperor does his best to avoid war, or defer it. But will he succeed??"[3] From his youth, national consciousness and trust in the order of the state had become for Sommerfeld virtually axiomatic foundational political principles. In Aachen, he had given expression to his civic sensibilities by entry into the National Liberal Party.[4] Given his political and ethical convictions, he felt an obligation to the "good German idealism," as he wrote in a newspaper article on the occasion of Bismarck's 90th birthday. Idealism was, "Fulfillment of duty and self-discipline, application of one's whole strength to the position assigned us, setting aside of personal comfort and self-indulgence, commitment to the greater good, contempt for incompetent

1 To Johanna, July 26, 1914.
2 Berghahn, *Weltkrieg*, 2003.
3 To Johanna, July 31, 1914.
4 Personnel file, BayHStA, MK 35736.

M. Eckert, *Arnold Sommerfeld: Science, Life and Turbulent Times 1868-1951*,
DOI 10.1007/978-1-4614-7461-6_7,

pretense, faith in the force of virtue—all this is what we understand by the good German word 'idealism.' And let us take these qualities, which are native to us, as the particular stamp of our people, in the political arena as well as the more broadly spiritual . . . And was not Bismarck himself the most splendid example of an idealism of ends joined to a healthy realism of means?"[5]

And it was in the matrix of this sensibility, too, that he experienced the weeks and months following the outbreak of war in August 1914. After a brief holiday in the mountains of Berchtesgaden, he returned to Munich without his family to assist in the changes taking place at the University in the wake of the war's outbreak. Preparations had to be made for trainloads of the wounded. "Write to me how many beds we can provide when we too are called on to take in wounded," Sommerfeld inquired of Johanna, who lingered with the children at the vacation house. "Much cheering, and extra editions of the papers," he wrote describing the atmosphere in Munich.[6]

Of course the question of his own participation in the war presented itself too. At 45, he no longer had to worry about being called up to military duty. His colleague the Romance language scholar Karl Vossler, 4 years his junior, had reported for duty, and been sent home again. "About my own situation I can't yet tell you much," he wrote on August 26. He speculated that as a reserve officer, he might make himself useful to the army drilling recruits.[7] No one was in a hurry, however, to order the Herr Professor to report for military duty on some barracks square. Even 2 months later, Sommerfeld didn't know whether to ready himself for military duty or prepare his lectures for the coming semester. "Judging from my personal experience at general headquarters, it seems no great store is set on my usefulness," he wrote to Karl Schwarzschild. "If I am to be left at home, it's just as well since I've never felt myself to be militarily strong." Schwarzschild did military service as commanding officer of a weather station in Belgium, which came as a great relief to Sommerfeld. "It would be a great shame for *you* to be sent to the front." He also made it known that he did not share the euphoria with which many professorial colleagues had experienced the outbreak of the war. "For me, the political future lies in utter darkness; I don't share the happy optimism of your brother-in-law Emden, whom I do often find edifying. Even your modest proposal to dispatch Belgium doesn't quite make sense to me. I think it would very seriously encumber us."[8]

Already at the outbreak of World War I, violation of Belgian neutrality had made Germany appear the aggressor in the eyes of world public opinion. Thereafter, Belgium became the object of unprecedented war propaganda. In an "Appeal to the Civilized World," 93 prominent German intellectuals aligned themselves in

5 Aachener Allgemeine Zeitung, April 4, 1905.
6 To Johanna, August 20, 1914.
7 To Johanna, August 26, 1914.
8 To Schwarzschild, October 31, 1914. SUB, Schwarzschild 743. Also in ASWB I.

solidarity with German militarism, denying the Entente's accusations of war atrocities committed by German troops in Belgium. The "appeal" included the signatures of respected scientists such as Felix Klein, Walther Nernst, and Max Planck. Among Sommerfeld's colleagues at the University of Munich, signers were Wilhelm Röntgen and Karl Vossler. Sommerfeld's signature is absent, though it is not known whether he refused it or simply was not asked; Karl Schwarzschild and Robert Emden, too, who unlike Sommerfeld did not think the invasion of Belgium burdensome, were missing from among the signers.[9] Shortly thereafter, more than 3,000 German university professors—and this time Sommerfeld among them—declared their affirmative position towards that which abroad was referred to disparagingly as "Prussian militarism." "The spirit of the German Army is no different from that of the German people, for the two are one, and we belong as well . . . This spirit is alive not only in Prussia, but is the same throughout the states of the German Reich. It is the same in war and in peace."[10]

Soon, Sommerfeld, too, no longer evinced any of that skepticism with which he had initially reacted to the war euphoria of his professorial colleagues. The war propaganda not only affected the broad public towards whom it was directed but poisoned the atmosphere among scientists as well. When on November 20, 1914, an article appeared in *Naturwissenschaften* in which Lorentz expressed publicly to Ernest Solvay, the Belgian patron of science, his "sympathy for the profoundly suffering people, represented so admirably by him," Sommerfeld was incensed. "Shouldn't Lorentz be just as scrupulous about the truth in the matter of inciting anti-German and pro-Belgian sentiments??" he wrote Willy Wien. One ought to "make it clear" to the editor of *Naturwissenschaften* "that it is inappropriate to print an article *in gloriam belgicam* in a German journal."[11] Shortly before, in answer to what he perceived to be the anti-German declaration by English physicists, Wien had authored an appeal culminating in the demand that "the unjustified scientific influence of the English be rejected." German physicists should publish their research in English journals only if there is reciprocation. In the citation of research literature, English authors ought "no longer be given greater consideration than our countrymen, as has often been the case."[12] Sommerfeld endorsed this appeal, although he was entirely unfamiliar with the declaration of the English physicists. No trace remained of the scientific internationalism that just a year before had brought Wien and Sommerfeld together with their colleagues from England and France at the last Solvay Congress in Brussels.

9 http://de.wikipedia.org/wiki/Manifest_der_93 (30 January 2013).

10 The original text with French translation and the entire set of signatures, organized by university and department, are to be found at http://publikationen.ub.uni-frankfurt.de/volltexte/2006/3235/pdf/A008838631.pdf (30 January 2013).

11 To W. Wien, December 25, 1914. SBPK, Autogr. 1/1253. Also in ASWB I.

12 Aufruf. DMA, NL 56, 005. Printed in ASWB I. Wolff, *Physicists*, 2003.

7.2 Return to Theory

In the months leading up to the outbreak of war, Friedrich and Ewald had shown that with the topic "X-Rays and Crystals," a rich gold lode had been opened.[13] Ewald had shared one of the two positions of assistant in the institute for Theoretical Physics with Wilhelm Lenz, who had become lecturer under Sommerfeld in February 1914. As second assistant, Friedrich was responsible for experiments in the institute basement. Now Friedrich and Ewald fulfilled their military obligations as X-ray technicians in military hospitals—much as the X-ray equipment which Sommerfeld donated to a military hospital in Munich.[14] Lenz served as a radio operator somewhere in Belgium.

Nonetheless, scientific research at the Sommerfeld institute did not come a complete standstill. Even if, for now, the "gold vein" of X-ray diffraction of crystals could not—experimentally, at least—be further mined, there was no dearth of subjects Sommerfeld could pursue even without assistants. In 1913, his *h*-discovery, on which in a joint publication with Debye he had drawn for the explanation of the photo effect, had ultimately proven wrong.[15] But this did not mean he would turn his back on quantum theory. In 1912, Debye, Max Born, and Theodore von Kármán had explained the temperature dependence of the specific heat of solids by quantizing the energy of lattice vibrations.[16] In a similar manner—so Sommerfeld thought—sonic oscillations in a gas could also be quantized. In 1913, at a quantum physics conference at Göttingen he had presented a theory developed from an idea of Lenz; he was soon forced to concede, however, that he could not proceed with it.[17] He handed another variant of this idea over to Alfred Landé as the subject of a doctoral thesis. Landé was to adapt the method to optics, that is, quantize the sum of the electromagnetic energy of the light waves and the resonating electrons. "My own interest in the question was in knowing whether the total energy belonging to the vibrational region $(v, v + dv)$is to be distributed by quanta, or whether perhaps the ether energy and the electron energy each behave quantally"—this is how Sommerfeld, in his testimonial on Landé's dissertation, specified his research interest.[18] In retrospect, Landé felt his doctoral thesis had dealt only with a matter Sommerfeld wished clarified, but that was of no particular interest to anyone else.[19]

13 Ewald, *Intensität*, 1914; Ewald/Friedrich, *Röntgenaufnahmen*, 1914.

14 To W. Wien, December 25, 1914. SBPK, Autogr. 1/1253. Also in ASWB I.

15 Debye/Sommerfeld, *Theorie*, 1913; Wheaton, *Tiger*, 1983, pp. 186–88.

16 Eckert/Schubert/Torkar, *Roots*, 1992, pp. 33–34.

17 Sommerfeld, *Probleme*, 1914; to Hilbert, October 14, 1913. SUB, Cod. Ms. D. Hilbert 379 A.

18 Sommerfeld to the Philosophical Faculty, 2. Sektion, 28 April, 1914. UAM, OC 1 40 p; Landé, *Methode*, 1914.

19 Interview with Landé conducted by Thomas S. Kuhn and John Heilbron, March 5, 1962. AHQP. http://www.aip.org/history/ohilist/4728_1.html (30 January 2013).

In light of the knowledge of quantum mechanics first developed in the 1920s and of quantum field theory built on its foundation, it is not surprising that early attempts at quantization were not always successful. From the viewpoint of 1914, however, a method such as Debye's, which had been so successful in one area, appeared exceptionally appropriate as the subject of doctoral work. Landé was not the only doctoral candidate Sommerfeld thus led to the field of quantum physics. Walter Dehlinger, who as a doctoral student of Debye had learned the method firsthand, took his degree with Sommerfeld in June 1914 with a dissertation on "Specific Heat of Two-Atom Crystals." Dehlinger was to generalize to two-atom crystal (such as salt and NaCl) calculations Debye had made only for the oscillations of single-atom crystal lattices. The reciprocal oscillations of the coupled atoms in every crystal cell also generate electromagnetic radiation ("residual rays"), so that two distinct physical phenomena were joined in one context. The heat motion of the atoms in question should lead to faster and slower oscillations if the mass of the two atoms differ significantly from one another and thereby appear also in the spectrum of the residual rays. As the particular merit of his doctoral candidate, Sommerfeld stressed that "Mr. Dehlinger has confirmed this conjecture, and has also investigated the context of the narrow spectrum with the frequency of the residual rays . . . He has thereby shown he is completely familiar with the methods of modern theoretical physics."[20]

As doctoral supervisor, Sommerfeld gained a feel for those problems and methods that were tied to the new approach to research from the work of his doctoral and habilitation candidates and a sense of whether it paid to pursue one or another path. Following the work of Lenz, Landé, and Dehlinger, Debye's "method of natural oscillation," as Sommerfeld called it, lost its attraction as a new approach to quantum theory. It had been useful with respect to heat radiation and specific heat, but beyond that no further areas appeared in which it might prove useful. Phenomena such as the emission of gamma rays or of electrons in the photo effect, for which Sommerfeld had developed the *h*-hypothesis, also eluded Debye's method.

New theoretical formulations were needed above all in the area of atomic spectra. In 1912, Friedrich Paschen and his doctoral candidate Ernst Back had published a discovery concerning the splitting of spectral lines in magnetic fields that focused new attention on magneto-optic effects. The phenomenon itself—the splitting of spectral lines in the magnetic field—had long been known, but with the discovery of the "Paschen-Back effect," it became clear that an understanding of it was still far off. In the simplest case, by application of a magnetic field, a spectral line was broken up into a group of three lines (a triplet). This seemed explicable if it were assumed that a spectral line was created by the oscillation of an electron around its position at rest. In the magnetic field, the oscillation is split into three parts: an undisturbed oscillation parallel to the magnetic field and two circular oscillations

20 Sommerfeld to the Philosophical Faculty, 2. Sektion, June 26, 1914. UAM, OC I 40 p.

each with opposite rotation.[21] In the context of his electron theory, Lorentz had formulated this concept in terms of mathematical equations and in 1902 was awarded the Nobel Prize, jointly with Zeeman, who had discovered the phenomenon. To be sure, this "normal" Zeeman effect soon turned out actually to be the exception. Far more frequently, "complex Zeeman types" were observed, with greater than triple splitting, such as quartets, sextets, and so on. Now Paschen and Back found that in the case of very strong magnetic fields, this "anomalous" Zeeman effect returns to the "normal" form. In editing the *Encyclopedia*, Sommerfeld had already encountered the problems of the Zeeman effect, but it was the Paschen-Back effect that first prompted him to attempt a theory of this phenomenon himself. For the three possible directions of oscillation of the electron, he set frequencies differing only slightly from one another. An anomalous splitting pattern occurs at first then in the magnetic field, since three normal Zeeman effects overlap. If the strength of the field increases to the point that the energy differentials between the fundamental frequencies become negligible, the normal Zeeman triplet appears as the splitting pattern.[22] He thereby challenged Woldemar Voigt, who had already presented a similar theory. This did not result in a struggle over priority, however. "So long as we have no theory of spectral lines, any theory of the magneto-optic effect will remain fragmentary," Sommerfeld stated in March of 1913.[23] By publishing a general description of his theory in the *Annalen der Physik*,[24] Voigt underscored his authority in the field of the magneto-optic effect, and Sommerfeld conceded that his own excursus into the magneto-optic effect could "clearly not in any way be compared to the Voigtian theory." He had wished "merely to illustrate the conditions prevailing in strong fields."[25] In June 1913, Sommerfeld lectured "On Complex Zeeman Effects" at the Munich Colloquium.[26]

A month later, the still scarcely known Danish theoretician Niels Bohr published a new theory of spectral lines, in which the electrons are, by a quantum rule, prescribed stable orbits around the atomic nucleus, and only leaps between such orbits lead to the radiation of electromagnetic waves.[27] Thereby, he gave a physical interpretation to the empirically discovered formula $v = N\,(1/n^2 - 1/m^2)$ for spectral series of hydrogen (v = the frequency of a spectral line, N = the Rydberg-Ritz constant ($= 2\pi^2 me^4/h^3$ according to Bohr's theory, m = the mass of the electron, e = the elementary charge), $n = 1, 2, 3 \ldots$ = the identifying number of the respective series, $m = n+1, n+2, \ldots$ = the running number within a series). For n = 2 this was the "Balmer" formula. "I have long had in mind the problem of expressing the

21 Kox, *Discovery*, 1997.
22 Sommerfeld, *Zeemaneffekt*, 1913.
23 To Voigt, March 24, 1913. DMA, NL 89, 015. Also in ASWB I.
24 Voigt,*Zeemaneffekte*, 1913; Voigt, *Ausbau*, 1913; Voigt, *Zeemaneffekte der Spektrallinien*.
25 Sommerfeld, *Theorie*, 1914.
26 Lecture delivered June 25, 1913. Physikalisches Mittwoch-Colloquium. DMA. 1997-5115.
27 Hoyer, *Introduction*, 1981.

Rydberg-Ritz constant in terms of the Planck *h*," Sommerfeld wrote to Bohr, after he had read his work. "I spoke about this to Debye several years ago. If, for the time being, I remain somewhat skeptical with respect to the atomic models generally, the calculation of these constants unquestionably represents a great accomplishment." Bohr's theory seemed to him of particular interest with respect to the current discussion of the Zeeman and the Paschen-Back effects. "Do you intend to apply your atomic model to the Zeeman effect also?" he asked Bohr. "I had wanted to take up this question."[28]

When Stark reported his discovery that he had split "spectral lines into distinct components by means of an electrical field" at the end of 1913,[29] the subject of spectral lines moved even further into the spotlight. On December 10, 1913, "The New Stark Effect (with Demonstration)" was the subject of an experimental physics lecture at the colloquium of the Munich physicists. One month later, Epstein presented Bohr's atomic model to the Munich physicists at the Sohncke Colloquium. On May 27, 1914, Lenz and Sommerfeld reported on "Theoretical Aspects of the Stark Effect According to Bohr and Voigt" at the Wednesday Colloquium, and on July 15, 1914, Bohr himself was invited to deliver a lecture "On Bohr's Atomic Model, Specifically the Spectra of Helium and Hydrogen."[30]

Sommerfeld, Voigt, and Bohr were not alone in feeling challenged by the new discoveries to advance theories of atomic spectra. The Berlin physicist and President of the Imperial Physical and Technical Institute, Emil Warburg, who had participated in the first Solvay Congress in 1911 and who was interested equally in theoretical and experimental physics, was the first to endeavor to describe the splitting of spectral lines in an electrical field with Bohr's model of the atom. After him, Schwarzschild attempted—without recourse to Bohr's theory—to formulate the theory of the Zeeman and Stark effects. Agreement with the measurements, however, eluded him.[31] Thus, it was no coincidence that Schwarzschild was regularly an interlocutor with Sommerfeld in these matters, with whom he exchanged theoretical ideas about spectral lines. When Schwarzschild wrote him from his military station in Belgium with further details of his theory of the Zeeman effect, Sommerfeld was pleased that his service was proving "so idyllic" that it allowed him time for science. He asked Schwarzschild to continue this epistolary shoptalk. "Next semester, I'll lecture on the Zeeman effect and spectral lines, and I can make good use of it."[32]

At first, Sommerfeld, like Schwarzschild and Voigt, approached the theory of spectral lines without recourse to quantum concepts. To be sure, the conception of electrons as balls oscillating on springs seemed unrealistic to him. But one might—he wrote Schwarzschild—reinterpret this conception without altering the

28 To Bohr, September 4, 1913. NBA, Bohr. Also in ASWB I.

29 From Stark, November 21, 1913. SBPK, Autogr. I/292.

30 Physikalisches Mittwoch-Colloquium. DMA. 1997-5115.

31 Warburg, *Bemerkungen*, 1913; Schwarzschild, *Bemerkung*, 1914; Schwarzschild, *Aufspaltung*, 1914.

32 To Schwarzschild, October 31, 1914. SUB, Schwarzschild. Also in ASWB I.

equations at all. It is possible, "to rewrite everything Voigt has described quasi-elastically in magnetic terms."[33] Instead of a complicated examination of the many coupled oscillations of electrons such as Voigt had performed, one ought to examine the reciprocity between each single electron and the magnetic field. Though he arrived at no finalized results, his grasp of the difficulties inherent in any classical theory of the Zeeman effect was thereby unquestionably enhanced.

Aside from Schwarzschild, he discussed these questions also with Paschen, who wanted to get the bottom of the "anomalous Zeeman types" experimentally. As Zeeman and Lorentz, around the turn of the century, had directed their attention experimentally and theoretically to the Zeeman effect, Paschen and Sommerfeld now went to work on the anomalous Zeeman effect. "Enclosed is a compilation of all the Zeeman types," Paschen began a long letter to Sommerfeld in December 1914, in which he entrusted the results his assistants had, up to their mobilization in the war, "obtained in a years-long work," and which had not yet been published. "Unfortunately, Herr Back has, since the start of the war, disappeared without a trace, so that I must still delay publication of the types until his fate is known. But since these types have with Back's consent also been shared with Voigt, I can let you have them as well." In the course of many pages of his letter, he presented a kind of empirical resume of the anomalous Zeeman effect. "These things are the foundation of the complicated anomalous Zeeman types. It is regrettable that Zeeman himself, who continues to write popular books on this subject, has not understood this."[34] Sommerfeld responded to Paschen's letter by return mail with more detailed explanations of his and Voigt's theory, which in turn led Paschen to confide to Sommerfeld his plans for future spectroscopic measurements. His enthusiasm was dampened only by qualms that now might "actually not be the time for such peacetime occupations . . . We are enlisted in all sorts of military work here, and several among us are thinking of getting into uniform to join the fight. The superior force of the enemy demands defense to the last man."[35]

Sommerfeld, too, was torn between his enthusiasm for the physics of spectral lines and the nervous tension spread by newspaper reports of events of the war. Since the military authorities had no use for him, he carried out the intention he had expressed to Schwarzschild by giving a special lecture course on "The Zeeman Effect and Spectral Lines" in the winter semester of 1914/1915. He must have taken the occasion to study thoroughly the extensive data obtained from measurements in laboratories and photographs of star spectra and brought by various formulas—more or less at random—into an ordering system that assigned the various chemical elements to spectral series.

Several of these series, however, fell outside the framework. The English spectroscopist Alfred Fowler had observed a spectral series in a mixture of helium and

33 To Schwarzschild, November 30, 1914. SUB, Schwarzschild.

34 From Paschen, December 15, 1914. DMA, HS 1977-28/A, 253.

35 From Paschen, December 21, 1914. DMA, HS 1977-28/A, 253.

hydrogen that did not conform to the well-established law of the Balmer formula. The American astronomer Edward Charles Pickering had discovered this same unusual series in star spectra. Both assumed these were cases of a hydrogen series. "I enclose a calculation of the Pickering series of hydrogen and note that in a Geissler tube containing helium and hydrogen, Fowler obtained potential lines at high potential corresponding closely to Pickering's values," Paschen wrote Sommerfeld concerning this curious series. "Based on Bohr's theory, Fowler thinks it possible that these are helium lines. Because of the divergence of Pickering's wavelengths it is not impossible, on the other hand, that Pickering's star lines are not identical with Fowler's, and are hydrogen lines. The question remains open, and is being addressed here by one of my students, though he is currently in military service. Meanwhile, I would like to report to you that according to our experiments too, Fowler's lines are probably helium lines, but that we must leave the possibility open that Pickering's lines are different, and belong to hydrogen. We will publish the proofs only when all the experiments have been concluded. Unfortunately, on account of the war, this cannot happen at present."[36]

What had at first been thought an unusual hydrogen series was attributed by way of Bohr's theory to ionized helium. Following this reinterpretation, it was not necessary to assume a new series law for the lines discovered by Fowler and Pickering; the formula $v=4N\ (1/n^2-1/m^2)$ where $n=3$ or $n=4$ sufficed; only the constant 4 which was owing to the doubled nuclear charge of helium distinguished this formula from the formula for the hydrogen series. According to the interpretation of the hydrogen formula and the derivation of the Rydberg constant, this was a further vindication of Bohr's theory. Even if several proofs were still missing, as Paschen indicated in his letter, the evidence spoke unequivocally for Bohr's conception. Whoever attended Sommerfeld's special lectures during this first winter of the war would, like Sommerfeld himself, have been convinced that further successes could be achieved on this path: "This semester, I have been reading about Bohr, and am interested in this so far as circumstances of the war allow," Sommerfeld wrote to Willy Wien at the conclusion of these special lectures. "Today's 100,000 Russians are, to be sure, more welcome even than Bohr's explanation of the Balmer series. I have very nice new results as well."[37]

7.3 Letters from the Front

Whatever new results he may have obtained regarding Bohr's theory, he did not yet reveal. Nor was Sommerfeld's attention in 1915 focused exclusively on such "peacetime occupations." His reference to the 100,000 Russians was to dispatch from the

36 From Paschen, February 7, 1915. DMA, HS 1977-28/A.253.

37 To W. Wien, February 22, 1915. DMA, NL 56, 010. Also in ASWB I.

eastern front about the destruction of the Russian 10th Army. "The total loot from the winter battle in Masuria has grown so far to 7 generals, over 100,000 men, over 150 heavy artillery pieces, and as yet untold numbers of smaller equipment including machine guns," said the report of the army high command on February 22, 1915.[38]

Apart from official war reports, in which news was mixed with propaganda, Sommerfeld learned directly of the events in the various theaters of war. But the letters from his students and colleagues hardly offered up triumphal reports of victory. "The war has already quite terribly decimated the group of my special students," Otto Wallach wrote him in January 1915.[39] Directors of institutes like Wallach and Sommerfeld, who had not been called up for duty and who were in contact with their students at the front via military mail, thus became informal communications centers. When the publishers of the *Physikalische Zeitschrift* decided to inform their readers about the war experiences of German physicists, they commissioned Max Born, who in 1915 was acting as substitute editor, to gather the relevant information from the Directors of the institutes of physics at the German and Austrian academic and technical universities. "Our top priority is to find which colleagues are on active duty in the field and where they are stationed," Born wrote Sommerfeld in February 1915. He asked for news concerning any who had received the Iron Cross or been distinguished in other ways. Besides this, "obituaries and photographs of those who have died 'a hero's death' in service to the Fatherland should be published. By such publication, the gratitude due defenders of our homeland is given voice; at the same time, however, we let the world at large know that, like all German science, physics too stands in solidarity with the Fatherland in its time of adversity and peril."[40]

The letters from the front not only contained news the *Physikalische Zeitschrift* wished to present but also informed their correspondents about personal impressions of the teacher-student relation, attitudes towards the war, and other things. Whereas heroism and enthusiasm for the war were meant to be fed the public, these letters more often spoke of resignation and fatalism.[41] The same day the army high command issued the announcement to the world concerning the "100,000 Russians," Otto Blumenthal, who was serving with a reserve foot-artillery regiment, wrote to Sommerfeld: "Life goes on here in a merely vegetative sort of way; I may prove useful agriculturally."[42] Military service came increasingly to be experienced as a spiritual wasteland. "I would be really happy to be able to study in peace once more," a student wrote Sommerfeld from his operational base in the Vosges.

38 Amtliche Kriegs-Depeschen, http://www.archive.org/stream/amtlichekriegsde02contuoft#page/n0/mode/2up .(30 January 2013).

39 From Wallach, January 21, 1915. DMA, NL 89, 014.

40 From Born, February 2, 1915. DMA, NL 89, 059. Cf. also "Übersicht über die Kriegsbeteiligung der Deutschen Physiker," in *Physikalische Zeitschrift*16, 1915, pp. 142–45.

41 Watson/Porter, *Motivation*, 2008.

42 From Blumenthal, February 22, 1915. DMA, NL, 89, 059.

"One sorely and continuously misses intellectual activity here. Perhaps the Herr Professor would have the kindness to send me from time to time any small publication from the *Annalen* or the like."[43] Sommerfeld was happy to comply with this wish. "Resting behind the front after many strenuous days, I thank the Herr Professor heartily for the intellectual nourishment sent me," the student wrote back.[44] One day later, Sommerfeld received a letter from the front from Wilhelm Hüter, who had taken his doctorate under him in 1911, and spent the spring of 1915 in the trenches on the western front. This letter, too, bespeaks a "longing for that far-off intellectual realm of physics," which in the cruel reality of trench warfare must have seemed as though removed somewhere in another world.[45]

Ludwig Hopf, who had taken his habilitation in Aachen shortly before the war, spent the first year of the war in a motor vehicle depot in Aachen. "I help requisition rubber, and for the most part sit in an office. My time is pretty well occupied, so that I can do little at the technical university, and scientifically next to nothing. But after all, that is not ultimately the point; I hope soon, in a better time, my ravenous appetite for physics can soon be sated; I wouldn't at all mind living to see peace in my current situation—I certainly won't win any prizes as a warrior." In his habilitation, Hopf had tried to solve the turbulence problem formulated by Sommerfeld in 1908; however, he came to the conclusion that the problem itself apparently eluded so sophisticated a mathematical analysis. "The turbulence still sleeps its sleep of war," he wrote Sommerfeld after further, fruitless efforts. "Were I ever to succeed with this, it would mean more to me than an Iron Cross."[46]

From time to time, Sommerfeld's students at the front even had occasion to satisfy their "thirst for physics" with subjects drawn from their immediate surroundings. "Yesterday and today countless grenades and shrapnel shells were fired at enemy flyers and missed their mark, which we can observe very well from here," Lenz reported from the western front. "From what I have seen up to now, it is only by chance that an airplane is actually hit."[47] Physicists and mathematicians were still not being systematically enlisted to address technical military issues, but that was soon to change. "Yesterday morning, I was at the Artillery Board of Examiners," Sommerfeld wrote home in April 1915 from a sojourn in Berlin, during which he was presumably consulting on various military-scientific problems. "A gentleman from Siemens-Halske invited me to observe their war production—a quite marvelous thing."[48] He also paid a visit to the Imperial Physical-Technical Institute on this occasion, although here, as he explained to his wife, it was "peacetime physics

43 From Lang, May 1, 1915. DMA, NL 89, 059.
44 From Lang, May 10, 1915. DMA, NL 89, 059.
45 From Hüter, May 11, 1915. DMA, NL 89, 059.
46 From Hopf, November 13, 1915. DMA, NL 89, 059. Eckert, *Birth*, 2010.
47 From Lenz, undated (presumably March, 1915). DMA, NL 89, 059.
48 To Johanna, April 24, 1915.

generally" that was being done.[49] En route to Berlin, he had also stopped over in Göttingen, where he discussed various military-scientific subjects with Prandtl, who directed a research laboratory for aerodynamics at Göttingen. Thereafter, he corresponded with him on questions "concerning the fall of bombs in water and air," as well as "sonic problems" that presumably related to fixing the position of submarines.[50] Meanwhile, at the front, too, there were nascent aspirations that as physicists, "one might put one's scientific capabilities to good use in the service of the larger issues," as Lenz wrote Sommerfeld in May 1915.[51] On his own initiative, he had focused on technical problems of radio transmission[52] but had received little encouragement from his superiors. "I have to assume the nexus of war and science is more to be achieved where you are than out here," he lamented on the lack of military initiative at the organization for military-scientific research.[53]

For the moment, military science was merely a peripheral issue for Sommerfeld, too. In a letter to Willy Wien around this time, he wrote that he had heard "very interesting things" from the latter's cousin Max Wien, in Berlin. Under the umbrella of a new "Technical Radio Division" (Tafunk), Max Wien had begun organizing experts in the field of radio for the exigencies of the war. Sommerfeld, too, numbered in this circle of experts, since in his 1909 theory of the spreading of electromagnetic waves he had taken up the question how dissemination of waves was dependent on the composition of the ground. But in the summer of 1915, no very great urgency attached yet to this "military physics."He had also "hit on an interesting formulation of the Stark effect based on Bohr's theory of the hydrogen lines," he wrote in the same breath.[54] Sommerfeld had still not published anything about his elaboration of Bohr's theory. He must nonetheless have dropped hints of it in letters to one or another of his students in the trenches. "The spectral questions too will surely lead to nice new results," speculated Lenz in May 1915.[55]

General relativity took its place beside the elaboration of Bohr's atomic theory among the subjects filling Sommerfeld and others with enthusiasm during this 1915 summer of the war. "This semester, I have lectured on relativity, ultimately in the sense of Einstein's latest Berlin work, and am very enthusiastic about it, almost as much as by Bohr last semester," he wrote Schwarzschild at the conclusion of the summer semester. He had acquainted his audience with the theory without "Einstein's dreadful tensor formalism."[56] Hopf too learned in this way about the latest advances in Einstein's theory. This is the crowning achievement of Einstein's years

49 To Johanna, April 20, 1915.
50 To Prandtl, May 9, 1915. GOAR 2666.
51 From Lenz, May 17, 1915. DMA, NL 89, 059 .
52 From Lenz, February 20, 1915. DMA, NL 89, 059.
53 From Lenz, undated (presumably, early May, 1915). DMA, NL 89, 059.
54 To W. Wien, May 3, 1915. DMA, NL 56, 005. Also in ASWB I.
55 From Lenz, undated (presumably early May, 1915). DMA, NL 89, 059.
56 To Schwarzschild, July 31, 1915. DMA, NL 89, 059.

of labor, he wrote Sommerfeld. "Now in my limited free time I sometimes break my head over this. Has your simplified presentation already appeared? The calculation of examples will surely make it much easier; I have struggled a bit with rotating bodies, but all the same, I won't come to clarity about any of this before we have peace."[57]

In these weeks, Einstein achieved the breakthrough with his general theory of relativity. "Have you seen Einstein's work on the precession of the perihelion of Mercury, in which he correctly obtains the observed value from his latest theory of gravitation?" Schwarzschild gushed in a letter to Sommerfeld shortly before Christmas, 1915. In the meantime, he did not let his transfer to an artillery unit in the Vosges prevent him from formulating an elaboration of Einstein's theory. Where Einstein had merely sketched an approximate solution, Schwarzschild achieved a strict solution. "The planetary motion and the perihelion of Mercury come out virtually as in Einstein. It is wonderful that these agree. How goes it with the Zeeman and Stark effects, and with your mathematical physics generally? To the tune of much artillery fire up on Hartmannsweilerkopf, our life goes along peacefully here on the plain."[58] This enthusiasm for physics, which could take such flight amid the war, contrasted starkly with the existence the reality of daily life imposed on most of the physicists in the various theaters of war. "For the moment, our attention is focused mainly on warm beds, food, and minimal shooting, so that naturally systematic idiocy makes great strides forward," a student of Sommerfeld's wrote back to Munich. Added to this, his fear of "gradually [losing] contact with profession and colleagues."[59]

7.4 Elaboration of Bohr's Atomic Model

In letters throughout the year 1915, Sommerfeld had repeatedly given students and colleagues to understand that he had elaborated Bohr's atomic theory, without indicating how or based on what results he had done so.[60] At first, his interest seemed focused entirely on the magneto-optical effects (". . . next semester, [I will] lecture on the Zeeman effect and spectral lines . . ."). Presumably, he was interested in transferring the harmonic electron oscillations of classical theory to Bohr's orbit (". . . to rewrite in magnetic terms everything Voigt expresses quasi-elastically . . ."). On January 16, 1915, he delivered a lecture at the Wednesday Colloquium on "The number of decompostions of hydrogen in the Stark effect."[61] This formulation of the subject suggests that in a spectral line he did not—like Bohr—see *one* leap from one to another orbit, but rather, provided no external disturbance occurred, *several*

57 From Hopf, November 13, 1915. DMA, NL 89, 059.
58 From Schwarzschild, December 22, 1915. DMA, NL 89, 059. Also in ASWB I.
59 From Fritz Eckert, November 25, 1915. DMA, NL 89, 059.
60 Nisio, Formation, 1973; ASWB I, pp. 429–493.
61 Physikalisches Mittwoch-Colloquium. DMA, 1997-5115.

leaps between equivalent circular and elliptical orbits. By the application of an electrical field, however, the various orbits were variously affected, so that the frequencies corresponding to the orbital leaps were no longer equally large, and appeared in the form of neighboring spectral lines. "The large number of components increasing with the number of the Balmer line that Stark observed in his precise analysis argues for this conception of the Stark effect," Sommerfeld reasoned in his first publication on this subject.[62]

If he already had this basic idea for the elaboration of Bohr's model by January 1915, why did he let nearly another year go by before publishing it to the scientific community? "Last semester, I hit on an interesting formulation of the Stark effect based on Bohr's theory of the hydrogen lines. Only carrying out the idea remains to be done," Sommerfeld wrote Willy Wien in May 1915.[63] In an electrical field, the electrons would no longer move in closed circular and elliptical orbits around the atomic nucleus. His inability to transfer Bohr's quantum formulation of such nonperiodic orbits was the reason he gave for his delay in the presentation of his elaboration of Bohr's theory to the Bavarian Academy of Science on December 6.[64]

But even at this juncture he was unable to explain the Stark effect, so that this difficulty must not have been the only reason for his delay. Presumably, he wished to await further evidence of the correctness of the Bohr model before complicating the picture with quantized elliptical orbits. Sommerfeld refereed just such a piece of evidence on May 12, 1915, at the colloquium, "The Rotational Momentum of Magnetization According to Einstein/de Haas."[65]The effect could be counted as evidence that, as Bohr had asserted, a rotation of electrons unconnected with an emission of energy actually occurred in the atom. When Bohr, bolstered with new experimental results, defended his theory against its critics in August 1915, he cited the Einstein/de Haas effect in material support of his quantum formulation.[66] In this essay, Bohr enumerated a string of other works that supported his theory, among them, a paper by Walther Kossel (1888–1956), a member of the Munich physics circle. Following his studies at Heidelberg and a doctorate under the supervision of Philipp Lenard in 1911, Walther Kossel had come to Munich, where in 1913 he became assistant to Jonathan Zenneck at the Technical University. Nonetheless, he felt more strongly allied with the Sommerfeld circle: "Our mutual student Kossel"—so wrote Lenard in a letter to Sommerfeld—would like "to remain near you, which he (as I know) values very highly."[67]

In Munich, Kossel made the characteristic X-radiation his research topic. In 1914, it occurred to him during the course of scrutinizing experimental data on

62 Sommerfeld, *Theorie der Balmerschen Serie*, 1915, p. 449.

63 To W. Wien, May 3, 1915. DMA, NL 56, 005. Also in ASWB I.

64 Sommerfeld, *Theorie der Balmerschen Serie*, 1915, p. 426.

65 Pysikalisches Mittwoch-Colloquium. DMA, 1997-5115.

66 Bohr, *Quantum Theory*, 1915, S. 397. Also in NBCW 2.

67 From Lenard, September 25, 1913. DMA, HS 1977-28/A, 198. Heilbron, *Kossel-Sommerfeld Theory*, 1967.

the absorption of X-rays in diverse elements that in the context of the Bohr model, a quite similar ordering of the X-ray spectra resulted, just as in the realm of visible light, for instance, it had appeared in the spectral series of hydrogen.[68] The radiation of an X-ray line could be conceived as a transition from an outer to an inner orbit. If an electron is removed from an inner electron ring, the gap can be filled by an electron from an outer ring. The energy difference between electron rings corresponds to the energy of the X-ray emitted. Thus, the frequencies of the spectral lines in the realm of X-ray could be conceived as the difference between each of two stationary conditions. "It will be seen that these relations correspond exactly to the ordinary principle of combination of spectral lines": Thus, Bohr summarized the essence of Kossel's observations.[69]

Reading this work, Sommerfeld may have felt Bohr had begun harvesting the fruits from his own garden. In any case, it provided the impetus finally to publish his elaboration of the Bohr theory as he had conceived it months earlier—even if he was still unable to explain the Zeeman and Stark effects, as he had hoped to do in the summer of 1915. Just at the start of the following winter semester, in a colloquium lecture on "Bohr's Recent Work," on November 27, 1915,[70] he revealed that, occasional digressions into military physics notwithstanding, he was following Bohr's every step Argus-eyed. Aside from Kossel's observations, Bohr threatened to exploit other ideas as well on which Sommerfeld based his theory: In February 1915, Bohr had already spoken of a possible relativistic elaboration of the theory that could explain deviations from the Balmer series formula. To be sure, he had carried out this calculation only for circular orbits and found that the observed deviations—a doubling of certain lines—could not be explained. In this connection, however, he suggested the doubling of lines could also be explained by noncircular orbits and announced that he would address such questions in a future publication.[71]

Presumably, Sommerfeld used his colloquium lecture on November 27, 1915, as a rehearsal for what he shortly thereafter presented for publication to the Bavarian Academy of Sciences and had already let be known among the narrower circle of his colleagues. "Yesterday, I presented a paper on the Balmer series to the Academy," he wrote Willy Wien on December 5, 1915. "In Würzburg recently I told you about the quantized elliptical orbits; in the meantime I have further developed this."[72] From the start, he had kept Einstein informed too: "I will now study both of your papers," Einstein replied on November 28, 1915.[73] That this related to the elaboration of the Bohr model is clear from Einstein's next letter, enclosed with which he returns Sommerfeld's manuscripts with the notation that Planck was just then

68 Kossel, *Bemerkung*, 1914; Kossel, *Bemerkung II*, 1914.

69 Bohr, Quantum Theory, 1915, p. 414. Also in NBCW 2, pp. 411–412.

70 Physikalisches Mittwoch-Colloquium. DMA, 1997-5115.

71 Bohr, *Series Spectrum*, 1915, pp. 334–35. Also in NBCW 2, pp. 379–80.

72 To W. Wien, December 5, 1915. DMA, NL 56, 010.

73 From Einstein, November 28, 1915. DMA, HS 1977-28/A, 78. Also in ASWB I.

working "on a similar problem. . . He too is engaged with spectral questions."[74] Planck had just presented a paper to the German Physical Society on "The Quantum Hypothesis for Molecules of Several Degrees of Freedom," which also presented an elaboration of the Bohr model and led to virtually the same results, albeit from a different starting point and with a different objective.[75]

What did this elaboration consist of? In his theory of the hydrogen spectral series, Bohr had assigned circular orbits to the electron of a hydrogen atom at certain discrete distances from the nucleus, on which the electron, he conjectured, emitted no electromagnetic waves. Only in a leap between such orbits was radiation supposed to be emitted or absorbed. According to the quantum law $E=h\nu$, the energy differential should correspond to the frequency ν of a spectral line. The energy of the electron in the nth radiation-free orbit was, according to Bohr, proportional to $1/n^2$ (with the proportional constant N= Rydberg-Ritz constant $=2\pi^2me^4/h^3$), so that the energy differentials between two orbits with an index n and m yield the series formula $\nu=N(1/n^2-1/m^2)$. Sommerfeld explained in his first Academy paper however that this theory was incomplete, since it did not provide for any elliptical orbits, which, according to the law of electrostatic attraction between nucleus and electron, were after all equally possible. He wanted to remedy this defect and thereby also show the way to an explanation of the spectral series of other elements, for which the hydrogen series formula no longer held. "This is explained according to the conception presented herewith, in that in the Balmer series a number of series coincide, but each of their lines arises in a number of different ways, not only in circular motion, but also in elliptical orbits of certain eccentricities."[76]

In order to admit elliptical orbits beside the circular orbits, Sommerfeld elaborated the quantum formulation. The circuit of an electron in an elliptical orbit around the nucleus can be described in terms of two motions: First, the radius vector between the nucleus and the electron rotates in one revolution by 360°; second, the radial distance changes from a minimum distance to a maximal distance and back again. The rotational and radial motions completed with a rotational angle φ and a distance r can be described by so-called phase integrals, $\oint p_\phi d\phi$ and $\oint p_r dr$, where p_ϕ indicates the angular momentum and p_r indicates the momentum in the direction of the radius vector. To calculate these values, which have the dimension of an action, Sommerfeld formulated the quantizing conditions: $\oint p_\phi d\phi = nh$ and $\oint p_r dr = n'h$. For the energy of the electron, he obtained the same expressions as Bohr, with the difference that in place of one quantum number there now appeared the sum of two quantum numbers. Sommerfeld's spectral formula read

$$\nu = N\left(\frac{1}{(n+n')^2} - \frac{1}{(m+m')^2}\right).$$

74 From Einstein, December 9, 1915. DMA, HS 1977-28/A, 78. Also in ASWB I.

75 Eckert, *Plancks Spätwerk*, 2010.

76 Sommerfeld, *Theorie der Balmerschen Serie*, 1915, pp. 425–26.

Since the sum of Sommerfeld's two quantum numbers could in the end be replaced by Bohr's one quantum number, what appeared mathematically as an unnecessary complication was from the viewpoint of physics nonetheless fundamentally distinct from the Bohr model. Bohr assigned a spectral line unambiguously to the leap from one orbit to another; in Sommerfeld's elaboration, to every circular orbit, there corresponded an energetically equivalent elliptical one, so that a spectral line could come about in quite a different way. "In every case," Sommerfeld summed up his results, "a hydrogen line appears in our formulation as a quite complicated overlapping of different and discrete events."[77]

In the second paper, which Sommerfeld presented to the Munich Academy on January 8, 1916, following the Christmas vacation, the extent of this elaboration was made evident: Only if the electron, speeding around the atomic nucleus, is attracted to the nucleus exactly according to Coulomb's law (i.e., with a force proportional to $1/r^2$), is its energy on equivalent circular and elliptical orbits the same. Any divergence from this law leads to different energies and thereby also to a proliferation of spectral lines. Since an electron in an elliptical orbit—like a comet around the sun—can approach the nucleus very closely and thereby attain extremely high velocities, Sommerfeld calculated the orbital motion according to relativity theory. "Whereas in classical mechanics the energy of the $n+n'$ different circular and elliptical orbits, that belong to the same value of $n+n'$, correspond exactly with each other, in consideration of the variable electron mass for these $n+n'$ various orbits, it turns out somewhat differently in each case. The respective spectral lines, or more properly, the respective term of the spectral line diverges—corresponding to the $n+n'$ generational possibilities—into a system of $n+n'$ neighboring lines or terms, that is in the case that $n+n'=2$ into a doublet, in the case that $n+n'=3$, into a triplet, etc."[78]

Pursuant to these results, new vistas opened for experimental spectroscopy as well. Paschen had reported to Sommerfeld in December 1915 that a spectral line of ionized helium that had at first been assigned to hydrogen exhibited a curious structure. He conjectured "that the term $4N/2^2$, corresponding to the second orbit, must be even more complex and split up. So pursuing this line of research would be very interesting. Were it not for the war, this would already be well under way."[79] Paschen found Sommerfeld's theory, which made concrete predictions for this as yet little investigated fine structure of spectral lines, "very impressive."[80] What had been thought to be experimental deviations proved the confirmation of theoretical predictions. "So, 'deviation' may be theoretically required! A good theory ignores nothing!"[81] According to Sommerfeld's formula,

77 Sommerfeld, Theorie der Balmerschen Serie, 1915, p. 448.
78 Sommerfeld, *Feinstruktur*, 1915, pp. 466–67.
79 From Paschen, December 12, 1915. DMA, HS 1977-28/A, 253. Also in ASWB I.
80 From Paschen, December 27, 1915. DMA, HS 1977-28/A, 253. Also in ASWB I.
81 From Paschen, December 30, 1915. DMA, HS 1977-28/A, 253. Also in ASWB I.

the fine-structure splitting was proportional to the fourth power of the nuclear charge. Since according to Kossel's conjectures the characteristic X-ray spectra could be interpreted as transitions of electrons from outer to inner orbits, the splitting of the X-ray lines from the heavy atoms ought to be easiest to prove. "I show that for all elements from Z=20 to Z=60, where observations have been made $\Delta v/(Z\text{-}1)^4=\Delta v_H! \Delta v$=the vibration differential of the X-ray doublet, Δv_H=the vibration differential of the hydrogen doublet"—thus Sommerfeld described this implication of his theory in a letter to Schwarzschild.[82] The doublets of the X-ray spectra could be conceived "as virtually an image enlarged by a factor of $(Z\text{-}1)^4$" of the as yet experimentally undemonstrated fine-structure splitting in hydrogen. In fact, in a dissertation completed in 1915, a doctoral student of the Swedish X-ray spectroscopist Manne Siegbahn had registered the doublets of numerous elements in the *K*-series of X-ray spectra (transitions from the second lowest to the lowest level) in which Sommerfeld saw his theory "exactly confirmed."[83] Kossel had lectured on this in the colloquium 2 days before Christmas, 1915.[84] Sommerfeld wrote happily in a letter to Willy Wien, "Now the hour has also struck for a genuine theory of the Zeeman effect, in which the nature of the doublets is recognized as different orbits."[85]

Although its application to the Zeeman and Stark effects was still several months in the future, the theory and the experiments it inspired made rapid progress. "Your last letter has made the solution of the 4686 picture puzzle possible," Paschen wrote Sommerfeld in March 1916 concerning a spectral line assigned to ionized helium which, without regard to the fine structure, was described as $v=4N(1/3^2-1/4^2)$ and thereby corresponded to a leap from the fourth to the third Bohr orbit. According to Sommerfeld's theory, however, the initial orbit split into four, and the final orbit into three orbits of slightly differing energies, so that altogether 12 electron leaps were possible. Measurements confirmed Sommerfeld's theoretical predictions about the fine structure of this line. Even if he could not prove all 12 components, the wavelengths of the components discovered corresponded to the values Sommerfeld had computed. "Your theory is almost completely correct," Paschen found. "One can't very well ask for more."[86]

Sommerfeld saw his activity confirmed theoretically too. Planck's "Structural Theory of the Phase Space" led to quite similar results, as Sommerfeld wrote to Schwarzschild: "From such different points of departure, and by such different modes of thought (Planck, cautious and abstract—I, somewhat reckless and ready

82 To Schwarzschild, December 28, 1915. SUB, Schwarzschild. Also in ASWB I.
83 Sommerfeld, Feinstruktur, 1915, pp. 460, 494, Fig. 3.
84 Physikalisches Mittwoch-Colloquium. DMA, 1997-5115.
85 To W. Wien, December 31, 1915. DMA, NL 56, 010.
86 From Paschen, March 28, 1916. DMA, HS 1977-28/A, 253.

to plunge into experimental observation) exactly the same results!"[87] Planck had written him shortly before that this had been just a small diversion for him, which he intended not to pursue, "since with you, the problem is now in the best of hands."[88] Sommerfeld then wrote to Willy Wien, who worked jointly with Planck as editor of the *Annalen der Physik*, that he next wished to publish in the *Annalen* "in a more refined form" the theory he had presented at the two lectures to the Academy. "It will interest you that Planck's quantization of the phase space corresponds exactly to my formulations. But Planck's explanation of the Balmer series is hideous and fundamentally different from mine."[89]

Sommerfeld's elaboration of the Bohr theory occasioned admiration in several theaters of the war too. "I devoted myself at once to a thorough study of it," Lenz wrote in a long letter from the front in northern France, thanking him "for sending your wonderful work on spectral lines." From Sommerfeld's formulas, Lenz derived the series law that must replace the hydrogen spectral formula in the case of relativistic calculation, and, he was convinced, "must also assume a simple form. Since this law is not explicitly given in your work, I have worked it out myself."[90] In this calculation, he also introduced the quantity $\alpha = 2\pi e^2 / hc$, which later gained fame as the "Sommerfeld fine-structure constant."

Schwarzschild's reaction, too, was full of enthusiasm. "Your work on spectral lines is a huge step forward," he wrote in thanks for the Academy paper sent him. Then, in four pages of his letter, he sketched out for Sommerfeld a more general method, oriented towards celestial mechanics (Hamilton-Jacobi formalism with angle-action variables), that yielded the same results. He extolled this procedure: From this we can derive "a compelling formulation of the Stark effect and the Zeeman effect as well."[91] Five days later he informed Sommerfeld on a postcard from the front that his procedure was also consistent with the Planck phase space quantization. "Are you convinced yet as to how it goes with the Zeeman and Stark effects? I'm in seventh quantum heaven!"[92] As he replied, Sommerfeld was "unfamiliar" with the celestial mechanical methods Schwarzschild had employed. But he, too, was ready with something new: With a view towards a theory of the Zeeman effect, he described to Schwarzschild how by means of a phase integral the slope of the orbital plane of an electron could be expressed in a quantum formulation.[93]

87 To Schwarzschild, February 19, 1916. SUB, Schwarzschild 743. Also in ASWB I.
88 From Planck, January 30, 1916. DMA, HS 1977-28/A, 263.
89 To W. Wien, February 10, 1916. DMA, NL 56, 010. Also in ASWB I. On Planck's theory, Eckert, Plancks Spätwerk, 2010.
90 From Lenz, March 7, 1916. DMA, NL 89, 059. Also in ASWB I.
91 From Schwarzschild, March 1, 1916. DMA, HS 1977-23/A, 318. Also in ASWB I.
92 From Schwarzschild, March 5, 1916. DMA, DM.NL 89 059. Also in ASWB I.
93 To Schwarzschild, March 9, 1916. SUB, Schwarzschild 743. Also in ASWB I.

Fig. 17: From the front in northern France, Wilhelm Lenz participated in the elaboration of atomic theory. In a letter from the front, he presented Sommerfeld with a derivation of the relativistic generalization of the hydrogen series formula, which expressed the significance of the fine-structure constant more clearly than Sommerfeld's original derivation (Courtesy: Deutsches Museum, Munich, Archive).

He was, however, not able thereby to explain the complex splitting in the anomalous Zeeman effect. In the meantime, he had turned the Stark effect over to Paul Epstein, who planned to write his habilitation on it. During the war, as a Russian citizen, Epstein was subject to the regulations for enemy aliens and had been placed under arrest; he was, however, allowed to stay at Sommerfeld's institute. The motion of an electron in the central field of the atomic nucleus with an overlapping homogeneous electrical field could be mathematically described as a borderline case of the astronomical "two-center problem," where one center, as its mass increases, moves to infinity. For this reason, just as Schwarzschild had done, Epstein made use of the methods common in celestial mechanics, which again proved superior to Sommerfeld's original procedure. And Schwarzschild and Epstein reached their goals virtually simultaneously. He had "been able to deal with the Stark effect without difficulty and entirely unambiguously," Schwarzschild wrote to Munich on March 21, 1916. Sommerfeld replied 3 days later: "Yesterday, your interesting letter arrived with the formula for H_β, and today, here comes Epstein with the general formula, which also describes H_α exactly, contains the lines still missing in H_β, and is unproven only for H_γ, H_δ. In the case of H_β, your formula is a special case of Epstein's, of course. Yours still lacks a fundamental viewpoint. Epstein will directly publish a provisional notice in the *Physikalische Zeitschrift*. Later, he plans to take his habilitation in Zurich with this work. He should write to you himself directly. He has of course attended my lectures on spectral lines, etc."[94] Epstein's and Schwarzschild's works were published at almost the same time. Epstein saw in his theory of the Stark effect "a new, persuasive proof of the correctness of the Bohr atomic model" that ought to convince even "our skeptical colleagues."[95] Schwarzschild found it "remarkable how extraordinarily close to the observed relations one [had been able to come] in this first more rigorous implementation of quantum theory using Bohr's formulation."[96]

Sommerfeld had sent the theory published in the Academy papers to Niels Bohr, who at this time was still in England and at once discussed these new things with British colleagues. "I do not believe I have ever read anything that gave me so much pleasure," Bohr wrote back, via Copenhagen, where his brother Harald translated his reply into German and forwarded it to Munich. Momentarily, enthusiasm for the atomic theory appeared to triumph over all the discord of the World War. Even Ernest Rutherford was extremely interested, Bohr confided to his German colleagues. He had himself just completed a reformulation of his theory for publication, but after studying Sommerfeld's Academy paper had in the last minute withdrawn it. In it, he had sought to use an idea of Paul Ehrenfest in Leiden ("On Adiabatic Transformation") as a basis for a new quantum formulation.[97] Sommerfeld had no need to puzzle over this idea for long, for Ehrenfest himself, shortly thereafter in a seven-page letter, described for him how he regarded the connection between Sommerfeld's "Quantizing of the Bohr Ellipses" with his "Adiabatic Hypothesis": "Every quantum permissibly motion form of the motion I (with respect to impermissible) is transformed by adiabatic reversible influence into a quantumly permissible motion II (with respect to impermissible)." With this hypothesis Ehrenfest had, 3 years earlier, successfully explained certain properties of the hydrogen molecule. "If only I could tell you about this whole adiabatic question in person!" Ehrenfest wrote, wishfully conjuring up the war's end.[98]

The success of the Bohr-Sommerfeld atomic theory caused a sensation also at Göttingen. In 1915, Debye, the star among Sommerfeld's students, had assumed directorship of the Physics Institute there. On June 3, 1916, he presented a paper to the Göttingen Academy on "Quantum Hypotheses and the Zeeman Effect," in which he showed that the celestial mechanical formalism in the theory of the Stark effect by Epstein and Schwarzschild was useful even when—as in the case of the

94 To Schwarzschild, March 24, 1916. SUB, Schwarzschild 743. Also in ASWB I.

95 Epstein, *Theorie des Starkeffekts*, 1916, p. 150.

96 Schwarzschild, *Quantenhypothese*, 1916, p. 564.

97 From Bohr, March 19, 1916. DMA, HS 1977-28/A, 28. Also in ASWB I.

98 From Ehrenfest, undated (presumably late April, 1916). DMA, HS 1977-28/A, 76. Also in ASWB I.

motions of electrons in a magnetic field—there were no parallels with astronomical processes. He was thus competing with his former teacher, who was just then readying for publication his work on the spatial quantizing of the electron orbit, already alluded to in his letter to Schwarzschild.[99]

"In contrast to Debye," however, Sommerfeld still did not recognize any accurate theory of the Zeeman effect, as he wrote to Ehrenfest.[100] What mattered to him in the first place was to connect the theory as closely as possible to the experimental results that were available by the summer of 1916. Thus, in the presentation of his theory "in a more refined form," for the *Annalen der Physik*, the formalism borrowed from celestial mechanics was not front and center—as in the work of Schwarzschild and Debye; rather it was the agreement with spectroscopic measurements he had obtained primarily from Paschen's institute at Tübingen. On May 21, 1916, Paschen wrote him that his measurements were "everywhere in nicest agreement with your fine structures." He had gotten agreement "down to exactly a few 1/1000 A°E." "Without your theory these results would not have been obtained, because the strongest lines cannot be interpreted and exploited as mixtures of components, while the weak, isolated lines are sharply defined."[101] Four weeks later, as Paschen was writing up his results for publication in the *Annalen der Physik*, he reiterated his assurances that his material was now "well worked through in all aspects, and agrees with the theory entirely. Whatever deviations remain I can certainly ascribe to observational errors."[102] On June 30, 1916, he reported to Munich: "Today, I've sent my work to Wien, and will send you the proofs. It is all quite satisfactorily and, I may say, also convincingly explained."[103] In response to an objection of Sommerfeld's, a week later he wrote once more: "Since all the numbers of the calculation have to be changed in the last decimal place, I'll have the manuscript returned to me, and will rewrite the number tables."[104] Coordinated thus, Paschen and Sommerfeld sent off their comprehensive works to the *Annalen der Physik*, where they appeared in successive issues in September 1916.[105]

7.5 Success

Planck saw to it that even long before the appearance of the "more refined" publication, Sommerfeld's theory had been duly noted among the circle of Berlin physicists.[106] Einstein had already been "delighted" by the first Academy paper.

99 Debye, *Quantenhypothese*, 1916; Sommerfeld, *Theorie des Zeemaneffektes*, 1916.
100 To Ehrenfest, November 16, 1916. Leiden, Museum Boerhaave. Also in ASWB I.
101 From Paschen, May 21, 1916. DMA, HS 1977-28/A,253. Also in ASWB I.
102 From Paschen, June 20, 1916. DMA, HS 1977-28/A,253. Also in ASWB I.
103 From Paschen, June 30, 1916. DMA, HS 1977- 28/A,253.
104 From Paschen, July 8, 1916. DMA, HS 1977- 28/A,253.
105 Paschen, *Bohrs Heliumlinien*, 1916; Sommerfeld, *Quantentheorie der Spektrallinien*, 1916.
106 From Planck, May 17, 1916. DMA, HS 1977-28/A,263. Also in ASWB I.

"A revelation!"[107] Then, in August 1916, he wrote, "Your spectral investigations take their place among my greatest experiences in physics. Only through your work is Bohr's idea completely convincing. If only I knew what kind of tiny screws God is using here!"[108]

This encomium from the mouth of the greatest theoretical physicist of his time demonstrated to Sommerfeld that he had more than met the expectations raised 10 years earlier when he had been appointed successor to Boltzmann in Munich. "The Munich school of mathematical physics has certainly become one of the first and best in the world!" Röntgen, too, declared.[109] If further proof were necessary that he was now among the pantheon of his discipline, Sommerfeld received an enquiry in January 1916 whether he might accept an offer of the chair in theoretical physics at the University of Vienna: that place—of all the venues in which the unstable Boltzmann had been active—that had most taken on his aura.[110]

Boltzmann's former Viennese chair was vacant because Friedrich Hasenöhrl, the professor's successor following his suicide in September 1906, had in turn died in the World War. At the outbreak of war, Hasenöhrl had enlisted at once. He was killed in October 1915 in trench warfare in the mountains of the South Tyrol when a grenade exploded nearby. He had brought increased prestige to the renowned Viennese chair and in just a few years had built up his own school, which produced physicists such as Hans Thirring, Ludwig Flamm, Erwin Schrödinger, and Karl Herzfeld.[111] Following Hasenöhrl's death, too, there was great concern that the chair be filled by a prestigious successor, as Sommerfeld learned from the mathematician Wilhelm Wirtinger, Dean of the Philosophical Faculty and chairman of the search committee who was dealing with the search for Hasenöhrl's successor. Wirtinger knew Sommerfeld from their years together as students of Felix Klein at Göttingen. He let Sommerfeld know that Einstein and Laue were also being considered for this position and that both he and a segment of the faculty laid the greatest stress on "maintaining this chair in *German* hands."[112]

This made it clear the appointment was to be regarded in a political light, as well, and like other matters with a political dimension, opinions differed among the Vienna faculty about which candidates—given these prerequisites—deserved preferential consideration. The matter of the appointment remained in limbo more than a year. "I am no longer Dean, and am thus free to express myself," Wirtinger wrote Sommerfeld in March 1917. Negotiations with Einstein and Laue had come to nothing. The physicists on the search committee would have placed the Polish

107 From Einstein, February 8, 1916. DMA, HS 1977-28/A,78. Also in ASWB I.
108 From Einstein, August 3, 1916. DMA, HS 1977-28/A,78. Also in ASWB I.
109 From Röntgen, January 6, 1916. DMA, HS 1977-28/A,288.
110 From Wirtinger, January 19, 1916. DMA, NL 89, 019, folder 5,5. Eckert/Pricha, *Berufungen*, 1984.
111 Bittner, *Geschichte*, 1949, p. 193.
112 From Wirtinger, January 27, 1916. DMA, NL 89, 019, folder 5,5.

theoretician Marian Smoluchowski and Debye on the list. But Wirtinger had entered a petition "to put you [Sommerfeld] *unico loco* on the list; Smoluchowski on the other hand, not at all." The majority of the committee had accepted this petition, although in a minority vote the physicists had insisted on Smoluchowski. Just why Wirtinger, along with the majority of his colleagues, was against Smoluchowski, he explained by writing, "that Sm. is a Pole, and declares himself as such. I can imagine that in the German Reich there is no true appreciation of what this means to us [in the Austro-Hungarian Empire]. For you can always count on your government's being *German*, whereas for us "German-ness" constitutes a chip the government bargains with in various difficult circumstances to accommodate the other nationalities, so that we have to protect it ourselves." The ministry had not yet come to a decision in this matter. "In this situation, I scarcely need assure you that it would be especially pleasing and a source of great satisfaction for us to attract you, and thus assure that Boltzmann's chair remain in German hands. It will of course be some time before the government acts." He closed his report with the importunate appeal, "to have a little patience with these negotiations."[113]

Sommerfeld had no wish to leave Munich. Nonetheless, he gave the Austrian Minister of Culture to understand that he "would take a potential call to the University of Vienna seriously into consideration," should such an offer eventually be made him. "Of course, given the expense of living in Vienna, my financial situation would accordingly need to be substantially improved."[114] Sommerfeld was then left hanging for several months, only in the end to be informed that, "with reference to the current negotiations carried on by the Finance Administration in the matter of filling the vacancy of Full Professor of theoretical physics at the University of Vienna, funds are not foreseeably available to provide for an increase in the official income you currently enjoy at Munich."[115] The Vienna "offer," Sommerfeld later wrote Hilbert, had been "merely a nonbinding inquiry from the faculty and the Ministry." "The matter concluded with the notification that I am too expensive for them. After that, of course, I couldn't squeeze another cent out of Munich."[116] In the event, though, the matter was not yet concluded. The faculty followed Wirtinger's proposal and placed Sommerfeld as the only candidate ("unico loco") on the list. Sommerfeld was thereupon invited to negotiations at the Ministry of Culture in Vienna, where the way was cleared to offer him a higher salary than Munich's after all. "The situation pleased me," Sommerfeld wrote Hilbert following the visit to Vienna. "When the Bavarian Minister offered me an increase in salary more or less equal to the Vienna offer, I of course decided in favor of Munich."[117]

113 From Wirtinger, March 14, 1917. DMA, NL 89, 019, folder 5,5.
114 To the Austrian Minister of Culture, July 25, 1916. DMA, NL 89, 019, folder 5,5.
115 From Cwiklinski, January 11, 1917. DMA, NL 89, 019, folder 5,5.
116 To Hilbert, March 13, 1917. SUB, Hilbert 379 A.
117 To Hilbert, July 10, 1917. SUB, Hilbert 379 A.

Subsequently, a new search was instituted in Vienna. Smoluchowski was killed in the war in 1917. Ultimately, Gustav Jäger, former assistant to Boltzmann, was given the position as successor to Hasenöhrl. In Munich, Sommerfeld's decision to remain was rewarded with an increase in his yearly salary of 3,000 Marks and with the "title and rank of a Royal Privy Councilor," as on July 13, 1917, the Bavarian Minister of Culture informed him: "May your Honorable Self perceive in this supreme token of favor an outward sign that Bavaria knows to value your talents, and gladly sees you retained in service to the unitary university."[118] As "Privy Councilor," Sommerfeld now numbered among the notabilities of the Bavarian state—which also often entailed social obligations. "Last Monday I had to lecture on the atomic model at the Polytechnic Association in the presence of the King!" he wrote Willy Wien in October 1917.[119] On other occasions, he was called on to serve as expert witness by the Ministry of Culture. For Oscar von Miller, too, who, in creating the German Museum, liked to get the reassurance of Privy Councilors from Science and Engineering, Sommerfeld became in these years a sought-after advisor and exhibition planner. Sommerfeld, together with a university colleague from physical chemistry, also carried out with pleasure the Museum founder's wish in working out—even in the final year of the war—a concept for an exhibition space: "The Structure of Matter."[120]

Sommerfeld enjoyed recognition and success in yet other ways. That he was now one of the giants in his field was confirmed for him in Stockholm, for example, when the Swedish Academy of Sciences asked him for a proposal for the awarding of the Nobel Prize in physics for the year 1918. He recommended "the quantum theory of Max Planck," in the following order: first, the discovery, then the discoverer, in explicit recognition that subsequently still other Nobel Prizes should be awarded for quantum-theoretical applications. But it seemed to him "impossible to crown a product of the quantum theory, for example, that certainly merits the Nobel Prize, the enormously fruitful investigations of Bohr, before the creator of the entire quantum field generally has been honored with the Nobel Prize. That bestowing this honor should occur just now is fully justified, since it is through the quantum theory of the spectra and the atom that the fundamental nature of Planck's thought is established beyond all doubt."[121] As he put this sentence to paper, the thought that he himself might win a future Nobel Prize must surely have occurred to him. In October of 1918, at the request of the Secretary of the Bavarian Academy of Sciences, who wished to place his name in nomination for the 1919 Nobel Prize, Sommerfeld suggested the wording for his own nomination, although he found it "somewhat unusual" to extol himself in this way. However, he

118 From Knilling, July 13, 1917. DMA, NL 89, 019, folder 5,5.

119 To W. Wien, October 24, 1917. DMA, NL 56, 010. Also in ASWB I.

120 Schirrmacher, *Atom*, 2003; Eckert, *Atommodelle*, 2009.

121 To the Nobel Committee, December 20, 1917. Stockholm, Academy of Sciences, Nobel Archive. Also in ASWB I.

obviously overcame these "nice scruples," since his suggested wording was "factually speaking, not altogether unjustified." Then doubt overtook him once more: "Actually it really is nonsense that I should draft your text!" he wrote the Academy Secretary. "Perhaps you should drop the whole matter after all!"[122]

Already at that time, the Nobel Prize was esteemed by physicists as the highest recognition of success in their scientific field. But this recognition was to be denied Sommerfeld. Some years later, Sommerfeld unburdened himself in writing in a moment of utter bitterness and resentment over the "scandal" that year after year he had been passed over for the award of the Nobel (see p. 319). But in the year 1917, when neither Planck, nor Bohr, nor Einstein had yet been chosen for the Nobel Prize, this sentiment was still alien to him, all the more so, as he was not otherwise lacking for recognition. In November 1917, the Academy of Sciences at Göttingen elected him Corresponding Member.[123] This was followed half a year later by his election to the Vienna Academy of Sciences.[124] The Prussian Academy of Sciences recognized him with the Helmholtz Prize for his work "Towards a Quantum Theory of the Spectral Lines." The award of this prize, with an endowment of 1,800 Marks, had—as Sommerfeld was informed on January 25, 1917—"been publicly announced in today's ceremonial session in celebration of the birthday of his Majesty, Emperor and King, and the anniversary of King Friedrich II."[125] This distinction for Sommerfeld came to the notice of colleagues in the Netherlands too. "Your results belong among the most beautiful ever achieved in theoretical physics," Lorentz wrote in congratulations after he read the news. "Who could have imagined, even a few years ago, that relativity mechanics would deliver the key to deciphering so many secrets."[126] Hilbert, too, in his letter of congratulations on the Helmholtz Prize, said that ultimately, "It is the joy in the forward progress and in the thing itself that is so wonderful!"[127]

7.6 Military Physics

His elaboration of Bohr's atomic theory in the middle of the war might leave the impression that Sommerfeld had withdrawn entirely into the ivory tower of science. But this impression is misleading. In April 1915 technical military problems were brought to his attention by colleagues as they arose in the technical divisions of the army and the navy, and this proved not merely a momentary detour but the

122 To Goebel, October 8, 1918. München, Archiv der Bayerischen Akademie der Wissenschaften, estate of Karl Ritter von Goebel.

123 From Runge, November 10, 1917. DMA, NL 89, 020, folder 6,2.

124 From Lecher and Wirtinger, May 28, 1918. DMA, NL 89, 020, folder 6,2.

125 DMA, NL 89, 020, folder 6,2.

126 From Lorentz, February 14, 1917. DMA, HS 1977-28/A,208. Also in ASWB I.

127 From Hilbert, February 18, 1917. DMA, HS 1977-28/A,141.

start of military research that after 1917 became increasingly a central rather than an incidental focus. He was not alone in this. At other universities, military research was also mixed more and more into the professorial routine.[128] From the viewpoint of the history of science, one can for the first time speak of a "disinhibition of the scientist himself to be a resource for the prosecution of the war."[129] New weapon systems, such as submarines and airplanes, which lent this war literally a "new dimension,"[130] presented natural scientists and engineers with a virtual cornucopia of new challenges.

Although Sommerfeld was not called up for duty as a reserve officer to a technical division, his involvement in military research did not long remain a merely private concern. In 1916, the Berlin ministerial official Friedrich Schmidt-Ott, in collaboration with Fritz Haber and Walther Nernst, established an organization whose function was, "through the cooperative work of the country's best scientific minds in collaboration with the best military minds, to promote development of scientific and technical aids to the prosecution of the war." The Kaiser Wilhelm Stiftung für kriegstechnische Wissenschaft (Kaiser Wilhelm Foundation for Military-Technical Science), or "KWKW," as this organization was called by its acronym, was funded from the endowment of a private donor. Nonetheless, with a board of trustees comprising ministerial officials and military officers, headed by the Prussian Minister of War, this foundation was a supremely official body.[131]

The KWKW began its practical work with lists of research subjects from various militarily relevant fields that had been directed to the Ministry of War from the army and the navy and then referred on to committees of experts for further processing. In January and February 1917, the Ministry of War, by way of the Rectors of the universities and technical colleges, informed the professors with the expertise to do work on one or another subject. Sommerfeld, as a long-time acquaintance of Nernst, was personally contacted and requested to take on the area of "theoretical investigations in the field of radiotelegraphy" as a member of a committee of physics experts.[132] Sommerfeld agreed at once. His area of responsibility was designated as "Theoretical Consideration of Practical Antenna Forms, Gyroscope Theory."[133] As a medium of communication between airplanes, ships, and submarines, "radiotelegraphy" was, during World War I, a still largely untested technology. In the army, Max Wien, scientific head of the technical department of the radio troops (Tafunk), was responsible for this area; the navy also had a department, in the torpedo inspectorate at Kiel, that pursued research in radio technology and, in the person of Heinrich Barkhausen, had at its disposal the talents of a university

128 Berg/Thiel/Walther, *Feder und Schwert*, 2009; Busse, *Engagement*, 2008.
129 Ash, Wissenschaft – Krieg – Modernität, 1996, p. 71.
130 Trischler, *Räumlichkeit*, 1996.
131 Rasch, *Wissenschaft und Militär*, 1991, pp. 94–96.
132 From Nernst, February 22, 1917. DMA, HS 1977-28/A,241.
133 Rasch, *Wissenschaft und Militär*, 1991, p. 100.

professor with the relevant experience.[134] Just how significant gyroscope theory was for military technology was reflected in Nernst's appeal that Sommerfeld visit him in Berlin at the next opportunity so they could "discuss the trajectory of rifled mortar shells."[135]

Wireless telegraphy and gyroscope theory were areas with which Sommerfeld had long been familiar. In the matter of gyroscope theory, he could rely on Fritz Noether, who, still a student under his supervision in 1910, had edited the last volume of *The Theory of the Top* and who had assisted also in the new revised edition of 1913.[136] At the start of the war, Noether had been sent as a soldier to northern France but was later transferred to Berlin, where presumably he worked on ballistic applications of gyroscope theory in Sommerfeld's commission. As they were shot, mortar shells were given a spin on their way through the barrel which, while it provided greater stability, also set them into a motion known in gyroscope theory as "conical oscillation" which led to deviations in their trajectories. The issue, as Sommerfeld indicated in a report to the KWKW under the heading "Mortar Ballistics," was "to eliminate or diminish the conical oscillations of mortar shells by some modification in their construction, and to achieve an optimally smooth trajectory."[137] This report revealed nothing explicitly about Sommerfeld's collaboration with Noether, but after the war, Noether, utilizing an unpublished manuscript by Sommerfeld, brought out a dissertation on "projectile oscillations" which, so far as the mathematical theory was concerned, must have presented the results of this KWKW commission.[138] Whether this theory was of any real military utility is doubtful. Several years later, Carl Cranz, who represented "physics" on the ballistics committee of experts, wrote in a textbook on ballistics that, in their treatment of the problem "by way of complex integration," Sommerfeld and Noether "offered their readers high intellectual pleasure," even though their results were not consistent with observed ballistic deviations.[139]

From surviving sources, it is possible to reconstruct Sommerfeld's military research only piecemeal. Apparently, the files of the KWKW, together with the stores of army archives, were largely destroyed during World War II.[140] Sommerfeld's report to the KWKW is among the few documents that offer an insight into any of the individual projects commissioned by this organization. In it, Sommerfeld has indicated only sketchily the scope of the research project "Mortar Ballistics," whose inception went back to an assignment from the army's artillery testing commission.

134 Krauß, *Rüstung*, 2006, p. 123.
135 From Nernst, March 2, 1917. DMA, HS 1977-28/A,241. For other applications see Broelmann, *Intuition*, 2002, pp. 295–310.
136 From Noether, July 12, 1913. DMA, HS 1977-28/A,246.
137 To the KWKW, undated [March, 1918]. DMA, NL 89, 019, folder 5,6. Also in ASWB I.
138 Noether, *Berechnung*, 1919.
139 Cranz, *Ballistik*, 1925, p. 358.
140 Rasch, *Wissenschaft und Militär*, 1991, p. 83.

Considerably more comprehensive was his service to the military in the area of wireless telegraphy. Aside from the army's *Tafunk* in Berlin, this research was commissioned primarily by the torpedo inspectorate of the navy at Kiel. There, a separate department was responsible for underwater telegraphy, although at the start of the war in 1914 it was regarded as insignificant, and began only in the fall of 1915 to pursue more extensive research on underwater communication, primarily via sound waves.[141] Sommerfeld's area of activity concerned the propagation of electromagnetic waves in water. In his report to the KWKW, he headed his work for the navy "Streamlined Telegraphy." This project had taken form in May 1917 "in a conversation with Prof. Barkhausen." Its aim was "to create a communication medium from submarine to submarine or from land to submarine through slowly alternating current." For the absorption of an alternating current signal in a medium, the product of the electrical conductivity σ and the frequency v is decisive (in accordance with the theory, the depth of penetration is proportional to $1/\sqrt{\sigma v}$), so that in sea water, due to its high conductivity, lower frequency signals are transmitted more efficiently than those of higher frequency. Experiments, however, determined "a high multiple of the computed critical wavelength," which led Sommerfeld back to an involvement of the less absorbent ground. Consequently, he made "Streamlined Telegraphy in shallow water" the subject of his own investigation. In this, he was able to calculate the spread of the signal according to the template of his work on the wave spread in wireless telegraphy from the year 1909, "where air in the former case corresponds to the ground in this, the ground in the former, to water in this. Just as waves spread in the air in the former case, and dissipate in the ground, so here the alternating currents spread in the ground, and penetrate the water only through a kind of skin effect." Other calculations were made for "Streamlined Telegraphy" respecting "the so-called alternating current compass, by which submarines were to orient themselves by means of a cable carrying an alternating current." Another subject of calculation was whether a wire extended from and dragged behind a submarine or a closed wire loop was the more advantageous antenna form. "The last investigation is also under consideration for a question posed to me on November 9, 1917 by the Imperial Naval Office," Sommerfeld wrote in his KWKW report. "How can the directional precision of wireless signals be improved, that is, how can the angular space capable of receiving the perceptible intensity be reduced. My efforts in this regard are still not concluded."[142]

He had long been acquainted with this aforementioned problem of the directivity of antennas. In 1911, he had supervised a doctoral thesis on the subject (see p. 169) and had written a separate report for the KWKW "presented graphically,"[143] since

141 On the Organization of Underwater-Telegraphy to 1916 see "Inspektion des Torpedowesens. Kriegstagebuch," dated from September 28, 1916, in BA-MA, RM 27 III 29.

142 To the KWKW, undated (March, 1918). DMA, NL 89, 019, folder 5,6. Also in ASWB I.

143 Ibid.

there were very controversial conceptions of it. Such questions were addressed in letters from the front too. In April 1916, Sommerfeld's prewar assistant, Lenz, was immediately put to work "in the office of the head of field telegraphy at the high command" on the practical effects of the theory and became—as Sommerfeld wrote Hilbert in March 1917—his most important contact with regard to "military problems of wireless telegraphy."[144] Sommerfeld discussed these matters with Max Wien as well. He was at the Tafunk in Berlin in April 1918, where his "new theory of directivity" sparked interest primarily with respect to applications for the air force. "So I will enlist Buchwald in the discussion," Wien wrote Sommerfeld in advance of the meeting.[145] Eberhard Buchwald, an experimental physicist from Breslau, had briefly joined the Sommerfeld circle in Munich in 1911 and thereafter counted himself one of his students though he had not taken his doctorate under Sommerfeld. In 1917, as a lieutenant of the reserve in the "Radio Airmen's Test Division, Döberitz," he was familiar with experiments in which both the directivity of an antenna hanging from an airplane and fixing the bearings of airplanes from a ground station were to be investigated. Sommerfeld displayed a lively interest in these tests. At the end of July 1918, he again spent a week in Berlin, to inform himself on the spot about the application of his theory. He had "worked in radio for 2 days in the province of Mecklenburg" and was "very amicably" received by the officers of the testing station, he wrote home.[146] To his daughter, he wrote that on one occasion he even flew along in an airplane, an adventure that ended fortunately without serious consequences. "In the end, we experienced motor failure and made an emergency landing, which came off without harm to either human or machine."[147]

In this case, too, theory contributed little to military practice. "Technical difficulties preclude the application of the theory," Sommerfeld once wrote concerning his efforts on behalf of the Torpedo inspectorate at Kiel, which explains why, with respect to this area of military research too, he was "not really satisfied."[148] Although after the war Buchwald and his colleagues, in a report on the "Experiments in Directivity and Fixing Position" at the airfields in Mecklenburg during the last year of the war, thanked Sommerfeld for "suggestions relative to a discussion of all these experiments," nonetheless, they also conceded that the investigations had yielded hardly any practical results.[149] A textbook-style presentation from 1920 on "Radio Telegraphy for Airplanes," wrote that "The theoretical results of Sommerfeld and his students were not even thoroughly tested on the ground; the situation in three dimensional space was left altogether unresolved." Not even in the tests at

144 From Lenz, April 19, 1916. DMA, NL 89, 059; to Hilbert, March 13, 1917. SUB, Cod. Ms. D. Hilbert379 A.

145 From M. Wien, April 20, 1918.

146 To Johanna, July 30, 1918.

147 To Margarethe, August 2, 1918.

148 To Johanna, August 6, 1917.

149 Baldus/Buchwald/Hase, *Geschichte*, 1920.

Mecklenburg could agreement between "calculation and experiment" be spoken of. "Nor could this have been expected, since the simplified assumptions underlying the calculations did not correspond to the conditions of the experiments."[150] A textbook on "Wireless Telegraphy and Telephony" from the year 1923 concluded that "So far as the question of directional transmission is concerned, the problem remains unsolved."[151] In none of these postwar presentations directed at professionals was the theory of the directivity of transmitting antennas initiated by Sommerfeld and further elaborated by his doctoral student Hoerschelmann commented on, even in an appendix. It would be many years before the gulf between theory and practice in this field was bridged.[152]

But Sommerfeld made himself useful to the German war effort also in other ways. For example, his Munich colleague Walther von Dyck, head of a "study commission" on the Flemization of the University of Ghent, enlisted him into the politics of war aims in occupied Belgium.[153] The university was to be reopened in 1916 as a Flemish university, an eventuality boycotted by Belgian professors. The plan of the German occupiers threatened to founder on a dearth of university teachers prepared, given such an omen, to assist in breathing new life into the University of Ghent. When Sommerfeld attempted to make acceptance of a professorship at Ghent palatable to a colleague from neutral Holland, he was given to understand that he would thereby be making no friends among the Flemish. "I don't think it is really understood in Germany what the true viewpoint of the Flemings is. Though they want to achieve the Flemization of one of the Belgian universities, specifically Ghent, they want to control their circumstances themselves, as Belgians, and not be dictated to by foreigners."[154]

This response proved effective: Sommerfeld forwarded the letter to Wilhelm Wien as a warning against ill-considered measures and doubtless to other colleagues, although he was not thereby moved to withdraw from other propaganda efforts promoting German war aims.[155] In January 1918, he agreed to a leave of absence "to lecture at the front in Tournay,"[156] where he and other German professors boosted the morale of the troops with lectures "on peacetime physics," as he wrote home. He also exploited this "shot-in-the-arm lecture" to acquaint himself with the practical problems of implementing wireless telegraphy, as well as to pay a visit to the University of Ghent.[157] This visit was "very valuable" for Dyck, and he

150 Niemann, *Funkentelegraphie*, 1921, pp. 308 and 326.

151 Lertes, *Telegraphie*, 1923, p. 143.

152 Eckert/Kaiser, *Nahtstelle*, 2002.

153 Hashagen, *Walther von Dyck*, 2003, pp. 503–536.

154 From Kamerlingh Onnes, June 6, 1916. DMA, HS 1977-28/A,160.

155 To W. Wien, June 15, 1916. DMA, NL 56, 010.

156 To the Rectorate of the University of Munich, December 22, 1917. UAM, E-II-N Sommerfeld.

157 To Johanna, January 9, 1918. Also in ASWB I.

counted on Sommerfeld's "testifying back home as to what has been achieved there."[158] Sommerfeld fulfilled this desire only too eagerly. In an article for the *München-Augsburger Abendzeitung*, he wrote that he had felt "exalted . . . to know of a place on old Germanic soil, at which formerly only the French language had been heard, but that had now been won back for German science." The "establishment of Ghent" had been the "most effective and promising move of German policy in Belgium," to promote which "the common roots of Germanic culture" had been evoked.[159] The value the German occupiers placed on the Flemization of the University of Ghent became clear several months later when, in the context of retreat from the western front, a withdrawal from the University of Ghent, seen as the "backbone of German Flemish policy" was also weighed. An internal memo of October 3, 1918, from the German civil authority in Belgium noted that "With the disappearance of the Flemish university, given the symbolism it surely bears, the entire German policy in Belgium is finished."[160]

7.7 An Emotional Roller Coaster

For Sommerfeld, the last year of the war was a time of intense activity. His "shot-in-the-arm lecture" blended nationalistic propaganda, enthusiasm for atomic theory, and military physics. In April 1918, he accepted an invitation from the "German Ladies Association of the Red Cross for the Colonies, Province of Württemberg" to deliver a popular lecture on "The Development of Physics in Germany since Heinrich Hertz." "Large audience, at least 1,000 people," he wrote his wife afterwards.[161] From there, his lecture tour proceeded to Brussels, where he evoked "great enthusiasm" from a "rather small audience" with a professional lecture on ballistics. An excursion to general headquarters ("Perhaps I'll even catch a glimpse of H[indenburg]'s and L[udendorff]'s coat-tails!")[162] presumably served the purpose of elucidating problems of wireless telegraphy, in tandem with Lenz. One week later, he was in Berlin, where he debated on the subject of the directivity of antennas at the Tafunk, and, at the celebration of Max Planck's sixtieth birthday, in his testimonial speech voiced his enthusiasm for quantum theory. "Berlin is exhausting, but all the same quite nice," he reported to his wife concerning this visit.[163]

158 From Dyck, January 19, 1918. Quoted in Hashagen, *Walther von Dyck*, 2003, pp. 529–530.

159 Literaturbeilage der *München-Augsburger Abendzeitung*, February 26, 1918. Cf. also, Sommerfeld, *Besuch*, 1918; Sommerfeld, *Besuch in Gent*, 1918.

160 Quoted in Hashagen, *Walther von Dyck*, 2003, p. 533.

161 To Johanna, April 14, 1918. Sommerfeld, *Entwicklung der Physik*, 1918.

162 To Johanna and Ernst, April 17, 1918.

163 To Johanna, April 29, 1918.

Fig. 18: Arnold and Johanna Sommerfeld in 1917 (Courtesy: Deutsches Museum, Munich, Archive).

These journeys were also something of an emotional roller coaster for Sommerfeld. From a political viewpoint, the last year of the war was quite depressing for him, for military success continued to elude the German army. The fact that nonetheless he experienced the visit to Berlin in late April 1918 as "quite nice" had more to do with the latest advances in atomic theory, achieved precisely among his own Munich school, which he reported to assembled colleagues at the Planck celebration. He answered the question why, of the many transitions possible between circular and elliptical orbits according to the fine-structure theory, only a few actually occurred. In the emission of radiation, the atom is coupled with the surrounding "ether," Sommerfeld explained; therefore, one had to carry out the quantization of "the coupled system, atom + ether." Assuming the radiation occurred in the ether rather than in the atom, the energy differential between final state and initial state in the atom had to correspond to the energy of an ether oscillation. "It is not the atom that oscillates, but the ether." In that case, however, not only the energy

Fig. 19: Adalbert Rubinowicz contributed the "selection rule" to Sommerfeld's theory that in the transition of electrons in the atom, the quantum number can change by only one unit (Courtesy: Deutsches Museum, Munich, Archive).

but also the angular momentum must be conserved. "Whenever the momentum of an electron changes at a transition in the atom, the same change has to be transferred to the ether." A selection rule follows from this, whereby the quantum numbers of the transition in the atom can change by just one unit. Aside from this, statements about the polarization of the spectral lines were obtained.[164] Sommerfeld did not take credit himself for this success but acknowledged the contribution of Adalbert Rubinowicz, who had occupied the post of assistant at the University of Czernowitz, and following the war-related closing of that university, had found a temporary home at Sommerfeld's institute in Munich. When Sommerfeld reported the reactions of the Berlin physicists, Rubinowicz felt as though he had been knighted. Especially pleasing was Einstein's reaction, who "had judged as 'excellent' the idea of employing the law of conservation for the derivation of the selection and polarization rules."[165]

164 Warburg/Laue/Sommerfeld/Einstein, *Geburtstag*, 1918, pp. 20–22.
165 A. Rubinowicz: Zur Geschichte meiner Entdeckung der Auswahl- und Polarisationsregeln. AHQP,OHI 1419/4.

After the Berlin physicists, Sommerfeld shared his pleasure over this "reconciliation of quantum theory and wave theory" also with Bohr. "The wave phenomenon exists only in the ether, which obeys Maxwell's equations, and operates quantum-theoretically like a linear oscillator, with indeterminate eigenfrequency v. The atom supplies to the wave phenomenon only a certain quantity of energy and momentum as substance of the wave phenomenon. But it has nothing directly to do with the oscillation. The ether takes its frequency from the energy according to your hv-frequency condition, its polarization from the momentum." Shortly before, Bohr had sent him the first part of a dissertation, "On the Quantum Theory of Line Spectra," in which the same results were derived from—as Sommerfeld expressed it—an "interesting comparison of classical and quantum theoretical emission for large quantum numbers." Compared to Bohr's "principle of correspondence" (as this method soon came to be called), Rubinowicz's formulation seemed to Sommerfeld, however, to be "in terms of physics, more instructive."[166]

For several weeks, enthusiasm for atomic physics took precedence over everything else. The "reconciliation of quantum theory and wave theory" seemed to Sommerfeld the long-sought solution to the dilemma of the wave-particle dualism. Shortly thereafter, news from the laboratory of Manne Siegbahn in Lund occasioned euphoria. On June 1, Sommerfeld reported to the Bavarian Academy of Sciences on certain lines in the X-ray spectra (K_β), which, according to his fine-structure theory, ought to show a doublet splitting—and quoted in this connection from a letter from Siegbahn who just a few days earlier had confirmed that in the course of closer scrutiny of his photographic plates, he had in fact found this dual nature of the lines in question.[167] Three days later, he wrote Siegbahn how he understood this doublet: "If an electron jumps from the M-ring to the K-ring, the effective nuclear charge diminishes . . . So the nuclear attraction is reduced, and thus the L-ring expands. So in calculating the energy balance between the initial and final states, a contribution of the energy of the L-term has to be included, notwithstanding the fact that nothing is altered in the number q of its electrons. . . But now the L-ring is doubled—circular or elliptical—and therefore different in its relativistic contributions." Pursuant to his theory, a dependence on the nuclear charge is demonstrated for this doublet splitting that would also be revealed on Siegbahn's plates. "You see from this we are dealing here with a very interesting and surely demonstrable new consequence of the fine structure theory. The particular interest lies in the fact that the relative size of the doublet . . . allows the particle density q of the L-ring to be determined directly and very exactly. We have thus taken a significant step forward in understanding the structure of the atom. I would be very grateful for your assistance by informing me of your measurements."[168] The K_β doublet was

166 To Bohr, May 18, 1918. NBA, Bohr. Also in ASWB I, pp. 595–596. On Bohr's formulation of the principle of correspondence, see NBCW 3, "Introduction," notably pp. 3–8.

167 Sommerfeld, *Feinstruktur der K_β–Linie*, 1918, p. 372.

168 To Siegbahn, June 4, 1918. Stockholm, Academy, Siegbahn. Also in ASWB I.

for several weeks the subject of lively correspondence between Lund and Munich. Sommerfeld wrote Einstein that Siegbahn had given him "a beautiful confirmation of an initially unexpected result concerning X-ray spectra. . . K_β is a double line; the *L*-ring expands at the transition of an electron from the *M*-ring to the *K*-ring, and thereby shows its dual nature." He went so far as to suggest that Einstein "seek specific funding from some Berlin patron for a lecture series. . . As the first lecturer, I propose, perhaps, Siegbahn or Bohr."[169]

In light of such plans, the war seemed in these summer months of 1918 to have receded far into the background. At the same time, though, bitter battles were raging on both the southern and western fronts. In July, the Austro-Hungarian Army bulletin announced the withdrawal of Imperial troops from Albania; German bulletins spoke of French advances.[170] The major German offensive initiated on the western front in the spring had not brought the hoped for breakthrough. Now, in light of a French-British-American counteroffensive, it seemed ever less likely that Germany might still win the war. "I have various scientific plans, but am first and foremost fully absorbed with military problems," Sommerfeld wrote Willy Wien in August 1918. "As the military horizon seems recently to have clouded over again, this focus is also more gratifying than the purely scientific." The juxtaposition of war and peace in his scientific work provided him satisfaction and qualms of conscience in equal measure. "Rubinowicz's observations about the polarization of spectra, about which you will have read in outline in my Planck lecture, seem very promising to me. Additionally, I am still working on the X-ray doublets and the doublets of the alkalis. When will we finally be able to address these things in clear conscience?!"[171] From Berlin, where he was working on his "military problems" with the army radio technicians, he wrote his wife, "It would all be quite nice if only the news from the west were better."[172] Three days later, he concluded a birthday letter to his daughter with his good wishes "for you, for us all, and above all for the fatherland."[173] When Rubinowicz reported new "oddities" from his researches in atomic theory, he replied with a note of regret that "on account of all sorts of military work," he himself had "not been able to do a thing in quantum theory."[174]

169 To Einstein, undated (June, 1918). AEA. Also in ASWB I.

170 Official war communiques, http://www.stahlgewitter.com/18_07_10.htm (30 January 2013).

171 To W. Wien, undated (August, 1918). DMA, NL 56, 010. Also in ASWB I.

172 To Johanna, July 30, 1918.

173 To Margarethe, August 2, 1918.

174 To Rubinowicz, September 2, 1918.

8 The Quantum Pope

With the end of the war and the attendant revolutionary turmoil, the ride on the emotional roller coaster became more dizzying still. The appointment at Uppsala offered to Willy Wien in November 1918 seemed to Sommerfeld like a call to desertion. He would have experienced it as "humiliating" had Wien accepted and abandoned his country.[1] The position taken by Einstein, who believed in the "new era" and was in this respect, too, an exception among German professors, was totally incomprehensible to him. "May God preserve your belief!," Sommerfeld concluded a letter to Einstein. "I find everything unspeakably miserable and stupid. Our enemies are the greatest liars and rogues, and we the greatest morons. Not God, but money rules the world."[2] To a Viennese colleague, he confessed that he saw the "German future as black on black . . . It is as though our misfortune and our revolutionary slogans have wiped out every feeling of self-respect and self-confidence."[3]

The grim political atmosphere stood in stark contrast to the optimism Sommerfeld felt for the future of his own field. In light of the exciting innovations in theoretical physics, Einstein and Sommerfeld readily set aside their political differences. In March 1919, following the murder of the Bavarian President, Kurt Eisner (1867–1919), when Munich was on the verge of proclaiming a soviet republic and government troops were fighting the Spartacists in Berlin, Sommerfeld sent the latest results of his research to Einstein with the request that he give a "little lecture" on them at the next meeting of the German Physical Society.[4] The minutes of the meeting of May 9, 1919 (2 days after publication of the draft Versailles Treaty) confirm that Einstein complied with this request. The paper by Sommerfeld and Kossel bore the title "Selection Principle and Displacement Law in Spectral Series."[5] This was not a case of a theory interlarded with equations, but of an empirically based conjecture. The spectra of sequential elements on the periodic table exhibit a notable peculiarity: Atoms with an uneven number of valence electrons display double lines (doublets, such as the yellow sodium lines); atoms with even valence show triple lines (triplets). The spectrum of an atom ionized by displacement of one of its electrons should assume the type of its neighboring element on the periodic table. "The spark spectra of an element are of the same character as that of the

1 To W. Wien, November 12, 1918. DMA, NL 56, 010. Also in ASWB I; on the political situation following WWI generally, Ullrich, *Revolution*, 2009.
2 To Einstein, December 3, 1918. AEA, Einstein. Also in ASWB I.
3 To Geitler, January 14, 1919. Also in ASWB II.
4 To Einstein, March 25, 1919. AEA, Einstein.
5 Sommerfeld/Kossel, *Auswahlprinzip*, 1919.

M. Eckert, *Arnold Sommerfeld: Science, Life and Turbulent Times 1868-1951*,
DOI 10.1007/978-1-4614-7461-6_8,

preceding element on the periodic table," Sommerfeld wrote describing the essence of this "Displacement Law." This was "a further step forward in bringing order to the tangled mass of spectroscopic facts."[6]

8.1 Atomic Structure and Spectral Lines

Scientifically, Sommerfeld saw himself at the summit of his career. Although mathematicians and physicists are conventionally thought to have passed the peak of their creativity as researchers by the age of 30, Sommerfeld, who had turned 50 on December 5, 1918, was in his explorations in theoretical physics ascending into the highest spheres. Although in correspondence with Einstein and other colleagues he gave vent to expressions of displeasure over the political situation, such passages came at the ends of letters whose principal topic was current physics, not politics. "For a certain purpose (a popular book on atomic models), I need a simple presentation of the foundations of quantum statistics," he had begun the letter to Einstein in which regarding the political situation, he thought "everything unspeakably miserable and stupid" and then went on in three pages to spell out his conception of quantum statistics and ask, "Do you agree with this?"[7]

Sommerfeld had sketched the outline of the "book on atomic models" 2 years before, following a lecture course on "Newer Experimental and Theoretical Advances in Atomism and Electronics," announced in the lecture catalogue as "popular, without mathematical elaboration," and which had clearly given him pleasure. "This semester, I have given a one-hour, popular course on atomic structure and spectral lines," he wrote Hilbert after the winter semester of 1916/1917, "to an audience of about 80, of which 12 were colleagues, primarily chemists, medical faculty, and philosophers, which I would also like to publish as a book. It was such fun that next semester, I will try to do the same with relativity, i.e., without mathematics, purely conceptually."[8] Subsequently he gave similar popular, 1-h lecture courses on "X-rays and Crystal Structure" (winter semester, 1917/1918), then once more "Atomism" (summer semester, 1918), and "Atomic Structure and Spectral Lines" (winter semester, 1918/1919). This must also have constituted part of his "lectures on peace-time physics" for the soldiers at the western front in January 1918.[9]

Even before the war's end, Sommerfeld had begun to bring this plan to fruition. "The last two weeks, I have been writing a popular book on 'atomic structure and spectral lines'—the text, for chemists, in the supplements also for physicists," he had written in a letter to Einstein in June 1918.[10] From time to time, he spoke also of a

6 To W. Wien, March 27, 1919. DMA, NL 56, 010.
7 To Einstein, December 3, 1918. AEA, Einstein. Also in ASWB I.
8 To Hilbert, March 13, 1917. SUB, Cod. Ms. D. Hilbert 379 A.
9 To Johanna, January 9, 1918.
10 To Einstein, undated [June, 1918]. AEA, Einstein. Also in ASWB I.

"semi-popular general presentation of the field"[11] or of a book "that would be comprehensible also to non-physicists."[12] In any case, Sommerfeld left no doubt that he intended to operate far afield from his own specialty. In March 1919, he completed the manuscript of the book "except for the last chapter"[13] and began negotiations with the scientific publishers Vieweg, Teubner, and Springer. He chose Vieweg, as he wrote Willy Wien. Teubner had been "not forthcoming." "Springer was very tempting, but I did not trust their business principles, and in the face of the current Jewish political mischief, I am becoming more and more an anti-Semite."[14]

He was letting it be known what he thought of the revolutionaries who shortly thereafter proclaimed a soviet republic in Munich and were lumped together by the press under the rubric "Jewish."[15] Otherwise, the political situation seems not to have exerted any influence on the progress of his work. "My book on the atom will shortly be printed," he informed Wien following the summer semester.[16] On September 2, 1919 he affixed his signature to the preface, and thanked Vieweg Publishers for having "upheld its ancient reputation splendidly through the most difficult of times." Aside from this allusion, and a passing reference to "university courses at the front" at which he had lectured on atomic models, it is impossible to detect from the book the historical circumstances under which it came into being. Instead, Sommerfeld let himself be carried away in rhapsodic statements about the subject of his book. "What we hear today in the language of the spectra is truly an ethereal music of the atom, a polyphony of integral relationships, a burgeoning of manifold order and harmony. The theory of spectral lines will forever carry Bohr's name. But another name will also be associated with it, that of Planck. All the integral laws of spectral lines and of atomism flow ultimately from quantum theory. It is the mysterious organum on which the spectral music plays, and in time with whose rhythm it rules the structure of atoms and their nuclei."[17]

The publisher took till the end of the year to produce and distribute the book. "I assume you and your institute, as well as several of your fellow workers have in the meantime received my book," Sommerfeld inquired of his Swedish colleague Manne Siegbahn in January 1920.[18] Although in his preface and in countless letters he repeatedly stressed the popular nature of his presentation, he was nonetheless very curious to know how his physicist colleagues would judge the work. He did not have to wait long. Runge, his delinquent encyclopedia author of the article on spectroscopy, was the first to thank him for his "handsome book." He had "already

11 To Swinne, December 25, 1918. DMA, HS 1952-3.
12 To Bohr, February 5, 1919. NBA, Bohr. Also in ASWB II.
13 To Landé, February 28, 1919. SBPK, Landé, 70 Sommerfeld.
14 To W. Wien, March 27, 1919. DMA, NL 56, 010.
15 See Ullrich, *Revolution*, 2009, Chap. 3.
16 To W. Wien, August 9, 1919. DMA, NL 56, 010.
17 Sommerfeld, *Atombau*, 1919, p. viii.
18 To Siegbahn, January 9, 1920. Stockholm, Akademie, Siegbahn.

studied its contents diligently" and found the presentation "very elegant." It did seem to him, however, that Sommerfeld had not maintained the popular character he had announced in the early chapters. Nonetheless, the book would "be an excellent introduction to the subject for many readers."[19]

In the professional press, *Atomic Structure and Spectral Lines* was very favorably received. It was to be welcomed "most heartily," wrote James Franck (1882–1964) in the *Naturwissenschaften*, that Sommerfeld had written this book midway in the research process and had not waited until he could present completely mature results. The presentation was "of such an immediate power that every reader who is interested in the natural sciences must feel himself swept away, following the author willingly into this new world whose opening up we owe in large part to his scientific intuition and that of his students." He likened reading to a mountain trek in which both practiced mountain climbers and novice hikers could profitably take part. "Even though many readers will not follow the leader all the way to the highest summits, there are nonetheless enough easily reached lookout points from which the view is well worth the effort."[20] The reviewer in the *Physikalische Zeitschrift* predicted that "this excellent work" would be "indispensable" for every physicist wishing to work in this area. The book was "extremely stimulating," but in the final chapters, where quantum matters are the focus, "not exactly easy to understand. This is due to the nature of the subject, and not a lack of expository ability on the part of the author."[21] Enthusiastic reactions came from abroad, too. "The book reads like a thrilling novel," wrote Zeeman from Amsterdam. He had found much there that, notwithstanding his intimate knowledge of the research involved, was still unknown to him, such as the Spectroscopic Displacement Law.[22] *Atomic Structure and Spectral Lines* had scarcely been published when the publisher had to issue a second printing. "I congratulate you on the new edition of your book," Planck wrote to Sommerfeld.[23] "Among all the nice reviews, this is certainly the most striking. . . Such a rapid reprinting" testifies best to the success of this work, Bohr wrote from Copenhagen.[24]

Atomic Structure and Spectral Lines was seen far and wide as a sign that physics was departing for unknown shores—and Sommerfeld, to extend the metaphor, came to be regarded as a captain who was setting the course. "Foreigners speak enthusiastically about the progress made by physics in Germany in spite of the

19 From Runge, January 12, 1920. DMA, HS 1977-28/A,298. Runge's encyclopedia article "Seriengesetze in den Spektren der Elemente" did not appear until 1925.

20 *Die Naturwissenschaften 8* (1920), p. 423.

21 *Physikalische Zeitschrift 21* (1920), pp. 223–224.

22 From Zeeman, January 16, 1920. DMA, HS 1977-28/A,380. Also in ASWB II.

23 From Planck, February 15, 1920. DMA, HS 1977-28/A,263.

24 From Bohr, November 8, 1920. NBA, Bohr. Also in ASWB II, pp. 85–86.

war," Paschen wrote to Sommerfeld. He had received letters from the USA full of admiration for the achievements of German physicists. "Einstein's papers and yours figure largely in the conferences there."[25]

8.2 The German Physical Society: Internal Strife

Sommerfeld, together with Planck and Einstein, was considered *the* representative of physics in Germany. Their scientific authority was such that over and above all political differences, the German physics community entrusted offices and functions to them that determined decisively the future course of the German scientific enterprise. Planck served as Permanent Secretary of the Physical Mathematics Class of the Prussian Academy of Sciences and in this capacity took part in the founding of the Emergency Organization of German Science (*Notgemeinschaft der deutschen Wissenschaft*), brought into being in 1920 primarily through the initiative of Fritz Haber, from which later the German Research Foundation (*Deutsche Forschungsgemeinschaft*, DFG) emerged.[26] Einstein directed the Kaiser Wilhelm Institute for Physics founded during the war, which to be sure existed initially only on paper, and like the Emergency Organization allocated resources for research to applicants.[27] In 1918, Sommerfeld took over the presidency of the German Physical Society (*Deutsche Physikalische Gesellschaft*, DPG), an office Planck and Einstein had occupied before him. The Berlin society of physicists founded in the nineteenth century evolved only gradually into a professional organization for physicists those outside Berlin also felt represented by. To judge from its name, the "Physical Society at Berlin" became after 1899 a "German Physical Association," but in practice, the Berlin physicists maintained control. The decisive transformation came following World War I, and Sommerfeld perpetually had to assume the role of mediator to keep physicists elsewhere from feeling patronized by "the Berliners."[28]

When Sommerfeld took up this office, he could not yet imagine what it entailed. "Yesterday evening, you were elected president by the executive board and the advisory council as well as the plenum of the German Physical Society, and, be it noted, with marked enthusiasm," Einstein had written him on June 1, 1918, and had added that this entailed "no duties." "If you happen to be in Berlin for a meeting of the Society anyway, you will chair it; otherwise you will be represented by one of the local board members (first in line, Rubens)."[29] As Director of the Physical Institute

25 From Paschen, June 16, 1920. DMA, HS 1977-28/A,253.

26 Heilbron, *Dilemmas*, 1986, Chap. 3; Szöllösi-Janze, *Fritz Haber*, 1998, Chap. 9; for a comprehensive survey of the founding of the "Emergency Organization," see Marsch, *Notgemeinschaft*, 1994.

27 Kant, *Albert Einstein*, 1996.

28 Richter, *Kämpfe*, 1973; Forman, *Support*, 1974; Wolff, *Konstituierung*, 2008.

29 From Einstein, June 1, 1918. DMA, HS 1977-28/A,78. Also in ASWB I.

of the University of Berlin, Heinrich Rubens (1865–1922) represented the patronizing attitude of the Berlin physicists the non-Berliners deplored. Only a few meetings were necessary for it to become clear Sommerfeld would not be able to discharge his duties as DPG President as casually as Einstein had represented to him. In the last months of the war, the pending reorganization could no longer be accomplished. "The constitutional debate is to be postponed until after the war," Sommerfeld wrote after a board meeting in August 1918 to Willy Wien in Würzburg, who vehemently opposed all patronizing from Berlin. Sommerfeld agreed that the interests of the DPG could no longer be "an affair of Berlin senior teachers," but sought also to mediate the strife. The Berliners were "not so bad" and would "gladly and sincerely meet the non-Berliners pretty much half-way." A splitting of the DPG, with which Wien was flirting, struck Sommerfeld as "quite unthinkable."[30]

After the war, the sentiments for a split erupted openly. Many industrial physicists no longer regarded the DPG as their representative, and in 1919 founded their own organization, the Society for Technical Physics (Gesellschaft für technische Physik).[31] To this contentious atmosphere was added the poor material conditions of the physics institutes following the war. The DGP hoped to remedy the situation with a "Support Fund for the Physical Institutes," and in November 1918, "the altered political circumstances notwithstanding" established a commission whose purpose was to solicit contributions.[32] Together with Einstein and Sommerfeld, the commission also included Rubens, Willy Wien, Franz Himstedt (1852–1933), and the Manager of the DPG, Karl Scheel (1866–1936). Despite this prestigious lineup, however, the "capital fund drive" consistently faltered.[33] The conferences on the DPG constitution made no progress, either. "Given the current conditions of the railroads, an assembly seems impossible in the foreseeable future," Sommerfeld wrote Willy Wien in March 1919 by way of consolation.[34] Two weeks later, the commission declared its efforts at industrial funding a failure. An industrialist had made it unmistakably clear "that furnishing adequate means was not the business of industry, but of the state."[35]

Despite these failures, at the next meeting of the board in May 1919, Sommerfeld was entrusted for another year with presidency of the DGP.[36] From then until Willy Wien was elected President in the fall of 1920, this office demanded of him a high degree of diplomatic skill. The decentralization of the Society was accomplished after several months by the establishment of "district organizations." But then the

30 To W. Wien, undated [August, 1918]. DMA, NL 56, 010. Also in ASWB I.

31 Forman, *Environment*, 1967, pp. 143–145; Hoffmann, *Etablierung*, 1987.

32 From Rubens, November 19, 1918; from Scheel, November 20, and December 8, 1918. DMA, NL 89, 018, folder 3,7.

33 From Scheel, March 20, 1919. DMA, NL 89, 018, folder 3,7.

34 To W. Wien, March 27, 1919. DMA, NL 56, 010.

35 From Scheel, April 13, 1919. DMA, NL 89, 018, folder 3,7.

36 From Scheel, May 11, 1919. DMA, NL 89, 018, folder 3,7.

disputes threatened to escalate with respect to another bone of contention. Some reorganization had to be instituted for the traditional Society journal, the *Verhandlungen der Deutschen Physikalischen Gesellschaft*, published by Vieweg-Verlag, who refused to continue distributing the journal, now bursting at its seams, free of charge to all the members of the DPG. The DPG board directed a commission to focus its attention on this problem, but its proposed solution proved incendiary as well. The *Verhandlungen* was to be divided into an "internal" and an "independent" periodical. The "internal" publication was to contain Society news and continue to be given out at no cost to DPG members; the "independent" publication was to have the character of a scientific journal and no longer to be distributed free of charge.[37]

On learning of these plans, Wien immediately saw in them competition with the *Annalen der Physik*, whose editorship he shared with Planck; he threatened to resign from the DPG. Sommerfeld was concerned too. Especially the title "Zeitschrift für Physik" ("Journal of Physics") which Vieweg wanted for the new periodical was problematic since it might be confused with the existing *Physikalische Zeitschrift*, published by Hirzel-Verlag. Vieweg stuck to his position "with such tenacity," as Planck wrote Sommerfeld, that he made "his acceptance of the entire contract dependent on [our] acquiescence to it."[38] Three days later Einstein wrote to Munich that "after prolonged back and forth," Vieweg had gotten his way. "Had you been here, you wouldn't have been able to do otherwise than we, nor would our colleague Wien. But from a distance, everything looks lop-sided and dubious, especially when it comes from those +[=damned] Berliners!"[39] Haber, who along with Einstein sat on the journal commission, added that there was consensus the *Annalen* should not be compromised. The *Zeitschrift für Physik* would be seen as bringing out new developments quickly, whereas the *Annalen* would retain "its role as archive of the important papers. . . Should fears arise that the *Annalen* might be harmed, we can best allay them, in my opinion, by supporting Prof. Wien's election as President of the Society at the conclusion of your term."[40] For his part, Planck assured Sommerfeld once more that he would leave "no stone unturned" to prevent Wien's departure. In the worst case, "a break with Vieweg [would be] preferable to a secession from the Society."[41]

Ultimately, however, "the Berliners" did not want to back away from the contract with Vieweg. Two days before Christmas, Haber and Planck wrote Sommerfeld that they expected nothing from renewed negotiations. Planck pleaded with Sommerfeld to impress on Wien "that for the time being at least he should refrain from his

37 From Scheel, November 3, 1919. DMA, NL 89, 018, folder 3,7.

38 From Planck, December 15, 1919. DMA, HS 1977-28/A,263.

39 From Einstein, 18. December 1919. DMA, HS 1977-28/A,78.

40 From Haber, December 18, 1919. DMA, NL 89, 018, folder 3,7.

41 From Planck, December 18, 1919. DMA, NL 89, 018, folder 3,7.

announced intention to resign from the Society."[42] But Sommerfeld did not feel up to it, all the more since he himself tended to side with Wien on this question. He must have been struggling inwardly, for in the draft of his response to Planck, he already declared his intention to resign as president of the DPG. Wien was annoyed. "Worst in my view are the business dealings of the Berliners," he wrote to Sommerfeld on December 23.[43] It may be that the Christmas festivities evoked more conciliatory feelings in Sommerfeld and Wien, for 2 days later Wien sent a telegram to Sommerfeld announcing that he would not withdraw immediately, but would await further developments.[44] This may have motivated Sommerfeld for his part to reverse his intention to resign. "Am planning to travel to Berlin," he wrote Wien on December 27. On the spot with his DPG board colleagues, he hoped to arrive after all at a compromise acceptable to all.[45] In the end, however, revision of the arrangement already agreed to did not come about. Final decisions would be reached only at the autumn meeting of the DPG, scheduled to take place in tandem with the next Natural Scientists Congress in September 1920, at Bad Nauheim.[46]

Yet another quarrel soon broke out. Reorganization of the physical journals touched the interests of competing publishers. In the short run, it was planned to merge the *Physikalische Zeitschrift* with the *Zeitschrift für Physik*, but after an initial round of negotiation between the publishers, Hirzel wrote Sommerfeld as DPG President that he no longer believed it possible to reach agreement in this matter with Vieweg. He would "never be prepared to cede to another publisher a journal built up over 20 years by dint of great financial sacrifice to the position of leading publication in physics, without any recompense whatsoever."[47] There was disagreement over the substantive organization, too. Debye, who up to now had been in charge editorially of the *Physikalische Zeitschrift* and had meanwhile moved from his professorship at Göttingen to the directorship of the Physical Institute at the ETH Zürich, did not wish to see his authority infringed on by Scheel, the intended editor of the *Zeitschrift für Physik*. He rejected a division of labor between experimental and theoretical physics. "I understand perfectly well that Prof. Debye chooses not to cede the authority of accepting experimental papers," Haber, as speaker of the journal commission, wrote in an effort to mediate. "Might I suggest that both editors be granted authority to accept experimental as well as theoretical papers, but that it be publicly requested that theoretical papers should preferably be sent to Debye, whereas experimental papers preferably to Scheel."[48] Debye and Hirzel, however, were not ready for any compromise. "After much deliberation and

42 From Planck, December 22, 1919. DMA, NL 89, 018, folder 3,7.
43 From W. Wien, December 23, 1919. DMA, NL 89, 018, folder 3,7.
44 From W. Wien, December 25, 1919. DMA, NL 89, 018, folder 3,7.
45 To W. Wien, December 27, 1919. DMA, NL 56, 010.
46 Forman, *Naturforscherversammlung*, 2007.
47 From Hirzel, April 14, 1920. DMA, MPGA, Debye.
48 From Haber, January 17, 1920. DMA, NL 89, 018, folder 3,7.

careful weighing of the issues, I have come to the decision to continue publishing my physical journal under one of the best editors it has ever had, and to spare no resources to maintain it on the high level it enjoys both at home and abroad," Hirzel announced to Sommerfeld.[49] The DPG board abandoned all hope of agreement between the publishers. "Merger of the two journals has unfortunately foundered with Vieweg and Hirzel, and it is because Debye wishes to retain editorship in Zürich," Sommerfeld wrote in advance of the Nauheim meeting to Wien, DPG President-elect.[50]

At Nauheim, though, even more was at stake. Though already in the fall of 1919, the DPG Board had approved decentralization through "district organizations." By establishing the "Munich District Organization" at the express wish of Berlin management in January 1920, the board had demonstrated to Sommerfeld that this was no mere lip service. Yet not everyone regarded this as liberation from Berlin dominance. The quarrel flared up anew over a publishing issue. In Lenard's opinion, the newly established review journal of the DPG, the *Physikalische Berichte*, through which physicists were to be offered abstracts of current publications, was too much under the control of the Berlin physicists. In April 1920 the managers of the DPG and the German Society of Technical Physics wrote jointly to Sommerfeld that Lenard wished to turn this into "a German" review journal. Lenard sought "above all to devote sufficient attention to the German literature in relation to foreign publication."[51] Sommerfeld attempted to mollify Lenard by assuring him that the management of the DPG would also invite non-Berliners to contribute to the *Physikalische Berichte*,[52] but this was not the only issue for Lenard. As he recorded in his diary of the summer of 1920, he had already "gathered together 12 gentlemen German enough to want to undertake making a German physical society out of the miserable international Berlin Physical Society."[53] The threat of a putsch was signaled when Stark, a member of Lenard's 12 adherents, turned to Sommerfeld with the request to hold the vote on "the disagreement between Berlin and the Reich" not in the general assembly of the Nauheim Congress, but "in a smaller circle of experts." Sommerfeld rejected this suggestion, however, citing the DPG's governing rules.[54]

In the wake of this correspondence, Sommerfeld sensed that Lenard, Stark, and their comrades in arms were serious in their plans for a putsch at the Nauheim Congress: "Please help me see to it that the Physical Society doesn't come apart at Nauheim," he wrote Debye in August. "The agitation of Stark and Lenard has to be countered."[55] At Nauheim, Lenard met with his "12 gentlemen" at the lecture to the

49 From Hirzel, April 14, 1920. DMA, MPGA, Debye.
50 To W. Wien, May 5, 1920. DMA, NL 56, 010.
51 From Mey und Scheel, April 28, 1920. DMA, NL 89, 018, folder 3,8.
52 To Lenard, May 7, 1920.
53 Quoted in Schirrmacher, *Philipp Lenard*, 2010, p. 233.
54 From Stark, July 23, 1920, Sommerfeld's draft reply. DMA, NL 89, 018, folder 3,8.
55 To Debye, August 6, 1920. MPGA, Debye. Also in ASWB II, pp. 78–79.

general assembly of DPG members, to agree ahead of the decisive vote how best to present their own concerns suitably to the plenary. They recognized an ally in Willy Wien as successor to Sommerfeld in the Chair of the DPG. Predictably, Wien was elected. Lenard and Stark were elected to the board, although both declined. "Of course, we cannot sit together with that outspoken enemy of the fatherland, the international Jew Einstein!" Lenard recorded in his diary as his rationale.[56] The putsch foundered. "On account of the Jews, and their Scheel, [they had] run up against resistance and rejection," Lenard noted to himself on a letter from Wien, whom he afterwards resented—only temporarily, be it said—because he had "improved nothing."[57]

As lame-duck DPG president, it required some effort for Sommerfeld to wield the authority of his office at the business meeting at Bad Nauheim. Nevertheless, discussions about the interests of the DPG were relatively innocuous compared to another quarrel stirred up likewise by Lenard and his followers that had reached its apex at Nauheim. DPG matters were not the issue here; rather, Einstein was, who had become the focal point of an unprecedented furor in the press after a British expedition to a solar eclipse in 1919 confirmed the predictions of his General Theory of Relativity regarding the deflecting of light rays by the sun's gravitation.[58] In Germany, enthusiasm abroad evoked a reactionary movement among anti-Semitic circles with which physicists such as Lenard aligned themselves. "Yesterday, we had here one of the 20 publicized protest demonstrations against the relativity theory," Laue informed Sommerfeld from Berlin on August 25, 1920. It had been organized by a "Union of German Natural Scientists," whose tone was set by a man by the name of Weyland whom one must classify "among the grafters." The tenor of the campaign was: "Einstein as plagiarist; anyone who supports relativity theory is a propagandist; the theory itself as Dadaist (this word was actually uttered!)." Together with Nernst and Rubens, he had immediately sent a letter of protest to all the Berlin papers, but "given the admixture of anti-Semitic politics already evident in the distribution of smear-sheets in the foyer of the hall," he was unsure they would be printed. He requested that Sommerfeld, "as President of the German Physical Society . . . introduce a counter resolution" at the Nauheim Natural Scientists Congress "in which either the German Physical Society, or better yet the natural scientists and physicians express their regret over this degeneration of scientific struggle."[59]

Einstein himself took a stance in regard to these attacks. Among physicists of international reputation, only Lenard had expressed himself as an "outspoken opponent of relativity theory," he wrote in the *Berliner Tageblatt*. Though admirable as a "master of experimental physics," Lenard had "made no contribution" to

56 Quoted from Schirrmacher, *Philipp Lenard*, 2010, p. 234.

57 Quoted from Wolff, *Konstituierung*, 2008, p. 378.

58 Fölsing, *Albert Einstein*, 1993, pp. 488–510.

59 From Laue, August 25, 1920. DMA, HS 1977-28/A,197. Also in ASWB II; Kleinert, *Paul Weyland*, 1993; Wazeck, *Einsteins Gegner*, 2009.

theoretical physics. His objections to the General Theory of Relativity were "of such superficiality" that he [Einstein] had deemed it unnecessary to refute them.[60] Sommerfeld wrote Einstein in reaction to these events that he had followed the "Berlin smears" against him with "genuine outrage" and would confer with Planck as to how to respond at Bad Nauheim. He hoped to elicit from the Congress an official statement of "strong defense against 'scientific' demagoguery, and a vote of confidence for you."[61] Einstein regretted that he had let himself be provoked into a public statement against Lenard at all. He hoped "some sort of clarification" would emerge from Nauheim to straighten things out again.[62] Before the Nauheim meeting, Sommerfeld undertook one more effort to take some of the sting out of the quarrel between Einstein and Lenard, but after Einstein's article in the *Berliner Tageblatt*, reconciliation was no longer possible. Indeed, he let Sommerfeld know he would reject even an apology from Einstein "with indignation."[63]

Thus, an open confrontation between Einstein and Lenard at Bad Nauheim was inevitable. All Sommerfeld could do at this point was to assure that it took the form of a scientific debate. The venue of the clash was the resort's large assembly hall, to which only Congress participants were admitted. The moderator was Planck, who was respected as a scientific expert by the overwhelming majority of physicists regardless of political orientation. The excitement that must have attended this debate is still palpable between the lines of the published report. Einstein's relativity principle could achieve general validity only by "dreaming up suitable fields," Lenard retorted at one point.[64] One Congress participant recalled that Planck successfully damped down "agitated digressions into the personal."[65] Lenard's diary entries reveal how uncompromisingly he had rejected all attempts at reaching an understanding: "I quickly take my leave, and hurry to the cloak room to get my hat and umbrella and leave. But then along comes Einstein right behind me to the cloak room and requests an exchange. I silently and forcefully shake my head. He, again, more urgently. I say: No, it cannot be done. You have already spoken publicly and *ex cathedra*. Then hat and umbrella finally arrive; I rush away and leave Einstein standing there."[66]

Following this contretemps, Sommerfeld wrote to Einstein's wife, who had accompanied her husband to Bad Nauheim: "I hope you have recovered from the Nauheim unpleasantness and so on, and are finally just as happy as I that the whole crisis has been passed with reasonable propriety. Naturally, principal credit has to go once more to your husband with his goodness and objectivity, qualities one cannot credit to his opponent Lenard."[67]

60 Quoted in Kleinert/Schönbeck, *Lenard*, 1973, p. 328.
61 To Einstein, September 3, 1920. AEA, Einstein. Also in ASWB II.
62 From Einstein, September 6, 1920. DMA, HS 1977-28/A,78. Also in ASWB II.
63 From Lenard, September 14, 1920. DMA, HS 1977-28/A,198.
64 *Physikalische Zeitschrift* 21 (1920), S. 649–675, hier pp. 666–667.
65 Fölsing, *Albert Einstein*, 1993, S. 526.
66 Schirrmacher, *Philipp Lenard*, 2010, p. 235.
67 To Elsa Einstein, October 7, 1920. AEA, Einstein.

8.3 Visiting Bohr

As President of the German Physical Society, Sommerfeld felt his responsibility to be not merely national, as in mediating between Berliner and anti-Berliner sentiments. The 2 years of his presidency fell in the period of transition between war and peace—a peace the German scholarly world experienced as deeply unjust. Under these circumstances, science became a kind of power substitute. No one gave clearer expression to this view than Fritz Haber, awarded the Nobel Prize in 1918 for the synthesis of ammonia, important in both war and peace. He spoke in 1921 of Germany's "intellectual great-power status" which it was vital, following the "collapse of the country as a political great-power" to emphasize.[68] "We know very well that we have lost the war, and no longer occupy a leading position politically or economically in the world," he wrote fully 5 years later to a Dutch colleague. "But we believe that scientifically, we still number among the peoples with a legitimate claim to be counted one of the leading nations."[69]

The bitterness grew not least out of the reorganization of international science, pushed through in 1919 on the initiative of the Entente. In lieu of the tradition-rich International Association of Academies, which before the war scholars of the leading scientific nations had long used as a framework for international activities, there was now an International Research Council, which excluded scientists of the former Central Powers from participation. This boycott seemed unjust even to most scientists in the neutral countries: Holland, Switzerland, Spain, and the Scandinavian countries.[70] "The conquests of German science will finally have to be universally acknowledged," Zeeman wrote in January 1920 in his congratulatory letter on Sommerfeld's just published book, *Atomic Structure and Spectral Lines*, announcing at the same time his solidarity against the scientific politics of the Entente.[71] In this matter, too, Einstein adopted an outsider role among the scientists in Germany: "By the way, it's just as well that these people be made to feel their dependence through a boycott on the part of foreign countries," he wrote to Lorentz. "This way, the residue of grandiosity and power-lust will be disposed of, which earlier, the economic boom had brought with it."[72] Nonetheless, Einstein, too, whether or not he wished it, was regarded abroad as a representative of German science.[73]

Next to Einstein, Sommerfeld was in the front rank of German scientists who traveled abroad after the war. His Swedish colleague Manne Siegbahn, with whom he had corresponded even during the war on X-ray spectra, sent him an invitation

68 Schröder-Gudehus, *Wissenschaft*, 1966, p. 181; Szöllösi-Janze, *Fritz Haber*, 1998, p. 545.

69 Quoted from Forman, *Internationalism*, 1973, p. 163.

70 Schröder-Gudehus, *Wissenschaft*, 1966; Kevles, *Camps*, 1971.

71 From Zeeman, January 16, 1920. DMA, HS 1977-28/A,380. Also in ASWB II.

72 Einstein to Lorentz, September 21, 1919. AHQP, LTZ-7; Forman, *Internationalism*, 1973, pp. 177–178.

73 Grundmann, *Akte*, 2004, p. 182.

to Lund for September 1919, which Sommerfeld regarded "as the first true dove of peace."[74] Nonetheless, he almost turned it down, as he wrote Siegbahn in July 1919, because something "inexpressibly painful" had befallen him. "A dear son of 15 years, full of hope, has drowned while swimming! I feared at the time I would have to cancel my visit to Lund, but at the urging of my wife I have decided to go ahead with it, and to absorb fresh vital energy from it."[75] The misfortune with "Ucki," as they had nicknamed Arnold Lorenz, born in 1904, had occurred during a family visit to the Ewalds in Holzhausen, on Lake Ammer. "We haven't even the prospect of recovering our beloved child's body," Sommerfeld wrote a colleague.[76] Germany's humiliating fate and the private tragedy appeared to him as a terrible parallel. In a letter to Debye, he harkened back wistfully to the time "when Germany was great and sound, when we were young, and our splendid Ucki was alive."[77] To his drowned son's namesake, he wrote in January 1921, "Life would be truly unbearable in the face of the continuing humiliations we must endure, were it not for the little bit of pleasure we get from the progress of science, which cannot be taken from us." He concluded the letter with a reference to the family tragedy. "Since our Arnold's death, who has now rested in the depths of Lake Ammer a year and a half, a pall has been cast over our family life; my poor wife especially is only slowly and with great difficulty regaining her energy."[78]

When despite this fateful blow Sommerfeld accepted the invitation to Sweden in July 1919, he had just been putting the finishing touches on *Atomic Structure and Spectral Lines*. Besides proof-reading the galleys, there were miscellaneous details concerning the results of recent research he wished to incorporate in the book. The paper he had authored jointly with Kossel, "Selection Principle and Displacement Law in Spectral Series," had just appeared, and this topic, too, he included in his book as an appendix to a chapter on optical series spectra. Paschen had just written him that he was planning new experiments on "your displacement hypothesis."[79] The Rubinowicz Selection Principle, which he also wished to present properly in his book, provided the theoretical basis for this hypothesis.[80] He also planned to lecture on the various topics of his book in Sweden. "It will accord with your intentions," he replied to the inquiry regarding possible lecture topics at Lund, "if I speak exclusively, or primarily, on the questions currently of particular interest to me, of atomic structure and spectral lines, including X-ray spectra."[81]

74 To Siegbahn, June 5, 1919. Stockholm, Academy, Siegbahn.
75 To Siegbahn, July 27, 1919. Stockholm, Academy, Siegbahn. Also in ASWB II.
76 To Geitler, July 2, 1919.
77 To Debye, August 6, 1920. MPGA, Debye. Also in ASWB II.
78 To Lorentz, January 5, 1921. RANH, Lorentz, inv.nr. 74. Also in ASWB II.
79 From Paschen, June 25, 1919. DMA, HS 1977-28/A,253.
80 To Geitler, July 2, 1919. Sommerfeld/Kossel, *Auswahlprinzip*, 1919, p. 244.
81 To Siegbahn, June 5, 1919. Stockholm, Academy, Siegbahn.

For Siegbahn, the invitation to Sommerfeld was also a special event. Although he had made a name for himself before the war in the new field of X-ray spectroscopy, his academic position had been only that of assistant at the Physical Institute of the University of Lund, under the direction of the renowned spectroscopist Johannes Rydberg (1854–1919). Following Rydberg's stroke in 1912, leadership devolved increasingly on Siegbahn,[82] but he was promoted to professor only after Rydberg's death in 1919. With the organization of the lecture series for which he invited Sommerfeld as well as physicists from all over Scandinavia to Lund, he underscored his claim to a leading role in Swedish physics. Sommerfeld's visit to Lund was the occasion also of considerable interest in nearby Copenhagen. Bohr arranged for Sommerfeld to receive an official invitation from the Danish organization for the advancement of science, "Danmarks Naturvidenskabelige Samfund." Every Dane interested in physics would look forward to his visit with great anticipation, Bohr wrote to Munich.[83] Finally, Sweden's most famous physicist, Svante Arrhenius (1859–1927) invited him to lecture also at Stockholm and Uppsala. Sommerfeld was already in Lund when he received this invitation and actually had made other travel plans, but not wishing to miss this opportunity to meet "my more northerly Swedish colleagues too," he rearranged his schedule so that the trip to Lund became an impressive Scandinavian lecture tour.[84]

Following so closely on the end of the war, the ability to travel abroad was no foregone conclusion. Sommerfeld thanked Siegbahn for his "forceful intervention" in procuring the necessary documents. So long as at the last minute no coal crisis or railroad strike were to thwart his plans, Sommerfeld wrote Siegbahn shortly before his departure, they would see each other in Lund on September 9. In case Siegbahn wished to pick him up at the railway station, he provided details of his stature and what he would be wearing. He was short ("1.65 m"), would be wearing "a light brown overcoat and black felt hat," and he would stick a visiting card in his hatband "as an identifying marker."[85] His innermost feelings about this journey, though, he revealed only in letters home. "I sail across the blue Baltic and think of you in love and sorrow," he wrote 1 day before his arrival in Lund to his wife, thinking of their drowned son and of how she might get over this loss. "I know you will pull yourself out of your despair because you must, because you must not make all four of us miserable." He recalled "Uckchen's" precociously distinct "little self-assured personality," which he urged her to adopt as her model. "We can both learn from his manner. I too have often weighed my life down with despair and self-doubt. He can serve as model to me too not to feel the national misery to the point of paralysis of my active life."[86]

82 From Siegbahn, November 20, 1918. DMA, NL 89, 013. Also in ASWB I; Kaiserfeld, *Theory Addresses Experiment*, 1993.

83 From Bohr, August 30, 1919. NBA, Bohr. Also in ASWB II.

84 To Arrhenius, September 19, 1919. Stockholm, Academy, Arrhenius.

85 To Siegbahn, September 3, 1919. Stockholm, Academy, Siegbahn.

86 To Johanna, September 8, 1919.

At Sommerfeld's arrival in Lund, however, Siegbahn quickly steered his guest's thoughts in other directions. Two years earlier he had attended Sommerfeld's lectures for 2 months in Munich. "I had no idea of this," Sommerfeld wrote home. Now Siegbahn was 32 years old, looking forward to a brilliant career, in which Sommerfeld, in his introductory lecture, wished him "conquest upon conquest." He was most graciously welcomed and at once "invited to a formal dinner with 12 gentlemen at Siegbahn's." Nor was he relegated to the anonymity of a hotel, but was housed with a colleague, where he entered into the family life and was spoiled with "outrageous quantities of excellent coffee."[87]

During the next 10 days, Sommerfeld gave six lectures at Lund to his physical colleagues, one lecture for the general student body, and two lectures for astronomers on Einstein's General Theory of Relativity. As a prelude, Siegbahn had scheduled a small conference that was opened with a talk by Bohr and "included around 40 gentlemen, among them people from Stockholm, Christiania [Oslo], and Copenhagen. . . My lecture, which was very elegant, began at 3:15," Sommerfeld reported to his wife. Except for Bohr, who spoke in English, the lectures and discussions were carried out in German. When he put these lines to paper, exactly 3 months had passed since the drowning of their son. "A sad day of memorial, on which we should have been together," he thought. "For me, the day has passed in the feverish activity of lectures. It is only now, late at night, that I am able to gather my thoughts in communion with yours."[88] Two days later he wrote his youngest son, 11-year-old "Didi," how it was for him just at the moment not to have to give a lecture. "Tomorrow and the day after, Saturday and Sunday, we are to have an excursion to the seaside, actually to the sound, an arm of the sea. Look it up on the map." His clothing and meals were also topics in his letters. He was being given "unbelievably much to eat," not only at the home of his host, but also at the numerous dinners to which he was invited. At the opening dinner, he had dressed formally, likewise the other gentlemen guests. They all had "an embroidered collar on their tuxedos," which was "the insignia of the doctorate here in Sweden. The tuxedo thus becomes a really handsome article of clothing."[89]

On September 20, he traveled to nearby Copenhagen to deliver two lectures.[90] On one of his days there, the Copenhagen Physical and Mathematical Society honored him with a "big dinner," at which in speeches given, "very sympathetic political references" were made, as he wrote home with some satisfaction. He was put up "in the finest Copenhagen hotel, the Palace Hotel, hot-water shower in the bathroom!" Most important to him, though, was his meeting with Bohr, who was nearly always at his side throughout the 3 days, and was "attentiveness itself." He "really became friends" with him, he wrote his daughter. He met Bohr's mother, too.

87 To Margarethe, September 9, 1919.

88 To Johanna, September 10, 1919. Also in ASWB II.

89 To Eckart, September 12, 1919.

90 To Bohr, September 16, 1919. NBA, Bohr. Also in ASWB II.

Fig. 20: The physics conference at Lund offered Sommerfeld and Bohr the opportunity to exchange ideas about the advances in atomic theory aimed at during World War I. Thereafter, Sommerfeld supported Bohr's plan for the establishment of a research institute for atomic matters at Copenhagen (Courtesy: Deutsches Museum, Munich, Archive).

"I couldn't resist telling the young Mrs. Bohr I was very happy to see Bohr in such good feminine hands, wife and mother. Both worry that he overworks himself, and both asked me to exert my influence on his colleague Knudsen to reduce his workload. Of course I did that. Bohr is just like Einstein, only better groomed and more refined." Martin Knudsen (1871–1949) directed the Physics Institute of the University of Copenhagen. At the Bohrs', he also met with the British mathematician Godfrey Harold Hardy (1877–1947), who had gained a reputation as a pacifist during the World War, and had come to Copenhagen to promote resumption of scientific exchange among the former wartime adversaries. Though Sommerfeld strove for the same goal, he did not share Hardy's pacifist convictions and avoided discussing politics with him. Otherwise, he enjoyed the excursions into the environs of Copenhagen, and the attentions of his hosts, who gave him no opportunity to spend money. "Much riding about in the car; in general a sudden return to the good old times before the war."[91]

Following this sojourn, Sommerfeld traveled on to Stockholm and Uppsala, where Arrhenius was no less attentive a host than Siegbahn in Lund and Bohr in Copenhagen. "Uppsala was also very beautiful, a noble seat of science, the oldest in

91 To Margarethe, September 24, 1919.

the north," he wrote of this final station on his journey. "Only, with the many new people and the constant lecturing, I was a bit tired now and then. But on the whole I feel refreshed."[92] Back in Munich, he felt himself armed "to overcome all the nastiness" that awaited him in "cold, gloomy Germany."[93]

Sommerfeld's visit was an event his hosts in Denmark and Sweden expected would also yield further benefit. In 1919, Arrhenius observed repeatedly that the Scandinavian countries which had maintained neutrality in the war could play a mediating role between the enemy war powers.[94] Sweden and Denmark now recognized an opportunity to profit from the investment of their neutrality in World War I. By taking up the cause of the scientists targeted by the Entente's boycott, they succeeded in bettering their heretofore peripheral position among scientific nations.[95]

Sommerfeld had incentives like these in mind in recommending that the Carlsberg Foundation in Copenhagen support Bohr's institute as a measure of international scientific politics. "The burdens of war and the unbearable conditions imposed by the peace have long made it impossible for Germany, which once generously supported experimental research at its numerous universities and technical universities, to carry out science as once it did. Not just Germany, but practically the whole of the European continent has been impoverished. More fortunate Denmark can step into this breach. She will do this all the more readily as—in the name of one of her most prominent sons—she will bring honor to herself. Professor Bohr's institute will serve not only Denmark's scientific progeny, but will become an international workplace for talented foreigners whose homelands can no longer provide the golden freedom of scientific work."[96]

At the time of Sommerfeld's visit, Bohr was not yet head of his own institute, and Copenhagen did not yet boast the reputation of a world center of atomic physics. He wished to build up an institute "for atomic matters" where "younger researchers both Danish and foreign" could work theoretically and experimentally, Bohr's brother wrote to Sommerfeld in the hope that a corresponding letter of support from Munich to the Carlsberg Foundation would further this plan.[97] "The idea of a great research institute at Copenhagen is splendid; I have tried to place it in a political context," Sommerfeld wrote Bohr, enclosing his "document" for the Carlsberg Foundation and requesting him "please, without reservation" to make any changes he deemed appropriate.[98]

In the event, his recommendation did not fail in its intended effect.[99] It was also no coincidence that shortly thereafter Adalbert Rubinowicz, a Sommerfeld

92 To Johanna, October 2, 1919.
93 To Bohr, October 26, 1919. NBA, Bohr. Also in ASWB II.
94 Widmalm, *Science and Neutrality*, 1995.
95 Lindqvist, *Center*, 1993.
96 To the Carlsberg Foundation, October, 1919. NBA, Bohr. Also in ASWB II.
97 From Harald Bohr, October 14, 1919. NBA, Bohr. Also in ASWB II.
98 To Bohr, October 26, 1919. NBA, Bohr. Also in ASWB II.
99 Robertson, *Early Years*, 1979.

student, came to Copenhagen to work with Bohr.[100] In the turmoil of the postwar years, Rubinowicz had a varied career. In 1918, he had returned to Czernowitz, which three times had been recaptured by Austria-Hungary during the war, but following the war's end had come under Rumanian control. The new government decreed Rumanian the language of instruction, which forced the majority of professors into neighboring German-speaking countries. In 1920, he accepted a professorship at the University at Laibach, which however had likewise lost its connection to the Austro-Hungarian monarchy, and did not offer Rubinowicz—whose sentiments were with Germany—the prospect of a permanent position. "Since Bohr is building up a new institute, you could take a travel leave there," Sommerfeld advised him shortly after he had assumed his position at Laibach.[101] Rubinowicz followed his advice and spent several months in Copenhagen before returning to Laibach, which was now called Ljubljana. In 1922, he again spent a short time in research with Bohr in Copenhagen and then took up an appointment in Poland as professor of theoretical physics at the Polytechnic University at Lemberg (Lvóv). Though he did not regard this position as permanent either, he failed in his goal of transfer to a professorship in Germany.[102]

8.4 A New Quantum Number

Rubinowicz was among the first of a small group of ambitious theoreticians who came to Copenhagen in the 1920s to conquer the new world of quanta and atoms at Bohr's institute. Like Rubinowicz, most of them had already completed their physics studies and used their time in Copenhagen to distinguish themselves with work at the leading edge of research. At the Munich "nursery of theoretical physics," as Sommerfeld had dubbed his institute in 1919, atomic and quantum physics following the war were likewise fields of particular emphasis, so that a certain rivalry between the two centers soon became evident. In Sommerfeld's eyes, Bohr was "a wonderful person," to be sure, and he felt "friendship not only scientifically but also personally" with Bohr and Siegbahn, as he had written Hilbert from Sweden.[103] But very soon their differing conceptions in matters of quantum theory came to the fore. Bohr wished to convince Sommerfeld of the advantages "of the general principle of analogy of the quantum theory," as he initially called the correspondence principle.[104] But Sommerfeld did not think this principle as

100 To Rubinowicz, 26. October 26, and November 1, 1919; from Bohr, November 19, 1919. DMA, HS 1977-28/A,28. Also in ASWB II.

101 To Rubinowicz, December 26, 1919.

102 To Else Rubinowicz, November 27, 1921; Robertson, *Early Years*, 1979, p. 158; Rubinowicz, interviewed by Théo Kahan and John L. Heilbron, May 18, 1963. AHQP.

103 To Hilbert, September 25, 1919. SUB, Cod. Ms. D. Hilbert 379 A.

104 From Bohr, July 27, 1919. DMA, HS 1977-28/A,28. Also in ASWB II.

fundamental as the viewpoint to which he and Rubinowicz subscribed. Even later, when he attributed greater importance to the correspondence principle, he did so not without qualification. "Nevertheless, I must confess I find the origin of your principle, which is so removed from quantum theory, awkward. I recognize nonetheless that it reveals an important connection between quantum theory and classical electrodynamics."[105]

For both Sommerfeld and Bohr, the issue was not just disagreement over a particular quantum matter, but rather a fundamental difference of research strategies that reflected the image of their separate institutes, and thereby touched on their reputations as trendsetters in the further development of atomic theory. Whereas Bohr gave primacy to the correspondence principle and used atomic spectra to confirm one or another statement of his theory, Sommerfeld read from the spectra the regularities that seemed to him fundamental for further action. After the spectroscopic displacement law, he discovered a "magneto-optical splitting rule" in the fine separations between spectral lines in the case of the anomalous Zeeman effect. In these separations he had found "something quasi-empirical," he had written to Runge, the authority in these questions, shortly before his trip to Sweden. Runge had long before discovered a relationship between the separations in the anomalous and normal Zeeman effects, which was known as "Runge's rule" (anomalous and normal separations stand in a rational relation to one another). Sommerfeld's "magneto-optical splitting rule" emerged from the combination principle, whereby each spectral line—even those separated in a magnetic field—can be represented as the difference between two terms. In the case of differences between fractions, as Runge's rule suggests, quite simple number relations follow for numerator and denominator. He had come to these regularities when Paschen had reported the separations of doublet and triplet lines, Sommerfeld wrote Runge.[106] Initially, it was entirely unclear what this "number mystery," as Rubinowicz expressed it, meant. Sommerfeld thought the word so trenchant that he used it as the headline of a publication on the "magneto-optical splitting rule."[107] "Is this not quantum music?" he wrote to Epstein, who had gone to Zürich as lecturer and soon thereafter received an appointment as professor of theoretical physics at the California Institute of Technology, in Pasadena. "Theoretically, of course, there's nothing to be done. But the empirical regularity is just as interesting to me as the explanation according to the model."[108]

The "number mystery" and the recently established "spectroscopic displacement law" could at first not be explained within the framework of the Bohr-Sommerfeld atomic model. They referred also to atoms dissimilar from hydrogen, whose spectra

105 To Bohr, November 11, 1920. NBA, Bohr. Also in ASWB II.

106 To Runge, August 16, 1919. DMA, HS 1976-31. Also in ASWB II.

107 To Rubinowicz, December 26, 1919. Sommerfeld, *Zahlenmysterium*, 1920; Sommerfeld, *Gesetze*, 1920.

108 To Epstein, October 26, 1919. Pasadena, CalTech, Epstein 83.

did not obey a simple Balmer series. Nonetheless, here too series were involved that ultimately had to be explained by an atomic model and ordered by quantum numbers. The Bohr-Sommerfeld model of 1920 provided for three quantum numbers: a radial quantum number for the distance of an electron from the nucleus, an azimuthal quantum number for the rotational motion in an orbit, and an equatorial quantum number for different spatial orbital planes. But even the simplest elements with more than one electron presented problems for this model. "I pause, perplex'd! Who now will help afford?" Sommerfeld wrote quoting Goethe's *Faust* in *Atomic Structure and Spectral Lines* when he came to discuss the neutral helium atom, in which two electrons orbit around the nucleus.[109] In the case of atoms with inner and outer electrons, he conjectured that the doublet and triplet terms were not the result of the azimuthal quantum number, since this corresponds only to an "outer rotation," but of an "inner quantum number" belonging to a "hidden rotation" in the interior of the atom.[110]

Aside from the vague allusion to a "hidden rotation," there was nothing that could connect the "inner quantum number" with a concrete physical conception. In the Sommerfeld atomic model, the radial, azimuthal, and equatorial quantum numbers were given a physical sense through three quantizing conditions for the orbit of an electron around the atomic nucleus. But there was no theory that would allow for similar integration of the "inner quantum number" into this model. It served only provisionally as an additional criterion to bring order to the plethora of observed and unobserved transitions between the doublet and triplet separated energy levels characteristic of the series spectra of many atoms. In the separations of the doublet and triplet structure in the anomalous Zeeman effect, Sommerfeld also saw these "hypothetical 'inner' quantum numbers" at work.[111]

But the hypothetical character and the absence of a conceptual model did not hinder Sommerfeld from elaborating the empirical basis. The spectroscopic displacement law and the regularities in the anomalous Zeeman effect were among the first topics he assigned as doctoral work after the war. In the displacement law he was primarily concerned with "the equivalence of the character of the lines (doublet or triplet system)" in sequential elements on the periodic table, not with a theoretical explanation.[112] In the anomalous Zeeman effect, too, there was much that was "initially theoretically incomprehensible," Sommerfeld conceded frankly in his report on a dissertation on this topic, but in this "currently still obscure area," it was most important to confirm "certain regularities" and thereby to perform "useful ground-laying work" for future theories.[113]

109 "Hier stock' ich schon, wer hilft mir weiter fort?" Sommerfeld, *Atombau*, 1919, p. 70.

110 Sommerfeld, *Gesetze*, 1920, pp. 231–232.

111 Ibid., p. 253.

112 Vote on the dissertation of Erwin Fues to the Philosophical Faculty, 2. Section, December 18, 1919. UAM, OC I 46 p.

113 Vote on the dissertation of Josef Krönert to the Philosophical Faculty, 2. Section, February 20, 1920. UAM, OC I 46 p.

Sommerfeld's former student Alfred Landé, who had done his habilitation under Max Born at the University of Frankfurt, also immersed himself in the new empirical regularities as a source for new information about the anomalous Zeeman effect.[114] While studying Sommerfeld's magneto-optical splitting rule, it occurred to him that by using the "inner quantum number," other empirical rules could be established. "The various 'inner' quantum numbers of a term will likely mean simply the total quantum numbers of the atom around its invariable axis in various spatial orientations of the valence electrons around the atomic nucleus," he conjectured on the physical meaning of this new quantum number in a letter to Bohr, with whom, like Rubinowicz, he had spent a research period.[115] "Bravo! You perform magic!" Sommerfeld enthused when he learned of Landé's progress. "Your construction of the doublet Zeeman types is splendid."[116]

For the time being, this construction drew its plausibility only from its agreement with the spectroscopic findings emerging from Paschen's laboratory in Tübingen. Ernst Back had chosen the anomalous Zeeman effect as his habilitation topic there and almost simultaneously with Landé had found the same empirical rules. Sommerfeld therefore "urgently" requested that Landé wait with publication until Back had completed his dissertation.[117] When Landé declined to follow this advice, he grew "seriously angry," as he wrote Max Born, Landé's mentor. "It is improper to publish the conclusions of the experimenter's work ahead of him," he wrote indignantly. In addition, Back was still "emotionally so exhausted by the war" and now needed peace and quiet. Landé was, "through his impatient ambition, on the point of harming him in his circles." He added that the new information was based on unpublished measurements from the Tübingen laboratory. The risk thus was "that Paschen will cease informing us of anything at all from his institute if we abuse his or Back's generosity."[118]

A few days later, Landé wrote to Sommerfeld "in the matter of Dr. Back" that he had been in discussions with the Tübingen experimenters, and "a disagreement between theory and practice [was] not to be expected." In addition, he had in the meantime elaborated his "Zeeman splitting rules."[119] The rivalry with Back apparently spurred him on to give a theoretical aspect to his construction of the anomalous Zeeman splitting, in distinction to that of Back. "Since Landé has derived the Zeeman types differently from Back, and has grounded them more fundamentally,

114 Forman, *Alfred Landé*, 1970.

115 Landé to Bohr, February 16, 1921. Quoted in Forman, *Alfred Landé*, 1970, p. 242. On the early visiting researchers at Copenhagen, see Robertson, *Early Years*, 1979, pp. 156–159.

116 To Landé, February 25, 1921. SBPK, Landé, 70 Sommerfeld. Also in Forman, *Alfred Landé*, 1970, pp. 208 and 249.

117 To Landé, March 3, 1921. SBPK, Landé, 70 Sommerfeld. Also in ASWB II; Forman, *Alfred Landé*, 1970, pp. 251–252.

118 To Born, March 8, 1921. DMA, NL 89, 025. Also in Forman, *Alfred Landé*, 1970, p. 257.

119 From Landé, March 17, 1921. DMA, HS 1977-28/A,192. Also in Forman, *Alfred Landé*, 1970, pp. 259–261.

he should publish his work," Paschen wrote to Sommerfeld. He implied, though, that he did not rate "Landé's speculations" very highly. "Ultimately, an important, well-grounded fact has greater value than any speculation, which may very well incite us to new conceptions, but which soon seem threadbare."[120]

Landé's "speculations" would, however, prove forward looking. The "inner quantum number" moved to the center of a future quantum theory of the anomalous Zeeman effect. Landé argued that if the "differences of magnetic interference energy" between the beginning and end states of an electron's transition are multiplied by a "*g*-factor," the anomalous Zeeman separations could be treated like the normal separations. This factor, put together from the different azimuthal, equatorial, and "inner" quantum numbers, contains all the classically incomprehensible aspects of the phenomenon.[121] When he and Back made new measurements of the anomalous Zeeman separations, Paschen soon had to concede that Landé's *g-factors* were brilliantly confirmed. For Paschen, though, Sommerfeld, as "father of the inner quanta and Landé's rules," was the true pioneer in this area.[122] The "inner quanta" seemed to him "the most important and felicitous," suitable "as a working hypothesis for the future . . . Landé rests his work on this. The combinations of terms follow from this, and everything comes out correctly."[123]

8.5 Teacher and Students

Sommerfeld could actually have taken satisfaction from the fact that the new quantum number he introduced had proven itself so well, but he resented Landé for not allowing Back to go first. On the substance of the matter, too, he had a few objections. "Several of his observations about quantum numbers seem quite crazy to me," Sommerfeld wrote Paschen.[124] He was referring to the circumstance that Landé sought agreement between theory and experiment only in that he fit half-integer values to a few quantum numbers. For the most part, Sommerfeld thought Landé's insights magnificent ("Bravo! You perform magic!"), but what Landé presented to him in March 1921 as the subject of his pending publication seemed to him still "not ripe for publication." One of his students ("1st semester!") had found the same, he informed Landé, but this result would not have been published.[125]

The student's name was Werner Heisenberg (1901–1976).[126] He had begun his studies in the winter semester 1920/1921 and had at once expressed the wish to

120 From Paschen, May 21, 1921. DMA, HS 1977-28/A,253.
121 Landé, *Zeemaneffekt*, 1921; Forman, *Alfred Landé*, 1970.
122 From Paschen, July 24, 1921. DMA, HS 1977-28/A,253.
123 From Paschen, September 13, 1921. DMA, HS 1977-28/A,253.
124 Draft of reply to Paschen, July 24, 1921. DMA, HS 1977-28/A,253.
125 Forman, *Alfred Landé*, 1970, p. 261.
126 Cassidy, *Uncertainty*, 1992; Rechenberg, *Werner Heisenberg*, 2010.

participate in the seminar and to prove his talents on the current issues of research. He had initially planned to study mathematics, but a preliminary interview with mathematics Professor Ferdinand Lindemann had ended in disappointment for the ambitious Heisenberg, so he had turned to theoretical physics. After nearly 25 years as an academic teacher, Sommerfeld knew how to deal with students who thought themselves special. As a rule, coming face to face with the difficulties of the subject alone was enough to assure that pretension did not too far exceed ability and to instill the understanding that even a very gifted student has first to concentrate on absorbing the fundamentals of physics. But another physical Wunderkind by the name of Wolfgang Pauli (1900–1958) had just persuaded Sommerfeld that it sometimes pays to stray from routine. Already in the winter semester of 1918/1919, Pauli had astonished Sommerfeld with his knowledge of the General Theory of Relativity. "A first semester student! His ability exceeds that of Debye by an order of magnitude!" Sommerfeld had written a colleague about Pauli's early steps under his wing.[127] Then he had entrusted Pauli with an assignment he had initially intended for Einstein: the article on relativity theory for the *Enzyklopädie der mathematischen Wissenschaften*. Pauli fulfilled this assignment in masterful fashion, and the article became an instant classic.[128]

Sommerfeld may have been thinking of Pauli when Heisenberg, another first semester student, wanted to join in discussion of the critical issues of research. "All right, you have an interest in mathematics; it may be that you know something; it may be that you know nothing; we will see," was how years later Heisenberg recalled Sommerfeld's reaction. The seminar had evolved into a kind of testing ground on which advanced doctoral candidates familiarized themselves with a topic at the leading edge of research, and where first semester students must have felt quite at sea. Without much introduction, Sommerfeld presented the newcomers to his seminar with spectroscopic measurements of the anomalous Zeeman effect. From these, Heisenberg was to determine the beginning and end states, between which the transitions in the atom observable in the spectral lines must have taken place. Since he was still barely familiar with the atomic theory of Bohr and Sommerfeld, Heisenberg approached this seminar assignment as a kind of number puzzle. Heisenberg recalled years later in an interview: "So after a very short time, I would say perhaps one or two weeks, I came back to Sommerfeld, and I had a complete level scheme. Then I came up with a statement which I almost didn't dare to say, and he was, of course, completely shocked. I said, 'Well, the whole thing works only if one uses half quantum numbers.' Because at that time nobody ever spoke about half quantum numbers; the quantum number was an entire number, you know, an integral. 'Well,' he said, 'that must be wrong. That is absolutely impossible; the only thing we know about the quantum theory is that we have integral numbers, and not half numbers; that's impossible.'" That Landé came

127 To Geitler, January 14, 1919. Also in ASWB II, pp. 46–47.
128 WPWB I, pp. 13–14 and 58.

to the same conclusion around the same time must have given Sommerfeld pause. Though he did not want Heisenberg to publish his results at once, he encouraged him to continue work on it and to seek a physically plausible explanation. In any case, Sommerfeld was very curious about the results to which Heisenberg's zeal for discovery was leading. "I should say that almost every morning, but at least every second morning, I was called in to Sommerfeld. I had to tell him about what I had tried in my own work, and what I thought about Landé's paper, and so on."[129]

In this situation, Sommerfeld reflected on the time before the Bohr atomic model, when Voigt had extended the old Lorentz theory of the normal Zeeman effect to a theory of the anomalous Zeeman effect, and in the effort to simplify this theory and use it on the Paschen-Back effect, he had nearly come into conflict with Voigt (see Chap. 7). With its many parameters, Voigt's theory had, after all, supplied a good description of the separation of spectral lines in a magnetic field. If Voigt's equations could successfully be transformed such that the combination principle, that is, the representation of the separation of lines as the difference between separated energy terms, was perceptible behind it, at least on this point an approach to the Bohr theory would have been reached. Up to now, all spectral lines could still be represented as the difference of two terms; if the combination principle was to be found also in Voigt's theory, one might be able to derive a quantum theoretical model from it. "While the Voigt oscillation theory gives us the separation of the lines, in the quantum theory, we have to ask about the separation of the terms," Sommerfeld argued. The result must have surprised even him, for the equations for the terms assumed a very simple form: "One recognizes here how much simpler and more uniformly the ultimate quantum theoretical description of the magneto-optical facts turns out than the original vibrational [description]." At the end of his "quantum theoretical reinterpretation," he did not fail to "thank my student, Mr. W. Heisenberg for his very successful collaboration on the whole problem of the anomalous Zeeman Effect."[130]

Sommerfeld's "reinterpretation" now supplied equations for the separated energy terms, in which in place of Voigt's frequencies, there were the azimuthal, equatorial, and inner quantum numbers. So long as the inner quantum number was based only empirically, however, it could not be connected to any model conception. Heisenberg took this step when he gave concrete expression to the "hidden rotation" that Sommerfeld had conjectured behind the inner quantum number: He distinguished the angular momentum of one or several valence electrons from the total angular momentum of the electrons in the inner orbits, which he imagined as a balled-up atomic core. At the beginning of the winter semester 1921/1922, his third semester of study, he gave Pauli an epistolary seminar lecture on the "atomics

129 Heisenberg, Interview by Thomas S. Kuhn and John Heilbron, November 30, 1962. AHQP. http://www.aip.org/history/ohilist/4661_1.html (31 January 2013)

130 Sommerfeld, *Umdeutung*, 1922, pp. 267, 269, 272.

of the anomalous Zeeman effect," in which he elucidated his model of the "double atoms," like sodium. These atoms in the first column of the periodic table consist of one valence electron and one "core," he explained to Pauli. "In its normal state (s term) the atom has a total momentum 1. This is divided (here it comes!) in mean time equally on the core and the electron. So mean momentum ½ + ½ ." Between core and valence electron, "a magnetic correspondence" is supposed to exist. By means of a longer calculation, he showed how with this concept he arrived at Sommerfeld's reinterpretation of "Voigt's equations."[131]

What Heisenberg discussed in this and other letters to the older Sommerfeld students Pauli and Landé went down in the history of atomic theory as the core model.[132] Although Sommerfeld was bothered by the half quantum numbers of the core model, he was so impressed by this third semester student of his that he permitted him to publish a paper about it in the *Zeitschrift für Physik*. Heisenberg's essay appeared in the same issue and adjacent to Sommerfeld's paper on the quantum theoretical reinterpretation of Voigt's theory, so that it must have been clear to any reader that this was a matter of the joint research of teacher and student.[133] "Heisenberg is a 3rd semester student, and enormously talented. I could no longer restrain his zeal to publish, and think the result so important that I agreed to its publication even though the form of the derivation may perhaps not yet be definitive," Sommerfeld wrote Bohr when he sent him Heisenberg's paper. "You will see that we ascribe the origin of the multiplicity of terms to the magnetic," he wrote, pointing out at once the essential difference from Bohr's conception.[134] Bohr, namely, explained the "multiplicity of terms," that is, the doublet and triplet character of atoms with more than one electron, not like Heisenberg as magnetic interaction between atomic core and valence electrons, but as the consequence of an asymmetrical distribution of electrical charge in the atom.[135]

The fact that with Heisenberg's model, Sommerfeld had not only to tolerate half quantum numbers, but also now had to accept a magnetic explanation beside the relativistic explanation of the doublet as he had presented it with his fine-structure theory, did somewhat dampen his enthusiasm. "It all works, but remains at bottom unclear," he wrote to Einstein. "I can only contribute to the technical aspect of quantum theory; you must devise its philosophy."[136] After Pauli's masterly encyclopedia article on relativity theory, Einstein at first saw Sommerfeld's charismatic pedagogical personality behind the Heisenberg core model. "What I particularly admire in you," he wrote in reply, "is that you seem to have conjured such a large quantity of

131 Heisenberg to Pauli, November 19, 1921. Reprinted in WPWB I.
132 Cassidy, *Core Model*, 1979.
133 Heisenberg, *Quantentheorie der Linienstruktur*, 1922.
134 To Bohr, March 25, 1922. NBA, Bohr. Also in ASWB II.
135 Cassidy, *Core Model*, 1979, p. 217.
136 To Einstein, January 11, 1922. AEA, Einstein. Also in ASWB II.

younger talent, as it were, out of the ground. That is something quite unique. You obviously have a gift for refining and stimulating the minds of your audience."[137]

Aside from Pauli and Heisenberg, other Sommerfeld students at the Munich "nursery" in these years were making a name for themselves as rising stars in the firmament of modern theoretical physics. For the first semesters following the war, Sommerfeld could rely on Lenz and Ewald, who were once more at his disposal as assistants and helped him keep the pedagogical enterprise afloat during the postwar and revolutionary turmoil. They aided Sommerfeld not only in the functioning of seminars and colloquia, but also by filling out the lecture program. In the "war-emergency half year, January–April 1919," Lenz gave a refresher course in electrodynamics and gave lectures for advanced students on quantum theory, kinetic gas theory (winter, 1919/1920), relativity theory (summer, 1920), and theory of heat radiation (winter, 1920/1921). In the "war-emergency half year," beginning of 1919, Ewald gave a refresher course in mechanics, and then elementary lectures on vector calculus and foundation for theoretical physics (summer, 1919 and 1920), and advanced lectures on the dynamics of crystal lattices (winter, 1919/1920), selected problems of electron optics (summer, 1920), and thermodynamic potentials (winter, 1920/1921).[138]

In the fall of 1919, the roster of the institute grew with the arrival from Vienna of Karl Herzfeld (1892–1978), who covered primarily the bridge area between physics and chemistry.[139] In 1921, Lenz and Ewald were appointed professors of theoretical physics at Hamburg and Stuttgart, respectively. Their successors as assistants to Sommerfeld were Adolf Kratzer (1893–1983) and Gregor Wentzel (1898–1978). Kratzer had come to Munich in 1918 as a casualty of war in October 1918 and had completed his doctorate in February 1920 with a dissertation on band spectra supervised by Lenz.[140] Wentzel had completed his doctoral studies in the summer semester of 1921 with a dissertation on X-ray spectra.[141] In this semester, he had "slaved" and was now "vacation ready," Sommerfeld wrote to Einstein. "In this semester, I've directed 4 doctorates (among them, Pauli), and one lectureship (Kratzer). I've paid for all of it in sweat."[142]

But Sommerfeld was also glad that such "active life" was once more manifest in Munich physics, as he wrote his former student Epstein, who had become a professor of theoretical physics at Pasadena. The "Röntgen Sleeping Beauty Institute" had

137 From Einstein, undated [shortly after January 11, 1922]. DMA, HS 1977-28/A,78. Also in ASWB II.

138 Lecture catalogues. http://epub.ub.uni-muenchen.de/view/subjects/vlverz_04.html (31 January 2013).

139 Herzfeld, unpublished autobiography [1971], typescript, Washington D.C., Catholic University of America, Archive, Herzfeld-Papers, box 2.

140 Vote on the Dissertation of Adolf Kratzer to the Philosophical Faculty, 2. Section, February 19, 1920. UAM, OC I 46p. Schmitz, *Adolf Kratzer*, 2011.

141 Vote on the Dissertation of Gregor Wentzel to the Philosophical Faculty, 2. Section, June 21, 1921. UAM, OC I 47p.

142 To Einstein, August 10, 1921. AEA, Einstein.

been wakened to new life since the fall of 1920 when Willy Wien had succeeded the long-ailing Röntgen; and now in his own institute "extraordinary people [were] once more" maturing. Wentzel was aiming at great strides forward in X-ray spectra, and "Kratzer is in charge of the bands [that is, band spectra]. The phenomenon, however, is Heisenberg, a 3rd semester student, who has mastered the model theory of the Zeeman Effect and the multiplicity of terms—as it seems, far beyond Bohr." All this was soon to be read in the third edition of *Atomic Structure and Spectral Lines*, which was about to be published.[143]

8.6 The Bible of Atomic Physics

In the first four editions of *Atomic Structure and Spectral Lines*, brought out almost annually, it is possible to read the extent to which Sommerfeld involved his disciples in the construction of atomic theory. The second edition followed upon the first so rapidly that there was not time for a thorough reworking. Only the mathematical additions and elaborations at the end of the book were revised; Sommerfeld had turned this assignment in large measure over to Pauli.[144]

In the third edition, however, Sommerfeld wished to present the latest discoveries comprehensively, necessitating a thorough revision. "My primary task was to support the order of the general series spectra," he explained in his foreword, dated January 1922. "I lay special stress on the introduction of the inner quantum numbers," he went on to point out, "and on the system of the anomalous Zeeman Effect." At the end of the chapter devoted to these innovations, he referred explicitly to Heisenberg's collaboration. Though he chose not to integrate the core model into his text because he thought it still somewhat speculative for a textbook, he did not want to ignore it and added it as an addendum to the chapter on series spectra. With respect to this section, he commented that Heisenberg had been able "to solve by way of a model" the puzzle of the multiplicity of terms. He also again thanked Pauli, Kratzer, and Wentzel by name for having "faithfully and knowledgeably" helped him with various parts of the book.[145]

At least by the time of this third edition, *Atomic Structure and Spectral Lines* had become the "Bible" and Sommerfeld the "Quantum Pope." In this new edition there was again "much that is new and valuable," Max Born wrote in gratitude when he thumbed through his freshly printed copy. "Today, it is the Bible of the modern physicist."[146] The mathematician Hermann Weyl (1885–1955) called it his "physical Bible."[147] One year later the experimental physicist Karl Wilhelm Meissner

143 To Epstein, February 12, 1922. CalTech, Epstein 8.3.
144 Sommerfeld, *Atombau*, 1921b, p. IX.
145 Sommerfeld, *Atombau*, 1922, pp. V–VIII and 496.
146 From Born, May 13, 1922. DMA, HS 1977-28/A,34. Also in ASWB II.
147 From Weyl, May 19, 1922. DMA, HS 1977-28/A,365. Also in ASWB II.

(1891–1959) dubbed the fourth edition as "the spectroscopist's Bible."[148] Paschen, too, who more than anyone else had explored the area of atomic spectra in his Tübingen laboratory, saw in it "the Bible of the practical spectroscopist." Paschen associated the sequential editions of the book with the memory of how he himself had been led to the quantum physical interpretation of the spectra. It was, "in the end, your work," he wrote Sommerfeld, in which he had found clarity on the subject, "and I believe the majority of spectroscopists will have proceeded similarly."[149]

Not everyone thought of Sommerfeld's role in the development of quantum and atomic theory in terms of quite such affecting gratitude. The title "Quantum Pope" was not entirely intended as a compliment. It originated with Paul Ehrenfest who, as Sommerfeld learned from Einstein, was annoyed that his contributions to the so-called adiabatic hypothesis had not been given appropriate recognition in the third edition of *Atomic Structure and Spectral Lines*.[150] "Ehrenfest is an obnoxious fellow; I've long known that he is angry with me," Sommerfeld wrote thereafter to his wife, referring to a letter from Ehrenfest to Einstein in which he spoke of "St. Sommerfeldus as the quantum pope."[151] The adiabatic hypothesis dealt originally with thermodynamics and related to alteration in states of a system that were not directly the result of heat exchange, but rather indirectly through the alteration of the conditions on which state of the system otherwise depends. In 1911, Ehrenfest had found a connection between the light quantum hypothesis and such adiabatic alterations of state and in the following years had made this into a quantum principle that helped Bohr substantially in his construction of atomic theory.[152] Although Sommerfeld devoted a long section to the adiabatic hypothesis, and credited Ehrenfest as its author, in the third edition he gave as reference only Ehrenfest's 1916 summary presentation of it and made no mention of its earlier history, in which Ehrenfest might certainly have been credited in the role of advance scout. Not until the fourth edition was Ehrenfest's years-long work on this quantum principle referenced.

Bohr, too, accorded Sommerfeld's new edition a somewhat cool reception. Without addressing the correspondence principle directly from which, in his view, Sommerfeld was still withholding an appropriate evaluation, he expressed his pleasure at seeing what he believed was a revision. He had "often felt himself scientifically isolated" in the attempt to work out the principles of quantum theory. It was "of course not to be expected that everyone would have the same view of everything." He was alluding to the part of the new edition in which by means of the inner quantum number Sommerfeld had brought order into the system of doublets and triplets in the series spectra and the anomalous Zeeman effect. Behind this,

148 From K. W. Meissner, December 2, 1924. DMA, NL 89, 011.
149 Von Paschen, January 27, 1925. DMA, NL 89, 012.
150 From Einstein, September 16, 1922. DMA, HS 1977-28/A,78. Also in ASWB II.
151 To Johanna, October 8, 1922.
152 Navarro/Pérez, *Paul Ehrenfest*, 2006; Pérez, *Adiabatic Theory*, 2009.

Bohr sensed physical assumptions that did not admit of a "unified conception of quantum theory," such as he had in mind. He expressed his thanks, however, for the "friendly spirit" in which Sommerfeld had honored his work.[153]

The third edition had hardly appeared when the concept of the inner quantum number won new confirmation from an unexpected quarter. In March 1922, Sommerfeld was invited to give talks in Spain, where he met in Madrid with the spectroscopist Miguel Catalán (1894–1957). Catalán had just returned from a research visit in England and told him of his investigation of the manganese spectrum, which he had carried out in the laboratory of Alfred Fowler (1868–1940) at Imperial College in London. Catalán showed Sommerfeld the "multiplets," as he called them, structures of five, six, and more lines, that broadened the known system of doublets and triplets and pointed to a very complex arrangement in the spectra of multi-electron atoms. Sommerfeld had studied his work with great excitement, Catalán recalled many years later, because he suspected therein a confirmation of his theory of inner quantum numbers.[154] Back in Germany, Sommerfeld first informed Paschen of the new developments from Spain, who at once initiated similar investigations at his own laboratory. "It all goes exactly according to Landé's schema," Paschen reported about new results at his laboratory in which the interrelation of multiplicity and Zeeman separation in the chromium spectrum was being investigated.[155] This interrelation described by the inner quantum number had up to then been confirmed experimentally only in the cases of doublets and triplets. Now, the higher multiplets could also serve as empirical basis for the inner quantum number.

In June 1922, Bohr was invited to Göttingen for a week of lectures. Physicists traveled there from all over Germany to hear firsthand about the latest advances in atomic theory. The major topic of this "Bohr Festival," as it soon came to be called, was Bohr's conception of the arrangement of the elements on the periodic table, published the previous year in an article in *Nature* under the title "Atomic Structure."[156] The goal of this "second Bohr atomic theory" was to distinguish the various arrangements of electrons, one from another, with quantum numbers. Sommerfeld's inner quantum number, however, was assigned no role. In this situation, the new multiplets were just the thing for Sommerfeld to justify the inner quantum number in response to Bohr. "I have a few new things concerning the 'inner quantum numbers' that I will bring along to Göttingen," he announced in a letter to Copenhagen 2 weeks before the Göttingen event.[157] At Göttingen, he gave a lecture "on line structures in the spectrum of manganese," in which, however, he only indicated that this could be explained by a suitable attribution of inner quantum numbers to the various electron levels. The comprehensive presentation of this

153 From Bohr, April 30, 1922. DMA, HS 1977-28/A,28. Also in ASWB II.

154 Sanchez-Ron, *Relations*, 2002, p. 11.

155 From Paschen, May 24, 1922. DMA, HS 1977-28/A,253.

156 Kragh, *Atomic Theory*, 1979; NBCW 4, pp. 19–20, 177–180.

157 To Kramers, June 1, 1922. NBA, Bohr.

attribution he reserved for a longer article submitted to the *Annalen der Physik* 2 months later. Almost in the same breath, Sommerfeld and Heisenberg presented another joint work on the intensity of the multiplet lines. The essence of these papers would secure the inner quantum number a firm place in the theory of multi-electron atoms. It was the decisive quantum number that distinguished the components of the different series terms from one another; physically, one could conceive them as characteristic magnitudes of "the total impulse momentum of the respective state of the atom," that is, the total angular momentum.[158]

Bohr's conception of the periodic table and Sommerfeld's interpretation of the complex spectra pointed the way to the future with respect to the description of multi-electron atoms. Sommerfeld determined to go into this more closely in the fourth edition of *Atomic Structure and Spectral Lines*. For now, he could content himself with the praise he garnered from all sides for the third edition. "The Spanish translation of my book also is nearly negotiated with a publisher here," Sommerfeld wrote his wife from Madrid.[159] The "also" was a reference to translations into French and English, for which the third edition served as the basic text. "Sommerfeld's masterly book" was awarded superlative praise especially in America.[160]

8.7 The Lusitania Medal

For German scholars in the years of the boycott, recognition from abroad transcended mere personal honor. Sommerfeld exploited interest in him personally to promote politically reconciliation with Germany abroad. Such an opportunity presented itself in September 1921, when William Frederick Meggers (1888–1966), director of the spectroscopic division of the National Bureau of Standards in Washington, paid him a visit on a European trip. Meggers made Sommerfeld the offer of publication in the *Journal of the Optical Society of America*. This organ of the American Optical Society was published by Paul Foote (1888–1971), Meggers's colleague in the spectroscopic division. "Mr. Foote was kind enough to solicit my submission to this journal, and wrote me that my prospective contribution would be translated into English," Sommerfeld wrote Meggers following his Munich visit. He expressed his thanks for the offer, although even the matter of language became a political issue. He would assent to translation of his essay into English only if the Journal of the Optical Society of America in principle did not print articles in other languages. "Should it be the case, however, that (like the *Astrophysical Journal*, for example) when occasion arises you also print articles in French, then it would be of great importance to me that you

158 Sommerfeld, *Linienstrukturen*, 1922; Sommerfeld, *Deutung*, 1923, p. 33; Sommerfeld/Heisenberg, *Intensität*, 1922, p. 132.

159 To Johanna, April 24. 1922.

160 Saunders, *Aspects*, 1924, p. 49. ("Sommerfeld's masterly book on atomic structure shows to what an extraordinary extent the theory has widened our horizon...").

permit the German language as well. You surely know that our enemies are not only suppressing use of the German language in the German territories taken from us, e.g., Alsace, Upper Silesia, but also seek to anathematize German as a cultural language wherever possible. Naturally I cannot accede to that."[161]

When he was informed that the *Journal of the Optical Society of America* in principle published essays only in English, the wind was in this respect taken out of his sails.[162] The paper he then sent to the USA was not on an optical topic, but was readily accepted nonetheless. Sommerfeld's rationale for sending this paper to America, he explained, was that he had been prompted to the subject by an article in the *Proceedings of the National Academy*; he wished only to clear up "a misunderstanding."

The publication in the *Journal of the Optical Society of America* served him also as an opportunity to win over Meggers to an entirely different concern that had been on his mind for several months. "Our political situation grows ever more difficult," he wrote, shifting gears from physics to politics. "After England betrayed us so ignominiously in Upper Silesia, and handed this thriving country over to Polish mismanagement, we are convinced that England, like France, desires our ruin. The means the two employ are different, to be sure. France behaves bestially, England, like a rogue. I do not know which is the more despicable." He enclosed a "short political article" in which he excoriated the war propaganda of the Entente and which he hoped Meggers would submit to an American newspaper to reprint.[163]

This article concerned an event that played a not insignificant role in the US' entry into World War I: the torpedoing of the passenger steamship Lusitania of the British Cunard Line by a German U-boat. The Lusitania had sailed on May 1, 1915 from New York bound for Liverpool, with nearly 2,000 people onboard. She also carried war supplies for England, even though the transport of such "contraband" presented the German side with a pretext for regarding her as an enemy ship. Over 1,200 people died at the sinking of the Lusitania off the southern Irish coast, among them more than one hundred US citizens.[164] Reports that subsequently turned up in the international press that the Germans had celebrated this barbaric act by striking a commemorative medal unleashed unprecedented public indignation. Sommerfeld considered these press dispatches war propaganda and felt it incumbent on him to clarify the facts of the case. "I believed I was safe in assuming that this medal was of English origin," he wrote in an article on "Causes—Effects! The Lusitania Medal," which was published in June 1921 in the *Münchner Neueste Nachrichten*. In the course of his research, however, he had to be convinced that such a medal actually existed; to be sure, the background and motives were quite

161 To Meggers, November 22, 1921. College Park, AIP, Meggers; on the linguistic quarrels following the First World War, see Reinbothe, *Wissenschaftssprache*, 2006.

162 From Meggers, December 9, 1921. DMA, NL 89, 011.

163 To Meggers, January 14, 1922. College Park, AIP, Meggers.

164 Preston, *Lusitania*, 2002.

different from those suggested in the foreign press dispatches. The medal had not been commissioned by any official agency, but was privately produced in a small quantity by a Munich satirist named Karl Götz (1875–1950) and was virtually unknown in Germany before attention was drawn to it abroad. One side of the medal depicted a sinking ship below the legend "No Contraband"; the other carried the legend "Geschäft über Alles" ("Business Over Everything") and depicted below it a skeleton in a ticket booth. The medal was meant, as Sommerfeld explained, "to castigate the hunger for profit of the Cunard Line, which had transported munitions on a passenger steamer. This had nothing to do with celebrating the loss of life of the victims; the medal expresses not satisfaction, but reproach." Sommerfeld left no doubt that he found such war medals in bad taste. But even more reprehensible seemed to him how "our enemies" were exploiting this for propagandistic purposes. "They have sought to make it appear as though this purely private enterprise were an expression of the mood of the German people or of the position of the German government, an expression of triumph over annihilated peaceable travelers."[165]

Sommerfeld's indignation would have been greater had he known the whole extent of the war propaganda made over this medal. In England, "a bronze reproduction of the medal in question [was] distributed in great numbers," the foreign office wrote angrily in 1917. "From this can be gauged how great a service the producer of this medal has done the enemy propaganda through its distribution."[166] Conversely, Berlin, too, thought propagandistically to exploit these obviously false dispatches in the foreign press about the Lusitania Medal. "His majesty has decreed," a diplomat of the Foreign Office wrote, "that the propaganda department of the Foreign Office pin the British lie appropriately on the foreign press."[167] Even after the war, the medal served propagandistic purposes: "by the way, the medal has recently reappeared on the Swiss market, and is selling well," the German envoy to Bern reported to Berlin. "In the Entente nations as well, reproductions of the medal are said to be circulating in great numbers, and are being exploited for propagandistic purposes with good success."[168]

It may have been these Lusitania medals distributed throughout Switzerland that occasioned Sommerfeld's newspaper article, for he mentioned in his letter to Meggers that he had taken to pen and paper only "after a conversation with Prof. Kunz." Jacob Kunz (1874–1939) was a native Swiss. He had studied briefly with Sommerfeld before accepting an appointment in 1908 as professor of mathematical physics at the University of Illinois at Urbana. But so long as Sommerfeld reached only readers of

165 Sommerfeld, *Ursachen*, 1921.

166 Foreign Office (from Radowitz) to the Reich Office of the Interior, January 25, 1917. BA, R 901 71964.

167 From Grünau (Legation Consul) to the Foreign Office, telegram, November 15, 1916. BA, R 901 71964.

168 German Mission in Bern (Köcher) to the Foreign Office, February 19, 1920. BA, R 707 14.

the *Münchener Neueste Nachrichten*, he felt his rebuttal was insufficient. “If you deem it appropriate,” he therefore asked Meggers, “it would be gratifying to me if you were to arrange for an American newspaper to reprint it (with my name).”[169] Although Meggers confirmed that the Lusitania Medal had been reported in the USA too, he did not accede to Sommerfeld's request to offer the *Münchner Neueste Nachrichten* article to an American newspaper. It had become known to everyone that the Lusitania had been transporting munitions and had thus endangered its passengers; all that ought not now be stirred up anew.[170]

Sommerfeld had also turned to Einstein with the request that he submit his article to an English newspaper, but Einstein gave him quite clearly to understand what he thought of that sort of rejoinder: “I frankly regret,” he wrote in response to the newspaper article Sommerfeld had sent him, “that you have written this.” In Germany, too, “terrible lies were told without retraction, and it would not be useful now in a joint effort to drag all the dirty laundry gathered during the course of the war into the light of day. In any case, I cannot offer you my assistance with this, and ask you in the interests of restoring international harmony to give up this fruitless matter. Everywhere in America and England, I have encountered sincere willingness to reach understanding, deep respect for Germany's intellectual workers, and admiration for your scientific work as well as sympathy for you personally. So, away with the old grudges.”[171] Sommerfeld struggled on undeterred, however, to enlighten other countries about this war propaganda, although he met with no success. “By the way, aside from you, my Lusitania article, has also been passed over in silence by the three neutral colleagues to whom I have sent it,” he wrote again later to Einstein.[172]

8.8 Karl Schurz Professor at Madison

Not only in neutral countries like Spain (where following his visit Sommerfeld had been elected member of the Academy of Sciences at Madrid), but also in the USA, many people were admiring of German science. At the University of Wisconsin at Madison, a guest professorship for German scientists had been established in 1911 by German-Americans honoring the 1848 revolutionary and later American statesman Karl Schurz (1829–1906). The professorship was offered every 2 years for one semester. Sommerfeld was the first Karl Schurz Professor after the war. He was to be invited for the winter semester of 1922/1923. The initiative was taken by Alexander Rudolf Hohlfeld (1865–1956), who had been Chairman of the Department of

169 To Meggers, January 14, 1922. College Park, AIP, Meggers.

170 From Meggers, February 24, 1922. College Park, AIP, Meggers. (“I am uncertain if any good will be accomplished by further discussion of this subject since everybody now knows that the ship carried munitions which involved the passengers in risk…”).

171 From Einstein, July 13, 1921. DMA, HS 1977-28/A,78. Also in ASWB II.

172 To Einstein, January 11, 1922. Jerusalem, AEA, Einstein. Also in ASWB II.

German Studies at the University of Wisconsin at Madison since 1904 and during World War I had fought passionately against anti-German sentiment in the USA.[173] Hohlfeld announced in the *Milwaukee Sonntagspost*, the organ of the German-American community in Wisconsin, that the invitation to "a representative of German science to an official position in public education" had been contemplated already at the end of 1921, "at a time, that is, when the general atmosphere in this country, especially at the universities, was even more unfavorable than at present."[174]

This was a remarkable event not only for German-Americans. Even the invitation to Einstein to visit the USA in 1921 had evoked the strenuous protest of the astronomer George Ellerly Hale (1868–1938), who had established the Mount Wilson Observatory, which Sommerfeld so admired, and numbered among the most influential spokesmen of science in the USA. During World War I, Hale had contributed decisively to the division of Axis power scientists from those of Entente nations into "enemy camps" and to the extension of hostilities in a "cold war of science" for several years thereafter.[175] "The appointment of Professor Sommerfeld marks the resumption of the professorship after the interruption of the war years," the American magazine *Science* wrote on August 11, 1922, in reporting to its readers the news that Sommerfeld would be coming to Madison as Karl Schurz Professor.[176] Even *The New York Times* reported the appointment under the headline "German Scientist Coming."[177]

Previously, the literary scholar Eugen Kühnemann (1868–1946) in 1912 and the economist Moritz Bonn (1873–1965) in 1915 had been appointed to the professorship. It was no coincidence that the position now fell to a theoretical physicist. At the beginning of the 1920s, quantum physics and relativity theory were relatively little known in America. But there were increasing calls for establishment of these subjects as areas of research at American universities. It was initially arranged for leading authorities from abroad to be invited to give guest lectures. After Einstein, Sommerfeld was the second German to whom this honor was extended.[178] Even the announcements in *The New York Times* and *Science* did not fail to mention that Sommerfeld would give courses on "Atomic Structure" and the "General Theory of Relativity."

The news that Sommerfeld would be coming to America spread like wildfire. He had not yet accepted the invitation to Madison, when he received another invitation. The spectroscopist Theodore L. Lyman (1874–1954) of Harvard University wrote to Sommerfeld that Meggers had informed him of his visit to the USA the following year, and he invited him to visit his laboratory. "During this visit I hope

173 Nollendorfs, *First World War*, 1988; Zenda, *Loyalty*, 2010.

174 Newspaper clipping in German Department Records, Series 7/14/4, University Archives of the University of Wisconsin, Madison.

175 Kevles, *Camps*, 1971; Kevles, *Physicists*, 1979, Chap. 10 ("Cold War in Science").

176 *Science, 56*, 1922, p. 166.

177 *The New York Times*, August 6, 1922.

178 Coben, *Establishment*, 1971; Holton, *Rise*, 1988; Sopka, *Quantum Physics*, 1988.

you will be willing to deliver three or four lectures, for which we will pay $100 per lecture."[179] In May 1922, when Meggers heard of Sommerfeld's invitation to Madison, he offered to brief him on circumstances in the USA and to arrange for other invitations like that from Lyman to Harvard. He also gave Sommerfeld a picture of the rankings of the American universities. There were two categories of universities, he informed him, private and public. Harvard, Chicago, and Columbia were private, the universities at Madison, Minnesota, and others were public. The University of Wisconsin at Madison was one of the best, he said in praise of Sommerfeld's host.[180]

At the end of June 1922, Hohlfeld came to Munich to sweeten the invitation to Madison. "This evening, I'm expecting a German-American who wants to abduct me for the winter to the land of dollars," Sommerfeld wrote a collegial friend. "I am supposed to be stationed at the Wisconsin-University as 'Karl-Schurz-Professor,' and from there make 'art tours' (to Harvard, for instance). Although I am not particularly eager, I also do not have the heart to decline—on political grounds."[181] Shortly after Hohlfeld's visit, the invitation was officially confirmed by telegram from the President of the University at Madison. It stipulated that, in consideration of an honorarium of $4,000, he was to deliver 6 h of lectures per week from September 18, 1922 to February 1, 1923.[182] Sommerfeld did not take long to decide and requested a leave of absence for the winter semester of 1922/1923, adding in his official application that he considered it his "duty" to accept the invitation.[183]

At a time of increasing monetary inflation, aside from his politically motivated sense of duty, the financial incentive, too, cannot have been insignificant. Before the war, a US dollar cost 4.20 Marks. After the war, the exchange rate of the dollar increased tenfold annually: from 42 Marks (January 31, 1920), to 420 Marks (October 3, 1921), to 4,430 Marks (October 21, 1922). Even if Sommerfeld could not imagine how the drastic fall in value of the Mark would further accelerate during his stay in the USA, it was clear already in the summer of 1922 that an honorarium of $4,000 and several hundreds of additional dollars income from lectures at various American universities represented a considerable emolument, far exceeding his annual salary as professor at the University of Munich: ca. 16,000 Marks by the end of the war.[184]

That in the inflationary time financial considerations were for Sommerfeld no trivial matter is reflected in an application he submitted to the Foreign Office at the end of July for an advance of $800 and 80,000 Marks to cover his travel expenses.

179 From Lyman, May 27, 1922. DMA, NL 89, 019, folder 4,1.

180 From Meggers, May 29, 1922. College Park, AIP, Meggers.

181 To Anschütz-Kaempfe, June 25, 1922.

182 From Birge, 3. und 5. July 3 & 5, 1922. DMA, NL 89, 019, folder 4,1.

183 To the Bavarian Ministry of Culture and to the LMU, July 4, 1922. DMA, NL 89, 004.

184 http://de.wikipedia.org/wiki/Deutsche_Inflation_1914_bis_1923 (31 January 2013); Eckert/Pricha, *Berufungen*, 1984.

For the cabin on his ship, he had already laid out "$50 = 27,000 Marks" and had made "payments of over $300" for other travel costs. In order to avoid the "exchange risk," he requested that the advance be granted him in dollars, "whether in currency or as credit. My travel costs will be increased because I am taking my son with me as travel companion and assistant."[185] The Ministry ultimately awarded him an advance of 250,000 Marks, but nothing in dollars. Instead, he was given a credit of at most $650 at the German Consulate General in New York.[186] Sommerfeld also received a supplement to his honorarium from his American hosts. "I am bringing along my eldest son, who will pursue his electrical engineering studies at Madison until Christmas," he wrote in justification of his higher expenses. Thereupon his American honorarium was raised by $500.[187]

Before his departure, he still had to arrange a number of things for the duration of his absence. He handed the Munich lecture and seminar operation over to his lecturer Karl Herzfeld. His assistant Gregor Wentzel was to oversee the other matters of his institute. He sent Heisenberg and three other advanced students to Max Born at Göttingen, who had recently written challenging him to have his students "quantize" now "to give you a little competition."[188] After completing his doctorate under Sommerfeld in 1921, Pauli had become Born's assistant and had thereby opened up the exchange in matters of quantum physics between Munich and Göttingen. There could hardly have been a better opportunity for Heisenberg to advance his unconventional education as a physicist. "I expect enormous things of Heisenberg, who is no doubt the most talented of all my students, including Debye and Pauli," Sommerfeld wrote his former student Epstein at Pasadena, who just at this time was involved in a competition with Born concerning perturbation theory. Born and Pauli had just devoted a seminar at Göttingen to the celestial-mechanical perturbation theory, in hopes of coming to grips at least mathematically by approximation with the multi-body problem, which was solvable exactly in only certain cases, and arose also with multi-electron atoms. Epstein had just published a series of articles on this subject. "Born/Pauli maintain they have a more convergent [perturbation theory]," Sommerfeld goaded, knowing full well that even the simplest atomic multi-body problem, the helium atom with a nucleus and two electrons, had up to now defied all attempts at solution. "So far, no one has had any reasonable idea what to do with helium, not even Kramers, despite laborious calculation." Sommerfeld's purpose in this letter was not shoptalk, however, but rather to point out to Epstein the possibility of a visit to California, where once again science, money, and politics ran together. "I would of course be eager to marvel at the astrophysical fairyland of Mt. Wilson, and the first-class research institution created by the industry of Prof. Millikan at Pasadena. Also discussions with Prof. Birge at Berkeley would appeal to

185 To the Foreign Office, July 31, 1922. DMA, NL 89, 025, folder Körperschaft (corporation).
186 From the Foreign Office, 10. und 30. August 10 & 30, 1922. DMA, NL 89, 019, folder 4,1.
187 To Slichter, August 13, 1922. DMA, NL 89, 019, folder 4,1.
188 From Born, May 13, 1922. DMA, HS 1977-28/A,34. Also in ASWB II, S. 117–119.

me, not to mention the Pacific coast. But it is clear that I will not offer myself as a German. So you must not take this letter as a broad hint, rather only as an outline of the existing possibilities, namely the temporal (between mid-January and the end of February), the pecuniary (I would have to have dollars enough not to have to assume personally the costs of travel and lodging), and the formal (I would have to receive an official invitation from the competent authorities)."[189]

Before Sommerfeld accepted an invitation to lecture, he wanted to be sure he would not be confronted with anti-German sentiments. Therefore, he stressed that he did not want to be coming "as a merely tolerated or even unwanted guest," as he wrote his long-time acquaintance from the Göttingen and Clausthal days, the mathematician William Osgood (1864–1943) at Harvard University. His acceptance of Lyman's invitation had to be dependent on its being made to him in official form. "In Germany it is thought that many Harvard professors were especially anti-German during the war," he wrote to Osgood concerning the mood among his Munich professorial colleagues. Several of them had advised him against a visit to Harvard University; therefore, he had to "insist that, for example, the President of the University agree to his lecturing at Harvard. I am convinced that Professor Lyman in this respect has already taken the necessary steps. Nonetheless, it would be reassuring to know from you that the President—if I should pay him an official visit—would receive me acceptably."[190] Osgood relieved him of these reservations: "Yes. Please come here. You needn't fear any untoward reception," he wrote back. "Our colleague Lyman is a very good man. If he has invited you to come here, he will certainly have arranged for you to be received in a fitting way. He would never have issued the invitation without first clearing it with the President." He did, however, concede that there had been anti-German sentiment at Harvard. He suggested further that neither Lyman nor the President would share Sommerfeld's conception of the war. But "give us too the opportunity of seeing the matter from the German point of view."[191]

On September 6, 1922, Sommerfeld and his son Ernst boarded the S.S. George Washington at Bremen for the crossing to New York. "It's unbelievable how the time passes in idleness," he wrote home from the ship. "You move leisurely from one chaise or lounge chair to another, and from the library to the smoking salon, or promenade along the covered walks. Unfortunately, I have been speaking more German than English, since my table-mates and nearer acquaintances are German-Americans." Already onboard ship, people wanted to hear firsthand from the famous physicist about the latest developments in this science. He had lectured to a small audience on relativity theory, "not very well," he added self-critically, "despite the fact that I spoke in German."[192] English was somewhat problematic for him.

189 To Epstein, July 29, 1922. Pasadena, CalTech, Epstein 8.3.
190 To Osgood, July 17, 1922. DMA, NL 89, 019, folder 4,1.
191 From Osgood, August 1, 1922. DMA, NL 89, 019, folder 4,1.
192 To Johanna, September 12, 1922.

"At the moment, my English is so bad," he wrote Epstein shortly after his arrival in Madison, "that for now I have to avoid political discussions; later on, though, I plan to speak on political questions with men I get to know better. Certainly, this possibility was decisive in my acceptance of the Wisconsin appointment."[193]

He had worked out an arrangement for his lectures that made accustoming himself to the English language easier: "There is a more advanced student for each of the two courses with whom I talk through the material in preparation," he reported to his wife after his first week of lecturing. "Of course I speak without notes."[194] He anticipated that the semester would be "quite comfortable," so long as he was not invited all too often to give courses and lectures elsewhere. But the invitations came in quick succession. First to reach him was a letter from Robert Andrews Millikan (1868–1953), President of the California Institute of Technology (Cal Tech) and Director of the Norman Bridge Laboratory there, who had been informed by Epstein of Sommerfeld's wishes. Millikan suggested that Sommerfeld lecture for 2 weeks at Pasadena, and 2 weeks at the University of California at Berkeley, and at Stanford. "I can at least guarantee to cover your expenses, and perhaps somewhat more."[195] Then Lyman reiterated his invitation to Harvard.[196] One day later, Kunz sent him an invitation to the University of Illinois at Urbana.[197] And this was not the end. He had already received "a great many invitations from elsewhere," Sommerfeld wrote in the next letter to his wife. He had accepted invitations to the three California Universities for the coming February. He would be able to accept an invitation to lecture at the Massachusetts Institute of Technology (MIT) at the same time he visits Harvard, and on this same trip to the East Coast, he wanted also to visit Columbia University in New York and the General Electric laboratory in Schenectady, to which he had been invited. He also wanted to honor invitations "in the closer vicinity (within 12 hours' train ride)," by which he alluded to short trips to Minnesota, Iowa, Ann Arbor, Milwaukee, and Cincinnati. Finally, Meggers had also visited him and had invited him to Washington. The prospect that his sojourn would be "quite comfortable" had now evaporated. In 2 weeks, he would begin the routine of "spending [the lecture-free rest of the week], from Thursday noon to Sunday evening on trains and lecture tours."[198]

For the time being, though, Sommerfeld found that he led "a much lazier life than at home." He enjoyed the fall in the university town situated between two lakes, took boat rides with Hohlfeld, walked from the university campus to nearby Picnic Point at the tip of a peninsula that stuck like a finger into Lake Mendota, and made music in the evenings at the home of one or another professor in

193 To Epstein, September 24, 1922. Pasadena, Cal Tech, Epstein 8.3.
194 To Johanna, October 1, 1922.
195 From Millikan, September 29, 1922. DMA, NL 89, 019, folder 4,1.
196 From Lyman, October 4, 1922. DMA, NL 89, 019, folder 4,1.
197 From Kunz, October 5, 1922. DMA, HS 1977-28/A,185.
198 To Johanna, October 8, 1922.

Hohlfeld's department. "I played a Mozart trio and a movement of Beethoven. In addition, I heard Weber, Schubert, Mozart, Brahms—in other words, have been altogether in the best German company," he wrote home after one of these musical evenings. His son, too, quickly made friends in Madison. "Ernst is away for the night, with the young physicists, staying in a hut by the lake called Black Hawk ... named for the last Indian chief. He returns at noon."[199] Ernst soon overtook his father in his knowledge of English. In his lectures, Sommerfeld felt up to the challenge, "my English is good enough for that. Conversation and comprehension go less well." But the warmth of his hosts banished all feeling of foreignness. At the Hohlfelds', he was already almost family. The professors from the German Department invited him on a picnic, "which is to say, dinner around a fire with lots of lovely German folksongs. Very atmospheric amid the lovely colors of the woods and summer temperatures." What he lacked in English fluency, he made up for with his piano playing. "Nearly every evening last week I played music, some solo, some with violin. The Hohlfelds had actually invited an audience." The anxiety that he would meet with anti-German sentiment in America proved unfounded. In the multi-page letters he sent home weekly and in which he held back nothing, only one "unfriendliness" appears: The invitation to the University of Chicago was withdrawn, without explanation. He suspected that Albert Abraham Michelson (1852–1931), "a real German hater," was behind it.[200] Born to Jewish parents in Strzelno in the formerly Prussian province of Posen, Michelson had emigrated with them to the USA in 1855. He numbered among an irreconcilable group of American scientists who rejected any cooperation with colleagues from the Axis powers after World War I.[201] An invitation to lecture at Northwestern University in Evanston, near Chicago, would almost certainly have led to an embarrassing situation, since it had been thought to invite Michelson and Henry Gordon Gale (1874–1942) from nearby Chicago as guests—an eventuality Sommerfeld managed to prevent. "I will tell you the reason for this"—viz., that "Professors Michelson and Gale" had just rescinded a planned invitation to him to Chicago, Sommerfeld wrote to Henry Crew (1859–1953), his host in Evanston.[202] On his return from this 3-day lecture tour, he described his experiences in detail. "Crew is a spectroscopist; my book was lying on his desk; I was like an oracle to him." Crew had also driven him to Chicago along the shore of Lake Michigan and shown him, among other things, a Goethe monument erected by German immigrants before the war. Later, sitting together in the evening, he had been able "to speak freely about the situation in Germany and the war." The following day he had been the guest of the Director of the German School in Milwaukee and other Americans of German background, "many

199 To Johanna, October 16, 1922.
200 To Johanna, October 22, 1922; from Kunz, October 17, 1922, DMA, NL 89, 019, folder 4,1.
201 Kevles, *Camps*, 1971, p. 50.
202 To Crew, November 13, 1922. Evanston, Northwestern University Archives, Crew Papers, Box 3, folder 2.

handsome, polished, educated, and German-minded women and sympathetic men. Our conversation: how can Germany survive the winter?? The contrast between all the enormous luxury here and the dark, winter-heavy background of the homeland is enormous."[203]

Nearly every letter to his wife also contained advice on how, in light of rampant inflation, she should handle the dollars he received for his lectures and transferred to a bank account in Switzerland. In this, the mechanic of his institute, Karl Selmayr (1884–1974), who was well versed in financial and business matters, also played a role. "Ask Selmayr to buy me a type-writer, preferably the same type as for the Institute, only best quality. No doubt as good a capital investment as the bicycles," he once wrote. "I'm also thinking of authentic carpets, if an occasion presents itself."[204] He also spelled out his income for upcoming lecture tours: "From California, I'll get $700, the trip costs $200; from Urbana, $200. I'll go to Minnesota just for the travel expenses. I declined an invitation to Ohio under similar conditions. You can see, I'm getting the Yankees to shell out."[205] Following the war, his institute mechanic had begun—with his boss's permission—to capitalize on the experimental X-ray structure research in the basement of the institute, by shipping lattice models of crystal structures on order (and payment) all over the world. Sommerfeld had brought a special trunk with a selection of such models with him to America in hopes "of making a tidy sum of dollars out of it for Selmayr."[206] On his return to Madison with $400 from a several days' lecture tour to the University of Michigan at Ann Arbor ("You see, the American 'cow' is a good milker!"), he advised his wife to rely entirely on his institute mechanic with respect to the investment of the dollars he had sent. "Just let Selmayr take care of things like that, for the typewriter too. Here, every snot-nose kid has such a thing; why not I? Capital investment."[207]

Sommerfeld did not spend Christmas with his son in Madison, but south of Chicago in the small university town of Urbana, where Kunz had invited him to lecture and also welcomed him as a guest in his home. Being so far from his family on this day made him melancholy. When he read his wife's Christmas letter, and paged through the leather-bound collection of her poems she had sent as a Christmas gift, he was overcome with emotion. "You could not have given me a more treasured gift: the most beautiful soul in beautiful form, the harvest of a quarter century. My eyes were still very red—and I did not even need to hide them from these good people as I came downstairs to breakfast and was warmly greeted," he wrote of his hosts and his mood on Christmas Day. He would have liked to read a

203 To Johanna, November 19, 1922.
204 To Johanna, November 12, 1922.
205 To Johanna, November 19, 1922.
206 To Epstein, September 24, 1922. Pasadena, CalTech, Epstein 8.3.
207 To Johanna, December 17, 1922.

Fig. 21: Sommerfeld spent Christmas and New Years 1922/1923 with his former student Jakob Kunz, who had become Professor of Physics at the University of Illinois, Urbana (Courtesy: Deutsches Museum, Munich, Archive).

few of his wife's poems aloud in the Kunz family circle, but she had forbidden him to do so. So he would, he promised her, "keep the gold-leather volume as a secret treasure, and safeguard it from the treasure-seeking Kunz eyes." His former student seemed to him like "a big child, who discovers each day anew the beauty of creation, and laughs at the follies of mankind. He seems more solid to me than in Munich, scientifically deeper as well, a more important talent than I had realized. And his wife is magnificent: fresh, active, smart." Between the lines it was also evident, though, that the many lectures of the past weeks had been taxing. "Kunz is extremely solicitous of my peace and quiet, and is very protective. This is truly a little German Christmas island amid the land of the prairies." Thus, he distinguished his stay in Urbana from the other lecture tours, which he designated "art tours"; he called this Christmas trip his "Kunz tour."[208] The pending departure of his son, who was soon to return to Germany, added to his melancholy. "He will be bringing a nice sum of money, which you must spend properly and quickly, in part

208 An untranslatable pun: "art tour" in German is "Kunstreise;" this visit was a "Kunzreise."

for yourself, in part for needy friends and acquaintances." Thus, he came to speak once more of the disagreeable problem of the "current financial catastrophe." Between October 1922 and January 1923, the exchange rate for US dollars climbed from 4,500 to 49,000 Marks, and the end of the inflation was not yet in sight. "I'm still in favor of buying carpets and other such valuables; we have the money; I have it here in America, or am sure of getting it soon."[209]

Sommerfeld had actually wanted to combine his stay at Urbana with a side trip to St. Louis ("they say here it's a 7-hour trip"), but he managed to save himself this trip, "since the man with whom I wanted to speak came to Urbana for 10 hours."[210] The visitor from St. Louis was Arthur Holly Compton (1892–1962). He informed Sommerfeld of his latest experiments, in which he had scattered X-rays onto graphite, and found that the wavelength of the scattered radiation was longer than that of the incident radiation. In a paper published shortly before, Erwin Schrödinger (1887–1961) had shown that the Doppler effect, familiar to physicists as a typical wave phenomenon (wave fronts of a sound or light source are perceived by an observer at rest as either compressed or stretched apart, respectively, as the source approaches or recedes), was not in contradiction of quantum theory. Could this explain the elongation of the wavelength observed by Compton?[211] Presumably, in the 10 h of his stopover in Urbana, Compton and Sommerfeld had puzzled over the various possibilities how his results, which soon came to be known as the "Compton effect," were to be interpreted. If it was a case of Doppler effect, the lengthening of the wavelength could be traced to the fact that the scattering electron underwent a repulsion in the graphite atom in the scattering process and was consequently moving away from the apparatus registering the scattered X-rays.

But quantitative analysis showed that theory and experiment were not to be brought into agreement in this way. In the scattering on the graphite atoms, the X-rays appeared to behave like colliding particles. Sommerfeld offered to apply himself to the theory of the effect, which Compton heartily approved, since as an experimentalist he was insecure in the rarified air of theoretical physics.[212] Sommerfeld planned a discussion of the Compton effect in the next edition of *Atomic Structure and Spectral Lines*.[213] For now, he was content to report this development to Bohr. "According to this, the wave theory of X-rays would finally have to be abandoned," he wrote to Copenhagen. He was "still not entirely sure" of the matter and did not feel authorized to publicize the results before Compton's paper appeared, but he wanted Bohr to be aware "that we can very possibly expect a quite fundamental explanation."[214] Bohr was not the only person Sommerfeld informed

209 To Johanna, December 28 & 29, 1922.
210 To Johanna, January 8, 1923.
211 Stuewer, *Compton Effect*, 1975, pp. 219–222.
212 From A. H. Compton, January 17, 1923. DMA, NL 89, 019, folder 4,1.
213 Sommerfeld, *Atombau*, 1924, pp. 52–56.
214 To Bohr, January 21, 1923. Copenhagen, NBA, Bohr. Also in ASWB II.

about the new effect. In his "art tours," he contributed significantly to the furor stirred by this news at other American universities and to making the Compton effect the subject of intensive research.[215]

8.9 California Impressions

The "Kunz tour" (cf. footnote 208) marked the end of the Karl Schurz Professorship. In January 1923, although Sommerfeld returned for a few days to Madison, this trip served only to bid his newly acquired friends and colleagues farewell and to prepare for the great journey to the West. "And tomorrow at noon, I will travel even further away from you," he wrote home on January 18. Subsequent letters would be written from California, where he was to give 2 weeks of lectures each at Pasadena (January 24 to February 7), and Berkeley (February 8 to February 22). "In sum, my work here has been fully appreciated," he wrote taking stock of his months as Karl Schurz Professor at Madison. "I leave behind many good friends."[216] His detour to Urbana remained an especially pleasant memory. "Everyone—students and professors alike—still speaks about your profound lectures," Kunz wrote him about the impression he had made there. "Several of the chemistry students expressed the wish that you could be here permanently. Even that old arch-fiend conceded: these lectures were a complete success."[217] The "arch-fiend" was Albert Carman (1861–1946), Head of the Physics Department. "Carman, devourer of Germans, was feeding out of my hand," Sommerfeld wrote his son, who had begun his journey back to Germany several days earlier. "In any case, neither the students nor the professors (including the President, whom I visited), displayed the least war-time animosity. My visit has not been in vain—unfortunately, though, the change in the American tone will not help us over the current crisis."[218]

Sommerfeld was alluding to the oppressive reparation demands of the victorious World War I powers, which led to critical developments in January 1923. On January 11, French and Belgian troops marched into the Rhineland to secure access to its coal production. "The situation is unpleasant, almost like July, 1914," Sommerfeld wrote in response to the news reports of these days. "What I know so far is this: The French in Essen and Gelsenkirchen, the Lithuanians in the Memel district, Russia versus Poland threatening, National Socialists in revolt in Munich, the Berlin government in passive resistance."[219] He thought the political situation so threatening that he considered interrupting his trip in order not to be separated from his family should the crisis escalate. But he determined to continue his journey after all. "Make

215 Stuewer, *Compton Effect*, 1975.
216 To Johanna, January 18, 1923.
217 From Kunz, January 16 1923.
218 To Ernst, January 8, 1923.
219 To Johanna, January 13, 1923.

your pilgrimage through the American lands in high spirits," Kunz counseled him, who for his part wished to "usher in a widespread movement of sympathetic feeling among American students. If every American student donated ½ a dollar annually for a German student, much could be achieved, and the American wouldn't even feel it."[220] "Kunz's letter was a real tonic for me," Sommerfeld wrote his wife. "He understands how to heal wounds. My situation is really not good. Reading the newspapers the last few years has always been painful, but here and now it is torture . . . It is hard for you to be alone now. But it is hard for me, too!"[221]

His next letter was begun in the "club room" of the "California Limited" on the journey through New Mexico. "Traveling in comfort like this is effortless, even if the trip takes 3 days and 3 nights—and it would all be very nice if I were not traveling further and further from you and from the German calamity." Two days earlier, he had interrupted his train journey for 24 h in Lawrence, to give a lecture at the University of Kansas. There, he had also had "several political conversations that, with the general annoyance with France, are now less problematic than before." Before his arrival in Pasadena, he made one more stop, at the Grand Canyon. "An unbelievable scenic structure of yellow-red-violet limestone debris, cut into the uniform high plateau, covered lightly with snow, the bed of the Colorado River 1,800 meters below." The beauty of nature eclipsed his torment over the German calamity for a few hours. Besides, it seemed pointless, given the hectic pace of developments in the crisis, to make "observations concerning the political future"; they would have been "long outdated" by the time his letter arrived in Munich. Nonetheless, he now "felt somewhat better" than during his last days in Madison. "I think the French are deservedly getting into trouble, and that the days of the Peace of Versailles are numbered. I also have the impression that passive resistance is better than active. Given our present circumstances, the moral support of America and England is assured." Arrived in Pasadena, the "retirement home of American millionaires," the situation in far-off Germany must have seemed as though in another world. "So here I am in subtropical California! Palm trees, roses, rubber trees, pepper trees, laburnum in full flower growing wild; orange trees loaded and over-loaded with fruit."[222]

The occupation of the Ruhr led to a deterioration of the economic situation in Germany. The German government's call for passive resistance among the people resulted in hobbling industrial production; the state was forced to print more money in order to pay the wages of the striking workers in the Ruhr region, which in turn further accelerated inflation. Meanwhile, Sommerfeld was in Pasadena amid wealth and luxury. "Of course everything in Germany is becoming even worse than it already was," Sommerfeld replied to a letter from his wife on February 1, 1923.

220 From Kunz, January 16, 1923.
221 To Johanna, January 18, 1923.
222 To Johanna, January 22–26, 1923.

"Will the domestic unity of the resistance hold, or will passive resistance devolve into active recklessness?" In the same letter, he described a convivial gathering among a "society of millionaires" at the home of a patron of the California Institute of Technology. "The group that gathered yesterday, could, I believe, buy up all of Munich." Sommerfeld himself was staying "very comfortably at the Millikans'." Here, too, German classical music was much prized. "Mrs. Millikan sings Schubert and Brahms very beautifully; I accompany her."[223]

From the scientific point of view, Pasadena was likewise something special. Here things were on "the highest level to be found in America," Sommerfeld wrote, preparing himself and his wife for what awaited him with Millikan, who was to be awarded the Nobel Prize that same year. "Here, there is to be a two-week long colloquy, and I cannot simply rest on my reputation as an oracle."[224] The astronomers and physicists from nearby Mt. Wilson Observatory attended his lectures, "a lot of people of the first order."[225] In keeping with the high expectations of his audience, Sommerfeld did not offer them quasi-popular presentations of "atomic structure," but chose such challenging topics as "Quantization in space, theory of magneton," and "Line structure of complicated spectra treated by the method of inner quantic numbers."[226]

With these, he was directing his attention to the spectroscopic research being carried out at Mt. Wilson and at Millikan's institute, where the spectra of stars were observed and compared with the experimentally established spectra of various earthly elements. The astrophysicists at Pasadena sought Sommerfeld's input with respect to the theoretical interpretation and put "their huge store of data most readily at his disposal, including the as yet unpublished," as Sommerfeld wrote enthusiastically to his wife. "The 14 days are hardly enough for all there is to see here."[227] He had already longed to visit "the astrophysical fairyland of Mt. Wilson,"[228] and his hosts at Pasadena gratified this wish only too gladly. During the first week of his stay, the weather was inclement, but relented at last, and the astrophysicists took him along one evening up the 2,000-m-high Mt. Wilson and all night long showed him their telescopes and spectroscopes. In the end, he was at least as enthusiastic about his 2-week-long stay in Pasadena as his hosts were. In addition, Millikan paid him even more generously than had been agreed to. They had been "delighted" by his 12 lectures; his wife and his institute mechanic could expect a fresh transfer of funds. Millikan had expressed his esteem in the form of a check "for $500 (instead of $400, as agreed); additionally, $110 for Selmayr."[229]

223 To Johanna, February 1, 1923.
224 To Johanna,. January 22–26, 1923.
225 To Johanna, February 1, 1923.
226 To Millikan, November 6, 1922. DMA, NL 89, 019, folder 4,1.
227 To Johanna, February 1, 1923.
228 To Epstein, July 29, 1922. CalTech, Epstein 8.3.
229 To Johanna, February 11, 1923.

At Berkeley, too, where Sommerfeld arrived on February 8, 1923, astrophysical spectroscopy was a prominent area of research, and nearby there was another world-famous observatory to be visited. "Tomorrow and the day after tomorrow, I will go to the Lick Observatory on Mt. Hamilton," Sommerfeld wrote home on February 16, 1923, about his plans for the weekend. In the evening, after his return, he continued his report. "So, now I'm back from Mt. Hamilton. Evening of the 18th. On the 17th, left here at six for the ferry, on top by noon; afternoon and evening, inspection. Awakened at 2:30 am to see Jupiter and Saturn through the world's largest telescope."[230] As he had done at Pasadena, at Berkeley he also delivered a program of lectures tailored to the spectroscopic interests of his hosts.[231] "I feel more and more equal to the demands made on me, in both lectures and conversation," he wrote his wife after the first lectures. Though the great interest in him personally was almost burdensome, he declined virtually no invitation. He was hit hardest by the reports of the political situation in Germany following the occupation of the Ruhr. "Yes. The newspaper is a hard nut to crack every morning after breakfast. I think no one can predict what will happen. But the position taken by the government and the German people makes me happy. If only the English were not such spiteful rogues, and the Americans such milquetoasts!"[232] Enthusiasm and depression were often mixed together on a single page of a letter. "California is wondrously beautiful, flowering fruit and almond trees, ocean and snow-covered mountains. A land of milk and honey. So, tomorrow the journey home begins! A strange feeling, after so much foreignness, to be once more journeying towards the homeland. How the homeland has recently been devastated. No! I read in the paper today that the French demand total submission, and the English are backing away again! The end of our time of sorrow is not yet in sight."[233]

8.10 Practical Spectroscopy

On March 1, 1923, Sommerfeld was to be at the National Bureau of Standards in Washington, D.C., to serve for 10 days as a consultant to Meggers and his team in the spectroscopic division. He had scheduled several stops along the way of his return from the West Coast as he had done on the trip there. "My stop at the University of Colorado, Boulder, was short and painless: a lecture with a dinner first, and an excursion by car the next day," he wrote about his first stop in Denver. "Then came Ames, Iowa, with 3 lectures, and a very warm reception. This last was manifested, among other things, in an honorarium of $100." On February 28, he

230 To Johanna, 16.-20. February 16–20, 1923

231 To E. P. Lewis, December 4, 1922. DMA, NL 89, 019, folder 4,1.

232 To Johanna, February 11, 1923.

233 To Johanna, February 16–20, 1923.

arrived in Chicago, where he opened a bank account with the First National Bank of Chicago, lunched with the German Consul General, and visited the Field Museum of Natural History.

Suddenly in sleet and fog, he felt the last few weeks in sunny California as almost unreal. In other respects, too, he had to adapt himself to a different environment in Washington. Working at the National Bureau of Standards, a government agency, was something different from a lecture invitation at a university. First, he was confronted with a bureaucratic ritual. "Today, I had to take an oath of allegiance to the Constitution of the U.S.A.," he wrote home. "I have been formally appointed for 10 days."[234] The "appointment" of a German professor so shortly after the war was something extraordinary for the officials of the National Bureau of Standards. But Meggers left nothing undone to ensure that Sommerfeld's stay was pleasant. "I meet Sommerfeld at 4:20," Meggers entered on his calendar for March 1, 1923. "Visit Senate and House." That evening they spent with Lyman Briggs (1874–1963), Director of the Engineering Physics Division and future Director of the National Bureau of Standards, at the elegant Cosmos Club, where during his Washington stay Sommerfeld was put up. "To Bureau at 9," Meggers noted the next day. "Downtown to get Prof. complete his appointment."[235]

The days in the American Capital were real working days for Sommerfeld. They were strenuous, but he also enjoyed the recognition he was accorded. "My stay in Washington is very satisfactory," he wrote in his next weekly letter home. "My work is highly valued. Everyone wants to tell me his stuff; they all scramble to get to me. It's a wonder I don't come to pieces. Absolutely obliging treatment. I'm able now to speak quite openly about politics with people. Everything here has a somewhat official veneer; here, my tuxedo is finally getting its due. I'm on open and friendly terms with Meggers and Foote."[236]

For Meggers and his team, Sommerfeld's visit was one of the most productive periods of their careers. In these years, spectroscopy was evolving into a science of ever increasing importance for industrial applications, and the National Bureau of Standards thereby garnered great renown in the USA. Already in the nineteenth century, one had learned to identify chemical elements by means of spectral lines and to employ this physical indicator in industry along with chemical analysis in evaluating the combination of substances. But before the 1920s, these investigations were limited to just a few spectra. The "complex structure" of iron and other multi-electron atoms in the mid-range of the periodic table, with thousands of lines, long remained inaccessible. Only when regularities could be read in them could a broad range of chemical substances be subject to spectral analysis. The

234 To Johanna, February 27 to March 2, 1923.
235 Meggers Diary, College Park, AIP, Meggers, Additions, Box 28.
236 To Johanna, March 9, 1923.

Fig. 22: In March 1923, Sommerfeld, seen here in front of the White House, spent several days in Washington as consultant at the National Bureau of Standards, where he briefed Meggers and his team in the spectroscopic division on the latest findings in atomic theory, and in return received valuable insights into the practice of spectroscopy (Courtesy: Deutsches Museum, Munich, Archive).

multiplets classifiable with Sommerfeld's "inner quantum number" proved to be the key to the attribution of the various chemical elements. Meggers and his team had devoted themselves to the experimental analysis of such spectra, which without Sommerfeld's preliminary work in atomic theory would have remained a hopeless undertaking. "Your visit to Washington has given great impetus to these investigations," Meggers wrote Sommerfeld when he reported measurements to him on elements such as platinum, titanium, zirconium, and uranium, which were now subject to spectral analysis.[237]

The contribution of atomic theory to this burst of activity was discussed also in the scientific publications of the spectroscopists from the Bureau of Standards. "The work of Bohr, of Sommerfeld, and of Landé, has inspired spectroscopists to

237 From Meggers, June 15, 1923. College Park, AIP, Meggers. See also Sweetnam, *Command*, 2000, Chap. 8.

attack more complex spectra, and in rapid succession one spectrum after another has yielded to systematic and logical analysis the secret of its structure," Meggers and his team reported in 1924 of their most recent work.[238] Even the newspapers reported what applications were thereby targeted. "Spectrum detects metal impurities," read a headline in the *New York Post* about this new trend in spectral analysis. The article illustrated this with an instance which seemed puzzling until it could be explained by the new method from the Bureau of Standards. A steamer had sunk following the explosion of its boiler. The boiler had been equipped with safety valves that were supposed to melt in case of overheating, but in the event described, the valve had failed. "Why?" the Bureau of Standards wanted to know. Through spectral analysis it was determined that the valve did not consist of pure tin, but contained traces of lead, zinc, and other metals that raised its melting point.[239]

In Germany, too, spectroscopy conquered new territory, thanks to the quantum theoretical interpretation of the complex structure of spectra advanced by Sommerfeld. "Multiplets in the Spectrum of Vanadium" was the headline of Otto Laporte's (1902–1971) July 1923 article in *Naturwissenschaften*.[240] As in the preceding publication of Sommerfeld's "On the Interpretation of Complex Spectra (Manganese, Chromium, and so on) According to the Method of the Inner Quantum Numbers," it was less a matter of mathematical theory than of ferreting out the regularities. For Laporte, occupying himself with the spectrum of vanadium was a kind of finger-exercise preparatory to interpretation of the spectrum of iron, which Sommerfeld had assigned him as the subject of his doctoral thesis and which was also the subject of comprehensive experimental investigations at the Bureau of Standards. Meggers wrote to Sommerfeld that he and his team were engaged in similar analyses. What looked at first like a rivalry between Munich and Washington, Meggers reconfigured into a cooperative venture that profited both theoreticians and experimentalists. As division of labor, Meggers proposed that experimental analyses, which required much experience in the evaluation of spectra, be carried out in Washington and that the interpretation of the resulting data be done in Munich. "Isn't that fair? We will be pleased for instance, to get your ideas on the inner quantum numbers and selection principles for Mo, Fe, Ti, etc.," he wrote, specifying his wishes.[241]

The collaboration of the Munich theoreticians with the American spectroscopists extended to the astrophysicists of Millikan's working group in Pasadena. Harold D. Babcock (1882–1968) from Mt. Wilson Observatory, for example, forwarded data to the Munich Institute on the Zeeman effect of the vanadium lines. "I would like now to request that you send me the Fe-spectrum from Mt. Wilson (along with the letter from Babcock)," Sommerfeld wrote Meggers in this

238 Meggers/Kiess/Walters, Displacement Law, 1924, p. 356.
239 *New York Post*, March 5, 1924.. News clipping, College Park, AIP, Meggers, Additions, Box 7.
240 Laporte, *Multipletts*, 1923.
241 From Meggers, June 15, 1923. College Park, AIP, Meggers.

connection. After Landé had supplied the theoretical-empirical explanation of the Zeeman effect for arbitrary multiplets, they had succeeded in identifying some multiplets in the titanium and vanadium spectra ("of which we have received the Zeeman Effects from Pasadena"). "So the comparison with Fe would be very interesting."[242]

One year later, Sommerfeld and Laporte would celebrate the success of this collaboration of "theoretical-empirical" quantum physics and practical spectroscopy with Laporte's dissertation. It interpreted the iron spectrum with its hundreds of lines. This spectrum, Sommerfeld asserted in his doctoral report, "was considered only a few years ago as hopelessly complicated. But new methods, proven in the series ordering of the simple spectra, made it possible to create order in the complex spectra as well. The introduction of an 'inner' quantum number helped here, through whose selection principles one determines which term can combine with which." After this dissertation, Laporte was "now the top specialist in questions of complex structure."[243]

It might seem superfluous to add that in the fourth edition of *Atomic Structure and Spectral Lines*, brought out shortly thereafter, Sommerfeld devoted a comprehensive chapter to the complex structure.[244] "Untangling the Complex Spectra, in Particular the Spectrum of Iron," ran the title of an article in the *Naturwissenschaften*, with which the spectroscopist at the Potsdam Astrophysical Observatory, Walter Grotrian (1890–1954), brought attention to the advances in his subject area.[245] What Grotrian called "Untangling," and Sommerfeld called "creating order," still did not mean an explanation in the sense of a physical mechanism. Nonetheless, the attribution of the various spectral lines to corresponding electron transitions in the atom was so persuasive, that there was scarcely any doubt about the numerical rules, on the basis of which this attribution was made. "Nowhere is the arithmetic character of quantum theory more simply and elegantly evident than in the complex structure of the series terms." Thus, Sommerfeld opened this chapter in *Atomic Structure and Spectral Lines*; only the physical interpretation remained in dispute. In the current state of atomic theory, it was thus "advisable to leave the model-related interpretation more or less open, and in substance to limit ourselves to the quantum-theoretical determination of the facts."[246]

242 To Meggers, June 30, 1923. College Park, AIP, Meggers.
243 Vote on the Dissertation of Otto Laporte to the Philosophical Faculty, 2. Sektion, July 26, 1924. UAM, OC I 50p.
244 Sommerfeld, *Atombau*, 1924, Chap. 8.
245 Grotrian, *Entwirrung*, 1924.
246 Sommerfeld, *Atombau*, p. 575.

9 Wave Mechanics

In retrospect it seems as though following World War I quantum physics intensified critically and in a revolutionary act in the mid 1920s freed itself from the many contradictions of the old quantum theory. The breakthrough is identified with "matrix mechanics," sketched out in the summer of 1925 in a solitary stroke of genius by Werner Heisenberg, and then established on a solid foundation by the "triumvirate" of Heisenberg, Pascual Jordan (1902–1980), and Max Born. Early in 1926, the old quantum theory was independently revolutionized in an entirely different fashion by Erwin Schrödinger with "wave mechanics," and shortly, the equivalence of matrix and wave mechanics was recognized. Since then, physicists have referred to the core of this new physics succinctly as "quantum mechanics."

In the meantime, historians of physics have elaborated this rough sketch of the history of quantum mechanics into a quite complex picture,[1] without much having changed in the widespread conception of a revolutionary upheaval. With an eye to Sommerfeld, who together with his students made a lasting imprint on the quantum mechanical formation of atomic theory, reservations about this conception are pertinent. Even if there is no doubt about the radical upheaval in physical thinking itself, the transformation that accompanied quantum mechanics seemed to Sommerfeld and to a number of his contemporaries to be not so much a revolution, as a necessary process of adaptation to continuously changing realities. "The new development represents not an overthrow, but a felicitous advance in what already exists, with many fundamental clarifications and with increased precision," Sommerfeld wrote in 1928 in the Preface to his *Wave Mechanical Supplement* to *Atomic Structure and Spectral Lines*.[2] He replied to a colleague in physics who had sent him a book on the foundations of quantum mechanics: "You take the revolutionary position; I, the evolutionary."[3]

Even in a period of evolutionary transition, there can be critical developments and changing, apparently mutually exclusive conceptions and paradigm shifts, such as are characteristic of scientific revolutions.[4] By contrast to truly revolutionary crises, though, in evolutionary developments, old and new can exist side by side for a long time, even if the concepts associated with them cannot be brought into agreement with one another. There is considerable evidence that the development

1 Jammer, *Development*, 1966; Mehra/Rechenberg, *Development*, 1982; Darrigol, *c-Numbers*, 1992; Rechenberg, *Werner Heisenberg*, 2010; Meyenn, *Entdeckung*, 2011.

2 Sommerfeld, *Ergänzungsband*, 1929.

3 To Arthur March, 12. January 12, 1931. DMA, NL 89, 025. Also in ASWB II.

4 Kuhn, *Structure*, 1962. On the criticism of the concept of revolutionary crises, see Seth, *Crisis*, 2007.

M. Eckert, *Arnold Sommerfeld: Science, Life and Turbulent Times 1868-1951*,
DOI 10.1007/978-1-4614-7461-6_9, © Springer Science+Business Media New York 2013

of quantum mechanics represented precisely such *evolutionary* critical processes. Physicists learned to live with contradictions, and they dealt with the situation in very different ways. Everyone who was actively involved in the developments that led to quantum mechanics lived through a period of processes of adaptation that in hardly any individual case were experienced as a collective revolutionary act. What is commonly labeled the "quantum revolution" was rather a process stretching over many years and experienced in quite different and individual ways at the quantum schools of Munich, Copenhagen, and Göttingen.

9.1 The Crisis of the Models

When Sommerfeld returned from the USA in April 1923, skepticism over model-related interpretations of quantum phenomena, which a year later he would express so clearly in the fourth edition of *Atomic Structure and Spectral Lines* in the chapter on complex structure of spectra, was already discernible. During his absence, Heisenberg and Born had calculated all possible orbits that one of the two helium electrons could describe when the atom was in an excited state. The paradigm for this lay in celestial mechanics with its methods of perturbation theory developed for planetary motion around a central star. But the hope of thereby calculating the energy states in the helium atom readable in the spectra in combination with quantum rules was not fulfilled. "The result of our calculation is negative," Born and Heisenberg wrote summarizing their model calculations.[5] Just a short time before, Born had been confident about the application of methods from celestial mechanics to quantum theoretical calculations of atomic models.[6] Heisenberg had even urged making this the sole topic of the seminar for the summer semester 1923 in Munich.[7]

But by the summer of 1923, all traces of this euphoria had vanished. Born and Heisenberg were not the only ones whose model calculations had proven to be failures. Pauli had struggled with similar calculations even earlier than Heisenberg. Considered an expert in the area of celestial-mechanical methods, he was asked by the publishers of an edition of the works of Karl Schwarzschild to contribute a paper on the applications of celestial-mechanical perturbation theory to atoms. "Since so much remains unclear about the theory for multi-electron atoms, however, this hardly fits together," he wrote Sommerfeld in June, 1923. "So there is really no justification for a physicist to undertake this work; an astronomer would be more appropriate." In light of this "breakdown of classical mechanics," it made no sense whatsoever to him to calculate the spectra of multi-electron atoms using methods of celestial mechanics. "This break-down is now hardly to be doubted, and it seems to me that one of the most important findings of recent years is that

5 Born/Heisenberg, *Elektronenbahnen*, 1923, p. 229.
6 From Born, January 5, 1923. DMA, HS 1977-28/A,34. Also in ASWB II.
7 From Heisenberg, January 15, 1923. DMA, HS 1977-28/A,136. Also in ASWB II.

the difficulties of the multi-body problem in atoms are of a physical, and not a mathematical nature." If the Born-Heisenberg helium calculations have failed, the cause is "certainly not that the approximation is insufficient," he commented on this most recent attempt at a model-related explanation of atomic spectra.[8]

Nonetheless, this did not represent a total abandonment of models. When Pauli sent his latest paper on the anomalous Zeeman effect to Munich a few weeks later, he conceded to Sommerfeld that although in his text he had "carefully avoided" any reference to models, he would not have gotten certain of his results "had I not been guided by model representations."[9] At this time, Sommerfeld was himself still not prepared to abandon model-related explanations. For example, to his former assistant Rubinowicz, he heartily recommended the helium model as a particular challenge to help lift him out of a depression.[10]

Model representations could not be entirely dispensed with, especially when spectroscopic findings were brought into relation with other physical phenomena. In 1920, Pauli had already pointed out that the spatial quantization introduced with respect to the Zeeman effect—that is, the quantization assumed in the Sommerfeld atomic model of 1916 of the inclination of the orbital plane in which an electron is in rotation around the atomic nucleus—could also elucidate the puzzle of the elementary magnets. According to experimental investigations carried out by Pierre Weiss (1865–1940), the smallest unit of magnetic moment of atoms or molecules was much smaller than the minimum magnetic moment an electron in its orbit around the atomic nucleus was supposed to have according to the Bohr atomic model. If, however, the orbital plane could adapt itself differently with respect to an external field, the "Bohr magneton" would amount to a multiple of the "Weiss magneton." When the spatial quantization of Otto Stern (1888–1969) and Walter Gerlach (1889–1979) was confirmed experimentally,[11] the search for the smallest possible magnetic moment moved to a new stage. In August 1923, Sommerfeld sent a brief notice to the *Physikalische Zeitschrift* in which he called attention to the fact that the latest spectroscopic findings on multiplets confirmed the magneton that had been expected according to the conception of the spatial quantization.[12] To be sure, the equation derived for the normal Zeeman effect had to be modified, because the multiplets of multi-electron atoms displayed an anomalous Zeeman effect in the presence of a magnetic field. In September 1923, Sommerfeld sent his theory of the magneton to the *Zeitschrift für Physik*.[13] In it, he was once more able to describe an "elegant regularity," which extended "beyond the

8 From Pauli, June 6, 1923. Geneva, CERN-Archive. Also in ASWB II.

9 From Pauli, July 19, 1923. DMA, HS 1977-28/A,254.

10 To Else Rubinowicz, August 18, 1923.

11 Friedrich/Herschbach, *Stern and Gerlach*, 2005; Schmidt-Böcking/Reich, *Otto Stern*, 2011, Chap. VII

12 Sommerfeld, *Magnetonenzahlen*, 1923.

13 Sommerfeld, *Theorie des Magnetons*, 1923.

area of the periodic table in question." But it presupposed the model representation that the electrons in the atom moved in different, variable orbital planes and thereby could adapt their angular momentum in an applied magnetic field variably. This did not please Pauli at all. "As you will see," he wrote Sommerfeld about his latest efforts on the theory of the anomalous Zeeman effect in July 1923, "I was so intimidated by the failure of all my model-related speculations that I have studiously avoided even the word impulse momentum [=angular momentum] in the paper."[14] For Heisenberg, too, model representations were necessary on the one hand, but on the other not really binding on physical understanding. He gave expression to this ambiguous conception, for example, when in December 1923, he reported to Sommerfeld about his efforts to deal with the Zeeman effect in the framework of his core model: "When one reflects in retrospect on what one has actually done, one sees clearly that none of the model representations really make sense. The orbits are real with respect neither to frequency nor to energy."[15]

Sommerfeld commented pointedly on this paradoxical situation half a year later when he wrote to Landé, "Recently, we have repeatedly had the experience that the arithmetic regularities go much further than would be expected from the model representations." Shortly before, Millikan had reported to him that he and his colleague Ira S. Bowen (1898–1973) had measured spectra in the range of ultraviolet light in "stripped atoms" (atoms from which all valence electrons had been blasted away in explosive spark discharges), which like X-ray spectra displayed a characteristic doublet nature, and also could be calculated with the same equation. Sommerfeld expressed his pleasure over the fact that "The relativity equation, far from being discarded or refuted, extends its validity to the optical domain."[16] This equation arose from fine-structure theory and explained the doublets as a relativistic effect, as opposed to the explanation of the optical doublets in alkali metals such as sodium, which were explained by the core model on the various orientations of orbits of valence electrons with respect to the atomic core. Thus, two different models of the doublet phenomenon stood in opposition to each other.[17] "The contradictions you and Millikan present are very serious," Paschen wrote Sommerfeld.[18] Sommerfeld, however, made a virtue of necessity. "This semester, I've lectured comprehensively on your and Bowen's work on the ultra-violet," he wrote Millikan towards the end of the winter semester 1924/1925.[19] He had "for the time being not [been able to] solve" the "serious contradiction," but he hoped for an elucidation soon from his assistant Gregor Wentzel, who was addressing this subject. Wentzel was unable to resolve the doublet puzzle definitively, however. "For me, the open

14 From Pauli, July 19, 1923. DMA, HS 1977-28/A,254.
15 From Heisenberg, December 8, 1923.DMA, NL 89, 009. Also in ASWB II.
16 To Landé, April 20, 1924. SBPK, Landé.
17 Forman, *Doublet Riddle*, 1968.
18 From Paschen, January 27, 1925. DMA, NL 89, 012.
19 To Millikan, February 9 1925. DMA, NL 89, 003. Also in ASWB II.

question of the 'relativistic' doublet is terribly unsatisfying," wrote Schrödinger too, "as you yourself keep stressing."[20] That Sommerfeld, without a certain model foundation, was nonetheless able to interpret the multiplets of the multi-electron atoms with the help of the inner quantum number seemed incomprehensible to Schrödinger. "How it was possible for you to infer these fundamentally so profoundly different regularities from really not a great wealth of evidence without an actual model, and based only on the sense of an analogy with classical theory, is still a mystery to me. I have slowly struggled to achieve clarity on the really quite complicated construction of these rules involving only integers, while you have incorporated these same rules into the observational data, so they now fit snug as a guard's uniform!"[21]

Pauli experienced the failure of model-related explanations as both crisis and incentive. He even praised Sommerfeld for the fact that the presentation of the complex structure of spectra in the fourth edition of *Atomic Structure and Spectral Lines* was "entirely free of model-related preconceptions":

> The model representations now find themselves in a difficult crisis of principle, which will, I believe, end in an even more radical intensification of the contradiction between classical and quantum theory. In particular, as it follows from the findings of Millikan and Landé concerning the representability of the optical alkali doublet by relativistic equations, the idea of specific, unique orbits of the electrons in the atom can scarcely be maintained. One now has the strong impression that with respect to all these models, we are speaking a language inadequate to describe the simplicity and beauty of the quantum world.[22]

Liberation from "model-related preconceptions" and the predilection for maximally simple, empirically based regularities led Pauli to formulate the eponymous "Pauli Exclusion Principle."[23] In a multi-electron atom, the quantum numbers used to characterize the energy level and makeup of the electron shells are ascribed to each individual electron—together with the rule that the quantum numbers of every electron must be different. In other words, every quantum state in the atom can be occupied by only one electron.

Viewed on its own, the passage quoted from Pauli's letter to Sommerfeld on the eve of the ground-breaking work on quantum mechanics would seem the revolutionary escalation of the model crisis. But in light of the reinterpretation of older concepts repeatedly necessitated by complex spectra and other phenomena, we see that this was merely one more process of adjustment in a far from concluded evolutionary development. Consciousness of "model-related preconceptions" did not carry with it renunciation of all model-based thinking. This was most clearly

20 From Schrödinger, March 7, 1925. DMA, HS 1977-28/A,314. Also in ASWB II.
21 From Schrödinger, July 21, 1925. DMA, HS 1977-28/A,314. Also in ASWB II.
22 From Pauli, December 6, 1924. DMA, HS 1977-28/A,254. Also in ASWB II.
23 Meyenn, *Paulis Weg I* and *II*, 1980 and 1981; Massimi, *Pauli's Exclusion Principle*, 2005.

exemplified in the work of two colleagues from the Ehrenfest Institute at Leiden, George Uhlenbeck (1900–1988) and Samuel Goudsmit (1902–1978), in which on the basis of Pauli's insights, they reinterpreted the vector model with which Landé, Heisenberg, and Sommerfeld had explained the magnetic splitting of spectral lines by the spatial orientations of different vectors of angular momentum. In the old vector model, a vector was attributed to the atomic core, which however entailed difficulties that continued to raise questions about this conception. Uhlenbeck and Goudsmit drew from this the conclusion that the atomic core had to be excluded from the otherwise very plausible vector framework model. Like Pauli in his formulation of the Exclusion Principle, they took up the burden of each individual electron which before had been borne by the core. To the electron, in addition to the three spatial degrees of freedom, they assigned a fourth, meant to represent "an individual rotation." The new degree of freedom was not understandable in classical terms (the rotation of a sphere on its own axis does not imply a new degree of freedom because it can be described classically by its three spatial coordinates), but now the core no longer presented problems. Thus, the electron took over "the still not understood property," argued the Leiden theoreticians, that before had been thought to belong to the atomic core.[24]

Hereby "spin" stepped onto the stage of atomic theory as an additional quantum phenomenon. Pauli stubbornly resisted the model-related interpretation of the new degree of freedom as intrinsic rotation, because for an object without spatial extension, this concept is actually meaningless. Nevertheless, the model took hold in the consciousness of physicists.

9.2 "We Believe in Heisenberg, but We Calculate with Schrödinger"

In the course of this evolution, Sommerfeld came to a position that might seem almost indifferent to the fundamental questions raised by quantum theory. "The difficulties in atomic physics that crop up ever more clearly these days seem to me to lie less in an excessive application of quantum theory, than in a somewhat excessive belief in the reality of the model representations," he commented on the state of research in the fall of 1924 to the Natural Scientists Congress convened that year at Innsbruck.[25]

The model crisis was not the only reason to undertake a critical review of classical conceptions such as the idea of the electron orbit in the atom. Another crisis arose from the wave-particle dualism. This crisis, too, had loomed for many years and compelled the physicists ever and again to adjust their theories to new empirical findings. "In light of this, the wave theory for X-rays would finally have to be

24 Uhlenbeck/Goudsmit, *Ersetzung*, 1925.

25 Sommerfeld, *Grundlagen*, 1924, p. 1049.

dropped," Sommerfeld had written Bohr in January 1923, after Compton informed him of the as yet unpublished results of his scattering experiments destined to enter history as the Compton effect.[26] "Whereas I formerly sought to uphold the wave theory for the pure propagation processes as long as possible, the Compton Effect forces me more and more to accept the extreme theory of light quanta," he wrote in the preface to the fourth edition of *Atomic Structure and Spectral Lines*, laying out his own process of reorientation on this question.[27] Up to now, it was still uncertain whether a theory developed in Copenhagen might succeed in interpreting the Compton effect also in terms of a wave concept.[28] When this was refuted experimentally in the spring of 1925, however, it could no longer be doubted that in the Compton effect, X-rays behave like particles. Since then, physicists have described the nature of light with a "this-as-well-as-that," even though wave and particle analogies are mutually incompatible.

For Heisenberg, too, quantum processes in 1925 were still to be understood by "model-related pictures of symbolic significance." He presented this view in a paper titled "On the Quantum Theory of the Multiplet Structure and the Anomalous Zeeman Effect," submitted in April 1925 to the *Zeitschrift für Physik*.[29] Two months later he authored the paper celebrated as the breakthrough to quantum mechanics, "On Quantum Theoretical Reinterpretation of Kinematic and Mechanical Relations."[30] In this paper the problem of the radiation of an electron in motion, in the simplest imaginable theoretical case in which the electron oscillates in only one direction, was formulated so that only experimentally observable quantities were taken into account and the familiar quantum laws were in force.[31]

Except at Max Born's Institute, where within a few months the new theory was further evolved to matrix mechanics and, at Cambridge, where a scientific loner by the name of Paul Dirac (1902–1984) was building out quantum mechanics in quite a different direction, Heisenberg's "reinterpretation" was at first met with reserve and skepticism. "Heisenberg has laid a big quantum egg," Einstein wrote in September 1925 to Ehrenfest at Leiden. "In Göttingen, they believe it; not I."[32] Even among Sommerfeld and his students, there was at first little enthusiasm. "Heisenberg's new quantum mechanics" first appeared only half a year after its publication as a colloquium topic at Munich.[33] Heisenberg had not exactly covered himself with glory at his doctoral exam following Sommerfeld's return from the USA in the summer of 1923 and had so annoyed his second reader (Willy Wien)

26 An Bohr, 21. Januar 1923. Kopenhagen, NBA, Bohr. Also in ASWB II.
27 Sommerfeld, *Atombau*, 1924.
28 Hendry, *Bohr-Kramers-Slater*, 1981.
29 Heisenberg, *Quantentheorie der Multiplettstruktur*, 1925.
30 Heisenberg, *Umdeutung*, 1925.
31 Rechenberg, *Werner Heisenberg*, 2010, Chap. 5.
32 Quoted in Fölsing, *Albert Einstein*, 1993, p. 644.
33 Lecture delivered February 19, 1926. Wednesday Physics Colloquium. DMA, 1997-5115.

that Sommerfeld was at pains to rescue his prize student from the disgrace of failure. Sommerfeld and Wien agreed that Heisenberg received an overall grade for his doctorate that just prevented the failure, being the average of the best grade from Sommerfeld and the lowest from Wien.[34]

Afterwards, Heisenberg had departed Munich as it were in flight, to continue his career with Born at Göttingen and Bohr at Copenhagen. Clearly, he—and doubtless Sommerfeld too—felt it as a token of ingratitude that he had sought refuge at competing quantum schools. In any case, when Sommerfeld sent him the fourth edition of *Atomic Structure and Spectral Lines*, Heisenberg expressed a measure of relief that Sommerfeld was apparently "not so terribly angry" with him.[35]

Presumably, the course Heisenberg had set out on with his quantum mechanics seemed to Sommerfeld a diversion from the recently so successful path of inductively deriving theoretical laws from the wealth of spectroscopic measurements. Almost all the work of Sommerfeld's students on quantum theoretical problems around 1925 dealt with such topics. Miguel Catalan, who had come to Munich in 1924 as guest researcher on a Rockefeller Fellowship, published papers together with Karl Bechert (1901–1981) on the structure of the cobalt and palladium spectra in the *Zeitschrift für Physik*. In May 1925, Bechert had completed his doctorate under Sommerfeld on the nickel spectrum.[36] Heinrich Ott (1894–1962), who had become assistant at the Sommerfeld institute after Wentzel, had addressed the "Problems of X-ray Spectroscopy."[37] Helmut Hönl (1903–1981), another doctoral candidate, focused on the problem of theoretically describing the intensity of spectral lines.[38] In the context of these papers, Heisenberg's "reinterpretation" seemed as though from another world. Limiting himself to one-dimensional electron motion made a comparison with experimental data impossible. On the other hand, Heisenberg's previous paper on multiplets more nearly fit the Munich tradition. And Sommerfeld had not let half a year go by before reacting to it. He considered it so important that he heartily recommended its closer study to the American spectroscopist and astrophysicist Henry Norris Russell (1877–1957).[39]

The "new quantum mechanics" first won adherents among Munich physicists when Pauli demonstrated how one could thereby treat the hydrogen atom.[40] "I too believe that one has to convert without reservation to Heisenberg's new mechanics," Sommerfeld now conceded after Wentzel, who was working with Pauli in

34 Rechenberg, *Werner Heisenberg*, 2010, p. 138.

35 From Heisenberg, November 18, 1924. DMA, HS 1977-28/A,136. Also in ASWB II.

36 Vote on the dissertation of Bechert to the Philosophical Faculty, 2. Section, May 18, 1925. Munich, UAM, OCI51p.

37 Vote on the dissertation of Ott to the Philosophical Faculty, 2. Section, July 12, 1924. Munich, UAM, OCI50p.

38 Vote on the dissertation of Hönl to the Philosophical Faculty, 2. Section, February 24, 1926. Munich, UAM, OCI 52p.

39 To Russell, July 3, 1925.Princeton, University Archive, C0045, box 63, folder 35.

40 Pauli, *Wasserstoffspektrum*, 1926.

Hamburg, had sent him Pauli's manuscript. To be sure, Sommerfeld still found missing the treatment of more difficult cases. Could Wentzel also derive "such from Pauli?"[41] He was already indicating what he hoped from the further development of quantum mechanics. It should explain what—from the fine structure of the spectra of hydrogen-like atoms to the complex spectra of multi-electron atoms—had heretofore been either derived from unrealistic models or merely sketched inductively.

A manuscript now burst onto the scene that Erwin Schrödinger had sent Willy Wien as editor of the *Annalen der Physik* with the request that he give it to Sommerfeld to referee. "An extraordinary mind, very well educated and critical," Sommerfeld had assessed Schrödinger in 1921, in recommending him for an appointment to the chair in theoretical physics at the University of Zürich (held previously by Einstein, Debye, and Laue).[42] Now, 5 years later, Schrödinger proved he was more than equal to the lofty standards that attached to the Zürich chair he occupied. He was well aware of the importance of the paper he had submitted to the *Annalen der Physik* under the title "Quantization as Eigenvalue Problem," for he wanted to know from Sommerfeld whether he "shared the very ambitious expectations" he himself had of it.[43]

Already in his initial reaction, Sommerfeld showed that Schrödinger's procedure appealed to him much more than Heisenberg's. "This is really terribly interesting," he wrote by return mail to Zürich. "I was just on the point of formulating a concept for lectures in London (this March) that was very much in the earlier key. Then your manuscript arrived like a thunder bolt. My impression is that your method is a replacement for the new quantum mechanics of Heisenberg, Born, Dirac." He conceded that it was still not clear to him how the one could be brought into harmony with the other, but he was "convinced that something entirely new will come of it that can set aside the contradictions that currently bedevil us."[44] To Pauli he wrote that same day that Schrödinger had obtained the same results from the hydrogen spectrum that Pauli had just calculated rather laboriously according to matrix mechanics, "but in a quite different, totally crazy way, no matrix algebra, rather as boundary value problems."[45]

Even before Schrödinger published his paper, the stage had been set for the competition between matrix and wave mechanics. "His way may not be so crazy," Pauli replied about Schrödinger's method, which he knew initially only through

41 To Wentzel, January 13, 1926. DMA, NL 89, 004.

42 To Edgar Meyer, 28. January 28, 1921. Zürich, University Archive, Papers of the Office of the Dean ALF. Also in ASWB II; Moore, *Schrödinger*, 1989, pp. 139ff.; Meyenn, *Entdeckung*, 2011, pp. 51ff.

43 From Schrödinger, January 29, 1926. DMA, NL 89, 013. Also in ASWB II; Meyenn, *Entdeckung*, 2011, pp. 170–172.

44 To Schrödinger, February 3, 1926. Carbon copy in DMA, NL 89, 004. Also in ASWB II; Meyenn, *Entdeckung*, 2011, pp. 173–175.

45 To Pauli, February 3, 1926. DMA, NL 89, 003. Also in ASWB II.

Sommerfeld's sketchy references.[46] Schrödinger published his paper in four "communications" in different issues of the *Annalen der Physik* of the year 1926. Subsequently, he gathered them together as a book with the title *Treatises on Wave Mechanics*, which appeared in 1927.[47] Schrödinger thought matrix mechanics "insupportable" and hoped it would soon "disappear. . . For I shudder at the very thought," he wrote to Wien, "of sometime down the road having to lecture to a young student on matrix calculus as the essential nature of the atom."[48] Pauli saw to it that among the matrix mechanicians at Göttingen, the rival theory from Zürich was thoroughly studied. "I believe that this paper numbers among the most important written in recent years," he wrote to Pascual Jordan on the appearance of Schrödinger's first "communication." "Please read it carefully and with reverence."[49] Although Schrödinger's wish to eradicate matrix mechanics entirely was not achieved, the physical equivalence of the two methods was soon demonstrated. Many of the problems in atomic physics studied heretofore could be solved by both methods. Preference thus fell to wave mechanics because its mathematical operations were simpler than those of matrix mechanics. Schrödinger's method was "far simpler and more convenient" than Heisenberg's, Sommerfeld wrote, praising wave mechanics on the occasion of his trip to England in March 1926. It employs "the language of the theory of vibrations."[50]

This language was familiar to every physicist. Opinions might differ radically as to the meaning of what Schrödinger's wave mechanics supposed was vibrating and propagating wave-fashion, but the mathematical formalism presented no fundamental difficulties. A vibrating string, a tuning fork, a vibrating membrane, the vibrating air in an organ pipe—every such system has, corresponding to its material properties, eigenmodes of vibration determined by magnitude and arrangement that can be found mathematically as the solution of an eigenvalue problem. When the underlying boundary conditions of the respective problem were given, the eigenfunctions (vibration forms) and eigenvalues (frequencies of the basic vibrations and the harmonics) could be calculated by means of a standardized process. The electron rotating around an atomic nucleus could, following Schrödinger, be represented according to the same formalism as a standing wave, whereby the eigenvalues corresponded to the energy terms that Bohr and Sommerfeld had calculated in a quite different manner 10 years earlier, and the quantum numbers revealed themselves as the indices of the eigenfunctions—in this case, spherical harmonics.

46 From Pauli, February 9, 1926. DMA, HS 1977-28/A,254. Also in ASWB II.

47 Meyenn, *Entdeckung*, 2011, pp. 176–208.

48 Schrödinger to W. Wien, February 22, 1926, quoted in Meyenn, *Entdeckung*, 2011, pp. 184–187.

49 Pauli to Jordan, April 12, 1926. quoted in WPWB I.

50 Sommerfeld, *Lectures*, 1926, p. 3. ("Schrödinger arrives at the same results as those obtained by the mechanics inaugurated by Heisenberg, but by a road that is presumably far simpler and more convenient [...] his treatment is expressed in the language of the theory of vibrations.")

Sommerfeld's lectures in England found a great reception. "I really believe everyone was very satisfied," he wrote home after his first week.[51] His host was the professor of physics at King's College, London, Owen W. Richardson (1879–1959), who 2 years later would be awarded the Nobel Prize for his work on thermionic emission of electrons in metals. But this visit was not exclusively about physics. Sommerfeld enjoyed the journey. "I have even tried playing golf (in Oxford), and once table-tennis (in London, at the Indian students' club, with Dasanacharya), often piano, e.g. in Edinburgh, accompanied song and cello," he reported to his wife 1 week later, when he rejoined his London hosts from lecture tours to other cities. "I have worked my way completely into the heart of the stout Mrs. Richardson."[52] Both Charles Galton Darwin (1887–1962), grandson of the famous biologist, and William Lawrence Bragg, who had invited him to Edinburgh and Manchester, respectively, appreciated his sociability. "Bragg is especially cordial with me, a good friend of Ewald," Sommerfeld wrote home from Manchester.[53]

"It is terribly kind of you to have promoted me in England already," Schrödinger wrote gratefully a few weeks later.[54] In his lectures in England, Sommerfeld had been content with outlines, but once back in Germany he immediately elucidated wave mechanics to the students and research colleagues of his school. "Here, we are closely studying Schrödinger's new Quantum theory, and estimate it very highly," he wrote to Richardson on the main topic of his seminar in the summer semester of 1926.[55] Towards the end of the semester, he invited Schrödinger to Munich so that he and his students could be introduced firsthand to the new theory.[56] In the process, a heated exchange arose with Heisenberg, who had traveled to Munich for this occasion, and was so critical of wave mechanics that even Sommerfeld began to waver again: "We've had Schrödinger here, together with Heisenberg," he wrote afterwards to Pauli, to whom he delegated the role of referee in this debate. "My general impression is that though 'wave mechanics' is an admirable micromechanics, the fundamental quantum puzzles are not in the least solved thereby."[57] These doubts were stirred up primarily by the lingering question what the concept of wave motion underlying Schrödinger's theory actually was. He corresponded with Einstein about this, too. "Of all the efforts to extract a deeper formulation of the quantum laws from the latest experiments, I like Schrödinger's best," Einstein

51 To Johanna, March 11, 1926.
52 To Johanna, March 19, 1926.
53 To Johanna, March 23, 1926.
54 From Schrödinger, April 28, 1926. DMA, HS 1977-28/A,314. Also in ASWB II.
55 To Richardson, June 12, 1926. Austin, Ransom, Richardson.
56 To Schrödinger, July 10, 1926. DMA, NL 89, 025. On July 23, Schrödinger spoke, on "Foundations of Wave Mechanics" and the following day on "New Results in Wave Mechanics." Physikalisches Mittwoch-Kolloquium. DMA, 1997-5115.
57 To Pauli, July 26, 1926. Geneva, CERN. Also in ASWB II.

wrote. "The Heisenberg-Dirac theories are admirable, certainly, but to me they don't exude the odor of reality."[58]

For advanced students like Hans Bethe, who had come to Munich from Frankfurt in the spring of 1926 to continue his studies under Sommerfeld's wing as a fifth semester student, this first encounter with wave mechanics remained an indelible memory. "We believe in Heisenberg, but we calculate with Schrödinger," as Sommerfeld put it, introducing his students to quantum mechanics.[59] Every participant in the seminar had to report on one subsection of the Schrödinger "communications" that had appeared by the summer of 1926. Thereafter, the seminar participants were prepared to write a doctoral dissertation on virtually any quantum mechanical topic.[60]

The first student to take up a doctoral dissertation at Munich using the Schrödinger method was Albrecht Unsöld (1905–1995). As Unsöld recalled years later, Sommerfeld initially proposed the wave mechanical treatment of the hydrogen ion, which had been the subject of Pauli's dissertation in the framework of atomic theory along the model of celestial mechanics 5 years earlier. "I soon saw that this was not going to work, and began to work with all sorts of more tractable spectroscopic topics," Unsöld recalled. Then, Sommerfeld became "really angry ... But when he then saw that I had found a number of new methods and theorems in the area of spherical harmonics, he graded the dissertation as summa cum laude."[61] In his commentary on the Unsöld dissertation, Sommerfeld stressed that it was "a characteristic of wave mechanics" that one could "put to good use" the mathematical methods of boundary value problems. Thus, Unsöld had "first derived the addition theorem of the spherical function" and demonstrated thereby "that the effect of the electron shells on external points exhibit simple spherical symmetry." From this, it had been possible to calculate the energy levels of the alkali and alkaline-earth atoms.[62] One year later, Unsöld's dissertation furnished the basis for an exhibit at the Deutsches Museum, where visitors could view "quantum mechanical atomic models."[63]

Around this time, at Schrödinger's request, Sommerfeld also arranged for Rockefeller Foundation grants for his former students Fritz London (1900–1954) and Walter Heitler (1904–1981) to allow them to pursue research on applications of

58 From Einstein, August 21, 1926. DMA, HS 1977-28/A,78. Also in ASWB II.

59 Bethe in Eckert/Pricha/Schubert/Torkar, *Geheimrat*, 1984, p. 8.

60 Bethe, *Sommerfeld's Seminar*, 2000.

61 Unsöld, personal communication, September 1, 1982.

62 Vote on the Dissertation of Unsöld to the Philosophical Faculty, 2nd section, December 11, 1926. Munich, UAM, OCI52p. Unsöld, *Beiträge*, 1927.

63 Eckert, *Atommodelle*, 2009.

Fig. 23: In 1927, Sommerfeld conceived this model of the gold atom for the Deutsches Museum. In place of electron orbits, there were, according to quantum mechanics, spatially distributed probabilities of an electron's being at that position, visualized in their ground state as spherical shells around the atomic core. The distance of each shell from the atomic core was calculated according to quantum mechanics and indicated an electron's positions of greatest probability; the thickness of the shells is proportional to the number of electrons in each state (Courtesy: Deutsches Museum, Munich, Archive).

wave mechanics.[64] They succeeded in elucidating the chemical relation between electrically neutral atoms by an interaction that had been classically inexplicable (the so-called exchange interaction), a quantum mechanical effect that illustrated forcefully the importance of the new theory for chemistry.[65]

Sommerfeld's institute in Munich also became a popular address for visiting researchers. American universities, primarily, used the study grants offered by the International Education Board of the Rockefeller Foundation and other support organizations to provide their students the opportunity of unrestricted research at one of the prestigious European scientific centers. The "traveling fellows" contributed significantly to the rapid spread of new scientific fields like quantum mechanics as widely throughout the USA as in Europe.[66] The first American grantees to

64 From Schrödinger, April 28 and May 11, 1926. DMA, HS 1977-28/A,314. Also in ASWB II; from Ewald, July 15, 1926. DMA, NL 89, 007; to Ewald, July 20, 1926. DMA, NL 89, 001; from Heitler, August 29,1926. DMA, NL 89, 009.

65 Heitler, Interview by John L. Heilbron, March 18, 1963. AHQP. http://www.aip.org/history/ohilist/4662_1.html (31 January 2013). Heitler/London, *Wechselwirkung*, 1927.

66 Sopka, *Quantum Physics*, 1988; Assmus, *Creation*, 1993.

come to Munich were the brothers Victor (1896–1985) and Ernst Guillemin (1898–1970). In 1923, Victor Guillemin had attended Sommerfeld's lectures at Madison and was so taken with them that he wished to delve deeper into atomic theory. During his stay in Munich, he witnessed the first debates about quantum mechanics. "Consequently quantum mechanics has been to me, not something I read about," he recalled years later; "I was 'there' when it was born."[67] In the summer of 1926, Linus Pauling also came as a visiting American student to Munich. Shortly before, he had completed his doctorate in physical chemistry at Cal Tech in Pasadena and had received a Guggenheim Foundation grant to study at Munich. "The exciting thing to me were the lectures Sommerfeld was giving on Schrödinger quantum mechanics and of course the seminars were devoted to it," he recalled years later.[68]

9.3 Electron Theory of Metals

According to the Pauli Exclusion Principle, every possible energy state in the atom can be occupied by at most one electron. If not only the electrons within an atom, but also the particles of a gas behaved according to this principle, the statistical distribution of particles across the different energy states of such a quantum theoretically "degenerate" gas was quite different from the determination reached by classical statistics for a normal gas. In 1926, on the basis of the Pauli Exclusion Principle, Enrico Fermi (1901–1954) and Paul Dirac established a new statistics—Pauli referred to it as the "Housing Authority" statistics.[69] In December 1926, he sent a manuscript, "On Gas Degeneration and Paramagnetism," to the *Zeitschrift für Physik*, in which he demonstrated in the theoretical case of "gas atoms with angular momentum" what the new statistics meant for the magnetic characteristics of such a gas: In an external magnetic field, according to "Housing Authority" statistics, all the particles could not line up the way, say, iron filings in proximity to a magnet would; only if, as a result of its reorientation, a particle acquired an energy state not already occupied by any other could it contribute to magnetization. "If the conduction electrons in the metal are regarded as an ideal degenerate gas," Pauli explained in transferring this conception to real circumstances, "we arrive on the basis of the developed statistics to at least a qualitative theoretical understanding of the fact that despite the presence of the electron's own magnetic momentum, many metals (especially the alkali metals) in their solid state show no or only very weak and roughly temperature-independent paramagnetism."[70]

In February 1927, Sommerfeld visited Pauli in Hamburg. When Pauli showed him the galleys of his paper "On Gas Degeneration and Paramagnetism,"

67 Victor Guillemin to KatherinSopka, 1972, quoted in Sopka, *Quantum Physics*, 1988, p. 2.41.

68 Pauling, Interview by John L. Heilbron, March 27, 1964. AHQP. http://www.aip.org/history/ohilist/3448.html (31 January 2013).

69 Pauli to Wentzel, December 5, 1926. In WPWB I.

70 Pauli, *Gasentartung*, 1927, p. 81.

Sommerfeld ventured the conjecture that other characteristics of metals could also be explained according to this paradigm. But Paul was less concerned with a theory of metals than with this as a test case of the new Fermi-Dirac statistics. Sommerfeld proposed that he should next apply this to the explanation of other characteristics of metals, Pauli recalled years later. "As I was not eager to do that, he made then this further application himself."[71]

Pauli's aversion to solid-state physics became legendary. Once, when an assistant planned to take up the theory of electrical resistance in metals, he reacted with the disparaging remark that this was "a filth-effect, and one shouldn't wallow in filth."[72] By contrast, electron theory of metals was quite to Sommerfeld's taste. The conception of an electron gas capable of moving freely between the atoms was a very old one. Unlike isolators, where electrons are bound to atoms, the electrical conductivity of metals appeared comprehensible only if the electrons were allowed freedom of motion. To be sure, this led to contradictions which brought the "free electron gas" into discredit. If the electrons could participate in the motion that showed up as an electrical current, this should be true for thermal motion as well, but the specific heat of metals is hardly distinct from the isolators, so that in this respect the electrons could not be assumed to be freely mobile.

Sommerfeld had long been familiar with this and other contradictions.[73] He found it attractive to investigate whether the "Housing Authority" statistics could also be made to account for the dilemma of specific heat and other characteristics of metals. Before he published anything on the matter, he familiarized himself and his advanced students with electron gas theory by way of a special course of lectures. He had always felt his way into new theories through this tried and true paradigm. For the summer semester of 1927, he therefore announced a special lecture course on "Structure of Matter." First, he explained the dilemma of specific heat: "Housing Authority" statistics provided that only a very few electrons could take part in thermal motion, so that the increase in specific heat was approximately only 100th that of isolators.[74] In the lecture hours that followed, he dealt with the ejection of electrons from metals (Richardson effect) and the phenomena accompanying contact between various metals (Volta effect). "During this semester" he had been "emphatically interested in the Fermi statistics and gas degeneration," he wrote Paschen towards the end of the semester. And there was "weighty evidence of the correctness of Fermi's degeneration formula."[75] Two weeks later, he sent a "presentation [of it] as broadly comprehensible as possible" to the editor of

71 Pauli to Rasetti, October, 1956. AHQP. ("[...] that one should make further application to other parts of metal theory like the Wiedemann–Franz law, thermoelectric effects, etc. As I was not eager to do that, he made then this further application himself.").

72 Pauli toPeierls, July 3, 1931. Quoted in WPWB II.

73 Eckert, *Elektronengas*, 1989.

74 Lecture notes of Unsöld, now DMA.

75 To Paschen, July 25, 1927. DMA, NL 89, 003.

Naturwissenschaften. "Using the new statistical methods of Fermi, my paper seeks to bring order to the age-old problem of the galvanic current, the Volta potential, thermal energy, etc."[76]

Following on this overture, electron theory of metals became, along with wave mechanics, a central focus of research at the Sommerfeld institute. Already in his first comprehensive publications on the subject, Sommerfeld referred to follow-up work being carried out by American fellowship recipients at his institute.[77] "I am very pleased with your two students Dr. Eckart and Dr. Houston," Sommerfeld wrote Millikan at Pasadena after these two had arrived at Munich in the fall of 1927 on Guggenheim Foundation grants. "I carry on the most interesting discussions with Eckart on fundamental questions of electron theory, and I admire the trenchancy and breadth of his observations. But Houston also proves himself admirably. He has taken up specific questions proceeding from my note on metal electrons energetically and with great success."[78] William V. Houston (1900–1968) had actually wanted to pursue research on spin, which became an increasingly challenging matter for quantum physics. But Sommerfeld had advised against this, he later recalled, and instead had given him the galleys of his application of quantum statistics to electron gas.[79] Sommerfeld had Houston work up a wave mechanical explanation of the mean free path lengths of the electrons in metal.[80] Shortly after the publication of Schrödinger's papers, Carl Eckart (1902–1973) had shown the equal validity of wave mechanics and matrix mechanics and had likewise had fundamental quantum mechanical problems in mind before coming to Munich. He too let himself be persuaded to pursue research on the electron theory of metals.[81] "Both gentlemen, Houston and Eckart, have been personally very agreeable, and have proven of direct utility to me," he wrote Millikan gratefully half a year later.[82]

The spark leapt across to other institutes as well. Sommerfeld never tired of proselytizing for the electron theory of metals as a promising area of future research.[83] This theory also offered the prospect of elucidating long inexplicable solid-state properties. Sommerfeld had at first merely replaced classical statistics with the Fermi-Dirac quantum statistics and otherwise had treated electrons as a free gas. But it was clear that this could actually be only a temporary solution. If the electrons in the atom obey the laws of quantum mechanics, this should also be true of their motion between the atoms in a crystal lattice. This quantum mechanical extension of electron theory was among the topics with which one could make a name as "a

76 To Berliner, August 6, 1927. DMA, NL 89, 001. Sommerfeld, *Elektronentheorie der Metalle*, 1927.

77 Sommerfeld, *Elektronentheorie der Metalle 1, 2*, 1928.

78 To Millikan, November 28, 1927. DMA, NL 89, 025.

79 Houston, Interview by G. Phillips and W. J. King, March 3, 1964. AIP.

80 Houston, *Leitfähigkeit*, 1928.

81 Eckart, *Elektronentheorie der Metalle*, 1928.

82 To Millikan, March 26, 1928. DMA, NL 89, 025.

83 Eckert, *Propaganda*, 1987.

modern physicist" at the end of the 1920s. To be "modern" was to be conversant with the new quantum mechanics, and the physics of solid-state phenomena offered a cornucopia of problems on which a contender for an academic career as a theoretical physicist could demonstrate his quantum mechanical expertise.[84]

The first of these new physics centers grew up at Leipzig, where in 1927, in the persons of Heisenberg and Debye, two Sommerfeld students were appointed to professorships in theoretical and experimental physics. Shortly thereafter, two other Sommerfeld students, Pauli and Wentzel, were appointed to professorships of theoretical physics at the ETH and the University of Zürich, respectively. The Munich "nursery" had thereby sprouted branches at Leipzig, Zürich, Stuttgart (Ewald), and Hamburg (Lenz), and as happens with subsidiaries of an enterprise, there was a lively exchange of knowledge and personnel among the branches of the Sommerfeld school. "So, you'd like to steal assistants? And naturally only the best!" The directors of the branches communicated in this tone when they needed to fill positions.[85] The founders of quantum mechanical solid-state theory—Hans Bethe, Felix Bloch (1905–1983), Rudolf Peierls (1907–1995), and others—began their careers in one of these branches and were occasionally transferred from one to another of them. The same was true for advanced students and recent doctorates in theoretical physics, who began their academic careers at the branches of the Sommerfeld school with a grant from the Rockefeller Foundation or some other granting institution. "I think it would be a nice idea," Heisenberg wrote Pauli on one occasion, "to establish a sort of physicists' exchange between Zürich and Leipzig."[86] In congratulating Sommerfeld on his 60th birthday, he coupled his best wishes with the hope that in Munich, Sommerfeld would "for a long time yet [sponsor] a nursery for physical babies as for Pauli and me at that time."[87]

9.4 The Planck Succession

Even though Sommerfeld's institute represented the "nursery" for the network of new quantum schools, the most prestigious chair to which a theoretical physicist could aspire was not Sommerfeld's, but Max Planck's at the University of Berlin. With this chair, Planck had assumed the legacy of Gustav Kirchhoff, who in 1875, as the first full professor of theoretical physics in Germany, had given this discipline the status of an independent field. As permanent secretary of the mathematical physics class of the Prussian Academy of Sciences, Planck exercised a significant representative function in addition to his university teaching activity. When the Solvay Congress of 1927 was being prepared in Belgium, to which for the first time

84 Hoddeson/Baym/Eckert, *Development*, 1987.
85 To Heisenberg, November 15, 1927. DMA, NL 89, 002.
86 Heisenberg to Pauli, August 1, 1929. In WPWB I.
87 From Heisenberg, February 6, 1929. DMA, HS 1977-28/A,136.

since the war physicists from Germany were to be invited, great value was put on Planck's participation. The topic was quantum mechanics. Aside from Planck, only Born, Heisenberg, and Pauli were invited from Germany—not Sommerfeld, which annoyed some of those invited. Planck judged himself, in contrast to Sommerfeld, "no longer among those on the leading edge of the development of quantum theory and in the front rank of those qualified to participate in the Congress."[88] Born also felt himself "quite taken aback" over the fact that Sommerfeld had not been invited, and thought "that your name should be at the top of the list of Germans invited to the new quantum congress."[89] In the event, the Belgians had not been prepared to invite more than four Germans: Born, Heisenberg, and Pauli represented unquestionably the front rank of the German quantum theorists, and Planck was invited because he—not Sommerfeld—would be recognized as the preeminent representative of German science.[90]

Nevertheless, Sommerfeld was hardly second to his 10-year senior Planck when it came to upholding the reputation of German science abroad. When Planck retired in 1926 at the age of 68, it was thus scarcely to be wondered at that Sommerfeld was at once thought of as his successor. Planck's chair was to be entrusted only to someone who, like Planck, could act as scientific spokesman. Already in the initial appointment deliberations, Sommerfeld ranked as the leading candidate; others named were Born, Hans Thirring, and Schrödinger; Einstein and Laue were also briefly considered, but withdrew their names from the list. Einstein did not wish to trade his position at the Academy, which offered him free pursuit of his own research, for a professorship that would burden him with teaching duties. And appointing Laue, who held the second full professorship in theoretical physics at the University of Berlin, would merely have evoked an additional succession debate. Thus, the first list of proposed candidates comprised "Sommerfeld, Born, Schrödinger."[91] With respect to Sommerfeld, there was no doubt even in ensuing deliberations of the Appointment Commission that he should be ranked in the top spot on the list. For the second and third spots, however, "after careful consideration" it was determined "that Schrödinger's physical achievements possessed an inherently more profound originality and a greater creative force" than those of Born. So Schrödinger was placed second and Born third. Heisenberg, as a representative of the younger generation, also came under consideration. He would "at some future date surely [number] among the first rank of researchers," but he was not yet to be entrusted with the role of scientific spokesman incumbent on the Planck successor.[92]

88 Planck to Lorentz, June 13, 1926. AHQP.

89 From Born, June 15, 1926. DMA, NL 89, 006. Also in ASWB II, S. 255.

90 Heilbron, *Dilemmas*, 1986, pp. 107–108.

91 Appointment Commission to the Philosophical Faculty of the Friedrich-Wilhelm-University, Berlin, June 18, 1926. UAB.

92 Minutes of the faculty meeting of November 18, 1926, and report of the Faculty to the Prussian Ministry of Culture of December 4, 1926. UAB.

Sommerfeld was informed of the imminent offer of appointment before receiving official notice of it from the Berlin Ministry of Culture. "The entire faculty has, as you are no doubt aware, decisively named you in the top position," Nernst wrote him in advance of the appointment.[93] At the time Sommerfeld received the announcement of appointment from the Prussian Ministry of Culture, he was on a trip to the Balkans. His wife, though, knew how she was to answer. "You have surely written to Berlin, as agreed," Sommerfeld wrote from Ragusa. "I will myself write first to Planck, and then after an appropriate interval to Berlin, that I will not come before the week after Easter."[94] He wished to negotiate with the Prussian Ministry of Culture the terms under which he would accept the appointment. Planck knew quite well how difficult it would be to pry Sommerfeld loose from Munich. He promised Sommerfeld he would do all he could to sweeten the appointment and closed his letter with the plea to Sommerfeld's wife "to exert her influence in favor of Berlin."[95]

What to Planck and the Berlin physicists was a hope appeared to Sommerfeld's Munich colleagues as a threat. The two mathematicians Oskar Perron (1880–1975) and Constantin Carathéodory (1873–1950) went at once to the Bavarian Ministry of Culture and on behalf of the faculty urgently requested the university overseer in the imminent negotiations to retain Sommerfeld in Munich. "I tried to make clear to him," Carathéodory reported to Sommerfeld later "what it would mean for the University if we were after all to lose you."[96] The President of the Bavarian Academy of Sciences wrote to Sommerfeld: "As pleased as I am that you have been accorded this recognition, I tremble in equal measure for Munich."[97] The Rector of the University, Karl Vossler, who was also a personal friend of Sommerfeld's, declared himself "prepared to take any step that would lead you to a favorable turn towards Munich. I am also convinced that the Senate would support me in whatever action would be appropriate to keep you in Munich." He invoked the cordial relationship between their families in averring that his wife and children would "suffer an irreparable personal loss by your departure. The farewell would be very heavy for us, and we have no intention of lightening the farewell for you and yours."[98]

Sommerfeld was probably determined from the outset to remain at Munich, but the Berlin offer presented him the opportunity of improving his position there. To achieve this, he had to negotiate with the Prussian Ministry of Culture an offer better than his current position, so that in subsequent negotiation with the Bavarian Ministry of Culture, he could in turn improve his Munich position.

93 From Nernst, March 19, 1927. DMA, NL 89, 019, folder 5,9.
94 From Windelband, March 24, 1927. DMA, NL 89, 019, folder 5,9. Also in ASWB II; to Johanna, March 30, 1927.
95 From Planck, April 7, 1927. DMA, NL 89, 019, folder 5,9. Also in ASWB II.
96 From Carathéodory, March 28, 1927. DMA, NL 89, 019, folder 5,9. Also in ASWB II.
97 From Gruber, March 28, 1927. DMA, NL 89, 019, folder 5,9.
98 From Vossler, March 30, 1927. DMA, NL 89, 019, folder 5,9. Also in ASWB I.

These negotiations extended over 2 months. Berlin was prepared to offer Sommerfeld a higher salary than Munich and to add a supplement for the rental of an apartment or the purchase of a house. In addition, he was given assurances of improved conditions at the institute, already considered insufficient by Planck.[99] At the negotiations over his remaining in Munich, Sommerfeld demanded above all the establishment of an associate professorship in theoretical physics. He was aware that, under prevailing financial circumstances, he could not count on fulfillment of this demand, but he wished on behalf of the University to put on record with the Ministry that such a professorship was to be instituted as soon as sufficient resources could be made available in Bavaria.[100] When he was assured of this, together with other improvements in his Munich position, he declined the Berlin offer. He gave as the principal reason for his decision "the much simpler working and living conditions, and the much better facilities of the Institute."[101] The Bavarian Ministry of Culture had "fulfilled entirely" his wishes, he wrote to Berlin, "not only with respect to my personal circumstances, but also with respect to the organization of the Institute and its pedagogy."[102]

Thereupon, the Planck succession was offered to the second candidate on the list, Schrödinger, who accepted after protracted negotiations over "virtually" the same conditions stipulated by Sommerfeld. "Differences: not quite the top salary, but 1,700 M less per annum," Schrödinger reported to his "advisor" Sommerfeld, who had fully briefed him beforehand on the inner workings at Berlin.[103]

9.5 "Not Sommerfeld, but Schüpfer"

At issue for Sommerfeld in his decision to remain at Munich was not just the increase in his salary and better equipment for his institute. Primarily, he did not wish to give up the successful pedagogical enterprise he had built up and been so personally involved in over the past two decades. "It seems to me doubtful that interaction with students in big and restless Berlin could be organized as intimately as at Munich," he wrote in an article for the *Süddeutsche Sonntagspost*. To be sure, it had not been easy for him, as an "old Prussian," to decline an appointment to Berlin, to "the city in which Helmholtz and Kirchhoff were active, where Planck and Einstein live, the center of German intellect and work." But he cherished the more informal Bavarian lifestyle, and the nearby mountains that offered him and

99 File notation, Prussian Ministry of Culture, April 29,1927. DMA, NL 89, 019. folder 5,9.

100 To the Dean, Philosophical Faculty, 2. Sektion, May 16. 1927. DMA, NL 89, 004. From the Bavarian Ministry of Culture, May 25, 1927 and June 20, 1927. DMA, NL 89, 019. folder 5,9.

101 To C. H. Becker, draft of an undated letter [between. May 31and June 11, 1927]. DMA, NL 89, 019. folder 5,9.

102 To the Prussian Ministry of Culture, June 11, 1927. DMA NL 89, 004. Also in ASWB II.

103 FromSchrödinger, July 14, 1927. DMA, NL 89, 019, folder 5,9.

his students opportunities for skiing, and lent his pedagogical enterprise a very personal character. Berlin "uses up its people quickly, whereas Munich, situated at the foot of the mountains, allows even the elderly to find refreshment and renewal."[104]

He already had very concrete plans for the associate professorship promised him in return for his declining the Planck chair. He had the outcome of his negotiations over remaining at Munich be given him in writing once more,[105] and wrote to Heisenberg, whom he had in mind for the associate professorship, describing his vision for the future of his "nursery." As a first step, he suggested "Please save yourself for the Munich associate professorship. You will thereby be entitled after a number of years to become my successor in the full professorship. Of course these are just my intentions, and are proposed without being binding on the faculty. But I see no obstacles in the way of their being carried out."[106]

Sommerfeld may have regarded the great esteem shown him during negotiations over the Planck succession as the expression of highest recognition on the part of Munich professors generally. In fact, many of his colleagues espoused radically different views. This became obvious when the next election for Rector was at hand in July 1927. Sommerfeld had put himself up for election as Vossler's successor, a move that stirred displeasure among anti-Semitic and right-wing circles. "The dissatisfaction with the current democratic-pacifistic and Jew-loving Rector, Dr. Vossler, is general," one might read in the *Völkischer Beobachter*, which sought at all costs to prevent "the Jew and Professor, Dr. Sommerfeld" from succeeding the hated Vossler.[107] That Sommerfeld, as the National Socialists later would concede, had no Jewish forebears down to his great-grandparents did not dampen the press campaign. He, like Vossler, was counted as Jew loving and liberal. Vossler had attracted the animosity of the right-wing circles when at the annual celebration of the establishment of the Reich, "standing on the grounds of the Constitution," as the *Vossische Zeitung* stressed, he had arranged for the black-red-and-gold national flag of the Republic to be raised alongside the black-white-red, which stood for the Kaiser's Reich. Unlike the *Völkische Beobachter*, the *Vossische Zeitung* maintained it would be "greatly" in the interests of the University of Munich "that the liberal era inaugurated by Vossler continue to prove its viability." To the liberal press, Sommerfeld was the guarantor of this tradition. The opposition candidate was a forestry expert by the name of Vinzenz Schüpfer, "whose scientific importance cannot in the least be compared to that of the famous physicist, Sommerfeld," as the *Vossische Zeitung* stressed.[108]

Sommerfeld lost with 50 votes to his opponent's 68. "Not Sommerfeld, but Schüpfer" ran the headline in the *Berliner Tageblatt*; the "scientifically insignificant,

104 Sommerfeld, *Berlin*, 1927.

105 From Goldenberger (Bavarian Minister of Culture), June 20, 1927. DMA, NL 89, 019, folder 5,9.

106 To Heisenberg, June 17, 1927. DMA, NL 89, 019, folder 5,9.

107 *Völkischer Beobachter*, July 8, 1927.

108 *Vossische Zeitung*, July 12, 1927.

Fig. 24: Procession of the Munich professors in 1926 on the occasion of the centennial celebration of the Ludwig Maximilian University (in 1826, the University was moved from Landshut to Munich). In 1927, Sommerfeld stood for election as Rector, but was defeated by "the right-wing sentiments of the professorial majority" (*Frankfurter Zeitung*) (Courtesy: Deutsches Museum, Munich, Archive).

but 'dependably nationalistic' forester Schüpfer" had defeated the "world-renowned physicist Arnold Sommerfeld."[109] "A victory for the party politicians," commented the *Frankfurter Zeitung* on the outcome of the election for Rector in Munich. Therein "the right-wing sentiments of the professorial majority [had] once more been documented." The election of Schüpfer was to be chalked up to the "Professors' table of German nationalists and like-minded adherents of the Bavarian People's Party."[110]

With Sommerfeld's defeat, the liberal era at the University of Munich represented by Vossler came to an end, even before it had properly begun. In an "address to Vossler," Sommerfeld paid tribute to his "extraordinary official service" and on behalf of his fellow signatories expressed the hope that it "would leave behind a lasting legacy in the history of the University." However, the notation "not sent" at the bottom of the hand-written draft implies that due to a lack of signatories, this declaration was stillborn.[111] Ultimately, Sommerfeld had to have been beset after all by regrets over his refusal of the Planck chair. "Now and then I am sorry not to have come to Berlin," he wrote to Einstein; "my dear Munich colleagues have certainly greatly annoyed me in the interim."[112]

109 *Berliner Tageblatt*, July 22, 1927.
110 *Frankfurter Zeitung*, July 21, 1927.
111 To Vossler, undated [ca. July, 1927].DMA, NL 89, 019, folder 5,9.
112 To Einstein, November 1, 1927. AEA, Einstein. Also in ASWB II.

9.6 The Volta Congress

Annoyance over politics at his university, however, was soon eclipsed by his enthusiasm for physics. "On Quantum Mechanics" and "Selected Questions in Wave Mechanics" were the titles of Sommerfeld's special lecture courses in the winter semester of 1927/1928 and the summer semester of 1928. Aside from the first efforts at a quantum mechanical solid-state theory, to which he himself had given the impetus with his electron theory of metals, quantum mechanics accounted in many other areas for a sense of breakthrough among theoretical physicists. In 1926, Born had for the first time applied quantum mechanics to collision processes and in this connection had given a new interpretation to Schrödinger's wave mechanics. It was not the electrons described by the Schrödinger equations that were spatially "smeared" like a wave, but the probability of finding them at this or that location.[113] In December 1926, Dirac and Jordan lifted quantum mechanics with a "transformation theory" to new abstract heights.[114] In March 1927, Heisenberg added fuel to the fire with his "uncertainty principle."[115] In less than 2 years, quantum mechanics had, as no theory heretofore, turned physics inside out with respect to both foundations and applications.

In September 1927, on the occasion of the hundredth anniversary of the death of Alessandro Volta (1745–1827), an international congress of physicists took place in Como, at which differing conceptions of quantum mechanics were exchanged for the first time in a larger context. At first, Sommerfeld suspected that politically motivated propaganda was behind the event. "I have been invited to a small conference of big shots in Como in 1927 in observation of the Volta centennial," he wrote James Franck. "I have serious reservations about attending because I assume the Italians will not forego the opportunity of making it political and trotting out Mussolini."[116] As in the microcosm of the Munich Rector's election, in the larger picture, too, science was not isolated from politics. An international conference in Italy, where the fascists had just taken power, seemed to Sommerfeld a chess move by Mussolini to make his politics internationally presentable. Before he accepted the invitation, therefore, he wanted to know whether he was alone in his reservations. "Have you also been invited, and what do you plan to do?" he inquired of Laue. "Is Planck going? It would be a good thing if we could agree on a common course of action."[117] Franck, Laue, and Planck, who like Sommerfeld had received invitations to Como, advised in favor of attending. "If the Italians do something tactless, it only reflects back on them," Laue replied.[118]

113 Beller, *Quantum Dialogue*, 1999, pp. 39–49.
114 Mehra/Rechenberg, *Development*, 2000, Part 1, Chap. I.5; Janssen/Duncan, *Transformations*, 2009.
115 Rechenberg, *Werner Heisenberg*, 2010, Chap. 8.
116 To Franck, July 20, 1926. DMA, NL 89, 001. Also in ASWB II.
117 To Laue, July 22, 1926. DMA, NL 89, 002.
118 From Laue, July 27, 1926. DMA, NL 89, 010.

Accordingly, Sommerfeld put his reservations aside and accepted the invitation. The Volta Congress brought the leading physicists from the recently adversarial states together for the first time since World War I. What Sommerfeld had thought would be "a small conference of big shots" turned out to be a congress of more than 60 participants from over a dozen countries (Denmark, Germany, England, France, India, Italy, Canada, Holland, Austria, Switzerland, Spain, the Soviet Union, the USA)—among them prominent physicists such as Niels Bohr, Arthur H. Compton, Hendrik A. Lorentz, Ernest Rutherford, and Robert A. Millikan. The lectures delivered at this congress covered physics in its broadest scope. At Como, Sommerfeld presented his current results in electron theory of metals.[119] A number of other lectures also dealt with solid-state physics. But what made the congress an unforgettable event for its participants were the discussions of the interpretations of quantum mechanics. Bohr's concept of the "principle of complementarity" was the subject of vigorous debate that continued well beyond the conclusion of the congress.[120]

As with his lecture tour in England in March 1926, Sommerfeld used the Volta Congress as an opportunity to socialize with colleagues over and above purely professional substance. He befriended the Russian physicist Jakow Frenkel (1894–1952) and after the congress traveled with him through southern Italy. Frenkel shared Sommerfeld's interest in the electron theory of metals and later made important contributions to solid-state physics. "My traveling companion here and for Sicily is Frenkel, a physicist from St. Petersburg," Sommerfeld explained to his wife.[121] They must have embarked on their joint trip through southern Italy quite spontaneously, for at one point they found themselves temporarily in financial straits such that Sommerfeld was compelled to ask his "illustrissime amice," Tullio Levi-Cività (1873–1941) in Padua, to help them out of their difficulty by sending money.[122] "October 1, and I am still in Naples! How is it going to work out for me to get home?!" Sommerfeld reported the impromptu extension of his trip to his wife. "First Pompeii, and along with it, Vesuvius, on horseback (!)," he enthused concerning his latest travel experiences. "I may go broke again today. But it was a great experience. The trip down into the crater (which entailed a supplement of 15 Lire), fabulous: every other minute a thunderous eruption of water and sulfur vapor. A terrific fumarole [. . .] Next morning, to Capri, Blue Grotto in a small bark, instead of the steam-boat company with the herd."[123]

119 Sommerfeld, *Elektronentheorie der Metalle und des Voltaeffektes*, 1927.

120 Beller, *Quantum Dialogue*, 1999, Chap. 6. See also http://www.science20.com/don_howard/revisiting_einsteinbohr_dialogue (31 January 2013).

121 To Johanna, September 29, 1927.

122 To Levi-Cività, September 28, 1927. Rom, BANL, Levi-Cività.

123 To Johanna, October 1, 1927.

9.7 A *Wave Mechanics Supplement*

Two months after the Volta Congress, in Brussels at the fifth Solvay Congress, there were lively discussions between Bohr and Einstein on how quantum mechanics, which could not be understood in the terms of classical physics, should be interpreted. But it was not just questions of interpretation that made the Volta and Solvay Congresses extraordinary events in the history of physics of that time. From solid state, nuclear, and astrophysics to chemistry, new applications of quantum mechanics opened up, giving the field ever more the appearance of a huge construction site for theoretical physicists, but one lacking a comprehensive plan underlying the whole and with no indication what new buildings were going up. In this situation it was no wonder that soon a demand for overarching surveys arose. The interest was "so great that a report concerning the current state of research would perhaps be in order," Rudolf Seeliger wrote in November 1927 to Sommerfeld. Seeliger had completed his doctorate under Sommerfeld's direction in 1910 and now taught theoretical physics at the University of Greifswald. As coeditor of the *Physikalische Zeitschrift*, Seeliger was familiar with the current publication focus of his colleagues and the scientific publishing houses. He had been requested from many different directions "to present a comprehensive report about wave mechanics," so he passed these interests along to his former teacher. In addition, a colleague planning to write a book about X-ray spectra had recently written him to inquire whether "in light of the rapid developments" Sommerfeld's exposition of this subject in the latest edition of *Atomic Structure and Spectral Lines* "was still valid and supported by you, or should be revised."[124]

Three years after the appearance of the fourth edition of *Atomic Structure and Spectral Lines*, the idea of a new edition must often have occurred to Sommerfeld. Even if quantum mechanics in his opinion was not a revolution, but only a further step in the evolution of atomic theory, this step was nonetheless so important that it had to be properly presented in a new edition. But since Sommerfeld had kept the previous edition "entirely free of all model-related preconceptions" (at least insofar as its most important part, the complex structure of spectra, was concerned), it was initially less a matter of correcting the pre-quantum-mechanical presentation, than of amending fundamental theoretical principles. The pre-quantum-mechanical conceptual system was perfectly adequate to embrace the laws of the spectra themselves, as he had described them in the first four editions. Although according to the Pauli Exclusion Principle and the introduction of spin, that which before had been intended for the atomic core with the inner quantum number had to be transferred to each individual electron. But this changed little in the empirically established laws formulated in terms of the concepts of the old Bohr-Sommerfeld atomic model. For example, in a 1928 book on spectra, the astrophysicist Walter Grotrian expressed the view that "even with the current state of theory, there need be no

124 From Seeliger, November 20, 1927.D MA, NL 89, 013.

reservations" against these conceptions that have actually been superseded by quantum mechanics, "so long as one maintains clarity that the Bohr electron orbits are to be regarded merely as illustrative aids to conceptualization, and not as reality."[125]

Given this background, Sommerfeld abstained from a revised presentation of the largely empirically based spectral laws he had described so comprehensively in the fourth edition and concentrated fully on quantum mechanics. There was, in other words, to be no fifth edition of *Atomic Structure and Spectral Lines* (for now, at least), rather just a "wave mechanical supplement" to the fourth edition. He quite intentionally employed the term "wave mechanical," and not "quantum mechanical," increasingly the general usage, "because in practical application, Schrödinger's methods are clearly superior to the specifically 'quantum mechanical' methods." In addition, he wished "as much as possible to restrict [himself] to concrete questions." He would discuss the "principal questions of uncertainty and observability" only peripherally.[126]

Like the various editions of *Atomic Structure and Spectral Lines*, the *Wave Mechanical Supplement* was not the product of a lone act of writing in an ivory tower, but rather mirrored the research enterprise of the Sommerfeld school. Anyone who had taken a doctorate under Sommerfeld or worked as an assistant or lecturer at his institute between 1925 and 1928 could find his work in one or another subchapter or could bask in the knowledge of having sown critical seeds of various passages. In this way, for example, the "crystal interferences of electron waves," which had only shortly before been discovered experimentally and in 1928 formed the subject of Bethe's doctoral dissertation, became the contents of one chapter.[127] Unsöld, the "latest Wunderkind" of the Sommerfeld school, found himself immortalized in a chapter on the "Spherical Symmetry of the S-Terms."[128] Sommerfeld registered special thanks to his assistant, Karl Bechert, without whose "devoted assistance" he could hardly have produced this work.[129]

Unlike when 10 years before Sommerfeld had conceived the outline of *Atomic Structure and Spectral Lines* in the course of giving popular lectures on atomic models, the *Wave Mechanical Supplement* did not emerge from an effort to popularize wave mechanics. Mathematical expositions of spherical harmonics and Bessel functions, complex integration, and other such matters more likely to scare off theoretically less knowledgeable readers were not relegated to "Addenda and Supplements," but were on

125 Grotrian, *Darstellung*, 1928, p. VII.

126 Sommerfeld, *Ergänzungsband*, 1929, Vorwort.

127 Vote on the Dissertation of Bethe to the Philosophical Faculty, 2. Section, July 24, 1928. UAM, OC Np1928. Sommerfeld, *Ergänzungsband*, 1929, pp. 241–250.

128 To Finlay Freundlich, January 14, 1927. DMA, NL 89, 001. Vote on the Dissertation of Unsöld to the Philosophical Faculty, 2. Sektion, December 11, 1926. UAM, OCI52p. Sommerfeld, *Ergänzungsband*, 1929, pp. 101–105.

129 Sommerfeld, *Ergänzungsband*, 1929, Foreword.

the contrary an integral part of the presentation. Although Sommerfeld was primarily addressing theoretical physicists, he also wished to disseminate the theory beyond his own discipline. The first opportunity to do so came in May 1928, when he was asked to lecture to the German Bunsen Society for Applied Physical Chemistry, which met that year in Munich. The Bunsen Society was an association steeped in tradition, which prided itself on "exchanging the gold of new foundational ideas and results in the area of physics into currency for the use of chemical science and chemical engineering," as their president emphasized in his welcoming speech.[130] Sommerfeld did not wish to represent quantum mechanics to the Bunsen Society as a complete overturning of the atomic concepts with which chemists had just become familiar. "I will of course stress," he wrote to Friedrich Hund (1896–1997), who also had been invited to give a talk, "that much from the original models remains intact, namely quantum numbers, spectra, the periodic table, the Pauli Principle." Hund also might wish to dilute his scientific wine with "a little popular water," Sommerfeld counseled; he suggested a quantum mechanical interpretation of the chemist's formulas of valence bond.[131]

In his own lecture, Sommerfeld drew quite a clear picture of the actually rather unclear new atomic theory. Just as wave optics replaces geometrical optics when one transitions from coarse optical instruments to finer ones such as the microscope, it is necessary to supplant the "ordinary macro-mechanics" of our everyday experience with wave mechanics when dealing with things of atomic dimensions. But wave mechanics is also a statistical theory. It describes the behavior of "swarms of electrons" with laws like those we recognize for waves. These are not waves in space, however, but rather an "abstract something" that describes the probability of an electron's location. Sommerfeld explained this "something" with the example of a hydrogen atom, in which the square of the Schrödinger wave function shows the density of the charge cloud of the electron around the atomic core. "What does 'density of the charge cloud' mean?" he went on to ask, to preclude the false image of an electron with spatial extension. "We believe that the electron is a virtually point-form structure, and that its entire charge is concentrated in the smallest space." The density of the charge cloud indicates the probability that a given point-form electron is to be met with here or elsewhere. The old atomic model with its planetary orbits displayed a "disc symmetry"; the new theory offered a far more plausible explanation of how the electrical charge filled the space in the atom. The formation of molecules and the forces among ions in a crystal could also be explained satisfactorily. "In general, I have the impression," he concluded his lecture, "that the new theory addresses the needs of chemists in an especially felicitous, and actually a better way than the earlier conception of individual electron orbits."[132]

130 *Zeitschrift für Elektrochemie und angewandte physikalische Chemie* 34 (1928), pp. 425–426.
131 ToHund, February 29, 1928. DMA, NL 89, 002.
132 Sommerfeld, *Bedeutung*, 1928.

10 Cultural Ambassador

"When I travel abroad I feel I am not merely a private individual and globetrotter, but an ambassador of German culture in the realm of science." Thus Sommerfeld began his lecture on December 8, 1928, in Tokyo.[1] Tokyo was one of many stops on an 8-month world tour on which Sommerfeld carried out this self-imposed cultural mission. The idea of a world tour occurred to him after the Volta Congress in September 1927 when Millikan proposed a split guest professorship that would take him to the University of Chicago and the California Institute of Technology for the winter semester of 1928/1929.[2] Though that plan fell through, Millikan wanted at least to bring Sommerfeld to Cal Tech: "pasadena [sic] wants you definitely winter quarter twenty-nine," he telegraphed to Munich.[3] After brief deliberation, Sommerfeld accepted the invitation and announced that this time he would travel to America from the east, across the Pacific.[4] He may have been giving himself an unusual present on the occasion of his 60th birthday, which he would celebrate somewhere in Japan on December 5, 1928. Or, was he perhaps dodging festivities threatening him at home on this day? Bethe believed that "A major motivation for this trip was that he did not want to be in Munich on his 60th birthday."[5]

In any case, Sommerfeld's travel plans quickly made the rounds, assuring that invitations to lecture flooded in from many countries. The first invitations came by telegram from India.[6] Sommerfeld asked Chandrasekhara Venkata Raman (1888–1970), who invited him to lecture at the University of Calcutta, and Meghnad Saha (1893–1956), whom he had met at the Volta Congress in Como, to arrange a 4-week lecture and sightseeing tour through India for him.[7] In Japan, his former student Otto Laporte, now a professor at the University of Michigan in Ann Arbor, just then on a guest professorship at the University of Kyoto, provided the first contacts. Toshio Takamine (1885–1959) wrote Sommerfeld in March, 1928 that he had just from Laporte received "the glad tiding . . . that in the coming winter, there may be a chance for us to have the pleasure + honour of being visited by you, + if possible, to hear your lectures a few times in Japan."[8] To leave no doubt that he regarded his guest lectures as a cultural mission, Sommerfeld consulted the Cultural Division

1 Sommerfeld, *Entwicklung*, 1929b.
2 From Millikan, October 22, 1927. DMA, HS 1977-28/A,232.
3 From Millikan, November 25, 1927. DMA, HS 1977-28/A,232.
4 From Millikan, February 6, 1928. DMA, NL 89, 011. Also in ASWB II.
5 Preface to Eckert/Pricha/Schubert/Torkar, *Geheimrat*, 1984, p. 9.
6 From Raman, February 11, 1928. DMA, NL 89, 024, folder India.
7 To Raman, February 28, 1928. DMA, NL 89, 024, folder India.
8 From Takamine, March 19, 1928. DMA, NL 89, 019, folder 4,3. Also in ASWB II.

M. Eckert, *Arnold Sommerfeld: Science, Life and Turbulent Times 1868-1951*,
DOI 10.1007/978-1-4614-7461-6_10, © Springer Science+Business Media New York 2013

of the Foreign Office regarding his travel plans. He was advised to apply for a subsidy from the Emergency Organization of German Science (*Notgemeinschaft der Deutschen Wissenschaft*).[9] In addition, he would have to arrange for his replacement at the University of Munich for the winter semester of 1928/1929. His lecturer Heinrich Ott was to give his main lecture course. Karl Bechert, his second assistant, would take over the exercise classes. Advanced lectures would be given by Laporte, who wished to spend several months in Germany on his return from Japan in the summer of 1928 to visit his parents in Munich, before returning to the USA in January 1929.[10]

10.1 German Science on the International Stage

Shortly before setting forth on his great journey around the world, Sommerfeld was once more put in mind of the peculiar situation of German science in the decade following World War I: His Spanish colleague Blas Cabrera (1878–1945) informed him of plans to open the *Conseil International de Recherches* (International Research Council) to membership by German scientists. The *Conseil* had been established after World War I as a replacement for the International Association of Academies, which was dominated by the Central Powers.[11] To Sommerfeld, however, this international research council was a relic of the boycott against German science following World War I. Although Germany should certainly not remain excluded from this international scientific organization, the manner in which the exclusion of German science was lifted only occasioned fresh embitterment. Fritz Haber had for months tried vainly to negotiate a solution acceptable to both sides. Ultimately, only the discriminatory paragraph of exclusion was stricken, without ceding to Germany's wish for acknowledgement of its role as one of the leading scientific nations. The treatment accorded Germany was no different from countries such as Siam, Haber noted critically, while resigning himself to the fact that the time was not yet ripe for an equitable international organization of science.[12]

Sommerfeld shared his colleagues' embitterment. His opinion of the international research council was "not exactly flattering," he replied to Cabrera, for "the *Conseil*, born of the political hatred, costs a great deal of money, and has, so far as I am aware, no accomplishments to show for it." Sommerfeld thought the whole organization of this research council was misconceived. "Morocco, Egypt, Tunisia with independent representations! This is called democracy, but in reality it serves

9 To the Notgemeinschaft, May 1, 1928. DMA, NL 89, 020, folder 6,6.

10 To the University of Munich, March 2, 1928. UAM, E-II-N.

11 Von Cabrera, 7. August 1928. DMA, NL 89, 006. Also in ASWB II.

12 Szöllösi-Janze, *Fritz Haber*, 1998, pp. 588–590; generally on this topic, see Forman, *Internationalism*, 1973; Schröder-Gudehus, *Wissenschaftsbeziehungen*, 1990.

no other purpose than to support mad French claims." He thought the best course would be revival of the International Association of Academies, although it was clear to him that this was not to be realized. "It may be that for the sake of international courtesy, Germany will feel obligated to join the *Conseil*," he conceded, but on precondition that "the inane insults to German science" cease. "I write this in a great hurry shortly before my departure, and in my own name only," he wrote, to avert the possibility Cabrera would construe his statement as an official German position on this much discussed and hotly debated situation among diplomatic circles of the Foreign Office and the negotiators from the various academies. But he also made it clear that he was not alone in this opinion. "I believe, however, that many of my colleagues feel similarly. That the current status must be altered is clear to everyone. I myself hope that this alteration may come about in a spirit of friendship and mutual trust, but—in light of its whole pre-history— the *Conseil de Recherches*—does not seem to me to offer a suitable means to this end."[13]

The *Conseil International de Recherches* was converted into the International Council of Scientific Unions in 1931 and gradually shed its character of an Entente organization in opposition to the Axis powers.[14] But in 1928, the boycott imposed on German science at the instigation of the *Conseil International de Recherches* following World War I remained in place in general consciousness. To be reminded of this just a few days before his world tour gave further impetus for Sommerfeld to make this trip a cultural mission with the aim of restoring the reputation of German science.

10.2 Impressions of India

Outfitted with clothing for the tropics and accompanied by his assistant Bechert, Sommerfeld began his world tour on August 21, 1928, in Genoa aboard a steamer bound for the Suez Canal. Atypically, on this journey he kept a journal. "August 23, early, the Aeolian Islands, Stromboli, very picturesque," he recorded as the ship passed Sicily. "27th, the Canal, fabulously interesting," he wrote on entering the Suez Canal. He was fascinated by the waves generated by the ship which, once free of the ship, moved on as though on their own. He was reminded of the story of the "horse at the Scottish canal by Reynolds or Kelvin," he wrote under the heading "waves in the Suez Canal" in the "scientific portion" of his journal. He was referring to a particular wave phenomenon that had fascinated British physicists of the nineteenth century, and that he intended to analyze theoretically in the future.[15] "Red

13 To Cabrera, August 11, 1928. DMA, NL 89, 001. Also in ASWB II.
14 Greenaway, *Science* International, 1996.
15 Journal of the world tour.

Fig. 25: Karl Bechert and Sommerfeld aboard ship en route to India (Courtesy: Deutsches Museum, Munich, Archive).

Sea, weathered well; will not dock at Aden; heat now bearable; humidity reduced," he telegraphed several days later to his wife.[16]

Bechert accompanied him as far as India and made final corrections to the *Wave Mechanical Supplement*, galleys of which had been taken along on the great journey.

The first stage led through the Gulf of Aden and the Arabian Sea to Ceylon (now Sri Lanka). It was so sultry on the Red Sea that the passengers could not endure their cabins through the night. "Silent, sleep-walking figures in bath-robes on deck," Sommerfeld wrote describing the scene. "I lay several hours on a deck-chair, managing quite well, naturally completely soaked with sweat. Amazing one doesn't come down with rheumatism."[17] This stretch of the trip was not inspiring of pleasant memories. The Red Sea seemed to him a "God forsaken corner" that cost one of the ship's cooks his life. "Heat stroke at 40° [C.] below-decks. Bechert sat with him last night because the medical aide was himself totally exhausted. Burial at sea

16 To Johanna, September 1, 1928.
17 To Johanna, September 5, 1928.

with chorale, speech by a young missionary, and prayers by the Captain under the German flag—very moving."[18]

Once disembarked at Colombo, they journeyed through northern Ceylon. On September 9, they embarked on a steamer for the short crossing to the Indian mainland, where they continued their journey by train. "During the journey, Bechert has calculated the essentials of the Zeeman Effect. The countryside is well built-up and irrigated," Sommerfeld noted about the trip through southern India.[19] The goal of this stage was Madras (now, Chennai), where he was to lecture at the university. He lodged "in the fine English house of the Principal of the College," he wrote home. Although he was "eaten up at the moment by mosquitos," he felt both "personally and professionally" very well.[20] Bechert added to the report of this first stop on their cultural mission with impressions common to all travelers in southern India: "What we have seen: enormous temple complexes, great dark halls, excessively decorated towers, priests, monks, dark- and light-colored, black-haired people, great poverty, friendliness and hospitality, palms, palms, red sand and blue-green fields, brightly colored birds, blue and gold-brown mountains, bananas, rice, deep dark blue sea, and flowering gardens." The professor is doing well, he reassured Sommerfeld's wife, though up to now, with the heat, they had not been terribly diligent. "When just lazing about is exhausting in this great heat, it's surely impossible to work, don't you agree?"[21] Two days later, Bechert began his return trip to Munich to fill in for Sommerfeld in one portion of the course-work for the coming winter semester.

From Madras, Sommerfeld traveled on by train into the interior of the country to Bangalore, the capital of Mysore State (now Karnataka). Here, his cultural mission included a lecture to the South Indian Science Association on German universities and students, among other topics. The Maharaja's representative invited him to a tea party and chatted with him about Goethe. His host was an English physicist with whom he immediately felt at home. "In the evening, I played with my host—an Englishman—some Beethoven violin sonatas, and sang Brahms's 'Feldeinsamkeit' with his wife. All with windows wide-open and lively participation of mosquitos." Thus he described these manifestations of German culture in far-off India. The next day, he awoke with a fever. "I felt pretty awful, as hot as on the Red Sea, and asked for the doctor, an English military physician. He admitted me to his hospital this morning, where I lie in a pleasant pavilion, open on 4 sides, and am given all sorts of medicines to swallow."[22]

The fever came and went repeatedly, so that he remained in the hospital for 10 days. Malaria was suspected but not confirmed. "Pretty weak, and in need of sleep," Sommerfeld wrote in his journal when finally he was able to return to the home of

18 Ibid.
19 Journal of the world tour.
20 To Johanna, September 12, 1928.
21 Bechert to Johanna Sommerfeld, September 13, 1928.
22 To Johanna, September 18, 1928.

his host.[23] He had been touchingly cared for, he reassured his wife in his next letter home. Nonetheless, the lost time was annoying. "I'll have to cut my whole India program short." In addition, there was now anxiety about developments in the Physics Department at Munich, for he had learned from letters from home that Willy Wien had died after a gallstone operation. Sommerfeld feared that Johannes Stark, with whom he had long been at odds, would be appointed Wien's successor. "Call Schmauss and tell him that illness has delayed my letter, but that I will send it before departing here," he wrote his wife. He asked her to convey to the Dean of the faculty, August Schmauß (1877–1954), his distress that decisions might be reached in Munich before receiving his recommendations.[24] But Heinrich Wieland (1877–1957), his colleague in chemistry at Munich, assured him that they would await his opinion. The names of James Franck, Walter Gerlach, Gustav Hertz (1887–1975), and Robert Wichard Pohl (1884–1976) had been placed on an initial, provisional appointment list. "Differing opinions regarding the ranking of the top two candidates could be discussed per telegram." Stark was not being considered as a candidate by a single member of the faculty. "But we will bear in mind the danger that attaches to this name."[25]

Because of his stay in the hospital, Sommerfeld set out only after a 2-week's delay for the next stop along the way of his India trip. To arm himself for the long journey and the exertions attending it, he had "now also taken a boy," he wrote his wife, who was worried about his welfare. "He is unbelievably attentive and proper, knows exactly where every article of my clothing is, sews on buttons for me, steers me to the dining car, and waits in my compartment until I'm there again." Because of this, the long journey became for him the "pinnacle of comfort." It was very hot, to be sure, but bearable "if one is motionless." His "boy" was, incidentally, "a married man of 30-something." He was paid "about 40 M for 4 weeks. Of course I pay for his III class ticket, but nothing for his board."[26]

In Calcutta, Sommerfeld was received like a statesman. Raman, who had invited him for 3 weeks of guest lectures, greeted him by placing a floral-chain around his neck. The German Vice Consul also made an appearance at the railway station on Sommerfeld's arrival. Sommerfeld was put up "extremely comfortably" at the German Consulate. "A huge hibiscus tree is blooming in my bed-room. At night, large glow-worms come flying in. Continuous medium-hot greenhouse air," he described his new surroundings to his wife. "This afternoon another reception at the Residency College. In the evening, dinner at the German Embassy for 8 guests." The University of Calcutta bestowed an honorary doctorate on him, and three Indian scientific organizations, the Mathematical Society Calcutta, the Indian Association for the Cultivation of Sciences, and the Indian Academy of Science, inducted him as an honorary member.[27]

23 Journal of the world tour.
24 To Johanna, September 27, 1928.
25 From Wieland, September 10, 1928. DMA, NL 89, 019, folder 5,10. Also in ASWB II.
26 To Johanna, October 3, 1928.
27 To Johanna, October 10, 1928.

Fig. 26: In Calcutta, Sommerfeld was the guest of the discoverers of the "Raman Effect," K. S. Krishnan (*left*) and C. V. Raman (*right*) (Courtesy: Deutsches Museum, Munich, Archive).

As a physicist, too, Sommerfeld enjoyed his stay in Calcutta. "My book—the English edition that is—is known in the remotest corners of the country," he wrote, delighted over the familiarity of the Indian physicists with his work.[28] Over and above this, in Calcutta, he witnessed history-making experiments. "At the Institute, saw scattering, blue-green, in a block of ice," he inscribed in his journal following a visit to Raman's laboratory. The "Raman Effect," as it was soon to be named, denotes the scattering of light onto atoms and molecules, whereby the incident light is scattered back at a lower frequency specific to the scattering material. It had been discovered only a few months earlier. "Promise indirectly to propose Raman for the Nobel Prize," Sommerfeld noted to himself.[29] He told a reporter from the Indian newspaper *The Statesman* he felt privileged to be present at these latest experiments of Raman's, and he hoped to be able to make some contribution to the theoretical elucidation of this scattering effect. He characterized the effect as one of the most interesting discoveries of recent years.[30] Two years later, Raman was awarded the Nobel Prize for this work.[31]

From Calcutta, Sommerfeld visited other cities in northern India. On October 15, he was in Benares (now Varanasi), the religious center of Hinduism on the Ganges, to give a lecture at the Hindu University. The Chancellor of the University,

28 To Johanna, October 3, 1928.
29 Journal of the world tour.
30 Cited in Singh, *Arnold Sommerfeld*, 2001, p. 1491; Torkar, *Meeting*, 1986.
31 Singh/Riess, *Seventy Years*, 1998.

"a friend of Gandhi, strict Brahman," invited him on a river cruise on the Ganges and conversed with him "on Goethe, Haeckel, Spinoza, matter, and spirit."[32] The following day he inspected Sarnath, 10 km to the north, a historic city of early Buddhism. "Countless monastic cells, each with an image of Buddha and a small stupa," Sommerfeld wrote in his journal. It brought to mind Pompeii.[33]

He took the occasion of his lectures and talks at the various Indian universities and colleges to discuss the political situation with professors and students. "Everywhere, much sympathy for Germany. Admiration for our speedy reconstruction. All would like to study in Germany, but only if they have been to Cambridge can they find academic positions," he wrote, in criticism of the colonial dependency on England. "Indians unanimous in condemnation of the current system and in the demand for a position of respect within the British Empire."[34]

He experienced a particular insight into Indian-Bengal culture in his encounter with Rabindranath Tagore (1861–1941), resident in Santi-Niketan (now Shantiniketan) as spiritual head of a small scholarly and artistic community. Tagore had met Sommerfeld previously on a visit to Munich and was pleased now to be able to offer the professor from Germany the experience of "an Indian autumn's tranquility."[35] "Here, total stillness prevails around the 'poet,' as he is generally known," Sommerfeld enthused over his visit to Santi-Niketan. "Tagore is incredibly diligent in all aspects, as poet, musician, philosopher, and organizer of Indian education." His role in the cultural development of India could scarcely be overestimated. Tagore had "thrown their 'Sir' back in the faces of the English" and was striving for a "restoration of the decaying village life," though not like Mahatma Gandhi (1869–1948), whose politics of "non-cooperation" he rejected. Sommerfeld compared Tagore to Goethe, primarily because of his influence on the intellectual upper strata of Indian society.

Sommerfeld had actually wanted to visit Delhi, too, but abstained from the trip to the Indian capital which Raman had characterized for him as follows: "You will find there the monuments of many big empires now destroyed and the monuments of one more big empire not yet destroyed." Sommerfeld quoted this sarcastic description in a letter to his wife to illustrate the anti-British sentiment he constantly encountered. "Condemnation of the current governing methods of the British is universal among Indian professors."[36]

In light of this sentiment on the part of his host, it was not surprising that Sommerfeld was "under surveillance by the secret police," as the German Vice Consul warned him. He had noticed no sign of this, however, he noted in his journal under the heading "Political Items from Calcutta." From his many political

32 To Johanna, October 18, 1928.
33 Journal of the world tour.
34 Ibid.
35 From Tagore, October 15, 1928. DMA, NL 89, 024, folder Indien.
36 To Johanna, October 22, 1928.

discussions with his hosts, he concluded that most Indians desired independence—not through separation from England, however, but in the sense of self-governance, as had earlier been granted the "dominions" of the British Empire. Currently, India was obliged to import everything from England, "from matches to locomotives." There was only one technical research institution (in Bangalore). The criticism was widespread that not enough was spent on education. "Everything else seems peripheral. Great respect for the guru (teacher)."[37]

Such journal entries make it clear that in his cultural mission, Sommerfeld was no ivory-tower scholar, blithely singing the praises of German science, but oblivious to the sociopolitical situation in his host country. He registered very precisely the wants and the needs of his hosts and was open to instruction wherever the opportunity presented itself.

10.3 German Science at Chinese Outposts

On October 26 in Calcutta, after a 6-week stay in India, Sommerfeld once more boarded a steamer, bound this time for Rangoon (now Yangon) in Burma (now Myanmar). After a tranquil, 3-day ship's passage, a similar round of lectures and sightseeing awaited him. "Today, tea-party with various addresses, to which I naturally have to answer," he wrote to his wife after his arrival. "Early tomorrow, excursion to Pegu; in the evening, popular lecture: German and Indian universities; day after tomorrow lecture on spectral lines. In between, visits to institutes, hospitals, etc."[38] Actually, he would gladly have lodged aboard ship during the 3 days of his Rangoon stay, going ashore only to fulfill his lecturing obligations. But his English hosts would not forego putting him up in their home and spoiling him with all the comforts they were privileged with as colonial masters. In contrast, the overwhelming majority of the Burmese population lived in extreme poverty. A rickshaw driver earned "a few miserable rupees," Sommerfeld wrote, describing his impressions of Rangoon. "These drivers trot quite fast in the heat of the sun, and naturally die around the age of 30." He reported to his wife also that he had met a Buddhist monk, "born an Irishman, and previously a British officer! It is not unusual for the English to convert to Buddhism or Hinduism. It seems to be something in the air here."[39] From Rangoon, the journey proceeded to Penang and Singapore (now in Malaysia). Here, freed for a few days from lecture obligations, Sommerfeld could enjoy being a tourist, although this visit was not entirely private, either. He had been "often together with the German Consul General," he wrote home. He also met the American and French Consuls for dinner and lunch.[40]

37 Journal of the world tour.
38 To Johanna, October 29, 1928.
39 To Johanna, November 3, 1928.
40 To Johanna, November 12, 1928; Journal of the world tour.

His next destination was the Philippines. "One day out from Manila. I've been in bed the last two days; from lying on the top deck, I've picked up a disagreeable rheumatism and a bit of fever."[41] He wrote this to his wife during the passage across the South China Sea aboard the German steamship "Coblenz," 3 days after departing Singapore. By the time he reached Manila, however, he was fever-free again. He described the hotel in which he lodged as "very elegant, very expensive, very loud." From 1898 to 1941, the Philippines was a US colony, and Sommerfeld registered the contrast with the British colonies in his journal. Unlike India and Burma, one traveled through Manila "in a one-horse carriage . . . The Americans apparently do not tolerate the rickshaws drawn by humans, and have replaced them with nice pony-drawn vehicles."[42]

From Manila, the voyage continued across the South China Sea to Hong Kong, and from there along the Chinese coast north to Shanghai. "We have arrived: no more heat. We wear woolens. I'm also free of the fever, and slowly regain my appetite," Sommerfeld wrote by way of diagnosing his recovery from the trials of the tropics.[43] In Shanghai, lecturing duties awaited him once more. The first invitation came from the "Quest Society," a club for popular science enthusiasts who had asked to hear a lecture by Sommerfeld "on atomistics."[44] Another request had come from the Germanophone Tung-Chi University in Woosung near Shanghai. "In local German circles, your visit to Shanghai is eagerly anticipated," the German Consul General had written Sommerfeld. The Director of the Tung-Chi University had expressed the "wish for contact with you," and he conveyed this request "all the more since from the appearance of a prominent German scholar I anticipate a particularly lasting impression on the Chinese students, and may hope that thereby the German cultural influence on the Tung-Chi University will be valuably reinforced." This technical university, consisting of a medical and an engineering school, was "one of the most valuable German cultural efforts in China."[45] It was established in order "to assure Germany, the Germans, and the German spirit a commensurate role in influencing Chinese reform," as a German Consul General in Shanghai had formulated it following the Boxer Rebellion early in the twentieth century. Principally, the engineering school, opened in June 1914 under German direction, was intended to secure Germany a preferential position among competing European powers in the exploitation of the huge Chinese market. But the outcome of World War I had shattered these hopes. The Tung-Chi University passed to Chinese ownership, and the main thrust of German-Chinese relations was perforce relegated to the cultural realm. The University retained its German faculty and enjoyed the uninterrupted support of its—now Chinese—owners.[46]

41 An Johanna, November 15, 1928.
42 Journal of the world tour.
43 To Johanna, November 22, 1928.
44 From Herbert Chatley, August 16, 1928. DMA, NL 89, 019, folder 4,3.
45 From Fritz August Thiel, November 13, 1928. DMA, NL 89, 021, folder 9,6.
46 Bieg-Brentzel, *Tongji-Universität*, 1984; Steen, *Beziehungen*, 2006.

"First evening, lecture to the Quest Society; next evening, lecture at the Paulun Hospital; third day, visit to Tung-Chi University with address to the students beneath a picture of Sun Yat-sen, the current national hero, both latter addresses in German since the students of this university take their classes conducted in the German language."[47] Sommerfeld gave his wife this summary of his 3-day sojourn in Shanghai. In his journal, he registered once more what he had said in his address in "conclusion to the students": They were "privileged over millions of others in that they were being taught the best science by German instructors" and were thus "duty-bound to idealism."[48] By printing his address in both German and Chinese, the *Tung-Chi Medizinische Monatsschrift* (Tung-Chi Medical Monthly) was responsible for extending the effect of his mission to this "furthest outpost of German science and culture" well beyond the term of his visit.[49]

While Sommerfeld was carrying out his cultural mission in China, the pending appointment of a successor to Wien took a turn that caused him some concern. Sommerfeld wanted to see Debye, Franck, and Gerlach placed equally in the top spot, Gustav Hertz in the second, and Ernst Back in the third spot. The candidate list drawn up by the appointment committee, however, ranked only Debye and Franck equally in the top spot; Gerlach and Hertz ranked second and third.[50] "If the Ministry gets a refusal from Debye and Franck, then it will be easier for an offer to be made to Stark, than if—as I wished—we had clearly placed a man in the first spot whom we would get, namely Gerlach," as he explained his fear to his wife. Johanna Sommerfeld acted the role of intermediary between her husband and the faculty in the matter of this appointment. Even Johannes Stark was aware that Sommerfeld's wife could exert some influence. But in his attempt to ease his strained relation to Sommerfeld through his wife, he suffered shipwreck. That Stark should exploit his absence "to wear down" his wife outraged Sommerfeld. He was all the more pleased to see his arch enemy sent packing. "I would really love to have seen you, coolly, politely, and oh so innocently, telling Giovanni Robusto to get lost."[51]

10.4 Birthday in Japan

Sommerfeld departed Shanghai on November 29, 1928, aboard the S.S. Nagasaki-Maru bound for Japan. After a tranquil passage across the East China Sea, the steamer arrived the next day in Nagasaki, where Sommerfeld was welcomed by a delegation of Japanese physics professors. Following a brief stay, his journey

47 To Johanna, December 1, 1928.

48 Journal of the world tour.

49 Sommerfeld, *Entwicklung*, 1929a.

50 From Wilkens, October 31, 1928, and November 15, 1928, with the draft of a reply from Sommerfeld, DMA, NL 89, 019, folder 5,10.

51 To Johanna, December 1, 1928.

continued to Kobe. From here, he traveled by train to Tokyo. "The Japanese really know how to make one's life comfortable," he wrote a few days later from Tokyo.[52] Here too, the guest from far-off Germany was treated with extraordinary attentiveness. Yoshikatsu Sugiura (1895–1960), an employee of the respected Physical-Chemical Research Institute (Rikagaku Kenkyujo, RIKEN), who had been a guest researcher from 1925 to 1927 at Copenhagen, accompanied him everywhere and paid his expenses "on orders from above." "I dubbed him my finance minister," Sommerfeld wrote his wife.[53] Sugiura and an "adjutant" anticipated his every wish. "That I should celebrate my birthday in Japan was seen as a token of special favor on my part towards Japan. They have, however, declared the 6th my birthday, and made it almost a national holiday."[54]

On account of this misunderstanding, December 5, 1928, the actual date of Sommerfeld's 60th birthday, ran its course relatively uneventfully. Apart from the congratulatory telegrams that arrived at his Tokyo hotel from Europe, this day was for him nearly a normal workday. To spare his hosts the embarrassment of last minute rescheduling for the fifth all the festivities planned for December 6, he did not correct the misunderstanding and delivered the first of several 2-h lectures at the Empirical University of Tokyo on "Fundamental Questions of Wave Mechanics" according to plan.[55]

The next day, accordingly, Sommerfeld was "surprised" with a wide-ranging birthday celebration. Following his lecture, invited guests, including the German ambassador in Tokyo, Wilhelm Heinrich Solf (1862–1936), adjourned to a reception at the Sanjo Palace of the University. The birthday dinner was served in traditional Japanese style, presided over by Count Masatoshi Okochi (1878–1952), Director of the RIKEN, who coincidentally on this day was celebrating his own 50th birthday. "At dinner, Germans and Japanese guests were seated in alternation," Sommerfeld described the event to his wife. "Shoes off, of course, cushions in place of chairs, straw mats on the floor, chopsticks in place of knife and fork. I had already practiced with these, and proudly declined knife and fork. I only asked to have my cushion raised a bit, because I can no longer fold my legs under me comfortably. In front of us, cute little Japanese serving girls sit (or rather crouch) on the straw mat, chatter superficially with the guests, and bring the innumerable dishes, all of them served individually in lacquered bowls. High point of the affair: dance of two geishas, high art, extremely graceful, dance or theater, as you will. Of course speeches by Okochi and me."[56]

52 To Johanna, December 4, 1928; Ozawa, *Aufenthalt*, 2005.

53 To Johanna, December 24, 1928.

54 To Johanna, December 4, 1928.

55 Ozawa, *Aufenthalt*, 2005, p. 51. I gratefully acknowledge Michiyo Nakane for her transmission of Sugiura's Japanese translation of the texts of these lectures.

56 To Johanna, December 24, 1928.

After weathering the birthday festivities and one more lecture at the University of Tokyo, on December 8, 1928, Sommerfeld was guest at an event of the Japanese-German Cultural Society. He again delivered the popularizing lecture "On the Development of Atomic Physics in the Last Two Decades," which he had already given at Tung-Chi University, and here too met with great interest.[57] He had used "Ernst's observation that I went abroad as a German cultural ambassador" as an introduction, he wrote home.[58] This lecture remained a pleasant memory also because of an observation of his translator, who on this occasion had compared Bohr with Copernicus, and Sommerfeld with Kepler.[59]

It may be that in light of this comparison, Sommerfeld was reminded that he, in contrast to Bohr, had not been honored with the Nobel Prize. Three days before, on his birthday, he had confided to his journal: "Read letters and verses from home, sadly also notice about Nobel Prize."[60] Several days later he wrote his Munich colleague Heinrich Wieland, who that year had received the Nobel Prize in chemistry: "Hail and conquer! I congratulate your dear wife also on her famous husband. According to everything I know about you, I am persuaded the choice was well deserved. But to dispel all suspicion of false modesty, I must simultaneously note that it is gradually becoming a public scandal that I have still not received the Prize." In India, he had heard rumors that Bohr, "out of rivalry," was blocking the award of the Prize. He knew nothing about any such machinations, but he had already several times been on the short list. "Once, the Stockholm press had actually asked for my picture. In any case, it would have been the only right and proper thing, after Bohr received the Prize in 1922, for it to be given to me in 1923. The Royal Society, for example, made Bohr and me Fellows at the same time, as was fitting. So much for unburdening my heart, and for the sake of truth."[61]

But Sommerfeld had no time to sink into depression over the withheld Nobel Prize. His Japanese lectures were being eagerly awaited, and Sommerfeld took great pains not to disappoint these high expectations. At Kyoto, among his audience were the future Nobel laureates Shin-ichiro Tomonaga (1906–1979) and Hideki Yukawa (1907–1981), third year physics students, who preserved these lectures in memory as "unforgettable and superb."[62] In Kyoto, Sommerfeld also repeated his popularizing lecture "On the Development of Atomic Physics in the Last Two Decades." Tomonaga recalled that Sommerfeld spoke on this occasion also about the energy levels in hydrogen that could take up an electron. "Then the following happened: as he explained this, he ran around the podium. But because he was going backwards, he did not see the edge, and fell off. My teacher, Professor Tamaki,

57 Sommerfeld, Entwicklung, 1929b.
58 To Johanna, December 21, 1928.
59 Ozawa, *Aufenthalt*, 2005, p. 52.
60 Journal of the world tour.
61 To Wieland, December 13, 1928. DMA, NL 57.
62 Cited in Ozawa, *Aufenthalt*, 2005, p. 55.

Fig. 27: On an excursion to Hakone, Hantaro Nagaoka (1865–1950), the patriarch of Japanese physics, introduced Sommerfeld to the natural beauty of the surroundings of Tokyo. In this nature preserve, volcanic activity is everywhere in evidence (Courtesy: Deutsches Museum, Munich, Archive).

who was sitting in the first row, quickly picked him up. Then Sommerfeld, without much ado, went right on: 'Exactly as I just now fell down, the electrons, too, fall down from here to there.' I remember that he got the people laughing in the audience on his side."[63] Sommerfeld recorded in his journal simply, "Lecture to a big audience. English. Very good and very popular."[64] In Kyoto, he went to see the temples and the Imperial grounds, so that his stay in this city was a special experience of Japanese culture.

63 Ibid.
64 Journal of the world tour.

On December 17, 1928, Sommerfeld returned to Tokyo to observe experiments at the RIKEN several days before embarking on the ship's long passage across the Pacific to America, and especially to meet Nagaoka, patriarch of Japanese physics, who had returned early from a trip to Europe in order to greet Sommerfeld in his homeland. On the last day of his 3-week sojourn in Japan, Nagaoka accompanied him to Hakone, a locale in the foothills of Mt. Fuji, where he was able to get a final, lasting impression of the volcanic nature of Japan ("sulfurous air through the gorge of fumaroles, witches' kitchen, with hot-springs").[65] On parting, Nagaoka gave him the gift of an artfully decorated, bamboo walking-stick, "carved with a rat's tooth (!), depicting 100 Japanese faces, truly a work of art, signed by the artist," as Sommerfeld wrote his wife in his last letter from Japan. He felt "great reverence for the ancient history and culture of the country."[66]

10.5 Visiting Professor in Pasadena

Sommerfeld passed the Christmas Season and the New Year aboard a Japanese steamship from Yokohama bound via Honolulu for the U.S. west coast. "Now there is a Christmas tree (with electric lights) in the dining room, and a maple tree (artificial leaves with cotton snow), decorated with cherry blossoms, Japanese paints, etc." he wrote home about the unusual circumstances of his Christmas observance. The passage was stormy, "the entire sea grey and white with spray; it's barely possible to write. Many are seasick; not I."[67] He recorded in his journal that he spent most of his time in letter-writing ("20 letters and numerous postcards").[68] Aside from his correspondence, he penned a longer article on his impressions of India for a Munich art journal.[69] Regarding his stay in Honolulu, where he went ashore for a few hours, he had little to report: "Hawaiian girls dancing, to the accompaniment of fatsocs."[70]

Although a 3-month stay in the USA still lay ahead, he experienced the crossing of the Pacific as the first leg of his return home. "The Japanese haven't trisected the master," he wrote musing on the previous weeks. He reviewed with amusement several situations that had befallen him among the many honors bestowed on him in Asia. At his being named an honorary member, an Indian mathematician had, "in grim earnest," analyzed his mathematical papers so conscientiously that he had said in rejoinder, "I can't know how a frog feels during its own vivisection.

65 Ibid.
66 To Johanna, December 24, 1928.
67 To Johanna, December 25, 1928.
68 Journal of the world tour.
69 Sommerfeld, *Reiseeindrücke*, 1929.
70 Journal of the world tour.

But I must say I felt quite alright during this friendly vivisection."[71] He was constantly asked for autographs with maxims such as "what is most important for research." In such situations he had delivered himself of bits of wisdom like "Onward and upward" or "Integral p dq = n h."[72]

With his arrival in San Francisco, that portion of Sommerfeld's world tour during which he felt his role was as a scientific missionary came to a close. Six years earlier on his first visit, the American physics profession had already rendered him great respect. The California Institute of Technology in Pasadena, where he would spend the next 2 months as visiting professor and meet his former students Epstein, Pauling, and Houston as colleagues, was on the way to becoming a center of modern physics that had no cause to shy away from comparison with European universities. More than anywhere else in America, Sommerfeld felt at home here, and he was immediately reinforced in this feeling by his hosts. "Today at noon I will be with the Millikans, in the evening, with the Paulings," he wrote home shortly after his arrival in Pasadena. "Have also already been with the Houstons; last evening to the theater on invitation from Epstein." In addition, he was staying in an idyllic apartment at the Faculty Club, "with a view of palms and fruit-bearing orange trees, and in the back, a view of the blue mountains."[73]

Also, his duties as visiting professor had more in common with his familiar teaching regime in Munich than with his function as cultural ambassador in Asia. His teaching load comprised four 1-h lectures weekly and participation in the colloquia. The subject of his lectures corresponded broadly to what he had written in the just published *Wave Mechanical Supplement*, so that little preparation was required. "Here in California, life is made easy for one in every respect," he wrote Rubinowicz. "To be sure, I have not only my lecture courses, but have also to speak at all sorts of meetings, in English of course."[74] On the social level too, Pasadena had something to offer. "Yesterday there was a faculty dance. Quite nice and easy," he wrote home 2 weeks after his arrival. "I even danced, in spite of the jazz music. Last week, I heard very good music, string quintet, at the home of a friend of Epstein's, a professional violinist; I'll go again next week. A week ago I played with Pauling's trio. I had to speak at a society lunch about India and Japan, ½ h. Also, a colloquium lecture in addition to the usual lectures. Next week I have to speak to a similar society in Los Angeles. But it is good that each day I have several quiet hours to myself to gather my thoughts –not as it was in India and Japan, and will be to a greater extent in America after March 15." This latter reference was to the numerous lecture invitations for the last weeks of his U.S. visit that he had received in Pasadena, and that would require careful travel planning for the period following his visiting professorship. He also had to devote not a few of his quiet hours to

71 To Johanna, December 25, 1928.

72 To Margarethe, undated [around December 27, 1928].

73 To Johanna, January 13, 1929.

74 To Rubinowicz, January 15, 1929.

composing letters of thanks for the numerous birthday greetings and especially for the festschrift for which thirty of his students had in his honor written articles on "Problems of Modern Atomic Physics."[75]

In the weeks of his Pasadena stay, the question who would be his new experimental physics colleague on his return to Munich was also resolved. Debye, placed first on the appointment list together with Franck, had withdrawn his candidacy, since he had only shortly before been appointed at Leipzig. An appointment at Munich so soon after taking up his post at Leipzig would have clashed with the understanding among the Ministries of Culture of the various states according to which there was to be no recruitment within 2 years of an academic chair's being taken up.[76] Thereupon, the offer had fallen to Franck, who placed conditions regarding the improvement of the Munich Institute's outfitting that the Bavarian Ministry of Culture would not accept.[77] "The offer has gone to Franck, as I hear," Sommerfeld wrote his wife in Munich at the end of January. "I greatly value him as a colleague. But Gerlach would have been better."[78] Sommerfeld feared that in the end, Stark would after all come under consideration and asked Franck to accept the appointment. But Franck did not feel he could do that, as he explained to Sommerfeld in a long letter, since the Ministry was unprepared to meet his demands.[79]

At the same time, Stark complained that Sommerfeld had been blocking his appointment as Wien's successor. He had been informed by a person he declined to name, "that you are the ultimate and decisive author of the candidate list for appointment to the Wien chair. Since this list excludes me, it is tantamount to an official discrediting of my person and my scientific achievements. You must understand that I will defend myself against this discrediting, and intend to make public my viewpoint on the scientific grounds you have adduced."[80]

Sommerfeld replied coolly to Stark that he could not respond to anonymous innuendoes, and that the appointment list had not at all been drawn up on the basis of his recommendations.[81] "I don't foresee any good ending here," he wrote his wife. "Ultimately, we'll have to go to Berlin or to America after all."[82] His mood soon improved again, however, when the Dean informed him per telegram that the appointment had now gone to Gerlach. He wrote his wife that he had immediately telegraphed Gerlach: "Accept unconditionally." And Stark could "Go jump in the lake with his polemical threats."[83] Gerlach did not accept the appointment right away, however, but went first to the Ministry to negotiate further. This was "pure

75 To Johanna, January 20, 1929. Debye, *Probleme*, 1929.
76 From Debye, 21. December 21, 1928. DMA, HS 1977-28/A,61. Also in ASWB II.
77 From Wieland, January 19, 1929. DMA, NL 89, 019, folder 5,10.
78 To Johanna, January 27, 1929.
79 From Franck, February 5, 1929. DMA, NL 89, 019, folder 5,10.
80 Form Stark, January 30, 1929. DMA, NL 89, 019, folder 5,10. Also in ASWB II.
81 To Stark, February 18, 1929. DMA, NL 89, 019, folder 5,10. Also in ASWB II.
82 To Johanna, February 17, 1929.
83 To Johanna, March 3, 1929.

Fig. 28: Together with Pauling and his family, Sommerfeld visited the "Painted Canyon" (in the picture, Pauling's wife and "little Linus") (Courtesy: Deutsches Museum, Munich, Archive).

theater," Johanna wrote her husband in Pasadena after Gerlach informed her of it. All the same, Gerlach had given her the impression that he would accept the appointment in the end. "Then you can exhale. But his appointment is to begin only in October."[84]

But the back and forth of the Munich appointment question was not enough to seriously dampen Sommerfeld's sense of well-being in Pasadena. On one weekend, the Pauling family took him along on an excursion to the desert. They slept under the stars in the "Painted Canyon."

"It was warm in the sleeping bag, in spite of the night's being quite cool. In the morning, we climbed around a bit, little Linus mostly on big Linus's back. Long car trip back through endless orange and lemon groves, blue mountains, snow-capped in part, up to 3,600 m. high, well-tended villages, wonderful roads. All of it very pleasant."[85]

84 From Johanna, March 21, 1929.
85 To Johanna, February 10, 1929.

To the "cordiality of Pasadena life" belonged also his regular weekly meetings following dinner at the Faculty Club for a game of bridge with the Swiss astrophysicist Fritz Zwicky (1898–1974), who had come to Cal Tech in 1925. These evenings were reminiscent of his childhood days in Königsberg when he had played whist at home "with father, Aunt Minchen, and Ochen, who always played incorrectly, which invariably annoyed father." It also became a pleasant custom to be picked up by a violin virtuoso for musical evenings. "In the course of my time here, we have played through all the Beethoven and Schubert violin sonatas. I have also often played piano at the social gatherings I'm frequently invited to."[86]

Scientifically, too, he felt thoroughly at home. His lectures on wave mechanics met with great approval. "The students here (mostly older, and very sensible) are already beginning to turn to me with their troubles and their discoveries," he wrote his wife after the first two weeks of lectures.[87] He had Vieweg Publishers send 20 copies of the *Wave Mechanical Supplement* to Pasadena, a number, however, that was insufficient to meet the demand of his audiences.[88] He was most pleased by Schrödinger's reaction to his book, which reached him at Pasadena: "What you have done here is once more—like the main volume—something only you could accomplish. You are the master builder, creating a whole for which the rest of us merely supply the building blocks, often enough so crudely hewn that you must chisel them skillfully, when you don't actually prefer setting a stone of your own in the place where the one supplied won't fit."[89] This reaction from the architect of wave mechanics quickly took the sting out the letter he had received the same day from Stark threatening a continuing and fruitless argument. "My joy over this was greater than my annoyance over Stark," he wrote home.[90]

The great demand for the *Wave Mechanical Supplement* also quickly engendered the desire for a translation into English. At Cal Tech alone, students ordered 60 copies of the German edition.[91] "My Wave Mechanics is already supposed to be translated into English," Sommerfeld wrote with pleasure. Additionally, he had "been calculating diligently."[92] He was referring to the problem of explaining quantum mechanically the generation of X-rays at the braking of electrons. He had first taken up this question on the passage from Japan to America and therefore referred to it as his "Pacific problem."[93]

86 To Johanna, February 24, 1929.
87 To Johanna, January 20, 1929.
88 To Johanna, February 3, 1929.
89 From Schrödinger, January 29, 1929. DMA, HS 1977-28/A,314. Also in Meyenn, *Entdeckung*, pp. 462–464.
90 To Johanna, February 17, 1929.
91 To Johanna, February 3 and March 10, 1929.
92 To Johanna, February 10, 1929.
93 To Johanna, January 27 and February 3, 1929.

10.6 The Second American Tour

In the last 6 weeks of his U.S. visit, Sommerfeld was much in mind of his experiences in 1923 when at the conclusion of his Karl-Schurz Professorship at Madison, he had crossed this enormous country from coast to coast, delivering lectures here and there along the way. "Yesterday, the Grand Canyon during the day, for the second time in my life," he wrote his wife on March 15 en route of his 4-day train trip on the "California Limited," which brought him to Chicago at the conclusion of his visiting professorship in California. Taking leave of Pasadena, as at the end of his visit 6 years earlier, was not easy for him. "The boys," as he called the students in his lecture courses, had arranged a big party in his honor at which they bade farewell with an original theater-piece, two Beethoven trios, a Grieg sonata, Chinese music, and all sorts of culinary delicacies. "Everyone was royally entertained in the broadly informal atmosphere."[94] One day later, "near the Missouri," he wrote to a colleague, "You can just imagine how interesting India and Japan were, and how attentively everyone has provided me the best and most comfortable their countries have to offer. But this last impression is the greatest: southern California with its natural beauty and amazing progress is remarkable, and the people there are marked with an unusual measure of optimism, cheerfulness, sociability. The competence of the kind of people who have settled there preclude their degenerating into hedonism."[95]

In Chicago, Arthur Holly Compton and Carl Eckart prepared a cordial welcome for him. "The Comptons were charming to me, took me to hear a good quartet, and also arranged for music at home," he wrote his wife.[96] Sommerfeld did not mention the "German hater" Michelson, who had disinvited him on his first visit to the USA, and this time too had acknowledged him only with a brief perfunctory greeting appended to a letter from Compton.[97] During his visit of only 4 days, Sommerfeld also nearly missed seeing Heisenberg, who delivered guest lectures shortly after his at the University of Chicago. Although he traveled in the opposite direction, Heisenberg, like Sommerfeld, used the occasion of this invitation to make a world tour, whose overture were the Chicago lectures that began in April, 1929. Scientifically, too, Heisenberg stressed different things. While Sommerfeld touted the advantages of Schrödinger's method in his Pasadena lectures, Heisenberg sought to spread the "spirit of Copenhagen."[98]

From Chicago, Sommerfeld's journey continued to Ann Arbor, where Laporte, newly minted as a professor in the physics department, welcomed him, then on to

94 To Johanna, March 15, 1929.

95 To Grimm, March 16, 1929. DMA, HS 1978-12B/172.

96 To Johanna, March 23, 1929.

97 From A. H. Compton, May 4, 1928. DMA, HS 1977-28/A,54.

98 Rechenberg, *Werner Heisenberg*, 2010, p. 631; Heisenberg, *Prinzipien*, 1930, pp. V–VI; from Heisenberg, March 28, 1929. DMA, HS 1977-28/A,136.

Madison, where the many invitations from old friends and acquaintances precluded much rest.[99] After a brief stop in Columbus, Ohio and Pittsburgh, Pennsylvania, the next destination was Philadelphia. "From the 9th floor of a fine hotel in the large and fine city of Philadelphia," he began his next weekly letter to his wife on the stationary of the "Bellevue Stratford." "This is actually the preferable form of hospitality. One is put up in a hotel, has no obligations, signs the check at breakfast and other meals, and leaves everything else to one's hosts. It was the same at the Athletic Club in Pittsburgh. In Columbus, Ohio, I was actually on the 14th floor, and the spring wind whistled around the windows at night as at Sudelfeld." His host in Philadelphia was the Director of the Bartol Research Foundation of the Franklin Institute, who also enjoyed some renown as a cellist, and invited Sommerfeld to musical evenings in his home. An evening spent at a concert by the Philadelphia Orchestra, world famous at that time under the direction of Leopold Stokowski (1882–1977), made for an air of relaxation. "Philadelphia is almost peaceful."[100]

The last portion of the trip was hectic again. In New York, he had to settle the taxes due on honoraria he had received for the lectures delivered in the USA, book passage for his return to Europe at the offices of North German Lloyd, and deliver a lecture at Bell Laboratories. Then he attended a conference in Tuxedo Park, NY, at the invitation of the legendary American physicist, banker, and patron of science Alfred Lee Loomis (1887–1975). "Mr. Loomis, who has hosted 30 overnight guests and 110 others who came today for the lectures, is the American Anschütz," Sommerfeld wrote, comparing Loomis to the German inventor of the gyroscopic compass, and patron of the arts and sciences, Hermann Anschütz-Kämpfe (1872–1931). The laboratory of this "American Anschütz" in Tuxedo Park achieved legendary status for the development there of microwave radar during World War II. Already in the 1920s, though, a particular aura surrounded Loomis. "He is a man of Wall Street and a physicist, by preference," Sommerfeld thought.[101]

Both his visit to Tuxedo Park and his subsequent stay in Washington, where his lecture to the National Academy of Sciences formed as it were his official scientific farewell performance in the USA, were very pleasant experiences for him. He was not dealing with a lay audience here but with the elite of American physics. He had been occupied with his "Pacific problem" just at the time these invitations to Tuxedo Park and Washington had reached him in California. Thus, he had proposed "production of X-rays according to wave mechanics" as his lecture topic, thereby putting the pressure on himself to work up his provisional calculations of the X-ray bremsstrahlung into a demonstrable theory.[102] In Washington,

99 To Johanna, March 29, 1929.

100 To Johanna, April 7, 1929.

101 To Johanna, April 15, 1929. Conant, *Tuxedo Park*, 2002; on Anschütz-Kämpfe, see Broelmann, *Intuition*, Chap. 4.3.

102 From Loomis, with the draft of a reply from Sommerfeld, February 11, 1929. DMA, NL 89, 019, folder 4,4.

this was also his lecture topic. He regarded it "as the fruits of my Pacific and Californian muse."[103]

Therewith, Sommerfeld's "Pacific problem" garnered some notice even before he had published anything about it. Helmuth Kulenkampff (1885–1971), an experimental physicist at the Technical University of Munich and regular participant in the Sommerfeld colloquium, had provided the initial impetus. Shortly before the beginning of Sommerfeld's world tour, Kulenkampff had carried out a string of precise measurements of the distribution of directions of X-ray bremsstrahlung, in which he had used extremely thin aluminum foil as anticathodes, in order to eliminate the secondary effects (diffusion and multiple scattering) that arise with normal anticathodes. With these measurements, he had confirmed that Sommerfeld's 1909 derivation for the classical radiation of a straight-line braked electron was in its essence correct.[104] Now, the challenge that presented itself to Sommerfeld was to elaborate the theory such that it confirmed the earlier results according to quantum mechanics as well.[105]

At first glance, it might not seem that Kulenkampff's experiments and Sommerfeld's efforts at explanation constituted a very great challenge. If the classical theory was already capable of describing the bremsstrahlung in these experiments well, what need was there for a quantum mechanical explanation? In truth, though, more was at stake than just a newer derivation of a classical theory. Treating the process of absorption, emission, and scattering of electromagnetic radiation quantum theoretically presented great difficulties. Already before quantum mechanics, much effort had gone into reinterpreting X-ray bremsstrahlung quantum theoretically.[106] "Beginning work on bremsstrahlung, in light of discussions with Sugiura," Sommerfeld had written about his "Pacific problem" in his journal one morning after a hot bath, shortly before the ship's docking at Honolulu. "Looks promising but complicated."[107]

In the earlier quantum theory, the problem had been to calculate the energy loss of an electron that was first approaching an atom on an energy-rich hyperbolic course and then moving away from it on an energy-poorer one. The energy difference corresponded to the energy emitted by the X-ray bremsstrahlung. Wave mechanically, the incident electrons could be pictured as a smooth wave that is scattered onto the atom. The decisive magnitude of the electromagnetic wave radiated in this scattering process, Sommerfeld argued, was that of the "matrix element" corresponding to the electrical dipole moment that has to be calculated from the product of the amplitudes of the incident and reflected electron waves,

103 To Johanna, March 10, 1929.

104 Kulenkampff, *Untersuchungen*, 1928, p. 629.

105 Sommerfeld, *Production*, 1929a and 1929b.

106 Kramers, *Theory of X-ray Absorption*, 1923; Wentzel, *Quantentheorie des Röntgenbremsspektrums*, 1924.

107 Journal of the world tour.

multiplied by the distance from the scattering center and integrated over the whole space. That, at any rate, was how he presented his "Pacific problem" to the National Academy of Sciences. Although he could reinforce his basic idea with several calculations, he left out the complete mathematical implementation. "I must leave that for a fuller paper to be published later in the *Annalen der Physik*."[108] Only in this paper—which, be it said, did not appear for another 2 years—did the whole complexity of the "Pacific problem" manifest itself.[109]

Sommerfeld was already onboard the steamer in New York for the return passage to Germany when a telegram reached him from Washington announcing his election as a nonresident member of the National Academy of Sciences.[110] His US colleagues, chief among them Millikan as incumbent Secretary of the Academy, could not have made Sommerfeld a more wonderful parting gift.[111] From Tokyo he also received a token of highest esteem. "Your visit to Japan marks an event in the history of the development of mathematical physics in Japan," Nagaoka wrote him in flowery language. Sommerfeld's lectures had "had no doubt an effect of balmy dew falling on the tender leaves beginning to sprout."[112]

In the light of so many tributes, the exertions of his world tour receded into the background. At home once more, Sommerfeld set himself the task of accepting every opportunity that arose to report on his travel experiences. Above all, he showed himself to have been deeply impressed with India.[113] At a meeting of the Bavarian District Association of the German Physical Society, he praised "the great scientific activity prevailing in India, particularly in the school of C. V. Raman in Calcutta." In Japan also he had seen impressive examples of physical research. He showed his assembled colleagues photographs with "Kikuchi lines," a diffraction pattern obtained through multiple scattering of electrons in crystals, which Seishi Kikuchi (1902–1974) had discovered shortly before his visit at the RIKEN in Tokyo.[114] He reported on his world tour also to the Bavarian Academy of Sciences, and to the "Casual Ones," a tradition-rich Munich society of scholars and artists which had admitted Sommerfeld to its ranks 3 years earlier.[115] "What was most edifying to those of us present, and made us truly proud," the historian of the

108 Sommerfeld, Production, 1929b.
109 Sommerfeld, *Beugung*, 1931.
110 From Millikan, April 24, 1929. DMA, NL 89, 020, folder 6,3.
111 To Millikan, April 25, 1929. Millikan Papers, Pasadena, Archives of the California Institute of Technology, 42.17.
112 From Nagaoka, May 3, 1929. DMA, NL 89, 019, folder 4,3. Also in ASWB II. The letter continues, "We are ever anxious to reap rich harvest of science in the Far East by tightening the band of connexion between the scientific circles of Germany and Japan in course of time. I must call for your help in fulfilling this ardent desire."
113 Sommerfeld, *Reiseeindrücke*, 1929.
114 Sommerfeld, *Physik in Japan*, 1929.
115 Sommerfeld, Bericht, 1929; Rohmer, *Zwanglose Gesellschaft*, 1937.

"Casual Ones" noted, "was the highly distinguished reception accorded Counselor Sommerfeld as a representative of German research, which is certainly not accessible to any random number of people in that far-flung region of the globe, which all too readily we had imagined as scientifically backward."[116]

10.7 Critique of Positivism

Sooner or later, the "Globetrotter," as the Casual Ones had dubbed him, needed nevertheless to reacclimate himself to daily life at home. His greatest concern was over the still orphaned Wien Institute. Gerlach's appointment appeared a done deal, but so long as this position remained unoccupied and Gerlach was still in Tübingen, Sommerfeld feared that ultimately Johannes Stark would after all be made the offer. Only when in June 1929 he received an inquiry from Tübingen whether he could recommend Stark as successor to Gerlach there was the situation clear. "The light and dark sides of Johannes Stark are generally known," he wrote to Tübingen. "Since I fought strenuously against his candidacy for Munich, just as strenuously as he sought to push it through, it would be inappropriate for me to recommend him to Tübingen."[117] Debye was pleased along with Sommerfeld "over having averted the great danger."[118]

Thereafter, Sommerfeld was once more at peace and able to concentrate on his subject. With a talk at the German Physicists' Day in Prague he showed that, for all his predilection for concrete problems, he was not indifferent to questions of fundamental principles raised by quantum mechanics. The trend towards the fundamental was evoked by the physicists of the "Vienna Circle," primarily by the theoretical physicist Philipp Frank (1884–1966), teacher at the German University at Prague, who opened the meeting with a programmatic lecture on the meaning of the "current physical theories" for epistemology. Richard von Mises (1883–1953) spoke on the causality principle and its statistical interpretation; the causality principle had been called into question by quantum physics. Frank and von Mises hoped for broad acceptance among the assembled physicists for the "scientific philosophy" of the Vienna Circle, which they sought to develop further with reference to the new discoveries of modern physics as the legacy of the positivism represented by Ernst Mach (1838–1916).[119]

116 Chroniken der Gesellschaft der Zwanglosen, 1924–1931, here, p. 152. BSB, manuscripts, Cgm. 8026(13a.

117 To the University of Tübingen, June 11, 1929. DMA, NL 89, 030, folder Gutachten.

118 From Debye, June 21, 1929. DMA, NL 89, 007. ". . .über die Abwendung der starken Gefahr" involves an untranslatable pun on Stark's name. "Stark" means "strong" in English.

119 Stöltzner/Uebel, *Wiener Kreis*, 2006, pp. X–XV.

But in the latest discoveries of physics, Sommerfeld saw no reason to revive the Mach positivism. Although in introducing quantum mechanics Heisenberg had directed his attention to observable magnitudes, this was not the essential distinction to the pre-quantum theories. "The error of the older quantum theory," Sommerfeld argued in his lecture, "was not the introduction of unobservable magnitudes, but excessive faith in classical mechanics. Wave mechanics, which so splendidly corrected this error, introduces unobservable magnitudes on a far greater scale than the old quantum theory." Nor did he regard the causality principle as being called into question by quantum mechanics; it needed only to be freed from the confining corset into which eighteenth century mechanics had forced it, and elaborated with respect to the principle of "finality." "The causality of the 20th century must not limit itself to the initial state, but must take the end-state into consideration as an equally determinative moment." This had already been made evident before quantum mechanics in the form of the spectroscopic combination principle, whereby the frequency of a spectral line is, after all, determined simultaneously by the difference of the respective energy values of initial and end states. The indeterminism presumably evoked by quantum mechanics was an inevitable consequence of the wave-particle dualism. Herein, Sommerfeld saw the true, philosophically meaningful discovery of the newer physics—and wave mechanics had found the appropriate formulation of it. To be sure, this dualism had not yet been reconciled. Sommerfeld did not believe that "this would be possible in the physical arena. . . More likely, perhaps, through some sort of philosophical synthesis." Perhaps, he concluded his lecture, it would 1 day be possible with a dualistic worldview to grasp "the infinitely more difficult, infinitely more delicate, but never to be evaded question of the collaboration of mind and body. Fortunately for our generation, much still remains to be done on the firmer ground of real physics."[120]

A few months later, speaking in Vienna, though Sommerfeld sounded a note of sympathy with the Mach positivism, he nonetheless carried wave-particle dualism into the field as the decisive argument against it: "According to the positivist conception, the dual nature of the electron, juxtaposed to the dual nature of light, means nothing less than the assignment of two different ways of describing related empirical facts. Is that all there is to be said, then? Is no remainder left over? Does not the conjecture suggest itself that this dualism in the area of physics is in some way related to the dualism that runs through our entire lives, the dualism of mind and matter, of I and not-I, of body and soul? . . . The scientific worldview of the Vienna Circle may be inclined to brush aside such questions as insubstantial. I do not believe the human spirit will be content with so dismissive a solution."[121]

120 Sommerfeld, *Bemerkungen*, 1929.
121 Sommerfeld, *Elektronentheorie*, 1930.

In a lecture in April 1930 in Würzburg, "On Clarity in Modern Physics," Sommerfeld again stressed the wave-particle dualism as the true challenge for the modern understanding of the natural world. Quantum theory did not permit precise predictions about the magnitudes coupled in the Heisenberg uncertainty relation; it has demonstrated an "insurmountable limit, beyond which the exact space-time description becomes illusory." This uncertainty concerns only our mental images, not the physical facts that can be determined experimentally. "Philosophy," he said, alluding to recent discussions with the Vienna Circle, "will cautiously follow after, and will ultimately, having overcome temporary difficulties, only gain thereby." The Würzburg lecture was published in the *Unterrichtsblätter für Mathematik und Naturwissenschaften*, the organ of high school teachers in Germany, and gained wide international readership as well through a reprint in *Scientia*.[122]

In this lecture, Sommerfeld had not identified the philosophy of the Vienna Circle by name. But it was clear that, with his reference to "mental images," and the call for a finalistically elaborated causality principle, he was distancing himself from Mach's positivism and its adherents. When Moritz Schlick (1882–1936), active since 1922 as successor to Ernst Mach at the University of Vienna, entered into this discussion, Sommerfeld expressed his pleasure "over the tolerant attitude and willingness to understand" with which the representatives of the Vienna Circle had received his critique. Although ultimately the Vienna Circle also saw the necessity of an accommodation of philosophy to the discoveries of modern physics, Sommerfeld believed the facts of physics tended to reinforce his conception. "I am not a dogmatist in the religious sense, but I am a dogmatist when it comes to the laws of nature. I cannot abide the Mach 'principle of the untidy laws of nature,' the Uncertainty Principle notwithstanding. Einstein rejects it, too. He once said to me: 'all physics is metaphysics.'"[123]

10.8 Quarrel with Stark

Sommerfeld's debate with representatives of the Vienna Circle related to the philosophical conclusions that had to be drawn from the discoveries of modern physics. His lectures at Prague, Vienna, and Würzburg, however, also provided the substance of a quarrel that only superficially concerned the questions thus raised. Johannes Stark seized the opportunity to give the veneer of scientific dispute to his resentment against Sommerfeld after the dashing of his Munich hopes. Stark had long led a campaign against modern atomic theory. He conceived of the atom as a structure rotating around an axis, out of which electromagnetic energy was ejected in form of "quantum eddies" and transformed into "light eddies." Initially, his

122 Sommerfeld, *Anschaulichkeit*, 1930.

123 To Schlick, October 17, 1932. DMA, NL 89, 025. Friedl/Rutte, *Moritz Schlick*, 2007, pp. 317–319.

polemic had been leveled against the Bohr-Sommerfeld atomic model. Now he set his sights on quantum mechanics. His attacks peaked in 1930 in a series of essays in the *Annalen der Physik*.[124] Caricaturing the statistical interpretation of the Schrödinger wave function Ψ, he wrote that one could "intellectually, of course, construct such a swarming about of electrons according to such a law," but this construction would be in conflict with experience. "Frenzied motion of the sort characterized above" had never been observed. "In order to establish the space-time behavior of the electron in its atomic field causally according to the Sommerfeld interpretation of the Ψ-function, one would have to depart the realm of physics, and postulate the electron's consciousness of the Schrödinger equation and its capacity to behave accordingly."[125]

Here, Stark had arrived at a point that seemed to him the cardinal sin of the whole quantum enterprise since Bohr: the violation of the causality principle. In an entire printed page, he quoted what Sommerfeld had argued in his Prague lecture on the necessity of elaborating the understanding of causality in order to set out clearly the irrationality of including both initial *and* end states in the quantum theoretical calculation of a spectral line: "The Sommerfeld construction of a new causality means not only the dissolution of the concept of causality as we have known it, but also the blurring of our concept of time. It arises from the effort to explain the dependence of the frequency on the end state, based on the conception that the radiation of the frequency during the transition from an initial state occurs after an end state. It has not been experimentally demonstrated, however, that this conception comports with physical reality. Nor has it been demonstrated that this is the only possible physical conception."[126]

Sommerfeld was accustomed to annoyance with Stark. In 1909, during the quarrel over interpretation of X-ray bremsstrahlung, the scientific aspect of the controversy was still uppermost. By 1921, when Sommerfeld defended the Bohr atomic model against Stark's attacks, hardly anyone took the experimental physicist seriously—his recent Nobel Prize in experimental physics notwithstanding—when he presumed to express an opinion in the area of theoretical physics. To a colleague, who also felt pilloried by Stark's latest attacks, Sommerfeld expressed his "sense of comradeship at having been jointly insulted" and made it clear that Stark's article "had been written more on personal than on substantive grounds." Thus, he did not intend "seriously to reply" to the article; he had, however, recently reacted to "several of Stark's objections" in a lecture given at Würzburg.[127] He was alluding to the lecture "On Clarity in Modern Physics," which actually had nothing to do with his quarrel with Stark and as Sommerfeld explained in a footnote to the printed version of the lecture also in the spoken version contained nothing of it. But in the

124 Kleinert, *Axialität*, 2002.
125 Stark, *Axialität*, 1930, p. 717.
126 Ibid., pp. 718–721.
127 To Ronald Fraser, May 13, 1930. DMA, NL 89, 001.

printed version, he could not forbear pointing "to obvious misconceptions in Stark's exposition."[128]

This incited Stark to a further attack. In the *Unterrichtsblätter für Mathematik und Naturwissenschaften,* he warned high school teachers against the "dogmatism" Sommerfeld was spreading with his theories.[129] Even in the *Annalen der Physik,* Stark's writing was now unmistakably polemical. Sommerfeld had spread the "theory of the swarming electron" even at the Deutsches Museum. There, "presumably with his collaboration," models had been displayed "which were supposed to represent the spherical-symmetrical form of the cloud of swarming electrons around an atomic center."[130] After the publication of this article, Gerlach, newly minted colleague of Sommerfeld, brought to the attention of Eduard Grüneisen (1877–1949), the responsible editor at the *Annalen der Physik,* that Stark was here less concerned with the substance of the matter than with a personal quarrel with Sommerfeld. "I see now," Grüneisen wrote Sommerfeld apologetically, "that I would have done better to have at least sent you the proofs to give you the opportunity of marking the passages you wished to see changed." He admitted that he did not judge Stark's views as negatively as Gerlach. "Not because I am sympathetic to them—quite the contrary—but because, despite his curious ideas and gross lack of tact (whose victim I myself once was) he is an important researcher, whose opinion in scientific matters readers of the *Annalen* are interested in hearing." That Stark often struck the wrong note was one thing; but that his expositions "sprang from un-objective motives, and were insulting" was not something he was prepared to grant. Stark had just "misunderstood a great deal." Therefore, he proposed that Sommerfeld "set forth to the readership of the *Annalen* from your viewpoint the problems around which Stark's argument revolves."[131]

"Why would you assume that Stark does anything on substantive grounds?" Sommerfeld replied to the editor of the *Annalen.* "His rage against me stems from the fact that the faculty rejected him as successor to Wien. First, he took the occasion of my 60th birthday to try to get chummy with me, and went so far as to trouble my wife in my absence with such an attempt. Then, when he became aware that he was not on the list, he wrote me a crude letter. Now he dumps his whole opposition to the development on my quite innocent head. Incidentally, his knowledge of this development is based solely on a single lecture of mine, not from the original sources of Heisenberg, etc." Sommerfeld did not blame Grüneisen for publishing Stark's article, although passages such as the "references to the Deutsches Museum" had no business being there. "But let's not be at swords' points over that! I know that an editor is a much harassed man." At all events, he was still indecisive whether he should take up Grüneisen's invitation to counter Stark's accusations

128 Sommerfeld, *Anschaulichkeit,* 1930a, p. 165.

129 Stark, *Dogmatismus,* 1930.

130 Stark, *Axialität,* 1930, p. 677.

131 From Grüneisen, October 4, 1930. DMA, NL 89, 024, folder, Starkiana.

with an article in the *Annalen der Physik*. "In any case, I will do it less than thoroughly; otherwise you will receive another dozen responses from Stark, and as a member of the board, I must seek to avoid that."[132]

In the end, though, Sommerfeld did send an article that filled three printed pages to the Annalen der Physik, "Rebuttal to the Attacks of Prof. J. Stark." He declined to dispute causality in modern physics, or wave-particle dualism, with Stark because "Prof. Stark, in his attacks, is ignorant not only of the general evolution of theory since 1926, but also of the experimental facts of electron diffraction, that were discovered in tandem with the theory." On the other hand, he made it perfectly clear where he stood with regard to Stark's exposition directed to chemists of the "axial structure" of the atom. "There is no doubt about the spherical symmetry of the charge distribution of filled shells or the ground states of hydrogen, the alkalis, the noble metals, etc., unless one is prepared to abandon entirely wave mechanics with its innumerable consequences, which are indispensable to experimentation." In the case of chemical bonding, the matter was very complicated, but what had heretofore been known about it was not in conflict with quantum mechanics. Quite the contrary: Here, for the explanation of polar bonding, for example, one required the quantum mechanical exchange interaction. For nonpolar bonding, the spin-concept was indispensable. No physical theory could presume to explain the wealth of chemical facts; it could "treat only simple, typical cases." Stark had on several occasions raised the chemistry of carbon bonding as an example. But Sommerfeld would not allow the still unresolved problems in this area to serve as an argument against modern physics, for the structure of such complex molecules belonged properly in the realm of chemistry. "It is not the role of physics to seek to replace or improve upon this work. It can, however, contribute to the basic elucidation of the valence concept, just as it was able to shed light on questions of atomic structure and the periodic table. Anyone denying this has simply been uninvolved in the modern development."[133]

Any settlement of this dispute was out of the question, and after this rebuttal Sommerfeld put out of his mind the "Starkiana," as he had labeled the folder containing the unpleasant evidence of this quarrel. Although his colleague Georg Joos (1894–1959) from the University of Jena thought some example of "Stark's nonsense" should be exposed "with relentless severity" so that it would be obvious even to "people at a greater distance" what to think of the statements of the Nobel laureate on modern physics,[134] Sommerfeld declined his advice. It was clear to the majority of his colleagues that in this quarrel, Stark had once more put himself in the wrong. "Heitler and I have read and discussed your 'Rebuttal,'" Born wrote Sommerfeld. "We both thought it splendid in both tone and substance: factually sharp and yet polite. It is very good that you publicly take Stark to task for not

132 To Grüneisen, October 9, 1930. DMA, NL 89, 024, folder, Starkiana.

133 Sommerfeld, *Erwiderung*, 1930.

134 From Joos, November 20, 1930. DMA, NL 89, 024, folder, Starkiana.

having read the original sources."[135] A letter arrived from Zürich from Aurel Stodola (1859–1942), a retired professor of engineering who had a lively interest in modern developments in physics even while mourning those bygone days when it was still possible to understand natural phenomena with classical mechanics. Stodola saw no profit in the "attacks of the hothead Stark." Obviously Stark was "barking up the wrong tree, and, with his 'light eddy,' which supposedly collides with the atomic ion, is merely yearning for an explanation according to the old mechanics of force and impact, whose time (sadly) is now past."[136]

10.9 On the Road Again

While the quarrel with Stark in the *Annalen der Physik*, in the *Unterrichtsblätter für Mathematik und Naturwissenschaften*, and in correspondence with numerous colleagues was still stirring passions, Sommerfeld went traveling once again. "I intend to go to Odessa to the Russian Physics Day," he notified Rubinowicz of a visit.[137] Lemberg (Lwów), now in Poland, where Rubinowicz was a Professor of Theoretical Physics at the Technical University, was on Sommerfeld's travel route. To be sure, this trip did not go off quite so comfortably as his recent world tour. The day he spent with Rubinowicz in Lemberg had been "by far the most pleasant of the whole trip," he wrote after his return from Odessa, "for in Soviet Russia, every comfort is gone. Nonetheless, the trip to the Black Sea was interesting and sunny, almost tropical."[138]

In October, 1930, Sommerfeld traveled to Brussels, where he had been invited to the sixth Solvay Congress. Perhaps in compensation for his not having been invited to the Congress in 1927, he was now accorded the fitting honor of delivering the opening lecture on the theme of the Congress: "magnetism."

Sommerfeld used the opportunity to address the old question of the "magneton" from the perspective of spectroscopy in light of the latest findings.[139] By way of preparation, he had chosen the topic of his special lecture course for the preceding summer semester, 1930 accordingly. "I haven't yet properly begun working on the Solvay report, but I am lecturing on the topic," he wrote Pauli, who was likewise to read a major paper at the upcoming Congress and was an important consultant especially on the subject of magnetism. In the same letter he mentioned almost in passing that he was just in the process of "building a small house."[140] The Sommerfelds completed the move from Leopoldstraße to number 6 Dunantstraße, their future address bordering the English Garden, just shortly before the Solvay Congress.[141]

135 From Born, November 13, 1930. DMA, NL 89, 024, folder Starkiana.
136 From Stodola, December 14, 1930. DMA, HS 1977-28/A,331.
137 To Rubinowicz, August 14, 1930.
138 To Rubinowicz, September 18, 1930.
139 Sommerfeld, *Magnetismus*, 1932.
140 To Pauli, June 24, 1930. DMA, NL 89, 003. Also in ASWB II.
141 To Vieweg, October 17, 1930. Wiesbaden, Vieweg-Archive, Sommerfeld.

Fig. 29: Sommerfeld and Auguste Piccard (1884–1962) at the Solvay Congress, 1930, in Brussels (Courtesy: Deutsches Museum, Munich, Archive).

Not long after the Solvay Congress, Sommerfeld went abroad once more. The *Institut Henri Poincaré* in Paris invited him to give talks in April 1931 on wave mechanics.[142] Unlike 1922, when on his trip to Spain he had experienced France still as enemy territory and had taken no pleasure in a stop in Paris along the way, this time he refused to let himself be governed by resentment. "I must revise my judgment that Paris is not a beautiful city," he wrote home following an initial stroll through the city. Nor did he feel it an imposition to deliver his lectures in French. "I have apparently not committed any linguistic errors, and scarcely needed to glance at my lecture notes."[143] Langevin, who at the Solvay Congresses before World War I had been very friendly to him, now too was particularly attentive. "Most elegant luncheon at his house," he wrote about an invitation to Langevin's. "Afterwards, climb up the Eiffel Tower accompanied by his son and son-in-law, boat-ride along the Seine, sight seeing in old sections of the city." One evening, he was taken to the opera. To be sure, on this visit to Paris he felt there was not as much interest in his lectures as he had hoped. Not even Langevin attended his lectures. "That is typical. No one has the time."[144]

142 Sommerfeld, *Problèmes*, 1931.
143 To Johanna, April 21, 1931.
144 To Johanna, April 25, 1931.

On his return from Paris, a 2-month stay in the USA was on Sommerfeld's travel schedule. "I would like very much to come to Ann Arbor in 1931," he had written Herzfeld in Baltimore already in July, 1929. Together with Epstein in Pasadena and Laporte in Ann Arbor, Herzfeld represented the Sommerfeld school in America.[145] Annually since 1923, summer courses had been offered in which in an informal atmosphere, advances in theoretical physics were discussed. The ambitious level of the lectures and discussions soon brought these courses into high repute and made Ann Arbor, even for European experts in theoretical physics, a desirable destination.[146] In the summer of 1931, Pauli and Kramers were, with Sommerfeld, among the distinguished guests from abroad. "I am convinced we are going to have a very nice time together," Sommerfeld wrote, delighted.[147] Since the event in Ann Arbor began as early as June, he had to absent himself from half of the Munich summer semester.[148] Millikan also sent him an invitation to a congress of the American Association for the Advancement of Science, scheduled for a week earlier in Los Angeles and Pasadena, but Sommerfeld passed up this detour to far-off California because for lack of time.[149]

When Sommerfeld boarded the "Columbus" in mid-June 1931 to make the voyage across the Atlantic, he gladly eased once again into the daily life of a comfortable ocean steamer. "Wake up at 7:15. First, ½ hour exercise, strenuous but necessary. Then a hot bath in seawater (in the tub)," he wrote his wife describing the daily routine on the crossing. "Then breakfast with a lot of good coffee, grapefruit, and emphatically declining all meat dishes. In good weather, shuffle-board on the sundeck, a very nice way to exercise, with an elegant young American woman and two American gentlemen. Lunch around 1:00, often with caviar as a starter." In the afternoons, he retired to his cabin. As reading matter for the journey he had brought a biography of Napoleon by Emil Ludwig (1881–1948) and *The Apple Cart* by George Bernard Shaw (1856–1950), "A Political Extravaganza," as the subtitle declared. "I did not enjoy the latter as much as I usually do Shaw. The former is exceptionally well written, interesting, and—so far as I can tell—credible." In the evenings, etiquette was in force. "Dinner at 7:30 at the Captain's table; naturally I have to get myself up in a tuxedo for that."[150]

Once arrived in Ann Arbor, the terrific heat put him in mind of his world tour. But the physicists of the University of Michigan went to great efforts to make their guests' life as comfortable as possible. In the company of Laporte, Pauli, and Walter Colby (1880–1970), he was taken "swimming in an isolated lake. Water, lukewarm.

145 To Herzfeld, July 25, 1929. DMA, NL 89, 002. Also in ASWB II.
146 Schweber, *Empiricist Temper*, 1986, pp. 78–79.
147 To Pauli, June 24, 1930. DMA, NL 89, 003. Also in ASWB II.
148 To the Philosophical Faculty of the University of Munich, January 14, 1931. UAM, E-II-N.
149 From Millikan, February 7, 1931. DMA, NL 89, 011; to Millikan, February 25, 1931. DMA, NL 89, 025.
150 To Johanna, June 19, 1931.

We swam until dark," he wrote, describing his first impressions. "Many fire-flies, larger than our homegrown and secretive glow-worms; they fly as high as the tops of the trees." He lodged in a fraternity house and was taken at every opportunity on excursions by car. He was driven about even close by the university. And even Pauli, whose social graces often left something to be desired, behaved "very nicely." The following day, Sommerfeld added a few lines to the letter. It was becoming "ever hotter," but in the evening they had enjoyed a "lovely swim in a lake." And of course he was again requested to play the piano. "Last evening I played two Beethoven sonatas with a (rather mediocre) violinist at the fraternity house."[151]

On this visit, too, he found the American lifestyle very much to his taste. "Everything is organized splendidly in this country: in instruction, financially, socially (dress is unbelievably casual for the men, so that I almost constantly run around in sandals and an Indian shirt). Everything geared towards having as good a time as possible, and accomplishing things with minimal effort and trouble—exactly the opposite of us!"[152] But as in the years 1923 and 1929, "this time, too, the pleasure of America [was] soured by politics," as he wrote home on July 12, 1931. He was again confronted with "once more alarming news about Germany" in the American newspapers: "extreme financial emergency, great French outrage, threat of resignation by Brüning and Hindenburg, continuing crisis despite Hoover plan." Three weeks earlier, the American president had proposed that German reparations payments to France and German war debts to the United States be suspended for 1 year. Sommerfeld thought such a moratorium was "tremendous," even if Hoover "had not acted out of friendship with Germany, but rather to rescue American capital in Germany and overseas business."[153] Added to his annoyance over politics in the larger picture came his concern over things at the University of Munich, where National Socialism was spreading like an epidemic. He read in the newspaper that two universities had been closed on account of National Socialist student unrest. "Was Munich one of them? What has happened in the election of the Rector?" he wanted to know from his wife. "I am really quite anxious about what is happening at home!"[154]

There had in fact been riots a few days earlier at the University of Munich. The trigger had been the lectures of the liberal constitutional law scholar Hans Nawiasky (1880–1961). On June 26, 1931, National Socialist students at the University of Munich had mounted an initial protest demonstration which gave the *Völkischer Beobachter* the pretext of further inciting the students over the "Nawiasky scandal." On June 30, the leader of the National Socialist German Student Alliance publicly attacked Nawiasky in a speech delivered in the atrium of the University of Munich. In the venue where normally academic speeches were delivered, the Horst-Wessel Song now rang out. The agitated students screamed "Heil Hitler," "Death to Jews,"

151 To Johanna, June 30, and July 1, 1931.
152 To Johanna, July 8, 1931.
153 To Johanna, July 12, 1931.
154 To Johanna, July 8, 1931.

and "Death to Nawiasky," in response to which the administration thought its only recourse was to order the police to clear the university and close it down for a week.[155]

But the American newspapers reported less about the situation at the University of Munich than about the foreign and economic political situation in Germany. "What is happening in Germany? That is the fearsome question I can't escape in conversation and in my solitary hours," Sommerfeld wrote his wife on July 24, 1931. "The economic devastation is clearly bad." As on his first trip to the USA, when growing inflation in Germany had worried him, now, too, he sent dollars and checks home in order to stave off economic distress at least domestically. The new house the Sommerfelds had occupied for the past year accounted for additional financial worries. "The question is: should I use my Ann Arbor earnings, which by the way will be paid only at the end, to pay house-related expenses, or should I leave a part of them in the U.S.?"[156] Mixed in with his private financial worries was anger over French politics which Sommerfeld blamed for the looming failure of the Hoover moratorium. "Bitterness over the French outrages is general, especially strong in the London papers, as an Englishman told me today, nor do the American papers offer any excuses for the French tactic of extortion, and the malicious stalling of the Hoover plan. Shame on *la grande nation*!"[157]

Because of worries over money and politics, his experiences at the summer school in Ann Arbor receded into the background—at least in personal letters to his wife. What he did report related to musical evenings with his hosts or the farewell party he and Pauli hosted for the professors and students "with their girls" towards the end of the summer school semester. For this occasion, they organized a "colloquium on hyperphysics," at which Sommerfeld "presided ceremonially," and "many comic lectures" were delivered. "Music followed, provided by a professional pianist (German-American), then dance. Served: ice cream and punch (nonalcoholic). A great success, general satisfaction."[158] This was surely not the only party Sommerfeld and Pauli threw in Ann Arbor, and no doubt not all these convivial events were quite so nonalcoholic. This is illustrated by a mishap that occurred just at the start of the summer school. Pauli, probably not entirely sober, sustained a complex shoulder break. Because of prohibition, the consumption of alcohol could not be openly acknowledged, but from Pauli's correspondence we learn that they did not have to suffer under excessive abstinence. The official version of Pauli's accident that was given out was that he had slipped and fallen at a swimming pool, but at the place in the correspondence relating this event, we find an exclamation mark. However it was that Pauli had sustained his injury, on top of it he had to endure the derision of his colleagues. He ran around with his arm extended in a cast "like a traffic cop signaling," one participant in the course wrote. Pauli himself

155 Behrendt, *Hans Nawiasky*, 2006.
156 To Johanna, July 24, 1931.
157 To Johanna, July 30, 1931.
158 To Johanna, August 19,1931.

seemed to take pleasure in the general amusement at his appearance and later added with a dose of self-irony that this was the only time he ever extended his arm in a Hitler salute.[159]

For all the informality, Sommerfeld's lectures and the discussions about the current problems of theoretical physics did sap his energy somewhat, so that ultimately he longed for "the tranquility of the ocean voyage" on his return.[160] As on his earlier trips, he had devoted his courses at Ann Arbor to his favorite topics of the theoretical physics of those years, electron theory of metals and wave mechanics.[161] Such rapid strides were being made in these areas that he could not rely on the lectures he had worked out earlier. Walter Brattain (1902–1987), at that time still at the dawn of his career, later to be awarded the Nobel Prize in physics as co-inventor of the transistor, recalled many years later how impressed he had been by Sommerfeld's lectures, which dealt directly with the area (thermionic emission) in which he was working at the time at Bell Laboratories. "Several of us had interesting discussions with him on some of the current problems of thermionic and field emissions, of which the theoretical interpretation was still in doubt."[162]

10.10 Consolidation of the New Theories

The thermionic and field emission of electrons, to which the inventor of the transistor referred, belongs to the electron theory of metals, which Sommerfeld, using the Fermi-Dirac statistics on the free electron gas, had established in 1927 as a promising subsection of theoretical solid state physics. In the early 1930s, too, electron theory of metals was still a focus of research at Sommerfeld's institute. Sommerfeld himself left the working out of details mostly to his students and confined himself to publishing the results in the role of coauthor. For example, he gave the work on the thermoelectric and magnetic properties of metals over to Nathaniel Frank (1903–1984), a physicist from the Massachusetts Institute of Technology who had come to Munich in 1929 on a grant from the National Research Council (NRC) in order to bring his theory into line with the latest results of research in this area.[163] To William Allis (1901–1999) and Philip Morse (1903–1999), who in 1930 had likewise come as NRC Fellows to Munich, he gave over the working out of a wave mechanical theory of scattering of slow electrons on

159 Cited in WPWB 2, p. 84.

160 To Johanna, August 19, 1931.

161 Symposium on Theoretical Physics and Courses in Physics. Summer Session, 1931, June 29 to August 21. University of Michigan Official Publication, Vol. XXXII, Nr. 54, April 4, 1931, p. 9.

162 Brattain to Goudsmit, December 15, 1955. Quoted in Schweber, *Empiricist Temper*, 1986, p. 78.

163 From Frank, January 28, 1929; to Frank, February 8, 1929. DMA, NL 89, 022, folder 9,35; to Frank, November 27, 1930. DMA, NL 89, 001; from Frank, December 15, 1930. DMA, HS 1977-28/A,101. Sommerfeld/Frank, *Statistical Theory*, 1931.

gas atoms and declined immortalizing himself as coauthor in the publication of the theory. Here was a case of explaining a phenomenon entirely incomprehensible in the absence of wave mechanics: the "Ramsauer Effect," discovered 10 years earlier. Experiments had shown that the weakening of an electron beam passing through a gas was not to be brought into line with the classical conception of particle collisions. At very low energies, the cross-section drops below the value that would have been obtained according to the gas kinetic theory, as though the gas atoms became more permeable for slow electrons than for high-velocity electrons. "The fundamental idea and impetus for this work," Allis and Morse wrote at the end of their theory published in the *Zeitschrift für Physik* "comes from Professor Sommerfeld."[164] He was content to make the results worked out at his institute public in a lecture to the Berlin Physical Society.[165]

In these years, the need for a consolidation of quantum mechanics, particularly in its significance for the solid state theory, was discernible in manifold ways. Editors of compilations and handbooks kept turning to Sommerfeld in hopes of persuading him to undertake a survey of the newer physical theories. In the fall of 1929, for example, the editor of the *Handbuch der Radiologie* wrote Sommerfeld from the Leipziger Akademische Verlagsgesellschaft, asking for "a concise presentation of the conductivity of metals from the viewpoint of quantum theory." Sommerfeld declined since at the time he was occupied with reworking *Atomic Structure and Spectral Lines* for the fifth edition, which appeared in 1931. But he proposed that the editor turn to one of his students. "Mr. Peierls is in Zürich. Perhaps he would not be disinclined to undertake the conductivity of metals for you. Bethe is with me in Munich. I would not personally encourage him to take on this work since he already has enough to do. You are of course free to make him the offer."[166]

Rudolf Peierls had begun his studies with Sommerfeld at Munich, completing them with Heisenberg at Leipzig. Thereafter, he became Pauli's assistant at Zürich. "Dr. Peierls works on the theory of heat conduction in solid bodies," as Pauli characterized his research area in 1929.[167] Following his doctorate under Sommerfeld, Bethe was also something of an academic vagabond, and quantum mechanical solid state theory was the research area in which he, too, made his name. As long as they occupied no secure professorships, they had to support their candidacies for openings with publications that were as innovative as possible. Comprehensive surveys, such as the editor of the *Handbuch der Radiologie* sought, were time-consuming and, as a rule, offered little space for the presentation of original research. Thus, they represented rather an obstacle to the pursuit of their personal careers. Sommerfeld counseled his protégés therefore against taking up such offers from the scientific publishers, though they could certainly be lucrative.

164 Allis/Morse, *Theorie der Streuung*, 1931.
165 Sommerfeld, *Theorie des Ramsauer-Effektes*, 1931.
166 To Marx, October 19, 1929. DMA, NL 89, 003.
167 From Pauli, May 16, 1929. DMA, HS 1977-28/A,254. Also in ASWB II.

In 1931, when Springer Verlag instituted a search for authors for two constituent volumes of the *Handbuch der Physik* on quantum theory and solid state physics, Sommerfeld was once more the first person they consulted. Bethe had already published several papers 3 years after his dissertation that qualified him for a pending professorship, so that Sommerfeld had no qualms about arranging for him to prepare an article for the quantum theory Handbook volume. Adolf Smekal (1895–1959) wrote gratefully to Sommerfeld in April 1931 that he was most pleased by this arrangement with Bethe, who had already accepted. Smekal, who had taken over the editorship of these volumes of the Handbook for Springer Verlag, added a request for one further article. In the volume planned for solid state physics, electron theory of metals and the theory of ferromagnetism were to be consolidated into a "quantum theory of the metallic state." Smekal courted Sommerfeld as the actual founder of this area: "There could be no greater contribution to the profession and to the *Handbuch der Physik* than if you could see your way clear to making a definitive presentation of your work and the research related to it."[168] Sommerfeld, however, did not want to take on the burden of this work alone. Bethe appeared to him to be the most suitable author for this task, although he had already passed on to him the article for the quantum volume. He would—he stipulated in forwarding Smekal's letter to Bethe—"accept [the offer] only if you take on 90% of both the work and the honorarium. Article to be signed . . . by A. Sommerfeld and H. Bethe." He "absolutely did not" wish to persuade him to take this on, and even cautioned him against "too much scribbling."[169]

Bethe was on a Rockefeller grant in Rome working with Fermi when this offer reached him. "In and of itself, this would of course be very attractive to me," he wrote thanking Sommerfeld, "but like you, I am afraid I am loading myself up with too much 'scribbling.'" He wanted to devote himself entirely to research during his stay in Rome. On the other hand, both the subject "and the quite substantial honorarium" seemed thoroughly attractive. He could only take on the assignment—he decided, after weighing the pros and cons—if he were permitted to deliver the article, not as Smekal had wished by January 1, 1932, but by, "say, April,'32."[170] Smekal accepted this condition "so entirely" that Bethe—as he wrote Sommerfeld several weeks later from Capri—saw himself "honor-bound, as it were," to take on this Handbook article, too.[171] The expectation of completing two Handbook articles in a year proved illusory, however. On the agreed upon date of submission, "only one chapter of the 1st Handbook article [was] finished," as Bethe confessed to Sommerfeld in April, 1932. Smekal granted Bethe an extension until August 1, 1932, but even this period proved insufficient.[172] It often became apparent only in

168 From Smekal, April 17, 1931. DMA, NL 89, 013. Also in ASWB II.
169 To Bethe, April 18, 1931. DMA, NL 89, 013. Also in ASWB II.
170 From Bethe, April 25, 1931. DMA, HS 1977-28/A,19. Also in ASWB II.
171 From Bethe, May 30, 1931. DMA, HS 1977-28/A,19. Also in ASWB II.
172 From Bethe, April 20, 1932. DMA, HS 1977-28/A,19. Also in ASWB II.

the act of writing that one or another aspect needed to be more closely researched before it could be cast in the form of definitive textbook knowledge for the Handbook. The "scribbling" demanded many months more of intensive and not merely authorial work. In the end, though, all concerned could be satisfied with the result. Bethe's article on the "quantum mechanics of the one- and two-electron problems" in the first part of volume 24 of the *Handbuch der Physik*, published in 1933, had the scope of an entire book and became a classic of modern physics.[173] It served as a model for many subsequent textbooks on quantum mechanics. The same is true of the article on the "electron theory of metals" in the second volume. The agreed upon listing order of the authors ("A. Sommerfeld and H. Bethe") was retained, although Sommerfeld contributed only the 36 page introductory chapter with his semiclassical electron gas theory, while Bethe in 254 pages presented the quantum mechanical theory of the behavior of electrons in rigid bodies.[174] The theory had thereby acquired a "definitive presentation," a grateful Smekal wrote Sommerfeld in November, 1933. "It is a quite signal honor for the other contributors to this volume of the Handbook to appear in your company."[175]

The *Handbuch der Physik* was the most celebrated, but not the only medium driving the consolidation of quantum mechanics, and the burgeoning theoretical atomic, molecular, and solid state physics forward. The *Handbuch der Radiologie* of the Akademische Verlagsanstalt, the *Müller-Pouillet* textbook series of the Vieweg Verlag, and others kept this trend in view. As in Bethe's case, one or another adherent of the Sommerfeld school was recruited for such survey articles.[176] Sommerfeld also often assigned his doctoral students of those years topics intended to underscore the importance of modern theory for a broad range of physical phenomena through the application of quantum mechanics to problems of solid state physics. Herbert Fröhlich (1905–1991) was, for example, to handle the photo effect on metals. In the case of a single atom, the emission of an electron resulting from the irradiation of light could be "very naturally and easily described" with wave mechanics—thus Sommerfeld began his report on Fröhlich's dissertation—but in the case of electrons of metal, a corresponding treatment presented "quite substantial difficulties."[177] Fröhlich remained committed to theoretical solid state physics and contributed to its dissemination and consolidation.[178]

173 Bethe, Quantenmechanik, 1933.

174 Sommerfeld/Bethe, *Elektronentheorie der Metalle*, 1933.

175 From Smekal, November 28, 1933. DMA, NL 89, 013.

176 Peierls, *Elektronentheorie*, 1932; Nordheim, *Statistische und kinetische Theorie*, 1934; Nordheim, *Quantentheorie*, 1934.

177 Vote on the dissertation of Herbert Fröhlich to the Philosophical Faculty, 2nd section, July 22, 1930. UAM, OC-Np-1930. Fröhlich, *Photoeffekt*, 1930.

178 Fröhlich, *Elektronentheorie*, 1936.

Two other dissertations completed in 1931 at Sommerfeld's institute dealt in addition to the consolidation of wave mechanics with questions in the area of the "Pacific problem."[179] Otto Scherzer (1909–1982) was to treat the scattering of protons pursuing the method Sommerfeld had developed for the braking of electrons and to explain why in the experiments conducted heretofore no proton bremsstrahlung had been observed.[180] August Wilhelm Maue (1908–1970) was to work out how the earlier solution for the X-ray bremsstrahlung found by Kramers according to the correspondence principle differed from the wave mechanical. Since Kramer's theory had been adduced for astrophysical problems, Sommerfeld hoped that with the theory employed by Maue heretofore inexplicable inconsistencies between theory and observation in astrophysics could be cleared up.[181]

Sommerfeld demonstrated with these papers that his Institute remained a very productive "nursery" of modern theoretical physics even more than 30 years after its founding. Theoretical solid state physics got its decisive boost in the early 1930s, and many of the pioneering publications originated in the Sommerfeld school.[182] Following the discovery of the neutron in 1932, nuclear physics also blossomed into a new subfield of physics, and here, too, quantum mechanics was the key to theoretical understanding. The university physics institutes in Germany could not keep pace with the explosion of knowledge in theoretical physics, so that even prominent theoreticians like Bethe faced a bottleneck in openings for professorships. Nonetheless, Sommerfeld's Institute and its "branches" in Stuttgart (Ewald), Hamburg (Lenz), Leipzig (Heisenberg), and Zürich (Pauli, Wentzel) remained for a few years still productive venues of the new physics.[183] "Still"—because in 1933 with the "seizure of power" of the National Socialists came decisive changes, which brought about a slow and painful end to the Munich "nursery," and had as a consequence the decline of modern theoretical physics in Germany altogether.

179 Sommerfeld, *Beugung*, 1931.

180 Vote on the dissertation of Otto Scherzer to the Philosophical Faculty, 2. Sektion, November 27, 1931. UAM, OC-Np-1931/32. Scherzer, *Ausstrahlung*, 1932.

181 Vote on the dissertation of August Wilhelm Maue to the Philosophical Faculty, 2. Sektion, November 27, 1931. UAM, OC-Np-1931/32. Maue, *Röntgenspektrum*, 1932.

182 Hoddeson/Baym/Eckert, *Development*, 1987; Eckert, *Sommerfeld*, 1990.

183 Eckert, *Atomphysiker*, 1993, Chaps. 6 and 7.

11 Descent

On January 30, 1933, Adolf Hitler (1889–1945) was named Chancellor of the Reich. Up until the Reichstag elections on March 5, 1933, Sommerfeld may still have foreseen a relatively brief reign for "Adolf the Great," as he scoffed at the new Chancellor.[1] Hitler's predecessors, Franz von Papen (1879–1969) and Kurt von Schleicher (1882–1934), after all, had held office only a few months. The National Socialists had suffered substantial electoral losses in the last Reichstag election in 1932. On March 5, 1933, however, the NSDAP more than made up for the electoral losses of the previous year, and with their "Führer" as Chancellor, the ascendency of National Socialism nationally too now seemed certain. Ensuing events left no doubt that the National Socialists were putting their "seizure of power" palpably and comprehensively into effect across all areas of society.[2] Unless they sought refuge abroad, political opponents were locked into concentration camps or murdered. On April 1, 1933, through acts of harassment and willful destruction, it was made palpable to Jewish owners of companies, medical practices, and other facilities what awaited them in Nazi Germany unless they withdrew from business life.[3]

On April 7, 1933, 1 week after this "Jewish boycott," the race obsession of the Nazis was extended to public service and thereby also to the universities, through a "Law for the Restoration of the Civil Service."[4] "Civil servants not of Aryan ancestry are to be placed in retirement . . . Civil servants who, in light of previous political activity cannot offer the guarantee of wholehearted and unwavering support for the national state may be dismissed from service." To be classified "non-Aryan," one Jewish grandparent sufficed.[5]

11.1 Consequences of the New Civil Service Law

In April 1933, Sommerfeld was on a lecture tour in Great Britain which he concluded on May 1, 1933, with the James Scott Lecture to the Royal Society in Edinburgh.[6] He planned to travel to the USA in June to represent the German

1 To Johanna, 20. February 20, 1933.
2 Frei, *Anmerkungen*, 1983; Wirsching, *Das Jahr 1933*, 2009.
3 Ahlheim, *Antisemitismus*, 2011.
4 Adam, *Judenpolitik*, 2003.
5 For specific wording, see http://www.documentarchiv.de/da/fs-antijuedische-verordnungen.html (31 January 2013).
6 To Johanna, April 21, 23, 25, and 28, 1933; to Richardson, April 23, 1933. Ransom, Richardson. Manuscript of the "Scott Lecture" in DMA, NL 89, 021, folder 9.9.

M. Eckert, *Arnold Sommerfeld: Science, Life and Turbulent Times 1868-1951*,
DOI 10.1007/978-1-4614-7461-6_11,

Physical Society at the World's Fair in Chicago and to receive an honorary doctorate from the University of Wisconsin at Madison.[7] He had already applied for leave for half of the summer semester, but when he learned, on his return from Great Britain, of the new Civil Service law and its consequences for his students, these plans became moot. "You probably do not know that my mother is Jewish," Bethe wrote Sommerfeld a few days after promulgation of the law. And therewith, Bethe's prospects for a professorship in Germany evaporated. It was "surely not to be assumed," he added, "that anti-Semitism would diminish in the foreseeable future, or that the definition of 'Aryan' would be revised. So for good or ill I probably have to face the facts, and try to find a place somewhere abroad." As a lecturer, Bethe was integrated into Sommerfeld's pedagogical enterprise but held no appointment. A teaching assignment for the summer semester at the University of Tübingen was supposed to supply him a modest income at least temporarily, but he now watched even this possibility slip away. It was a puzzle how anyone there could know of his "congenital defect." Be that as it may, a letter which he found "almost insulting" from Hans Geiger (1882–1945), Professor of experimental physics at the University of Tübingen, made unmistakably clear to him that he was not welcome there.[8] Two weeks later, his teaching assignment was withdrawn, as Paul Ewald, Bethe's future father-in-law, reported to Munich.[9]

In 1932, Ewald had been elected Rector of the Technical University of Stuttgart. In this capacity, only a few days after the promulgation of the Civil Service law, he took part in a conference of rectors at which there must have been a discussion of the "Jewish Question." The "Jewification" of the universities—said the Rector of the University of Berlin—had been facilitated because "many bolts had been left open that could have been thrown shut."[10] Although most of the rectors were against the law and a protest vote against it had actually been debated, Ewald recalled later, several of the rectors were closely aligned with the Nazis, and would have "immediately instituted a counter declaration." Thus, no unanimous protest was issued.[11] But Ewald did not wish to be made an enabler of Nazi politics. Back in Stuttgart, he wrote the Minister of Culture: "Since I cannot share the view of the national government with respect to the racial question, I request permission to vacate my office as Rector of the Technical University of Stuttgart, effective immediately, and so be removed also from the office of Pro-rector." He sent this extract from his letter of resignation to the Senate of the TH Stuttgart, with a copy to

7 To the Bavarian Ministry of Culture, March 28, 1933. UAM, E-II-N.

8 From Bethe, April 11, 1933. DMA, HS 1977-28/A,19. Also in ASWB II.

9 Ewald to Johanna Sommerfeld, April 21, 1933. DMA, HS 1977-28/A,88.

10 Record of the Rector of the TH Munich, Richard Schachner, quoted in Heiber, *Universität*, 1994, p. 297.

11 Ewald, Erinnerungen anlässlich des 175jährigen Jubiläums der Universität Stuttgart, 1979. University Archive, Stuttgart SN1/35, p. 17.

Sommerfeld, together with the news that the Ministry had just issued him the "requested release" from the office of rector.[12] "What are we actually to do with all the people whose lives in Germany have been made impossible?" Ewald wrote thereafter to Sommerfeld's wife. He was not even sure whether he himself numbered among those affected. With a Jewish maternal grandfather, he was considered "non-Aryan," although his service in World War I might count as "service at the front," and in line with the Civil Service law offer him exceptional status. On the other hand, because his wife was Jewish, the four Ewald children were branded as "non-Aryan." "In any case, our children have not the slightest prospects in Germany, and whenever the opportunity presents itself, I have to get out."[13]

When Sommerfeld learned of these developments, he felt compelled to abandon his plans for the trip to America. His stay in Great Britain had been "interesting and pleasant," Sommerfeld's wife reported to her sister following her husband's return. "I cannot say the same of our situation here." Ewald and his mother had announced they would be coming to visit in the next few days, "upsetting, because this is a case of profound troubles." She reported also about Otto Blumenthal, who had been taken into "protective custody" in Aachen.[14] Shortly thereafter, Sommerfeld withdrew his application for the leave that had already been granted. "Decisive for me was the wish to participate in the organizational revisions at our university, and not to be absent from the pertinent faculty deliberations," he explained to the faculty his forgoing the trip to America.[15]

In the matter of the "organizational revisions," set in motion with bureaucratic thoroughness, however, the faculty actually had no say. On May 24, 1933, the Bavarian Ministry of Culture directed the Rectors of the universities to distribute packets of questionnaires to "all members of the teaching staff," to be filled out and returned together with the certificates of Aryan ancestry required by the Civil Service law.[16] According to this law, Sommerfeld himself did not belong to the targeted group. His grandfathers and grandmothers had all been members of the Protestant religion. But because the name "Sommerfeld" was thought to sound Jewish, in this case the authorities were extraordinarily thorough. "My Aryan ancestry was confirmed by certification of the Bavarian Ministry of Education and Culture of July 18, 1933," Sommerfeld wrote 4 years later beside his signature at the

12 To the Senate of the TH Stuttgart, April 20, 1933, copy to Sommerfeld. DMA, HS 1977-28/A,88.

13 Ewald to Johanna Sommerfeld, April 21, 1933. DMA, HS 1977-28/A,88. Eckert, *Paul Peter Ewald*, 2011.

14 Johanna to Helene Rhumbler, May 6, 1933. Felsch, *Blumenthals Tagebücher*, 2011.

15 To the Philosophical Faculty, 2. Sektion, of the University of Munich, May 13, 1933. DMA, NL 89, 004. Also in ASWB II.

16 Schemm to the Rectors of the three Bavarian State Universities, May 24, 1933, forwarded by Rector (Leo von Zumbusch) to the professors of the University of Munich. DMA, NL 89, 030, folder Hochschulangelegenheiten.

bottom of another questionnaire, to which he responded with the same statements as in his "certificate of Aryan ancestry" from the year 1933.[17]

The ministerial decree of May 24, however, concerned not only the appointed university teaching staff but also instructors and lecturers who like Bethe did not have the status of civil servants. Every questionnaire was sent via the Rectory to the Bavarian Ministry of Culture, where the documents furnished were further scrutinized. As a result, the faculties that were directly affected by the threatened loss of their "non-Aryan" colleagues were unable to intervene in the bureaucratic process. No doubt for this reason, on July 5, 1933, at the meeting of the Philosophical Faculty of this summer semester, the measures decreed by the Civil Service law were not even on the agenda.[18] By this time, on the other hand, after questioning by the Rector, a number of lecturers had already been dismissed "effective immediately." In accordance with the ministerial decree of July 21, 1933, finally, other non-civil service professors and lecturers, among them Bethe, were stripped of the right to teach.[19] Thereafter, Bethe's father wrote to Sommerfeld that in this matter he had been "amazed" most at the total lack of any collegial solidarity. "Ultimately, until he leaves, a lecturer is a colleague of the other university teachers, so that collegiality would have demanded that the Rector express gratitude for the previous contributions of those affected." To Sommerfeld, he expressed his "regret that you have lost a good colleague," and closed with the postscript: "This letter is not confidential!"[20]

Sommerfeld heard of similar developments at other universities. Ludwig Prandtl reported just 6 days after the promulgation of the new Civil Service law that his assistant, Wilhelm Prager, who had just received the offer of a teaching position in mechanics from the Technical University at Karlsruhe, had been "dispatched again by his new bosses just before taking up the post because of his Jewish sounding name. . . He is, incidentally of three quarters German ancestry!"[21] In none of these cases was the dismissal blocked by resolutions of the faculty, which is not to be wondered at insofar as the traditional system of self-governance of universities was in this summer semester of 1933 revised along National Socialist lines. The election of the rector and the senators, which was actually among the organizational measures Sommerfeld planned to have a hand in, was by decree of the Ministry of Culture "deleted from the agenda," as was succinctly noted in the minutes of the faculty meeting of July 5, 1933. For the election of the Dean, the faculty also awaited "further instructions from the Ministry."[22] Although professors were still free to offer suggestions, decisions emanated henceforth from the Ministry. The most

17 To the Rector (von Zumbusch) of the University of Munich, June 10, 1933. UAM, E-II-N; from Zumbusch, July 29, 1933. DMA, NL 89, 024, folder Nazizeit; Fragebogen, signed May 3, 1937, in Sommerfelds personnel file. BayHStA, MK 35736.

18 File 1d: Meeting of the Philosophical Faculty, 2. Sektion. UAM, OC-III-27.

19 Böhm, *Selbstverwaltung*, 1995, pp. 112–115.

20 From Albrecht Bethe, August 11, 1933. DMA, NL 89, 024, folder Nazizeit.

21 From Prandtl, April 13, 1933. MPGA, III. Abt., Rep. 61, Nr. 1538.

22 File 1d: Meetings of the Philosophical Faculty, 2. Section. UAM, OC-III-27.

important office in the university, the position of rector, was, in conformity with the "Führer Principle," to be held only by people who, as ardent Nazis, enjoyed the trust of the Ministry.[23]

Nonetheless, there were some measures that could be taken on the faculty level in one case or another that ran counter to the prevailing spirit. "In light of Professor Sommerfeld's explanation of the situation, the faculty and the Dean approve the granting of the doctoral degree to Mr. Romberg."[24] With this terse determination, the student Werner Romberg (1909–2003) was enabled to complete his studies in the summer semester of 1933. Later on, this would presumably not have been possible. Romberg was a member of a Socialist student group.[25] The University had previously withheld a prize for a theory of the "polarization of the light of canal rays" from him "because he lacked the necessary maturity." The actual reason for denial, aside from Romberg's socialistic convictions, was his distant relation to the Bavarian Prime Minister and Independent Social Democrat Kurt Eisner, who had been murdered in 1919. Sommerfeld nonetheless arranged for justice to be done his student by accepting the prize submission as a doctoral dissertation, making it possible for Romberg to complete his studies. Thereafter, he smoothed his path to emigration. Romberg was "well versed in wave mechanical calculations," he wrote in a letter of recommendation; in addition, he was "enthusiastic about going to Russia; he would integrate well into conditions there." To this allusion to Romberg's political convictions he added the admonition "to be somewhat cautious" in formulating any reply to Romberg "lest you endanger him somehow with the local authorities here."[26]

Worries about Romberg and other of his circle of friends and acquaintances whose academic careers in Germany the Nazi politics were making impossible became a constant burden for Sommerfeld in these weeks. Max von Laue, who took office in 1933 as Chairman of the German Physical Society, asked him and other directors of institutes for the names of those in their respective institutes who were "affected by the Civil Service law," in order to set relief measures in motion.[27] Sommerfeld named Bethe, Herbert Fröhlich, and Walter Henneberg (1910–1942), a recent doctoral recipient. Anyone who, like Romberg, had by this time still not completed a degree, and therefore did not qualify as a candidate by an employment bureau for a position in industry or abroad, was not included in the relief measures. "The names are most important to me," Paul Ehrenfest, who was supporting the initiative from Holland, wrote to Sommerfeld. "Because of the very great moral pressure, I would prefer that those affected not write to me directly."[28]

23 Böhm, *Selbstverwaltung*, 1995.

24 File 1d: Meeting of the Philosophical Faculty, 2. Section. UAM, OC-III-27.

25 Eckert, *Atomphysiker*, 1993, Chap. 7. Interview of Werner Romberg, October 8, 1985.

26 Carbon copy of an undated letter of recommendation to an unnamed physicist in the Soviet Union [after July 26, 1933]. DMA, NL 89, 030, folder Gutachten.

27 From Laue, May 21, 1933. DMA, NL 89, 024, folder Nazizeit. Also in ASWB II.

28 From Ehrenfest, May 21, 1933. DMA, NL 89, 024, folder Nazizeit. Also in ASWB II; to Ehrenfest, June 11, 1933. AHQP, Ehr; Eckert, *Atomphysiker*, 1993, Chap. 7.

Sommerfeld could empathize with Ehrenfest's wish, for in these weeks he was himself repeatedly subject to this "moral pressure." It was not only from Ewald and Bethe, to whom he was particularly close, that he learned firsthand what it meant to be suddenly relegated to limbo, outstanding personal accomplishments notwithstanding. Ludwig Hopf, who was among those affected by the Civil Service law in Aachen, described these feelings in especially vivid terms. "But being expelled from the only society one has known, out into empty space, is very painful. The human being is not a solitary creature, and these past days have taught me the true meaning of home, fatherland, and folk (not of course in the Nazi sense). To go abroad would constitute exile for me, and I would do this only under duress so that the children can once more have a home." When Hopf wrote this to Munich, he still hoped that as a former "frontline soldier" in World War I, he would fall under the rule of exception of the new Civil Service law. "And common sense tells me that in 3 months most of this will be repealed. A people that has complained for 14 years about the law of exception directed against it, after all, cannot itself impose laws of exception."[29] Several weeks later, he retrospectively described his state of mind in the first days as "terrible." Then he had forced himself to the insight "that we are no worse off than everyone else we took to be our fellow countrymen—they just haven't realized it yet."[30]

Towards the end of the semester, Sommerfeld fell ill and was forced to cancel his lectures. Now he himself was the focus of concern. "We all deeply regret your decision to break off your lecture course for reasons of health," 24 of his devoted students wrote. "From the bottoms of our hearts we wish you full recovery over the course of the vacation, and hope that next semester we may once more call ourselves your students."[31] The fact that Sommerfeld preserved this letter and filed it under "Nazi Period" hints that he blamed his illness on the nervous exhaustion the "cleansing" of the universities by the Nazis had caused him. He suffered primarily from insomnia and depression. At the beginning of the vacation, he sought recuperation in Lautrach, where Hermann Anschütz-Kämpfe, benefactor of the University of Munich, had placed his mansion at the disposal of the professors. But even in the idyllic setting of the Allgäu Alpine foothills, the nervous strain did not abate. He felt he made "poor company, because the nights have been bad again," and he had "suffered through it with the help of Compral [headache tablets], reading, etc.," as he wrote home to his wife. Only after several days did his condition improve. Depression gave way in his letters to more characteristic confidence—mixed with allusions to Nazi propaganda. "Paraphrasing the government announcements 'East Prussia free of the jobless,' I can report 'East Prussian free of sleeping pills for several days,' with mostly good success."[32] He was resorting only "very exceptionally"

29 From Hopf, May 24, 1933. DMA, HS 1977-28/A,148. Also in ASWB II.

30 From Hopf, June 28, 1933. DMA, HS 1977-28/A,148. Also in ASWB II.

31 From the "Members of the Physical Institute," July 22, 1933. DMA, NL 89, 024, folder Nazizeit.

32 To Johanna, August 7, 1933.

to sleeping pills and was feeling much better, he reported 1 week later.[33] In late August, he traveled to the vacation resort Zirmer Hof, a mountain hotel near Bolzano, to convalesce further. "Good weather, but today almost cool. The ascent on Sunday itself was bracing in tingling air. I sleep well both at night and in the afternoon," he wrote his wife. He asked for the galleys of the Handbook article on electron theory of metals to be sent to him, so that with respect to physics, too, he could look to the future.[34]

But even in the mountains of the South Tyrol, events over which Sommerfeld had lost sleep those previous weeks again caught up with him. "I understand only too well that you are very sad over present circumstances!" Max Born wrote him from a vacation spot in the Dolomites. According to the new Civil Service law, Born was "non-Aryan" and already in April 1933 "retired" from his Göttingen professorship. In May, he had left Germany and in August had requested a permanent leave from the Prussian Ministry of Culture. "Old man Planck," Born continued, "would have preferred to see me request leave for a few years, so I could return later. But that struck me as dishonorable. One cannot serve a state that treats one as a second class citizen, and one's children even worse."[35]

Many regarded the Nazi politics in the summer of 1933 as nothing more than an overreaction with which the new government was making a dramatic entrance. Planck had "spoken with the head of the government," as Heisenberg put it in a letter to Born in July 1933, "and received assurances that the government would not undertake any measures beyond the new Civil Service law that might encumber our science."[36] As President of the Kaiser Wilhelm Society and as "Permanent Secretary" of the Prussian Academy of Sciences, Planck felt duty-bound to mitigate at least the worst consequences of Nazi politics for science, without distancing himself from the ruling elite. Einstein had, "through his political behavior, himself made his continuing in the Academy impossible," Planck had commented on Einstein's resignation from the Prussian Academy of Sciences.[37] For his part, Einstein had the highest regard for Planck scientifically and personally but could not understand his loyalty to the Nazis. He "would not have remained as President of the Academy and the Kaiser Wilhelm Society even had he been a *goy*," he told an American colleague in 1934 speaking of Planck.[38] Schrödinger—though not himself affected by the Civil Service law—had also given up his professorship at Berlin and, as Einstein expressed it, "thrown their trash at the feet of the pirates."[39] In this case

33 To Johanna, August 14, 1933.
34 To Johanna, August 30, 1933.
35 From Born, September 1, 1933. DMA, NL 89, 006. Also in ASWB II; Szabó, Vertreibung, 2000, pp. 414-416.
36 Quoted in Hoffmann, *Max Planck*, 2008, p. 92. On Planck's visit to Hitler, see Albrecht, *Max Planck*, 1993.
37 Quoted in Hoffmann, *Max Planck*, 2008, p. 87.
38 Ibid.
39 Quoted in Meyenn, *Entdeckung*, 2011, p. 511.

as well, Planck tried to change Schrödinger's mind. "You will be as unhappy as I over Schrödinger's leaving Germany," Heisenberg wrote Sommerfeld at the end of the semester break. In the same breath, he expressed his excitement of the just discovered antiparticle of the electrons, the positrons, which had been theoretically postulated by Paul Dirac. "Let's hope you get enough of a break from politics in the winter semester to take pleasure in the positive electrons."[40]

11.2 A Deceptive Normalcy

But even after the summer of 1933, "rest" was out of the question. Karl Leopold Escherich (1871–1951), a zoologist and pioneer of National Socialism because of his early relationship with Hitler's party, was named by ministerial directive "Führer of the University,"[41] as the position of rector was now designated. The new rector enjoyed great respect among his professorial colleagues, certainly, and by no means exercised his office as a compliant accomplice of the Party. As would soon become evident, danger threatened less from the rector than from the "lecturer corps," an organization of ardent Nazi lecturers, assistants, and non-civil service lecturers, whose mission became the implementation of Party goals in the universities and, with the blessing of the Party, control of the entire academic enterprise.[42]

In the winter semester of 1933/1934, however, there was still no sign of this, so that the academic day-to-day ran almost as usual. That this normalcy was treacherous, however, Sommerfeld experienced in many ways. On December 19, 1933, Ludwig Hopf asked him "for a favor," not for him personally but for his son, who in 1932 as a student at the University of Munich had put his signature to a list of anti-Fascist students and who—though no longer a student at Munich—was now supposed to submit to a "disciplinary investigation." His father suspected this was to formalize his son's exclusion from the University, which might well be "a matter of complete indifference" to him, since the entire family sooner or later would be forced to emigrate. "It would not be a matter of indifference if, however, the police were then to proceed to move against Hans." He asked Sommerfeld to make discrete inquiries in the matter. If there was no threat of danger, he wanted "to let Hans quietly spend his Christmas in the bosom of his family; on January 3, he plans to travel to London. He will go at once only if you write that the University intends harsher punishment than mere expulsion."[43] Sommerfeld carried out the requested reconnaissance, which seemed to indicate that Hans Hopf was to be "administratively" excluded from study at the University of Munich, since he was

40 From Heisenberg, October 9, 1933. DMA, HS 1977-28/A,136. Also in ASWB II.

41 Literally "Leader of the University." But "Führer" had taken on the political connotation lent it by Hitler's assumption of that title.

42 Böhm, *Selbstverwaltung*, 1995, pp. 150–168.

43 From Hopf, December 10, 1933. DMA, HS 1977-28/A,148.

suspected of membership in the Communist Party of Germany. "Personally, I would not view this affair so tragically," the father wrote back, since "the boy is now to pursue his studies in England in any event." Of course, he bid him take into account, "the police might be ordered to refocus on Hans, and perhaps revoke his exit visa, or put his liberty in jeopardy." His son had doubtless been "Communistically active" and had perhaps even joined "some Communist aid organization, if only a charitable one. . . The question for us is just this: is there a direct line from the University court to the police?" Should this be the case, he wished to send his son to England at once. If yes, he asked Sommerfeld to send him a telegram with the message "Bon Voyage." If he was "simply being alarmist," he requested a telegram with the message "Merry Christmas."[44] The situation appears to have run its course innocuously, since there is no record of any police persecution. But the incident does demonstrate how what was in truth an "absurd farce"[45] might suddenly turn threatening given the new state of affairs.

In the daily life of the University, one grew accustomed to the juxtaposition of National Socialist and traditional academic routines. Admission to the university was governed by constantly changing decrees and laws. Students were trained through work duty, political indoctrination, and martial sports exercises in the ethos of Nazi ideology. From September 1934, "non-Aryan" students were by decree of the Ministry of Culture no longer allowed to matriculate in lecture courses. By the "Reich's Habilitation Order" of December 1934, prospective university teachers were required to attain the academic degree of Dr. Habil and at "lecturer training camps" to demonstrate their "personal aptitude and the suitability of their character as teachers at the universities of the National Socialist State."[46]

In practice, however, the academic climate was quite dependent on the particular individuals who functioned pedagogically as professors, assistants, and advanced students. Thus, a "non-Aryan" female student was with Gerlach's help still able in 1935 to complete her doctoral dissertation in experimental physics, even though professors were prohibited by decree of the Ministry of Culture "on their own initiative" to allow "non-Aryan" students to matriculate in their lecture courses. Without Gerlach's support, this student would surely not have been able to complete her dissertation. The other doctoral candidates in Gerlach's institute expressed their Nazi displeasure at her presence unmistakably and attempted to exclude her. By contrast to her experience there, she recalled the neighboring Sommerfeld institute as an "oasis."[47]

Sommerfeld also took on the task of finding places abroad for physicists driven from their positions in Germany whenever an opportunity arose. Asked by the Chairman of the physics department at Duke University in North Carolina to identify a young theoretician suitable for a possible appointment from a list of

44 From Hopf, December 16, 1933. DMA, HS 1977-28/A,148.

45 From Hopf, December 10, 1933. DMA, HS 1977-28/A,148.

46 Olenhusen, *Studenten*, 1966; Böhm, *Selbstverwaltung*, 1995, pp. 184–186.

47 Interview with Gertrude Scharff-Goldhaber, June 13, 1985.

émigré physicists then circulating in the USA, Sommerfeld recommended Ewald first. At 45, he was not exactly young but had a family with four children and his mother to support. In second place, Sommerfeld named the 50-year old Fritz Reiche (1883–1969), whose name was not even on the list. Only in third place did he name younger theoreticians such as the 35-year old assistant to Ewald, Carl Hermann (1888–1961), and the 34-year old Lothar Nordheim, who was ultimately appointed to the position.[48]

Though Ewald, Bethe, and other Sommerfeld students had emigration in view, only rarely was evidence of the attendant troubles and anxieties discernible in their letters to Munich. After the summer semester of 1933, Bethe traveled first for several months to Manchester, where he learned the methods of structural analysis of alloys in the laboratory of William Lawrence Bragg, and tried "to formulate the corresponding theory." In a multi-page letter, he lapsed positively into rhapsodizing over recent discoveries, from solid state to cosmic ray physics.[49] He expressed his expectation of seeing Sommerfeld again in the fall of 1934 at a congress on the physics of metals in Geneva.[50] From this correspondence, one might think that after the uneasy summer of 1933, the physicists had quickly returned to life as usual. But this impression is deceptive. When Sommerfeld traveled once more to the South Tyrol to recuperate at the Zirmer Hof after an outwardly uneventful semester, he took the opportunity to compose a long letter to Einstein. "It's been so long since we heard directly from one another!" he began, giving vent to his feelings. "And so much has happened since then! We saw each other last in Caputh in 1930. It seems as though the turmoil the world has lived in since 1914 just goes on." Following Hitler's "seizure of power," Einstein had not returned to Berlin from a trip to Belgium and had made the USA his new home. "Sadly, I cannot excuse my countrymen for all the injustice done you and many others, nor excuse my colleagues from the Berlin and Munich Academies." Sommerfeld referred to the fact that Einstein had not only been driven from Germany as a citizen by the Nazis but had also been expelled from the ranks of the academies by his own colleagues. Sommerfeld assured him that he was nonetheless still present in the physics lecture hall at Munich. "It might interest you to know that, as in previous years, I brought my winter lectures on electrodynamics to a conclusion in the four-dimensional, and topped that off with an introduction to the special theory of relativity. The students were enthusiastic: not one voice in opposition. The same with the summer course on optics, which I began with your optics of media in motion. Not once has mention of your name

48 From Edwards, November 24, 1933. DMA, HS 1977-28/A,75; to Edwards, January 22, 1934. DMA, NL 89, 015.

49 From Bethe, December 23, 1933. DMA, HS 1977-28/A,19. Also in ASWB II; from Bethe, February 25, 12. March 12, and April 12, 1934, DMA, HS 1977-28/A,19.

50 From Bethe, May 7, 1934. DMA, HS 1977-28/A,19. Also in ASWB II, pp. 412–414.

evoked protest. Please take this as evidence that the German student has long since tired of the intellectual tyranny to which a small group of 'leaders'[51] would like to yoke him, and that he longs for the pure air of the intellect." A reply was unnecessary, he added, since no letter from Einstein would ever reach him in Germany, and for the same reason he was writing this from Italy. Should Einstein wish to reply, he should write him at Geneva, where Sommerfeld would participate in a conference on the physics of metals in October and would be able to receive mail at the "discrete address" of a Geneva colleague.[52] In an unsent letter draft, he had formulated one additional sentence, which he later crossed out: "Incidentally, rest assured that, because of our rulers' abuse of the word 'national,' I am cured of the nationalistic sentiment I once so distinctly espoused. I no longer have any objection to the ruin of Germany as a power, and its assimilation into a peaceful Europe."[53]

Travel had been a perennial source of physical and spiritual renewal for Sommerfeld, but under the new political circumstances, travel abroad took on new significance. It enabled him at least for short periods to escape the increasingly stringent restrictions of the new rulers. "Geneva has been outstanding in every regard, emotionally and scientifically," he wrote home at the conclusion of the conference on the physics of metals.[54] On the return journey, he stopped over at Zürich to visit his friend and former colleague Robert Emden, who had immigrated to his native Switzerland after the Nazis had driven him from Munich. In Geneva, Sommerfeld met with Bethe and Peierls, who in their contributions to the conference laid out the ascendancy abroad of the quantum mechanical solid state theory.[55] He must have been painfully conscious on this occasion of the fact that this theory, developed in his Munich school, had now been driven into exile along with its best exponents. Heisenberg, Hund, and Debye still maintained a flourishing pedagogical and research enterprise at Leipzig, but elsewhere, modern physics was living a shadow existence. Sommerfeld wrote a rather wistful letter of congratulations to his colleague Ludwig Prandtl at Göttingen on his 60th birthday. Around this time, Prandtl's field of specialization, aerodynamics, was being strongly patronized by Hermann Göring's (1893–1946) Ministry of Aviation. "May you long be a refuge of research in our terribly devastated German science!"[56]

51 Sommerfeld's term is "Führer" in its plural form.
52 To Einstein, August 27, 1934. AEA, Einstein.
53 To Einstein, August 26, 1934. DMA, NL 89, 015. Also in ASWB II.
54 To Johanna, October 21, 1934.
55 Schweizer Physikalische Gesellschaft, *La théorie des électrons*, 1934.
56 To Prandtl, February 4, 1935. MPGA, III. division., rep. 61, nr. 1538. On the upswing in aerodynamics, see Eckert, *Dawn*, 2006, Chap. 8; on support for research under National Socialism in various sciences, see Maier, *Rüstungsforschung*, 2002; Flachowsky, *Notgemeinschaft*, 2008.

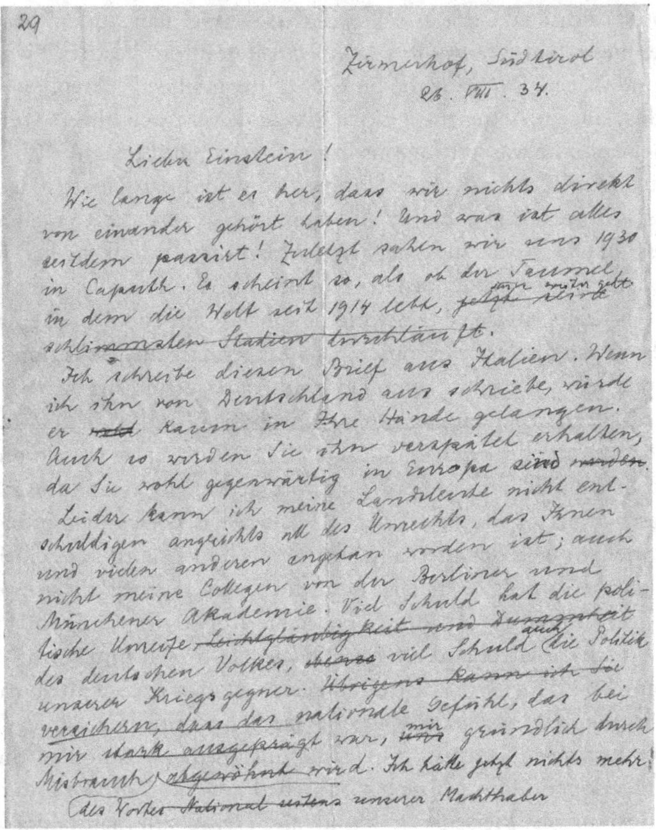

29

Zirmerhof, Südtirol
26. VIII. 34.

Lieber Einstein!

Wie lange ist es her, dass wir nichts direkt von einander gehört haben! Und was ist alles seitdem passiert! Zuletzt sahen wir uns 1930 in Caputh. Es scheint so, als ob der Taumel in dem die Welt seit 1914 lebt, [immer weiter geht] ~~schlimmeren Stationen~~ ...

Ich schreibe diesen Brief aus Italien. Wenn ich ihn von Deutschland aus schriebe, würde er kaum in Ihre Hände gelangen. Auch so werden Sie ihn verspätet erhalten, da Sie wohl gegenwärtig in Europa sind.

Leider kann ich meine Landsleute nicht entschuldigen angesichts all des Unrechts, das Ihnen und vielen anderen angetan worden ist; auch nicht meine Collegen von der Berliner und Münchener Akademie. Viel Schuld hat die politische Unreife ~~Leichtgläubigkeit und Dummheit~~ des deutschen Volkes, viel Schuld [auch] die Politik unserer Kriegsgegner. ~~Übrigens kann ich Sie versichern, dass das~~ nationale Gefühl, das bei mir ~~stark ausgeprägt~~ war, [mir] gründlich durch Missbrauch ~~abgewöhnt~~ wird. Ich habe jetzt nichts mehr! des Volkes ~~National~~ seitens unserer Machthaber

Fig. 30: First page of a draft letter to Einstein, which Sommerfeld penned at his vacation lodgings in the South Tyrol. Einstein had previously been compelled by his "colleagues" to withdraw from the Academies of Munich and Berlin (Courtesy: Deutsches Museum, Munich, Archive).

11.3 "Retirement" with Postponement

With this congratulatory wish, Sommerfeld was thinking not only of the fate of science in Germany but also of his own future and that of his Munich "nursery." On December 13, 1934, Hitler's cabinet had decreed a law "occasioned by the restructuring of German university education," which, together with the Reich's Habilitation Decree promulgated shortly before, established the personnel policy requirement that subordinated academic and technical universities even more strictly to the Nazi agenda. "Pursuant to this law, the civil service university teachers of the German Reich are to be relieved of official duties at the conclusion of the semester in which they complete the 65th year of their lives," read the first paragraph of this law, which came into force on January 21, 1935.[57] Most older professors did not exhibit the revolutionary fervor of their younger academic colleagues. Now, they were to vacate their teaching chairs to make way for younger professors who had been indoctrinated at training camps for lecturers with the correct ideological underpinnings for their impending careers. The "Law Pertaining to Retirement and Transfer of University Teachers," as it was officially named, brought an end to the long-standing academic practice by which professors determined their own individual and quite various retirement ages.[58] At the time this law came into force, Sommerfeld was 66 years old and thus among those affected. "Just now, when so many good theoreticians have been driven from Germany, we would have had need of your leadership in Munich for quite some time yet," Heisenberg wrote him when he learned of Sommerfeld's impending "retirement." "And I need not tell you how much the 'Sommerfeld school' has meant for our science. But politics follows its own laws."[59]

But Sommerfeld's colleagues at Munich were not so quick to give in. Heinrich Wieland immediately formulated a request that the faculty appeal to the Rector for an extension of Sommerfeld's employment. The mathematicians joined in this appeal. Sommerfeld's reputation as an internationally respected teacher and researcher justified the ruling of an exception. "Students stream in from near and far to study with this teacher who continues to function with undiminished creative powers. To retain him is most urgently in the interests of the young students who seek out our Alma Mater."[60]

The Rector forwarded the appeal to the Bavarian Ministry of Culture. When no response was forthcoming, the faculty secured the support of the lecturer corps, headed up until then by the astronomer Wilhelm Führer (1904–1974), an activist thoroughly imbued with Nazi ideology, who shortly was to continue his career in the Bavarian Ministry of Culture and then in the Berlin Reich's Ministry of Science, Education, and Popular Education.[61] Führer and his friend from student days Bruno Thüring (1905–1989)[62] would soon orchestrate the ignominious end of the Sommerfeld "nursery." In 1935, however, there was as yet no hint of this. The appeal of the faculty went with Führer's approval via the Rector to the Ministries of Culture in Munich and Berlin. There, however, sentiment ran against a ruling of exception for Sommerfeld. "Pursuant to the law, from the end of March, 1935, you are herewith relieved of official duties," the Minister wrote Sommerfeld. But since no successor could possibly be found in just a few weeks for the summer

57 http://www.documentarchiv.de/ns/1935/beamte_hschule_ges.html (31 January 2013).

58 Böhm, *Selbstverwaltung*, 1995, pp. 186–188.

59 From Heisenberg, January 18, 1935. DMA, HS 1977-28/A,136. Also in ASWB II.

60 Wieland to the Philosophical Faculty, 2. Sektion, February 6, 1935; Mathematical Seminar of the University of Munich to the Philosophical Faculty, 2. Sektion, February 9, 1935. UAM, E-II-N Sommerfeld.

61 Litten, *Astronomie*, 1992, pp. 237–238.

62 Ibid., p. 256.

semester, in the same letter, he lifted the relief of duties he had just decreed and entrusted to him "the care on a substitute basis of your previous teaching chair for the summer, 1935."[63]

So far as the matter of his successor was concerned, Sommerfeld had long since formed a firm opinion. In 1927, as a condition of his declining the Planck succession in Berlin, he had obtained from the Bavarian Ministry of Culture the promise that his institute would be granted an associate professorship, which he intended to fill with Heisenberg. "You will thereby be entitled after a number of years to become my successor in the full professorship," he had painted the future of his teaching chair to both Heisenberg and himself in 1927.[64] Then Heisenberg was appointed at Leipzig, while simultaneously Bavarian financial conditions had never improved to the point where the Ministry of Culture could have honored the promise of the associate professorship. Nonetheless, Sommerfeld had never backed off his position that Heisenberg should be his successor. When he informed Heisenberg of his imminent removal from his chair pursuant to the retirement law, he was simultaneously announcing that the long entertained plan would soon become a reality. Heisenberg also assured his teacher immediately that he would "make every effort . . . to uphold the tradition of the 'Sommerfeld school' should fate place me in this position." Since he was sure his political "past life" would play a role in the process, he at once informed Sommerfeld of the relevant facts. He had "taken part in the struggle against the soviet republic in Munich" and had aligned himself with the "Deutsche Freischar," a German youth organization. He had never joined a political party.[65]

Once the appeal of the faculty to exempt Sommerfeld from the retirement law "by resolution of the Reich's Minister of Culture" was denied, the question of his successor could no longer be postponed. On April 24, 1935, Sommerfeld presented to the faculty the proposal of the hastily convened appointment commission that Heisenberg should be placed first on the candidate list, Debye second, and Richard Becker (1887–1955) third. "The Dean asks the faculty whether it is in agreement with the proposal read out, and determines that it has the consent of the faculty." Nor is there any mention in the minutes of opposition on the part of the two representatives of the corps of lecturers present, Wilhelm Führer and Ernst Bergdolt (1902–1948).[66]

The following day, the virtually unaltered proposal—with Heisenberg and Debye together atop the list and Becker in the second spot—still lay on the Rector's desk.[67] It was nonetheless clear to Sommerfeld and the faculty that only Heisenberg

63 From Rust, carbon copy, prepared on March 23, 1935, conveyed on April 1, 1935. UAM, E-II-N Sommerfeld.

64 To Heisenberg, June 17, 1927. DMA, NL 89, 019, folder 5,9. See Chap. 9.

65 From Heisenberg, January 18, 1935. DMA, HS 1977-28/A,136. Also in ASWB II.

66 File 1d: Meetings of the Philosophical Faculty, 2. Section. UAM, OC-III-28. On Bergdolt, see Böhm, *Selbstverwaltung*, 1995, p. 402.

67 Senates Files. UAM, Sen-I-272; Litten, *Mechanik*, 2000, Chap. 1.2.3.

was in contention. The presence of the names of Debye and Becker is to be chalked up to the convention requiring that at least three candidates be put up. As full professor of theoretical physics at the Technical University, Berlin, Becker was as highly regarded a modern theoretician as Debye, with a sense for practical applications. Heisenberg was firmly counting on the appointment but hoped—as did Sommerfeld himself—that the Ministry would somewhat extend the postponement beyond a single semester. "It is also very important for theoretical physics in Germany that you continue to collaborate as long as possible, in order to bring up a younger generation. So, let's hope for the best!" Heisenberg wrote in June 1935 to Sommerfeld. He had also heard that a change was to be made at Munich in experimental physics as well. The Technical University of Berlin sought a successor to Gustav Hertz, and the rumor was that Gerlach had been chosen, Heisenberg informed Sommerfeld, "is anything known about this at Munich?"[68]

As two of the prominent experimental physicists in Germany, Gerlach and Hertz were the favored candidates already at the succession to Willy Wien. Führer suggested to the Rector that as long as this question remained unresolved, the succession to the chair in theoretical physics at Munich should also be deferred.[69] This was thoroughly in line with the wishes of Sommerfeld, Heisenberg, and the faculty. "Several signs indicate that I will be asked to continue my teaching this winter," Sommerfeld wrote to Leipzig. "I have told our Rector, though, that in that case the extension should be made till the end of my 68th year of life, so that both I and my students know where things stand."[70] The Rector forwarded the request to the Bavarian Minister of Culture and quickly received the reply that he had requested that the "Reich's and Prussian Minister for Science, Education, and Popular Education" in Berlin defer the replacement.[71]

The requested deferment was approved, but so far as the candidacy for the Sommerfeld succession was concerned, Berlin had other plans. Sommerfeld's dream candidate, Heisenberg, was supposed to take up the empty chair of Max Born at Göttingen as soon as possible. The administration of the Kaiser Wilhelm Institute for Physics, established in Berlin and soon to be opened, awaited Debye. So two of the three candidates on the Munich appointment list were unavailable. "Spoke yesterday with Bachér," Debye wrote Sommerfeld after a conference with the relevant official at the Berlin Ministry of Science. "He would like for 'the Munich people' to come up with a new list, because he wants Heisenberg for Göttingen."[72] A corresponding request was sent simultaneously to the Munich faculty. The appointment commission then assembled a new list but made unmistakably clear that Heisenberg remained their favored candidate. Finally, Heisenberg himself did

68 From Heisenberg, Jun3 14, 1935. DMA, HS 1977-28/A,136. Also in ASWB II.
69 Führer to Escherich, July 3, 1935. UAM, Sen-I-272.
70 To Debye, July 7, 1935. MPGA, (Debye). Also in ASWB II.
71 Boepple to Escherich, July 13, 1935. UAM, Sen-I-272.
72 From Debye, September 20, 1935. DMA, HS 1977-28/A,61. Also in ASWB II.

not want to go to Göttingen but to Munich. "A researcher of Heisenberg's standing will best develop his abilities, valuable equally for science and for pedagogy, if he is granted that position he himself deems most suitable for the direction of his work."[73]

At Göttingen, though Heisenberg was the favored candidate to succeed Born, here too other candidates had been put on the list (Heisenberg, Hund, Becker, Jordan) in case their first choice was unavailable. After some back and forth, the Ministry decided on Becker, who accepted the appointment, and assumed the Born chair at Göttingen at the start of the winter semester of 1936/1937.[74] The path to Munich now seemed clear for Heisenberg. Towards the end of the winter semester, the official then in charge of the matter at the Berlin Ministry of Science, Wilhelm Dames (1904–?),[75] inquired of the Rector of the University of Munich whether an appointment of Heisenberg for the next summer semester was desired. The Rector replied at once that Heisenberg, now as formerly, was the favored candidate of the faculty, and that "after conscientious consideration of all relevant factors" he concurred. "I might add that I have recently been informed by the Reich's Student Leadership that as a teacher, Heisenberg is also rated very positively by the National Socialist students at Leipzig."[76] Shortly thereafter, Heisenberg received the appointment to Munich. Effective April 1, 1937, he was to take up the Sommerfeld chair.[77]

11.4 "German Physics"

After the situation had hung so long in limbo, the appointment of Heisenberg now came quite suddenly. He preferred to take up the Munich position only after the summer semester. The Ministry acceded to this delay, and Sommerfeld continued for another semester as "substitute" in his own academic chair. Already before the end of the summer semester, however, new obstacles were put in the way of Heisenberg's Munich appointment. "I will have to continue teaching next semester, because at the moment, in deference to Lenard, the government dare not appoint H.," Sommerfeld wrote on July 7, 1937, to his colleague, Kasimir Fajans, now immigrated to the USA.[78]

73 Appointment commission to the Dean of the Philosophical Faculty, November 4, 1935. DMA, NL 89, 004. Also in ASWB II. (Friedrich Hund was first on the new list, Gregor Wentzel second, Ralph Kronig (1904–1995) and E. C. G. Stückelberg (1905–1984) both third, as well as Erwin Fues (1883–1970), Fritz Sauter (1906–1983), and Albrecht Unsöld as further possible candidates).

74 Rammer, *Nazifizierung*, 2004, pp. 49–54.

75 On Dames, see Lemmerich, *Angriff*, 2005, p. 216.

76 Kölbl to Ministerium, March 3, 1937. UAM, Sen-I-272.

77 Cassidy, *Uncertainty*, 1992, p. 371.

78 To Fajans, July 7, 1937. SBPK, Fajans, 46 Sommerfeld.

The reference to Philipp Lenard pointed to the source of these difficulties. Lenard, together with Johannes Stark, numbered among Hitler's early adherents. They were united in fanatic hatred of everything "Jewish." Lenard had collaborated actively as "comrade in arms in the spirit of the Führer for a renewal of Germany" and had taken a stand "clearly against the nonsense . . . that the Jew Einstein—supported by many, even non-Jewish physicists, like Planck—have constantly attempted to impose on the German people's established conception of nature," Lenard wrote in his memoire of the "time of struggle of the NSDAP."[79] In March 1933, in a memorandum to Hitler, he had offered his help in the Nazi reform and declared himself prepared "to examine the university proposals" for the ministries of culture "in the matter of personnel, to evaluate them, influence them, and in specific cases reject and replace them with alternatives."[80] As Nobel laureates, Stark and Lenard commanded considerable respect. In the early years of the Third Reich, they could get a hearing even with leading Nazis, so that their vendetta against modern theoretical physics did not remain in the realm of mere polemics but carried scientific-political consequences in its train. In 1936, Lenard gave his textbook the title "German Physics" and therewith also gave the ideology the banner under which the war against modern theoretical physics and its exponents still remaining in Germany, primarily Planck, Sommerfeld, and Heisenberg, was to be waged.[81]

No direct influence exerted by Lenard on the Munich appointment process is discernible in surviving files, but Wilhelm Führer who represented the lecturer corps on the spot and thereby the NSDAP soon made it clear that for his part a candidate other than Heisenberg was desired. "I have requested both Professor Stark and Professor Tomaschek to make suggestions regarding this chair," Führer put on record in April 1936. Rudolf Tomaschek (1895–1966), a student of Lenard's, was a short time later appointed at the Technical University at Munich. In light of Führer's inquiry, the lecturer corps took the liberty of presenting to the Ministry its own list of candidates for the Sommerfeld succession, in opposition to the faculty. Stark responded to the inquiry by proposing the appointment of Hans Falkenhagen (1895–1971), a physical chemist from Leipzig, to the Sommerfeld chair. Tomaschek introduced the name of Fritz Sauter into the discussion. The Berlin Ministry of Science, however, wished to hear none of these counterproposals. As Führer, who on October 1, 1936, had become officer in charge of university matters at the Bavarian Ministry of Culture, found in a filed memorandum, both Falkenhagen and Sauter were "out of the question" for an appointment at Munich.[82] This order came from Führer's superior, Rudolf Mentzel (1900–1987), who as a member of the SS with the support of Heinrich Himmler (1900–1945) rose directly to a leading position as scientific functionary of the Third Reich.[83]

79 Schirrmacher, *Philipp Lenard*, 2010, pp. 252 and 265.
80 Ibid., pp. 275–276.
81 Beyerchen, *Wissenschaftler*, 1982.
82 Memorandum from Führer, April 7 and May 26, 1936. UAM, Sen-I-272.
83 Flachowsy, *Notgemeinschaft*, 2008, pp. 172–174.

It is not surprising that the lecturer corps failed in their recommendation of these candidates, since Mentzel was a determined opponent of Stark. The fanatics flocking around Lenard and Stark, who, clearly tolerated, not to say sponsored, by leading representatives of the Party, looked for a forum in the Nazi press, were regarded as undesirable competition at the Berlin Reich's Ministry of Science. On January 29, 1936, the *Völkische Beobachter* carried an article under the headline "German Physics and Jewish Physics" that brought the issue to a head. Mentzel called on Heisenberg to submit a report on the position of physicists regarding the state of physics in Germany. Collaborating with Heisenberg were the widely known experimental physicists Max Wien and Hans Geiger, and in May 1936, they presented to the Berlin Ministry a memorandum signed by 75 physicists in which the necessity of modern theoretical physics was stressed, and the attacks of Lenard and Stark were judged to be harmful. The "Heisenberg-Wien-Geiger Memorandum," as it was designated, made clear that the great majority of physicists in Germany rejected the campaign instigated by Lenard and Stark.[84] But although Führer thereby suffered a setback and the appointment commission represented by Sommerfeld (which included Gerlach, Wieland, and the mathematician Constantin Carathéodory) was confirmed as the actually governing agency, the conflict smoldered on under the surface. The lecturer corps had not prevailed with their counterproposals, but they had served notice of their opposition to the faculty. That the Rector found it necessary in March of 1937 in his support of the faculty proposal to add "that Heisenberg [had been] assessed very positively as a teacher even by the National Socialist students at Leipzig" shows that he had to take the opposition of the lecturer corps into account. After all, behind the lecturer corps stood the National Socialist Lecturers' League[85] at Munich Party headquarters, so that the opinions expressed by individual fanatics within this circle could not be casually dismissed. Not even Sommerfeld believed the opposition had raised the white flag, as the remark "in deference to Lenard" in his letter to Fajans of July 7, 1937, reveals.

One week later, an essay appeared in the SS periodical *Das Schwarze Korps* that far exceeded in stridency previous attacks against modern theoretical physics. "Several of the most respected theoretical physicists were described as 'white Jews' and insulted," Sommerfeld complained to the Rector. "I was mentioned by name, as were Planck and Heisenberg . . . Prof. Stark [had] obviously provided material" for this attack. Sommerfeld requested that the Rector lodge a complaint with the Bavarian Ministry of Culture. "In the interests of the reputation of German science, the Department to which Prof. Stark reports should move to prohibit publication of that sort of expression in the press, and he should be called to account for the insulting article." Heisenberg intended to lodge an independent protest of the article. So far as Sommerfeld himself was concerned, Stark had "deprived [him of] national

84 Beyerchen, *Wissenschaftler*, 1979; Cassidy, *Uncertainty*, 1992, Chap. 18; from Geiger, Wien, and Heisenberg, undated [before May 19, 1936], DMA, NL 89, 024, folder Nazizeit.

85 NS-Dozentenbund. Nagel, *Dozentenbund*, 2008.

honor." Quite to the contrary, Sommerfeld took pride in the fact that on his travels abroad he had "benefited German science more than Prof. Stark could harm it with his intemperate attacks." This complaint was among the few documents he signed with the closing set phrase obligatory in official correspondence, "Heil Hitler."[86]

At the Berlin Ministry of Science, the Heisenberg appointment that had actually already been decided was reversed. The article in the *Schwarze Korps* made it evident that in his attack against Heisenberg and modern theoretical physics, Stark could count on support from the SS. "With respect to the appointment of Professor Heisenberg to the chair of theoretical physics at the University of Munich, I await further information from the outcome of an ongoing investigation," the Reich's Minister of Science informed the Rector of the University of Munich. At the same time, Sommerfeld was once more charged with filling in for the yet to be appointed occupant of his own academic chair.[87]

11.5 Science on the Sidelines

Thus, the Sommerfeld succession had officially, as it were, become politics. While power struggles among the functionaries of the various rival authorities of the Nazi Party and government apparatus presiding over science were being fought out behind the scenes, Sommerfeld tried to maintain the normal pedagogical and research enterprise as best he could under prevailing circumstances. "I imagine your life in beautiful Princeton as comfortable and idyllic, and sometimes long for the amiable and unproblematic 'Godsland,'" he wrote Einstein in January 1937 from Zürich, where the Swiss Physical Society was holding a meeting on solid state physics celebratory of its 50th anniversary.[88]

Lecture invitations to congresses abroad such as this were now also subject to approval from the ministerial bureaucracy in Berlin. Shortly before, Sommerfeld and several of his professorial colleagues had vainly opposed this infantilizing restriction. "It would compromise our situation in Germany were one obliged to answer an invitation from abroad with the caveat that one's acceptance was contingent on approval from above," they had argued, recalling the negative impression left by similar officious patronizing of Russian scholars by the Soviet government in Moscow on the occasion of a recent congress. Such decrees were "detrimental to the reputation abroad of German science."[89] Not only was the measure not rescinded, however, it was actually broadened in its application. A circular directive from the Reich's Minister of Science of March 1937 compelled German university teachers to supply an accounting of their travel abroad and to report to the foreign representation of the

86 To Kölbl, July 26, 1937. UAM, E II N Sommerfeld.
87 To the Rector, November 16, 1937. UAM, Sen-I-272.
88 To Einstein, January 16, 1937. AEA, Einstein. Also in ASWB II.
89 To the Rector, November 12, 1936. DMA, NL 89, 004.

NSDAP and other foreign offices. Apparently, the professors at first paid scarcely any attention to this "confidentially" promulgated decree, so that towards the end of the summer semester of 1937 it was once more forcefully restated to them.[90]

Sommerfeld also attempted to bring the "reputation of German science abroad" into play in the argument over his succession. After the attack in the *Das Schwarze Korps*, he turned to Ernst Freiherr von Weizsäcker (1882–1951), who held an elevated position in the foreign office, and who, as the father of Heisenberg's student Carl Friedrich von Weizsäcker (1912–2007), he assumed would also take a personal interest in the affair. Weizsäcker replied that he had already tried "to interest various relevant people from the perspective of the foreign office in the matter."[91] But for the time being, this diplomatic initiative had not proven effective. During the Christmas holidays of 1937/1938, Sommerfeld wrote resignedly to Einstein, "The politics of my sworn enemies, Giovanni Fortissimo [Johannes Stark] and Leonardo da Heidelberg [Philipp Lenard], who do not wish to grant me Heisenberg as successor, force me to continue teaching and looking after my now diminished flock." Advances, such as were being made in the USA in nuclear physics, were unimaginable in Germany. "The future looks grim for German physics. I have to console myself with having known its golden age, 1905–1930."[92] That he referenced nuclear physics in comparing American with German physics was not coincidental. In a 13-page letter, written after his immigration via England to the USA and his appointment at Cornell University in Ithaca, Bethe had described to him at what a furious pace physics was evolving there. At first, he had felt "like a missionary traveling to the darkest part of Africa to spread the true faith," but it was not half a year before he was converted from this heresy. The predominant research area there was in nuclear physics. "With the result that 90 percent of all work in America is done in this area . . . You can see in essence what I myself have done in the Physical Review and Reviews of Modern Physics. It is all about the nuclear."[93]

Although the future looked "grim," Sommerfeld did all in his power to maintain the pedagogical and research enterprise he had established in his institute over three decades. Even in the 1930s, famous theoreticians still went forth from the Sommerfeld school. His most important students included Bechert, Maue, Scherzer, and Heinrich Welker (1912–1981). Sommerfeld set himself and his collaborators a great challenge: the adaptation of the *Wave Mechanical Supplement* to the rapidly progressing development of quantum mechanics. The substance had grown so significantly that in the new edition, "breadth and contents

90 From the Rector, July 30, 1937. DMA, NL 89, 024, folder Nazizeit.

91 From Ernst Freiherr von Weizsäcker, September 30, 1937. DMA, HS 1977-28/A,360. Also in ASWB II.

92 To Einstein, December 30, 1937. AEA, Einstein. Also in ASWB II.

93 From Bethe, August 1, 1936. DMA, NL 89, 005. Also in ASWB II. Bethe/Bacher/Livingston, *Basic Bethe*, 1986.

Fig. 31: After his retirement in 1935, Sommerfeld was assigned for several semesters to fill in the as yet unoccupied position of his own academic chair. The photo shows him delivering a lecture in 1937 on the diffraction of X-rays on crystals (Courtesy: Deutsches Museum, Munich, Archive).

[constituted] a multiple of the earlier presentation," Sommerfeld explained in 1939 in the Preface to this new edition, which was no longer designated a supplement but as *Atomic Structure and Spectral Lines, Volume 2*. Many of the papers written at his institute during the 1930s had in one way or another to do with this work. Sommerfeld also gave it an especially personal touch in that he presented the methods of complex analysis in the treatment of wave mechanical problems, and thereby as it were erected a monument to the first scientific love of his career. In a chapter on X-ray bremsstrahlung, it also emerged clearly how much effort he had devoted to the "Pacific problem."[94] These works realized in the Munich "oasis" demonstrate that Sommerfeld and his students were still dealing with problems at the leading edge of research in theoretical physics even after the end of the "golden age" and obtaining results whose traces are discernible in the relevant scientific literature to this day ("Sommerfeld-Maue eigenfunctions," "Elwert factor," etc.).[95]

94 Sommerfeld/Maue, *Verfahren*, 1935; Elwert, *Berechnung*, 1939; Sommerfeld, *Atombau 2*, 1939, Chap. 7.

95 Haug/Nakel, *Process*, 2004.

11.6 The Seventieth Birthday

Since no decision about his succession had yet been reached nearly a year after the attack in the *Das Schwarze Korps*, Sommerfeld informed the Dean of his faculty that he no longer wished to fill in as substitute in his own academic chair, and that his two students, Maue and Welker, should be given teaching assignments to cover this function.[96] Now it became clear, however, that the lecturer corps (Bruno Thüring in particular in this matter—Führer had been transferred to the Bavarian Ministry of Culture) had already had their own favored candidate in mind for the Sommerfeld succession: the Cologne associate professor of applied physics, Johannes Malsch (1902–1956).[97] The "teaching fatigue of Counselor Sommerfeld" presents "a new opportunity now to move at last with determination towards a solution to the problem of this academic chair," the lecturer corps argued against the Rector, demanding that Malsch should be engaged to fill the vacant academic chair provisionally for the winter semester 1938/1939. But Malsch chose not to take on this assignment, as the Berlin Ministry of Culture informed the Munich Ministry, which in turn instructed the University to engage Maue to fill the chair.[98]

So, for the time being, Sommerfeld's institute was spared a radical transformation. But the politics of the Third Reich made itself felt in various ways nonetheless. In December 1938, the Dean of the faculty, botanist Friedrich von Faber (1880–1954), an ardent Nazi, in a memorandum designated "confidential," asked all members of the faculty to inform him whether they were (1) actively engaged in the NSDAP or any of its allied organizations; (2) working with the office of Rosenberg, Todt, Ahnenerbe, or other National Socialist organizations; (3) lecturing to Nazi organizations ("preferably providing lecture topics"); (4) working collaboratively with the Nazi press; (5) carrying on research "with respect to present times"; or (6) otherwise being of exemplary service to the state. This survey was supposed to "demonstrate" the commitment of university teachers "within as well as beyond the boundaries of professional duties to the Party and the state." In reply to this questionnaire, Sommerfeld wrote: "1 through 5, not applicable. To 6—On my travels abroad (several times to the U.S. and England, also to India, Japan, France, etc.), most recently on my trip to Italy in 1938 in which the Foreign Office took particular interest, I have worked with great success for the honor of the name of Germany."[99] Shortly before, to a questionnaire "concerning membership and activity in the NSDAP, its divisions, allied organizations, in the NSFK, in the Reich's Anti-Aircraft League, etc." he had reported membership in five NSDAP-affiliated organizations: NS-Public Welfare, the NS-Teachers' League, the NS-League of German Engineering ("that is, as a member of the German Chemical Society"), the

96 To the Office of the Dean, July 7, 1938. DMA, NL 89, 004.

97 On Malsch, see Litten, *Mechanik*, 2000, pp. 70–77.

98 To the Rector, October 15, 1938. UAM, Sen-I-272.

99 To Friedrich von Faber, December 22, 1938. DMA, NL 89, 024, folder Nazizeit.

Reich's Anti-Aircraft League, and the Reich's Colonial League. In the case of the NS-Teachers' League, he was apparently not exactly sure of his membership; to a similar questionnaire he had in March 1941 entered a question mark at this place.[100] "Affiliated organizations" such as the NS-Public Welfare, the NS-Teachers' League, or the NS-League of German Engineering showed high membership numbers; conclusions about the true sentiments of a member can hardly be drawn, since through the "Gleichschaltung"[101] of organizations, membership of predecessor organizations was automatically transferred to current membership, often without the members' knowledge—thus were many names attributed to the membership of the NS-Public Welfare. Many may also have been seeking to demonstrate their patriotic solidarity with their fellow countrymen[102] by joining these groups and thus in a relatively innocuous way counteracting the hostility of their Nazi neighbors, colleagues, or superiors.[103] Sommerfeld's membership in the five "affiliated organizations" can hardly be taken as a concession to National Socialism, however, for had it been he would presumably not have replied "not applicable" to the Dean's question concerning his commitment to Party and Nazi state.

A short time later, the lecturer corps made another attempt to call Sommerfeld's Certificate of Aryan Ancestry into question. "The suspicion [has] quite often and from many quarters been voiced that Counselor Sommerfeld is of Jewish ancestry," it was argued, with reference to entries in biographical compilations. The lecturer corps, however, was compelled to accept the fact that the "expert in racial research at the Reich's Ministry of the Interior" had "verified [Sommerfeld's ancestry] back to his great-grandparents" and had found "that the forebears of Prof. Sommerfeld are of Aryan descent."[104]

Under such circumstances, even Sommerfeld's 70th birthday on December 5, 1938, was swept up in the whirl of Nazi politics. Even the plan to honor the anniversary with a special issue of the *Annalen der Physik* could be realized only with "secondary political conditions."[105] German publishers no longer dared publish essays "by non-Aryan authors," Pauli wrote to Peierls, who had immigrated to England. In addition, there were "renewed newspaper attacks against Sommerfeld from the ranks of the Stark group" to be feared. The "Sommerfeld students living outside Germany" therefore decided to make the December issue of the *Physical Review* a Sommerfeld festschrift.[106] "With respect to the non-scientific secondary

100 Personnel Files, UAM, E-II Sommerfeld.

101 The process of "bringing into line" or "coordinating" of all organized entities with the ideology and aims of the National Socialist German Workers' Party (NSDAP). Organizations not "brought into line" were eliminated.

102 The German term is "Volksgenossen," which conveys a more National Socialist tone.

103 Feiten, *Lehrerbund*, 1981; Vorländer, *NSV*, 1988; Ludwig, *Technik*, 1974.

104 Bavarian Ministry of Culture to the lecturer corps, April 18, 1939. Personnel Files. UAM, E-II Sommerfeld.

105 Heisenberg to Pauli, July 15, 1938. WPWB II.

106 Pauli to Peierls, July 18, 1938. WPWB II.

conditions for authors established by the publishers," Pauli wrote Heisenberg, "I hope a growing number of authors will cease placing their work in the journals of such publishers, regardless of whether the authors number among the white or the black class of theoreticians."[107]

A boycott of the *Annalen der Physik* did not materialize, however. Besides, it was not just the "non-Aryan" Sommerfeld students who had already gone abroad who were affected. In a letter to Debye, editor 10 years earlier of the Sommerfeld festschrift in celebration of his sixtieth birthday, Ludwig Hopf expressed surprise that he had not yet heard of any plans for Sommerfeld's seventieth birthday. Debye replied that people at the *Annalen der Physik* were preparing a birthday issue, but that "no non-Aryan would be published" in it. Additionally, the Munich District Association of the DPG was preparing a gala, from which—Debye assumed—"Jews would also be excluded. . . I'm afraid I bring you no joy with what I write, but I think it best that you know the unvarnished truth . . . I can understand that you will be unhappy. You do of course always have the option of personally extending your congratulations to Sommerfeld on his birthday."[108] Of necessity, Hopf came to terms with his exclusion from the *Annalen* festschrift but insisted with respect to the DPG's gala in Sommerfeld's honor "to be treated like any other member." In the past, he had, as a dues-paying member, received an invitation to every DPG meeting. Since Debye was currently serving as DPG Chairman, he was the proper contact for Hopf in this matter. "In the event I am not invited, I will of course resign."[109]

Hopf did not receive an invitation. "Never could I have dreamed I would not be with you on your 70th birthday," Hopf later wrote in his congratulatory letter to Sommerfeld. "But fate has played us all curious tricks, and to my and your way of thinking being present on the birthday is after all a mere formality." To his birthday greeting he attached the wish: "May you for a few days forget all that is depressing!"[110] At the same time, the DPG was preparing the expulsion of Jews from its ranks. "Under the compelling prevailing circumstances, consistent with the Nürnberg Laws, membership of Jews of the German Reich in the German Physical Society [DPG] can no longer be countenanced," stated the document sent on December 9, 1938, to all DPG members and signed by Debye as DPG Chairman. Those affected were instructed to report their resignation from the DPG to Debye.[111] "The scrap of paper from the Physical Society was not so depressing," Hopf wrote once more to Sommerfeld, "but the name at the bottom gave me a slight shock."

107 Pauli to Heisenberg, August 15, 1938. WPWB II.

108 Debye to Hopf, October 18, 1938. MPGA, III. Abt., Rep. 19 (Debye), Nr. 377.

109 Hopf to Debye, October 19, 1938. MPGA, III. Abt., Rep. 19 (Debye), Nr. 377. On the history of the DPG in the Third Reich, see Hoffmann/Walker, *Physiker*, 2007, in this volume especially Wolff, *Ausgrenzung*, 2007.

110 From Hopf, December 3, 1938. DMA, NL 89, 040.

111 Wolff, *Ausgrenzung*, 2007, pp. 111–112 und Hoffmann/Walker, *Physiker*, 2007, Appendix, pp. 564–565.

This postscript to his birthday letter contained an additional bit of information which "would not have suited the joy of celebration." Friends, a married couple, had taken their own lives, "because they were no longer up to the calamity that had befallen them. Tragic, but perhaps they were right."[112]

Four weeks before, in the November pogroms Jews throughout Germany had been murdered, locked into concentration camps, and terrorized in other ways. The mathematician Arthur Rosenthal (1887–1959) apologized 3 weeks after Sommerfeld's birthday, because although on December 5 he had been residing "very near Munich," he had not been able to offer his congratulations in person since his residence had been "Dachau Concentration Camp." Up to the last, he had believed he could hold out "doing scientific work in complete seclusion . . . The events of November 10"—he was alluding to the so-called Kristallnacht—had taught him differently, and now he planned to emigrate as soon as possible. At his last meeting with Sommerfeld, though, he had been delighted to see "that you have remained in every respect the same as you always were." The allusion to the growing Nazi sentiment taking hold throughout the German populace was unmistakable.[113] "We have often spoken of you," Sommerfeld wrote back. The news that Rosenthal had been interned in a concentration camp during the November pogroms apparently spread quickly among the Munich physicists and mathematicians. "Your letter has relieved us of an anxiety. Its contents are being conveyed to your Munich friends."[114] Rosenthal immigrated shortly thereafter via Holland to the USA. Ludwig Hopf also managed to rescue himself and his family by immigration to Ireland, where, however, he died shortly thereafter.[115]

11.7 The Decision in the Succession Dispute

Under these circumstances, Sommerfeld's seventieth birthday was not, for him, a day of unalloyed joy. The essays published in the *Annalen der Physik* and in the *Physical Review* reminded him of the halcyon days of his Munich "nursery," but the events surrounding his birthday were not conducive to looking confidently to the future. And worse was still to come.

Since 1938, the lecturer corps had been dominated by Bruno Thüring, a fanatical Nazi and anti-Semite like his predecessors, who as coeditor of the *Zeitschrift für die gesamte Naturwissenschaft* had declared his commitment to the ideological campaign against modern theoretical physics in the realm of publication. On September

112 From Hopf, December 11, 1938. DMA, NL 89, 040.

113 From Rosenthal, 21. December 21, 1938. DMA, NL 89, 040.

114 To Rosenthal, December 23, 1938. I am indebted to Dr. Roland Rappmann (RWTH Aachen, University Library) for a copy of this postcard.

115 Siegmund-Schultze, *Mathematicians*, 2009, pp. 151–152; Eckert, *Atomphysiker*, 1993, pp. 165–169.

9, 1938, he had presented to the Rector his own list for the Sommerfeld succession, with Malsch as his top candidate. Wilhelm Müller (1880–1968), a professor of mechanics from the Technical University at Aachen, and Hans Falkenhagen ranked second and third, respectively. These latter names had been proposed to the lecturer corps by Stark and ranked on the list more as backup candidates. On December 29, 1938, the Dean of the faculty gave this list his backing. He maintained that Gerlach had in the meantime "come to the insight that the appointment of Heisenberg to Munich [was] impossible." Therefore, the faculty, "concurring with the lecturer corps," proposed that Malsch be appointed. Each of the three candidates was a "Party member." Malsch was "one of the few contemporary physicists" dedicated to restoring "the lost unity of experimental and theoretical physics."[116]

When Sommerfeld learned of this turn in the struggle over his successor, he asked Ludwig Prandtl for a report on Müller to be presented to the faculty. Müller's papers on fluid mechanics were "perhaps mathematically sound," Prandtl wrote assessing the Aachen Professor of Mechanics, "but in my view uninteresting, since he consistently evades all non-linear problems." This was no compliment for someone who had made fluid mechanics, in which the nonlinearity of the fundamental equations presents the essential challenge, his profession. Prandtl characterized Müller otherwise as "extremely formal," which, for someone named as a candidate in opposition to formalism in theoretical physics, was equally unflattering. "I am not aware of any observation by Prof. Müller on physics (the examination in physics for teaching at the advanced level, which he has undoubtedly passed, could hardly be considered a sufficient demonstration of his qualifications for a professorship)," Prandtl concluded his report.[117]

Thereafter, Müller seemed hardly to come under serious consideration as a candidate. When it came to Sommerfeld's attention, however, that the two other candidates, Malsch and Falkenhagen, were not being considered by the Berlin Ministry of Science for appointment to Munich, he asked for a conference with the Rector. Müller was "the very worst" on this list, he wrote in a note on this meeting. He insisted on the original appointment list and hoped the Rector would support it in Berlin in order to effect Heisenberg's appointment to Munich after all.[118] As Sommerfeld wrote to Heisenberg following this conference, the Rector preferred ultimately to leave the decision to the Berlin Ministry since "agreement between the viewpoints of the two reports in Munich [was] not to be reached. . . I will be very happy if, through your relationship with the staff of the SS, and using this letter, you can bring pressure to bear on the Ministry of Culture in favor of your succession. The difficulty lies in the fact that, according to the Dean, the leading officer in the Party opposes your appointment."[119]

116 Litten, *Mechanik*, 2000, pp. 83–89.
117 From Prandtl, October 28, 1938. MPGA, III. Abt., Rep. 61, Nr. 1538. Also in ASWB II.
118 Conference notes, February 27, 1939. DMA, NL 89, 019, folder 5,11.
119 To Heisenberg, February 28, 1939. DMA, NL 89, 019, folder 5,11. Also in ASWB II.

Heisenberg's parents and the father of "Reichsführer" Heinrich Himmler knew one another, and Himmler had taken Heisenberg under his protection from the attack in the *Das Schwarze Korps*.[120] Thereafter, smear articles from the hand of the SS had ceased. The *Das Schwarze Korps* "no longer accepts articles written in the vein of my previous one," Stark wrote to Lenard in April 1938.[121] Heisenberg was therefore optimistic that he could still turn the page in his favor. He had, he replied to Sommerfeld, "turned at once to the SS" and hoped "that the Reichsführer [will] still decide to overrule the Party's position on my appointment."[122] The situation is "quite favorable, although the politics is delaying all decisions," he wrote to Munich 4 weeks later.[123]

At the Ministry in Berlin, however, things in March 1939 were moving in a different direction. "Party Comrade Dames at the Reich's Ministry of Education [has rejected] the candidacy of Malsch," read a directive from the Bavarian Ministry of Culture to the leader of the National Socialist League of Lecturers at Munich Party headquarters. At the same time, however, Dames considered appointment of Heisenberg to Munich "under the prevailing circumstances as impracticable." Therefore, negotiations on the appointment were underway with Müller, as the second place candidate on the lecturer corps' list.[124]

Soon thereafter, the decision was made. "That Dames has now appointed the most impossible man on this list," Heisenberg wrote Sommerfeld during the Easter vacation of 1939, "may be explained as a gambit in expectation that nothing would come of it." He still hoped to have the upper hand "over the Workers' and Soldiers' soviets à la Thüring."[125] Sommerfeld had also not yet surrendered the cause of his succession. "The affair with W. Müller seems to be not so bad," he wrote his son, who was working at Telefunken in Berlin as an electrical engineer and patent expert. Debye had informed him that Müller had "already been dropped" by the Berlin Ministry of Science. "But the lecturers (Thüring) will stop at nothing to thwart any further development."[126] For Sommerfeld, the outcome remained open. "The barometer of my hopes and fears with respect to the succession goes up and down," he admitted to his son 2 weeks later.[127]

Soon, Heisenberg was able to report details of the tug-of-war going on behind the scenes between party and SS. "The appointment to your chair has become a purely political matter," he wrote to Munich. "I understand very well that Dames is pushing the appointment of Müller: he wants thereby to mobilize the counter-forces against the Lenard clique, and—should that not succeed—disgrace this

120 Cassidy, *Beyond Uncertainty*, 2009, S. 277–281.
121 Kleinert, *Briefwechsel*, 2001, p. 259.
122 From Heisenberg, March 3, 1939. DMA, HS 1977-28/A,136. Also in ASWB II.
123 From Heisenberg, March 30, 1939. DMA, HS 1977-28/A,136. Auch in ASWB II.
124 Litten, *Mechanik*, 2000, pp. 90–91.
125 From Heisenberg, April 9, 1939. DMA, HS 1977-28/A,136. Also in ASWB II.
126 To Ernst, April 13, 1939.
127 To Ernst, April 27, 1939.

group through Müller. The purely political question of Lenard's influence over the leadership of the lecturers is naturally of greater importance to him than the Munich professorship." His source at the SS had informed him "doubtless on instructions from Himmler" that discussions meanwhile between Himmler and the NS League of Lecturers at Munich Party headquarters had taken place. Himmler had supported his appointment to Munich, but the leader of the League of Lecturers, who was "probably supported by Hess," maintained the view that this was tantamount to a "loss of prestige for the Party." Himmler now was "unwilling to push my candidacy by force as it were because I might turn out to be a bad Nazi in Munich. Thüring and his cohorts would certainly be at pains to prove that the case, and Himmler doesn't want to disgrace himself through me in the eyes of the Party, so to speak." Himmler now wanted to help him "rehabilitate" himself with a different appointment.[128] This was to happen, as Heisenberg wrote in his next letter, through "an appointment to the University of Vienna." He had not committed himself to this plan, however, and quite apart from his own career "had expressed the urgent wish that efforts be made for a reasonable physicist to be brought to Munich, not Prof. Müller."[129] Sommerfeld also wished once more to lodge a complaint with the Berlin Ministry but was never able to reach any of the responsible authorities.[130]

The wishes of Sommerfeld and Heisenberg no longer affected the decisions already reached in any way. Müller was appointed at Munich effective December 1, 1939.[131] Exactly what the critical factor in this choice at the Berlin Ministry of Science ultimately was can no longer be determined unambiguously from the surviving files.[132] Much evidence, however, suggests that Dames, as Heisenberg conjectured, wished to disgrace the "Lenard clique" with Müller.[133] Shortly before, in a complaint from Stark addressed to the Minister, Dames himself had been numbered among "the Jewish-minded group around Heisenberg," since he had been "for some time the assistant of the now emigrated full-Jew James Franck at Göttingen."[134] Dames must also have felt secure in the support of his superior Mentzel regarding the decision for this appointment. Both were members of the SS. As Gerlach in a report for the denazification of Mentzel testified after the war, Mentzel had expressed to him his "disapproval" with respect to the appointment of Müller and had even asked that complaints be lodged against Müller, for "in this case one really had to expose the consequences of the meddling of the League of Lecturers."[135]

128 From Heisenberg, May 13, 1939. DMA, HS 1977-28/A,136. Also in ASWB II.
129 From Heisenberg, June 8, 1939. DMA, HS 1977-28/A,136. Also in ASWB II.
130 To Ernst, June 22, 1939; to Debye, July 12, 1939. MPGA, III. Abt., Rep. 19.
131 Litten, *Mechanik*, 2000, p. 95.
132 Ibid., pp. 95–104.
133 Eckert, *Deutsche Physik*, 2007.
134 Lemmerich, *Angriff*, 2005, p. 214.
135 Gerlach, Sworn Statement, December 13, 1948. DMA, NL 80, 290; Nagel, *Dozentenbund*, 2007.

12 Bitter Years

Following the appointment of "the worst imaginable successor," Sommerfeld wrote a colleague in December 1939 that he had "gone permanently into retirement."[1] Three months earlier, with the invasion of Poland, Hitler had unleashed World War II. The conjunction of the catastrophe on the large scale and the ignominious end of his Munich "nursery" on the smaller made for a depressing transition into retirement for Sommerfeld. Once again, as in the summer of 1933, he suffered from insomnia.[2] In 1939 at the neighboring Technical University of Munich, the "Lenard clique" also succeeded in installing a member of its ranks in the person of Rudolf Tomaschek in an academic chair. "It seems Munich is becoming the capital of the counter-movement in physics," Sommerfeld's former assistant Karl Bechert commented on this appointment. As professor of theoretical physics at the Justus Liebig University, Gießen, Bechert could only observe events unfolding in Munich from afar.[3]

The assumption of the Sommerfeld academic chair by a non-physicist was regarded as a scandal not just in the circle of Sommerfeld and his students. If the Reich's Ministry of Science was actually looking to establish a precedent with the appointment of Müller through which the harmful "meddling" of the Party ideologues of the League of National Socialist Lecturers was to be exposed, their calculation proved correct. For Sommerfeld, though, this was tepid consolation. To cope with the bitter struggle over succession to his academic chair, he set himself a new challenge in the form of publication of his lectures. In this project he saw the "faint possibility," of "preserving for the future the many (and for many) valuable personal observations of my lectures."[4]

12.1 The Scandal Intensifies

It seemed at first as though the representatives of the lecturer corps at the University of Munich and their sponsors at Munich Party Headquarters had scored a victory. When the philosopher of science Hugo Dingler (1881–1954), a fatherly friend of Thüring's and like him a fanatical Nazi and anti-Semite, read of Müller's appointment in the *Völkischer Beobachter*, he wrote to Thüring: "This is certainly a great victory, for which we have your toughness and your pure aspirations to thank . . .

1 To Paul Rosbaud, December 19, 1939. DMA, NL 89, 025.
2 To Ernst, September 10, 1939 and January 19, 1940.
3 From Bechert, December 30, 1939. DMA, HS 1977-28/A,12.
4 To Ernst, April 1, 1945.

M. Eckert, *Arnold Sommerfeld: Science, Life and Turbulent Times 1868-1951*,
DOI 10.1007/978-1-4614-7461-6_12,

Though I do not know Müller personally, I may permit myself to write him a letter of congratulations."[5] Dingler also introduced Müller to Ludwig Glaser (1889–?), a student of Stark's who, in anti-Semitism and revolutionary zeal for National Socialism, was in no way surpassed by members of the Munich lecturer corps. Glaser had done his habilitation under Stark in 1921 and had taught for several years as an associate professor at the University of Würzburg before being relieved of duties there in 1928. Thereupon, though he gained prominence as an SA activist (among other things at the Deutsches Museum, where in 1934 he had called for "cleansing" of the library of books by Jewish authors),[6] he never again took the stage of academic physics. "In order to offer him the opportunity to resume his academic career," Müller requested that the Dean of the faculty give Glaser a teaching assignment "in theoretical physics with a particular emphasis on the needs of engineering and military science." It was asked he be given the "rank and title of an associate professor."[7]

Müller sought thereby to preempt the objection that there were no lectures on theoretical physics at his institute—an objection that was in fact not long in coming. Even in retirement, Sommerfeld was unwilling to abandon the field to the "opposition" without a fight. In July 1940, he appeared at the Berlin Ministry of Science to report on the situation at his institute since Müller's appointment. Müller was "preposterous not only scientifically," but "personally as well," which he demonstrated "by producing a letter." Dames had responded with "head-shaking" and the ejaculation "outrageous," Sommerfeld later noted to himself about the conversation. On this occasion his suspicion that Müller's appointment was intended to "push the lecturer corps *ad absurdum*" was confirmed. "This goal has been achieved. But it is not flattering to have been the guinea pig in the experiment." When he reported the impending teaching assignment for Glaser to Dames, because the demands of the physics lectures were too much for Müller himself, he received the reply, "If Müller is not to give the principal lectures, the University is obliged to report this to Berlin."[8] Thereupon, Gerlach complained officially again to the Dean of the Faculty that Müller was "once more [teaching] no theoretical physics." Müller rejected Gerlach's complaint as "an unwarranted schoolmasterish interference," justifying himself with the contention that his lectures on mechanics included theoretical physics as well. At any event, his students were not confronted with "dogmatic and Talmudic physics," as had been taught in the Sommerfeld era. Doubtless, "Prof. Sommerfeld and his cohorts" were the true authors of the campaign against him. "This is ultimately a matter of a war against my ideological

5 Dingler to Thüring, December 8, 1939. Aschaffenburg, Hofbibliothek, Dingler estate.

6 Hilz, *Bildungsanstalt*, 2010, pp. 274–276.

7 Müller to the Faculty of Natural Sciences of the University of Munich, February 24, 1940. DMA, NL 89, 030, file Müller.

8 Transcript of a conference with Dr. Dames, July 16, 1940. DMA, NL 89. Also in Benz, *Arnold Sommerfeld*, 1975, pp. 183–184.

mission, which is and remains absolutely obligatory for me."[9] With a subsequent complaint addressed to the Rector, signed this time by Sommerfeld, Carathéodory, and Wieland,[10] the pressure on Müller was increased. But as Sommerfeld wrote thereafter to Gerlach, the Rector "chose not to do anything on his own." Dames, in turn, could engage actively only if he received "some prompting from Munich."[11]

That the tide began to turn for the Munich lecturer corps in 1940 was discernible from the fact that the Berlin Ministry simply ignored an objection raised by Bergdolt, who in his role as leader of the lecturer corps disputed Carathéodory's "guarantee of political reliability" in connection with intended travel abroad.[12] While the question of the lectures temporarily hung fire, annoyance arose over another matter. Based on a right traditionally granted emeritus professors, Sommerfeld had counted on office space being at his disposal and on access to the institute's library also in his retirement. In view of the shortage of space at the University, he volunteered to make do with space in the basement. Müller, however, denied Sommerfeld access to the institute and had the "full support" of the Dean of the Faculty therein.[13] The Rector and the Bavarian Ministry of Culture left it to Müller to reach an agreement with Sommerfeld on the question of office space. Sommerfeld's countermove to Müller was to offer to donate a portion of his private library to the institute, but Müller advised Sommerfeld in no uncertain terms finally to "draw the only possible conclusion" and to leave "the institute unmolested in future."[14] To Sommerfeld, this was an unprecedented affront. "I'll never set foot in the institute again because my successor is utterly shameless," he wrote his son. "I've been given a decent room elsewhere at the university, however."[15] He gave this news to a colleague just traveling to Berlin to take along "for use at the Reich's Ministry of Culture that my succession is an incredible public scandal, and that my successor has thrown me out of my own institute."[16]

Among the "opposition" too, this expulsion was the subject of conversation and celebrated as a token of Müller's effectiveness. "The institute is now permanently rid of Sommerfeld," Dingler informed Thüring, who meanwhile had been appointed at the observatory in Vienna. "Now he is even trying to palm off his superfluous books on Müller, who is however deaf on this point."[17] Now Müller

9 Müller to the Office of the Dean of the Faculty of Natural Sciences of the University of Munich, September 11, 1940. DMA, NL 89, 030, file Müller.

10 To the Office of the Rector, September 1, 1940. DMA, NL 89, 018, folder 3,14. Also in ASWB II.

11 To Gerlach, September 10, 1940. DMA, NL 89, 015. Also in ASWB II.

12 Hashagen, *Constantin Carathéodory*, 2010, pp. 22–23.

13 Friedrich von Faber to the Buildings Department of the University, February 8, 1940. UAM, E-II-N.

14 Cited from Litten, *Mechanik*, 2000, p. 107.

15 To Ernst, April 4, 1940.

16 To Grimm, April 8, 1940. DMA, HS 1978-12B/172.

17 Dingler to Thüring, March 10, 1940. Aschaffenburg, Hofbibliothek, Dingler estate.

and Glaser went on the offensive. In a 13-page document of complaint against Gerlach and Sommerfeld in October 1940, Glaser wrote that Sommerfeld had assigned his mechanic Selmayr "to remove incriminating files." The files in question would have documented how Sommerfeld had "promoted even foreign Jews" to fill academic chairs. "I just want to say that we old Party comrades and SA-men were too easy-going in 1933—otherwise back then we could easily have dealt differently with certain Jew-lovers and institutional hacks such as Gerlach and Sommerfeld."[18] Dingler informed Thüring in Vienna likewise about this intensification of the situation: "Today, Glaser was here. It is good that the battle has now broken into the open, so that the fronts are clearly drawn, although our poor Müller's nerves have much to endure."[19]

12.2 The End of "German Physics"

During this time, it was regarded as scandalous also at other universities, that a small group of fanatics should disparage creditable authorities in physics like Sommerfeld and Heisenberg. Since this was occurring with the obvious consent of the Nazi League of Lecturers and Munich Party Headquarters, Scherzer and Wolfgang Finkelnburg (1905–1967) of the Nazi League of Lecturers at Darmstadt attempted to persuade their Munich League comrades to withdraw their support of the fanatics. On November 15, 1940, a debate ensued at NSDAP Party Headquarters that was soon—borrowing from celebrated historical debates over religious disputes—characterized as "religious dialogue."[20] Six adherents of modern physics—Scherzer, Joos, Finkelnburg, C. F. von Weizsäcker, Otto Heckmann (1901–1983), and Hans Kopfermann (1895–1963)—confronted six representatives of "German physics"—Thüring, Müller, Tomaschek, Alfons Bühl (1900–1988), Harald Volkmann (1905–1997), and Ludwig Wesch (1909–1994). A functionary of the Nazi League of Lecturers assumed the role of moderator, advised by two physicists referred as experts by the Party, Herbert Stuart (1899–1974) and Johannes Malsch. As Scherzer later noted sarcastically, the discussions led to the "not entirely unprecedented piece of wisdom that perception is the root of knowledge of nature, that under the pressure of observation clear conceptions must from time to time be revised, and that when perception is unable to advance further, formal mathematical treatment is a welcome aid." In the end, all participants subscribed to the resolution that quantum theory and the special theory of relativity were "established and indispensable components of physics" and that "uninformed attacks on physics by the Party" should cease.[21]

18 Glaser: Accusation against Gerlach, October 21, 1940. DMA, NL 89, 030, file Müller.

19 Dingler to Thüring, October 6, 1940. Aschaffenburg, Hofbibliothek, Dingler estate.

20 Scherzer, *Physik*, 1965; Beyerchen, *Wissenschaftler*, 1982, Chapter 9; Hentschel/Hentschel, *Physics*, 1996, p. 341; Eckert, *Deutsche Physik*, 2007, p. 155.

21 Scherzer, *Physik*, 1965, p. 57.

When Sommerfeld learned of this outcome, he is said to have called it "weak and trivial."[22] Heisenberg, however, thought one should "be quite pleased" with the consensus arrived at: "That Thüring and Müller disappeared before the signing of the resolution is surely also quite satisfying. Thüring and Müller are perhaps the dumbest and thus the most fanatical adherents of the opposition, whereas the wily Tomaschek already sniffs the shift in the wind from above."[23] Laue, too, saw a "great triumph" in the outcome of the debate.[24]

As though to give the lie to this triumph, Müller brought out an article in the December issue of *Zeitschrift für die gesamte Naturwissenschaft* with the title "The State of Theoretical Physics at the Universities." "In the midst of the creation of the new Germany, and in the midst of the process of ridding our cultural life of Jews,[25] the internationalist and Einstein conspirator, Counselor Arnold Sommerfeld has been given the opportunity," Müller wrote polemically ignoring the compromise just arrived at, "of retaining his old seat as high priest—albeit stripped of its former luster—of theoretical physics."[26] At another colloquium, held shortly before, to which Stark had been invited, Sommerfeld had also been denounced. The speeches were published in 1941 in a pamphlet titled "Jewish and German Physics." "And only recently, Sommerfeld, chief propagandist for the Jewish theories, has been an academic teacher"—thus was Sommerfeld personally attacked. The polemic climaxed with the threat: "Let the Jewish-minded dogmatists be advised that their time in Germany is over. There is no longer any place for them in German physics."[27]

To counter the objection of his lack of competence in theoretical physics, Müller proposed around the same time that "applied mechanics" be added to the designation of his institute. A glance at the institute's library would reveal not only the "dogmatic mentality" of the previous occupant of the academic chair, but also the neglect of areas which, for example, "are of critical importance for the modern development of the air force and ballistics." Thus, he affixed his polemic against Sommerfeld to his intention of carrying on future research for the war. On March 11, 1941, his proposal was granted by order of the Bavarian Ministry of Culture. Henceforth Müller's academic address read "Institute for Theoretical Physics and Applied Mechanics."[28]

The new direction was immediately manifested in a collaboration arranged by Glaser with the BMW Company, for which the Göttingen Aerodynamic Laboratory (Aerodynamische Versuchsanstalt, AVA) furnished a wind tunnel and a variety of measuring apparatuses. No use was ever made of these devices, however, because in

22 Beyerchen, *Wissenschaftler*, 1982, p. 241.
23 From Heisenberg, December 4, 1940. DMA, HS 1977-28/A,136. Also in ASWB II.
24 From Laue, December 4, 1940. DMA, HS 1977-28/A,197. Also in ASWB II.
25 "Entjudung," literally, "un-Jewing".
26 Müller, *Lage*, 1940, p. 295.
27 Stark/Müller, *Deutsche Physik*, 1941, pp. 21–22, 30.
28 Litten, *Mechanik*, 2000, p. 105.

the summer Glaser was appointed to the "Reich's University of Poznan," which functioned as academic outpost of National Socialism in occupied Poland following the German invasion. Since at his departure from Munich to Poznan Glaser had left Müller with a mountain of unpaid bills, disputes flared that led within a short time to a break between Glaser and Müller. Thüring apologized to Müller for having advised bringing Glaser to the institute. "I really had the best of intentions. But he is and remains a psychopath."[29] Glaser deflected the blame for the unfinished business at Munich onto BMW and promised on his next visit to resolve "everything in person." At the moment, he was overburdened with organizational work in Poznan. "My weight is only 60 kg." He was currently preparing a lecture series on "The National Mission of the Natural Sciences as Introduction to the Jewish Question."[30] During the brief time of his activity at Munich, however, Müller had come to know Glaser as a colleague who did not shy away from malicious intrigues even among his own ranks. He would have to "revise completely [his] earlier view of Glaser," Müller wrote a colleague at Poznan, "and in the interests of the matter [I] urgently advise you and the responsible authorities of the Reich's University to exercise the greatest caution."[31] On a later occasion, he characterized Glaser as a "GPU spy-type"[32] and as a "public enemy." It was "high time that we render this person harmless, and stick him where he belongs. He has damaged our whole struggle, and sullied our flag."[33]

Sommerfeld was kept abreast of these developments by his institute mechanic, Selmayr, tied to him by a bond of friendship and faithful devotion. This did not long remain a secret to Müller, who tried all he could to rid himself of Selmayr.[34] He complained to the Rector that the institute mechanic worked actively "in the spirit of agitation as a tool of the Jew-lovers Sommerfeld and Gerlach."[35] Selmayr was eventually transferred to the neighboring institute for physical chemistry, but he was able all the same to photograph or transcribe a portion of the Müller correspondence in hopes of conclusively documenting the scandalous conditions. "Müller is a perfect idiot; the demands of his Munich position are simply piling up over his head, even though he has now also been made Dean (!!)," Sommerfeld wrote in October 1941 to Prandtl, when he learned from Selmayr about the dispute between Müller and Glaser. The wind tunnel Glaser had ordered, which now lay "abandoned in the University courtyard," made the scandal of Müller's appointment ever more manifest. Sommerfeld advised Prandtl to demand immediate reimbursement for the cost of the wind tunnel in the amount of 15,000 Marks.

29 Thüring to Müller, June 24, 1941. DMA, NL 89, 030, file Müller.
30 Glaser to Müller, September 29, 1941. DMA, NL 89, 030, file Müller.
31 Müller to Geisler, September 18, 1942. DMA, NL 89, 030, file Müller.
32 Müller to Geisler, January 4, 1943. DMA, NL 89, 030, file Müller.
33 Müller to Bomke, January 4, 1943. Transcript. DMA, NL 89, 030, file Müller.
34 Litten, *Mechanik*, 2000, pp. 107–110.
35 Müller to Wüst, June 23, 1941. DMA, NL 89, 030, file Müller.

Müller would not be able to pay this sum from the institute budget of 3,000 Marks. "The wind-tunnel could easily be his ethical and professional undoing. So, good luck on this praiseworthy executioner's mission!"[36] But Prandtl replied that Müller had already "very politely" asked approval to return the wind tunnel, so that it was not possible "somehow to use this matter against him . . . But putting something aside does not mean putting it away!"[37]

Ultimately, Prandtl proved a powerful ally in the fight against Müller and his kind. Already half a year earlier, when Sommerfeld informed him of his expulsion from his institute, he had composed a memorandum to Göring in which he condemned the attacks against Sommerfeld and Heisenberg in the strongest terms.[38] In November 1941, he informed Sommerfeld "that currently there was a new action brewing against the saboteurs of theoretical physics, in which my memorandum of April will also play a role."[39] He was referring to an initiative of Finkelnburg and Carl Ramsauer (1879–1955), who in their capacity as representatives of the German Physical Society in January 1942 presented a "Petition of the German Physical Society" to the Reich's Minister of Science, in which they alleged a decline of physics in Germany.[40]

Although Müller was not expelled from his chair because of these criticisms, he came under steadily mounting pressure. Copies of the "Petition of the German Physical Society" went to various people in the army and industry. "It seems that Müller is teetering," Sommerfeld wrote Prandtl in March 1942. "A shove from your side—demand for reimbursement for the wind tunnel, and at the same time informing the Bavarian Ministry of Culture—might do the rest."[41] Also in his capacity as Dean, Müller made himself so ridiculous that even his relationship to Walther Wüst (1901–1993), who had served as Rector of the University of Munich since 1941, deteriorated. As a high-ranking member of the SS, Wüst certainly had no argument with Müller's commitment to National Socialism. In the summer of 1942, Müller was on the point of independently requesting transfer to a technical university.[42] The very tenor of his letters now revealed how much he felt put on the defensive. "Recent events have taken such a toll on me, that I fear a total nervous breakdown," he wrote the Rector in June 1942. "Occasioned in part by my actions," Selmayr wrote in the margin of the copy of this document which he transmitted to Sommerfeld.[43] Also in a letter to the Göttingen Aerodynamic Laboratory, Müller

36 To Prandtl, October 10, 1941. MPGA, III. Abt., Rep. 61, Nr. 1538. Also in ASWB II.

37 "Aufgeschoben ist nicht aufgehoben!" From Prandtl, October 22, 1941. MPGA, III. Abt., Rep. 61, Nr. 1538. Also in ASWB II.

38 Reprinted in Vogel-Prandtl, *Ludwig Prandtl*, 2005, pp. 210–214.

39 From Prandtl, November 13, 1941. MPGA, III. Abt., Rep. 61, Nr. 1538. Also in ASWB II.

40 Eckert, *Deutsche Physik*, 2007; Hoffmann, *Ramsauer-Ära*, 2007.

41 To Prandtl, March 12, 1942. MPGA, III. Abt., Rep. 61, Nr. 1538. Also in ASWB II.

42 Litten, *Mechanik*, 2000, p. 130.

43 Müller to Wüst, June 28, 1942. DMA, NL 89, 030, file Müller.

portrayed himself as victim. "After the break in relations between Glaser and BMW, both of them cleared off and left me holding the bag," he wrote deflecting all the blame for the fiasco of the unused wind tunnel from himself. "I have been duped and compromised in an absolutely incredible manner."[44] Even among his equals, Müller was now considered a failure. "That Müller has essentially gone downhill at Munich is no longer in doubt," Thüring wrote at the end of the summer semester, 1942, to Dingler. "Deeply regrettable, but there is probably nothing more to be done. The human qualities of his character were just not up to the heavy tasks he was assigned."[45]

In November 1942, the National Socialist League of Lecturers hosted another debate on "German Physics," to which Müller was not even invited. This demonstrated that its proponents were totally at odds among themselves. "Arrayed against us, whose only representative I am here, is a broad front, its ranks closed," Thüring complained.[46] Heisenberg judged this debate as a "victory celebration."[47] It was not by chance that in a subsequent contest at the University of Munich between ideology and expertise, the latter carried the day. When a search was begun in 1938 for a successor to Carathéodory, the respective positions of the lecturer corps and the mathematicians in the faculty had hardened. In 1943, in the person of Eberhard Hopf (1902–1983), a mathematician of international standing was finally appointed who showed no predilection for Nazism.[48]

12.3 Political Misgivings

The depressing circumstances at his institute did not prevent Sommerfeld from continuing to observe several traditions that had grown dear to his heart over the decades of his teaching career. Chief among them was the colloquium, instituted in 1909, in which the latest findings of research in theoretical physics were discussed. When this tradition could no longer be observed in his own institute, he sought and found an alternative at the institute for physical chemistry. With Klaus Clusius (1903–1963), who had taken over this institute following Fajans's emigration, Sommerfeld had an ally in his struggle against Müller and the lecturer corps. "Since my retirement, I have been presenting a colloquium on theoretical physics together with my colleague Clusius at his institute, which is attended by many colleagues and students," Sommerfeld wrote in October 1940 to the Chairman of the German Physical Society, requesting funding to bring speakers from abroad.[49] The request

44 Müller to the Aerodynamic Laboratory, October 31, 1942. DMA, NL 89, 030, file Müller.
45 Quoted from Litten, *Mechanik*, 2000, p. 128.
46 Thüring to Dingler, November 2, 1942. Aschaffenburg, Hofbibliothek, Dingler estate.
47 Beyerchen, *Wissenschaftler*, 1982, p. 258.
48 Litten, *Carathéodory-Nachfolge*, 1994; Hashagen, *Constantin Carathéodory*, 2010, pp. 15–16.
49 To Steenbeck, October 26, 1940. DMA, NL 89, 018, folder 3,12. Also in ASWB II.

was denied lest it create a "precedent for similar requests from other institutes."[50] But the colloquium flourished even without the financial support of the DPG. Sommerfeld's reputation assured a brisk attendance. When a participant congratulated Sommerfeld in 1941 on the golden anniversary of his doctorate, he added the hope that another "golden era" for theoretical physics might soon dawn throughout Germany too. "Should it not come about so quickly after all, let us reasonable ones still stick together in our sequestered colloquium, consoling ourselves with the proverb 'Against stupidity, the gods themselves are helpless!'"[51] Carl Friedrich von Weizsäcker, for example, gave a lecture in this "sequestered colloquium" on "cosmology and the creation of the heavy elements."[52] Even when the war made travel difficult, Sommerfeld continued to invite participants to this "special colloquium."[53] In the DPG, too, Sommerfeld worked to see that theoretical physics did not entirely surrender its reputation. In 1938, he had assumed chairmanship of the committee in charge of awarding the Max Planck Medal, which up to then Planck himself had held. In 1942, Ramsauer, as DPG Chairman, wished to cede greater influence in the awarding of the medal to the Reich's Ministry of Science, "as a conciliatory coda," as it were, "to the whole quarrel over modern theoretical physics." But Sommerfeld argued that the Ministry should at most "be consulted" regarding the material of the medal, but not on the choice of a prize winner. This should be left solely to the Medal Committee.[54]

His great reputation also constantly brought Sommerfeld new invitations to travel abroad. On such occasions, it was clear how he was thought of by the political authorities, who had to give such journeys their assent. "Please inform me in detail whether there are doubts about the political reliability of the afore-named, and if so, on what facts these doubts are based," an NSDAP official in Berlin inquired of the local Party subsidiary "Biederstein" in Munich-Schwabing, for example, when Sommerfeld was planning a trip to Italy in August 1940. "Whether Sommerfeld is philosophically steadfast enough to be allowed to undertake a journey abroad cannot at the present time be ascertained," the answer to this inquiry read. Though the trip, in this case rather of a private nature, was approved, an eye was nonetheless kept on Sommerfeld. When in January 1941 the district leadership of the NSDAP Munich/Upper Bavaria received an inquiry from a publisher whether there were political or racial reservations about Sommerfeld, the reply came back that Sommerfeld should be "politically and scientifically most emphatically rejected."[55] Although Sommerfeld was retired, the University continued to

50 From Steenbeck, October 31, 1940. DMA, NL 89, 018, folder 3,12.

51 From W. Meißner, October 26, 1941. DMA, NL 89, 017, folder 2,4.

52 From Weizsäcker, January 28, 1942. DMA, HS 1977-28/A,359.

53 To Eucken, April 17, 1943. DMA, HS 1977-28/A,86; from Stueckelberg, December 26, 1944. DMA, NL 89, 013.

54 Beyler/Eckert/Hoffmann, *Planck-Medaille*, 2007, p. 231.

55 Files on Sommerfeld in the possession of the "Berlin Document Center," BAB, VBS 307, 8200002950.

include him in the category of those who had to submit information regarding their position vis-à-vis the NSDAP and its affiliated organizations. Conversely, the University was also consulted for information about Sommerfeld's political reliability. In March 1942, for instance, the General Consul of Italy submitted an inquiry with the University of Munich whether an invitation to Sommerfeld to deliver lectures in Milan could be approved. The Rector and the General Counsel of the University thereupon sought information from various authorities and informed the Consul "that serious objections had been raised against Prof. Sommerfeld's giving lectures at Milan." This "inquiry" revealed that "the SD, which had typically stood quite aside from the notorious physicists' quarrel , [had] raised objections against Sommerfeld."[56] Exactly what the SD (Sicherheitsdienst, "Security Service"), a division of the Secret Service in the Reich's Security Headquarters of the SS, suspected Sommerfeld of is not revealed by the surviving files. It may be that political ramifications were foreseen since the Milan invitation had been prompted by Giovanni Gentile Jr. (1906–1942), the son of the first fascist Minister of Culture in Mussolini's cabinet, who around this time had just fallen into disfavor. Gentile junior was a theoretical physicist at the University of Milan and an admirer of Sommerfeld's.[57]

Doubts about Sommerfeld appear in the end not to have been overly significant, for the trip to Italy was approved. But Gentile Jr. died unexpectedly in March 1942, shortly before Sommerfeld's arrival. In an article titled "Twenty Years of Spectroscopic Theory in Munich" in the autumn, 1942 edition of the Italian scientific journal *Scientia* (which presumably outlined the content of his Milan lecture), Sommerfeld referred to Gentile Jr. as "my young friend."[58] He had been particularly impressed by Gentile's recent work on quantum statistics, which seemed to him a very promising approach to the problem of the superfluidity of helium at very low temperatures.[59] "My lecture at Milan has been postponed until April 29, and is more and more assuming the aspect of a memorial in honor of the young Gentile," Sommerfeld wrote his wife from Rome. "I have visited his father here." He met also with Gian-Carlo Wick (1909–1992), who had assumed Fermi's chair at the University of Rome on his emigration and with whom he planned a joint theoretical-physical paper.[60] Sommerfeld and Heisenberg also invited Wick to Munich and Berlin to deliver lectures on cosmic radiation physics.[61] But the

56 Personnel files Sommerfeld, UAM, E-II-3187.

57 Bonolis, *Giovanni Gentile Jr.*, 2008.

58 Sommerfeld, *Zwanzig Jahre*, 1942, p. 123.

59 Sommerfeld, *Quantenstatistik*, 1942.

60 To Johanna, April 19, 1942.

61 To Heisenberg, June 17, 1942. Munich, Max-Planck-Institute for Physics; from Heisenberg, June 19, 1942. DMA, HS 1977-28/A,136.

planned publication by Sommerfeld and Wick did not materialize. It never went beyond shoptalk by letter, and even this was only sketchy, "otherwise, the poor censor will have to read too much!"[62] In 1943, Wick once more tried hard to bring Sommerfeld to Italy, but under the difficult wartime conditions, these plans could not be realized.[63] During his stay in Rome in April 1942, Sommerfeld met Wick's mother, Barbara Allason (1877–1968), "a charming lady, an author, who knows Ricarda Huch, and who raves about *Lotte in Weimar*," Sommerfeld wrote home.[64] He did not mention that as an anti-fascist she shared with Ricarda Huch (1864–1947) an aversion to Nazism. One year later, Sommerfeld visited "the 79-year old, still beautiful Ricarda Huch" in Jena,[65] and on this occasion, too, the conversation surely revolved not only around literary matters.

The reservations of the SD may well have related to another matter, too, that resulted in an entry in Sommerfeld's personal file in anticipation of the Italian trip. The German Embassy in Madrid had learned that Sommerfeld, in a letter to a colleague, had supported the return to Spain of Enrique Moles (1883–1953), a renowned chemist who, during the Spanish Civil War, had "sided with the Reds" and now lived in exile. Sommerfeld's letter had been "read in the presence of the Minister of Transport" at the Academy of Sciences in Madrid, the Reich's Minister of Science wrote to Walther Wüst, Rector of the University of Munich. "During the Spanish War," Moles had expressed himself "even in writing in favor of Russia and against Germany." Therefore, the Rector was advised to caution Sommerfeld "to distance himself in future from such recommendations and proposals."[66]

Except when Sommerfeld was—as in this case—expressly warned by the Rector, he remained unaware of such politically motivated reservations against him. Müller's accusations were passed on to him by Selmayr, but once the political higher-ups of the regime distanced themselves from the lecturer corps, he no longer considered these a serious threat. Presumably he knew nothing of the suspicions of the SD against him. What he himself thought of various figures of the Nazi power structure he never committed to paper. Occasionally, he hinted at his opinion rather *en passant*. Concerning an event at the Berlin Academy of Sciences "at which also our big boss Rust was in attendance," he wrote: "Since the meeting dragged on and on, I was more nearly asleep than awake."[67]

62 From Wick, August 28, 1942. DMA, HS 1977-28/A,366.
63 From Wick, 24. February 24, and March 22, 1943. DMA, HS 1977-28/A,366.
64 To Johanna, April 19, 1942.
65 To Buchwald, July 13, 1943. Danzig, Technical University, Atomphysik, Sommerfeld.
66 Rust to Wüst, March 6, 1942. UAM, E-II-3187. Also in DMA, NL 89, 024, folder Nazizeit.
67 To Buchwald, July 13, 1943. Danzig, Technical University, Atomphysik, Sommerfeld.

In one instance, however, it was unmistakably clear how the Nazi rulers were regarded in the Sommerfeld home. When Rudolf Heß, Hitler's representative, flew to Scotland on May 10, 1941, and wound up in English detention, Sommerfeld's wife composed the following poem "to Hitler."[68]

Now it begins! The towering fortress trembles
A cornerstone crumbles, timbers creak,
Sinister creatures slither through the walls.
What seemed green without is dead within!
The million-fold sorrows you have sown
Raise threatening hands in accusation,
And should sleep ever grace your chamber more,
In dreams, ghostly pale you'll vision the end!
Flight avails you nothing! Known everywhere,
Nowhere can you hide that liar's grimace,
Your Babel Tower cannot withstand time.
Those still cheering now will be your hangmen.
And as one day they speak of you "as though
Of the Black Death," come long ago to judgment,
You will join beneath their blood-red banner
The greatest fiends in the history of the world!

The background of the Heß flight remained hidden from the German public. What appeared a "flight" was presumably envisioned as the opening move of secret diplomacy by which, on the eve of the invasion of Russia, Heß was attempting to preempt a two-front war.[69] But even if Johanna Sommerfeld misconstrued the act of Hitler's deputy, the epithets with Hitler in mind ("liar's grimace," "hangmen," "greatest fiends") show manifestly how she—and presumably Arnold

68 "Als Rudolph Heß nach Schottland flog, an Hitler," May 13, 1941:
Nun fängt es an! Die stolze Zwingburg zittert
Ein Eckstein wich, es knistert im Gebälk,
Unheimlich Leben durch die Mauern schlittert
Was grün nach außen schien, ist innen welk!
Das Leid das du gesät millionenfach
Anklagend hebt und drohend seine Hände,
Und tritt noch je der Schlaf in dein Gemach
Siehst du im Traum gespensterbleich das Ende!
Dir frommt nicht Flucht! Die aller Welt bekannt,
Die Lügenfratze kannst Du nirgend bergen,
Dein Turm zu Babel hält der Zeit nicht stand,
Die heut noch jubeln, werden deine Schergen.
Und spricht man einst 'wie von dem schwarzen Tod'
Von dir, wenn lang verfielst du dem Gerichte,
Zählt man dich unterm Banner blutigrot
den größten Teufeln zu der Weltgeschichte!

69 Schmidt, *Rudolf Heß*, 1997.

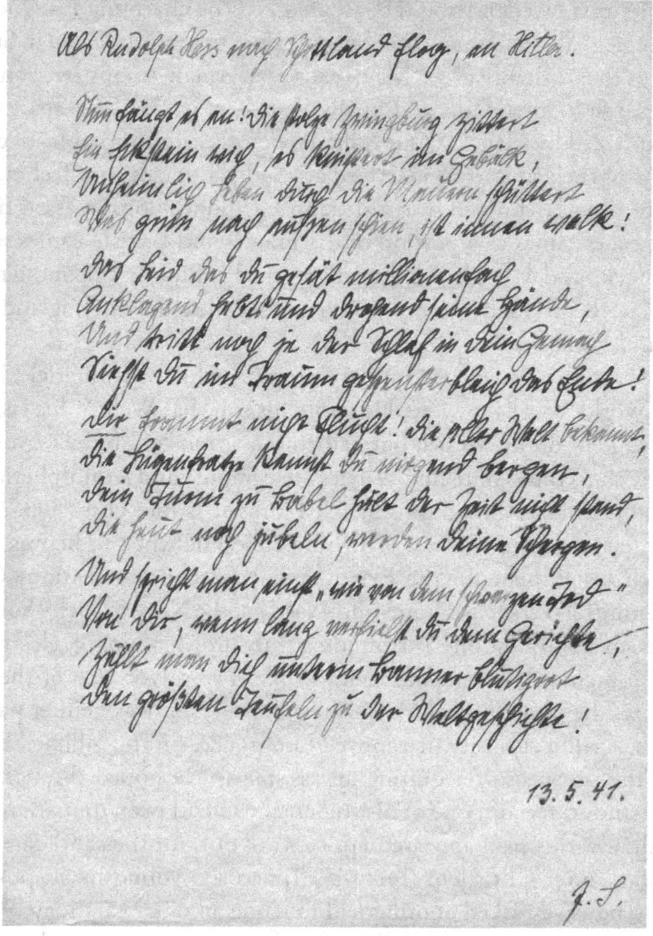

Als Rudolph Hess nach Schottland flog, an Hitler.

13. 5. 41.

J. S.

Fig. 32: Johanna Sommerfeld commented on the flight of Rudolf Heß to Scotland with this poem.

Sommerfeld—really felt about the rulers. Since these lines were kept secure within their own four walls, there was no danger they would be read by a third party—an ever-present consideration in any epistolary expression of opinion.

12.4 A Research Assignment for the Navy

Unlike his reaction during World War I, Sommerfeld's enthusiasm for the initial military "successes" of the Wehrmacht, which elsewhere in Germany had evoked great fervor, was limited. After the "General Government" was established in Poland, and the occupation of Denmark and Norway had begun, Sommerfeld hoped that "at least Sweden [would be] spared . . . The charming Lise Meitner, and also the great Siegbahn and many other people close to me are located in Stockholm, you know."[70] The "blitzkrieg" of May 1940, as German troops overran Belgium, Luxembourg, the Netherlands, and finally France, may well have sparked feelings like those expressed by Planck: "Our military victories are admirable, of course, the achievements of the army to be marveled at. And yet one cannot feel unrestrained joy when day after day human lives are sacrificed and millions in assets destroyed." Planck wrote these lines as a portion of the French and British armies were encircled at Dunkirk, and Hitler's Wehrmacht seemed militarily unconquerable. "My heart aches at the thought of my friends and colleagues in England, Norway, Denmark, and Holland."[71]

Nor was the commitment to "military physics" which Sommerfeld had exhibited during World War I the same. He wrote his son that Welker, his last assistant, who had been kept on by Müller, but had gotten a transfer on the pretext of "war-related" research to the "Gräfelfing Wireless-telegraphic and Atmospheric-electrical Experimental Laboratory," had thus finagled himself "a nice cushy job."[72] When in 1940 he himself was sounded by a Würzburg colleague whether he was "available" for a research assignment for the Experimental Telecommunications Command (Nachrichtenmittel-Versuchskommando, NVK) of the Navy in Kiel,[73] Sommerfeld's old enthusiasm for questions surrounding the propagation of electromagnetic waves was momentarily rekindled. "Meanwhile, I've solved one of the problems put before me," he wrote his son. Ultimately, however, it was neither patriotic call of duty nor scientific curiosity that moved him to take on the military assignment, but rather "the prospect of acquiring an assistant." He hoped through the Navy High Command to reclaim one of his students who had been drafted into military service.[74] Initially, this plan appeared not to work out, for the candidate chosen by Sommerfeld, Günter Christlein (1915–2008), was too young for a "position with deferment." The NVK asked Sommerfeld to name an older candidate, "because in future, extended deferments for younger employees are hardly to be expected."[75] Sommerfeld next proposed Fritz Renner (1907–1998), who had done his doctorate under him in 1937 and who had already been detailed to the NVK for other military research. In his dissertation,[76] Renner had also demonstrated virtuosic ability with complex integrals, which likewise offered useful qualifications for the military assignment at hand.

70 To Ernst, April 28, 1940.
71 From Planck, Ma23, 1940. DMA, NL 89, 012. Also in ASWB II.
72 To Ernst, February 22, 1940.
73 From Harms, February 12, 1940. DMA, HS 1977-28/A,132.
74 To Ernst, 20. April 1940.
75 From NVK, June 3, 1940. DMA, NL 89, 020, folder 7,2.
76 Renner, *Theorie*, 1937.

The NVK dealt with myriad radio-engineering developments for the Navy.[77] Sommerfeld's contacts, however, were mostly colleagues from Telefunken, who handled this research assignment for the NVK. He was supposed to work up "theoretical problems in the area of wireless telegraphy." One of these problems consisted in calculating the effect of the earth's curvature on the propagation of electromagnetic waves and their refraction in the upper atmosphere.[78] Sommerfeld left the calculations involved to Renner, his assistant, and limited himself to indicating the approach used in each case. Renner moved into a room in the Sommerfeld home and functioned also as secretary managing correspondence that piled up and as a helper to run errands. "Renner is terribly industrious and ambitious," Sommerfeld reported happily, "personally nice to have."[79] Since their grown children were out of the house, it was a great relief for Sommerfeld and his wife to have a strong young household companion around them. Renner soon counted as a member of the family. When life in Munich became ever more difficult in the later years of the war, Renner made all sorts of household purchases. During the last winter of the war, for example, he cited "war work" being carried out at the Sommerfeld home to justify acquisition of heating fuel at the "coaling station."[80] "He helps out assiduously in all household needs."[81] The "war work" also served as rationale to shield Renner from conscription to the front and to assure the Sommerfelds of household help.[82]

By contrast, the Navy seems not to have ranked the importance of their work to the war effort very highly, for Sommerfeld and Renner were put under no particular time pressure and were even permitted to publish portions of their results.[83] "Fortunately," he had not been tasked with "weighty" problems, such as the camouflage of U-boats, Sommerfeld wrote a colleague following a conference at Telefunken in Berlin; he had "only" been assigned "a more innocuous problem . . . which my assistant can basically handle."[84] In August 1944, in view of "impending postal restrictions," Sommerfeld asked his son Ernst to arrange for a certificate allowing him, with reference to his "armament work," to retain his telephone.[85] He clearly expected to achieve more through Ernst's contacts as patent attorney at Telefunken than from his own superiors at the Navy, who, had they regarded this as urgent, could easily have certified the necessity of his having a telephone.

77 On individual research projects, see BA-MA, M 697 E. On organization, see Krauß, *Rüstung*, 2006, pp. 165–174.
78 Transcription of a conference on August 30, 1940. DMA, NL 89, 020, folder 7,3.
79 To Ernst, September 13, 1940.
80 To Ernst, January 14, 1945.
81 To Ernst, February 3, 1945.
82 Interview with Fritz Renner, November 25, 1996.
83 Sommerfeld/Renner, *Strahlungsenergie*, 1942.
84 To Buchwald, July 13, 1943. Danzig, Technical University, Atomphysik, Sommerfeld.
85 To Ernst, August 12, 1944.

Fig. 33: Following an air raid in July 1944 in front of Sommerfeld's house on Dunantstrasse: Sommerfeld (with a pot serving as "helmet"); to his left, the chemist August Albert (1882–1951), who lived next-door. On the street (with swastika armband), a Party functionary ("block warden") from the neighborhood.

12.5 Lectures on Theoretical Physics

From time to time Sommerfeld found some interest in his "armament work" since it had put him back in touch with his first scientific love, viz., partial differential equations, which are central to all theoretical questions regarding electromagnetic waves. It did give him palpable pleasure to be able to report to his son that he had "nicely finagled" some question that had arisen in connection with his war assignment.[86]

His true ambition, however, lay not in solving mathematical problems in some physical application or another, but in the publication of the lectures which, during his decades-long Munich teaching career, he had developed, refined, and repeatedly adapted to the most up-to-date state of knowledge in theoretical physics. This plan for a textbook of theoretical physics based on his lectures was not a new one. Implementation had always been hampered, though, because Sommerfeld had shied away from the enormity of the task. "It's a great pity you don't wish to

86 To Ernst, May 20, 1943.

collaborate on the text-book in theoretical physics," a disappointed Planck had noted as early as 1909, referring to the proposal from the Leipzig Hirzel publishing house that Planck and Sommerfeld, as the outstanding representatives of theoretical physics, jointly bring out a textbook on the subject.[87] In 1906, Planck had brought out his *Lectures on the Theory of Heat Radiation* as a textbook and would gladly have pursued this direction together with Sommerfeld. But Sommerfeld saw too little common ground between his and Planck's lectures,[88] and Planck, too, had eventually to concede that such a textbook would be better written by a single author. When Hirzel repeated this proposal in 1924, Sommerfeld had no wish to take on obligations beyond *Atomic Structure and Spectral Lines*, "which constantly requires new, revised editions." The publisher even offered to take on the transcription of lecture notes, but Sommerfeld still declined. "The effort is far greater than you imagine; a secretarial assistant is not sufficient to the task." Besides, Hirzel had just brought out Planck's five-volume *Introduction to Theoretical Physics*. "I have often thought of publishing my lectures," Sommerfeld informed the publisher, "but now that Planck's lectures have been brought out by your press, I don't consider it urgent."[89]

"But putting something aside does not mean putting it away!" Throughout the entire 1930s, Sommerfeld had been occupied with adapting the *Wave Mechanical Supplement* to the rapid evolution of quantum mechanics, but when this work was completed in 1939, there was no longer any reason to postpone his ambition. Occupation with his lectures also helped him get over "the obscenity reigning in my former institute (I have no other word for it!)," about which he informed Prandtl in October 1941.[90] He must already have taken up the first volume of lectures, on mechanics, around this time, for a few months later he reported to his student Christlein, who had been sent to the Eastern Front in Russia, that he was "diligently" at work on publication of his lectures.[91] "What you write me from Munich seems, despite all the privations, like a swan song," Christlein wrote back. "I am especially thrilled about the publication of your lectures. You will be making a great gift to all Sommerfeld students, above all to those, like me, who for years have had to interrupt their studies."[92]

Many felt the same. "The former ways of getting access to your lectures—assuming one didn't happen to be in Munich—were really rather complicated," August Wilhelm Maue recalled. "I remember, for example, photographing lecture

87 From Hirzel, February 15, 1909. DMA, NL 89, 009; from Planck, February 24, 1909. DMA, HS 1977-28/A,263. Also in ASWB I.
88 Seth, *Quantum Theory*, 2004.
89 To Hirzel, 30. June 30, 1924. DMA, NL 89, 004. Also in ASWB II.
90 To Prandtl, October 10, 1941. MPGA, III. Abt., Rep. 61, Nr. 1538. Also in ASWB II.
91 To Christlein, March 5, 1942.
92 From Christlein, March 30, 1942. DMA, NL 89, 020, folder 7,1.

summaries with a cine-film camera."[93] The news spread among Sommerfeld students in many countries. Werner Romberg, who had narrowly escaped Stalin's "cleansing" in Russia in 1937 and had found a position in Oslo, learned of it in 1943 from the nuclear physicist Hans Jensen (1907–1973), who had come to Norway with the German occupation. "Several of my cherished and valuable notebooks from your lectures were lost in 1937 in my somewhat hasty exit from Russia," Romberg wrote to Munich in December 1943. "Now I dare to hope these will soon be available to me again in a newer form."[94]

At this time, *Mechanics* had already appeared, and *Mechanics of Deformable Bodies* was in progress as the second of the six planned volumes of lectures. But the war impeded production and distribution of the volumes according to schedule. At the Leipziger Akademische Verlagsgesellschaft, where the volumes were printed, the entire type-set composition of Volume II was destroyed in a bombing raid in December 1943. Pursuant to his son's advice, however, Sommerfeld was prepared for such contingencies of war and had had his assistant, Fritz Renner, prepare a copy. "So, not a total loss, just a delay!"[95]

In this case, the delay actually proved beneficial, since it gave Sommerfeld the opportunity, with the help of the experts at Prandtl's Kaiser-Wilhelm Institute for Fluid Dynamics, to correct several vulnerabilities in his initial presentation. Because of their manifold applications, from the hydrodynamics of ships to aerodynamics of airplanes, subspecialties such as boundary layer theory and turbulence theory had been caught up since the 1930s in feverish development. Applied mechanics, to which these theories so important in the engineering sciences belonged, had evolved into an independent discipline. In physics, scarcely any attention was paid the headlong advances occurring in the technical applications of fluid mechanics. "I have naturally simply adopted as is your reworking of my § on the boundary layer," Sommerfeld wrote gratefully to Prandtl, who had reformulated this portion for him. In the "especially problematic paragraphs on turbulence" and other portions, too, the Göttingen experts on fluid dynamics came to his aid.[96] In October 1944, Sommerfeld completed work on Volume II. Even amid the chaos of the last months of the war, it was printed and bound, although distribution was out of the question. "By now, it may be sitting somewhere in Moscow," suspected Sommerfeld, describing the status of his textbook project to a colleague in September 1945.[97]

93 From Maue, October 5, 1942. DMA, NL 89, 020, folder 7,1.

94 From Romberg, December 28, 1943. DMA, HS 1977-28/A,290. Also in ASWB II.

95 To Ernst, December 21, 1943.

96 From Prandtl, February 15, 1944, February 19, 1944, and February 20, 1945; to Prandtl, February 24, 1944, November 3, 1944, and January 31, 1945. MPGA, III. Abt., Rep. 61, Nr. 1538. Also in ASWB II.

97 To Jordan, September 8, 1945. SBPK, Jordan 606.

12.6 Relativity Theory Without Einstein?

Following the war's end, the Leipziger Akademische Verlagsgesellschaft was situated in the Soviet occupied zone. Sommerfeld gave the later volumes for publication to the Dieterich Verlagsbuchhandlung of Wilhelm Klemm, who had moved from Leipzig to Wiesbaden. But since the Akademische Verlagsgesellschaft maintained its interests in the textbook, a confusing duplication of West- and East-German editions ensued.

Apart from these superficial matters, it is hard to detect in the work itself any evidence of the historical circumstances under which it came into being. The correspondence, though, makes correspondingly more evident the toll taken by the "zeitgeist" on the textbook author in the year 1942. And it was none other than Heisenberg, who had been targeted as no one else by the attacks of "German physics," to whom Sommerfeld referred in noting that even after the attacks had ceased, theoretical physics could no longer be taught as it had been before 1933. In the mechanics manuscript, "in the section on relativity theory the name of Einstein is cited very often," Heisenberg wrote. An acquaintance, "who is politically active and has been of much valuable assistance to us," had expressed the wish "that at this point somewhat greater deference could be paid to the zeitgeist." Heisenberg thought it should be enough to know "that the special theory of relativity is correct, and that it would have come into being even without Einstein. I usually negotiate these sensitive political questions now by stressing the factual correctness of the special theory of relativity, entirely independent of its historical development."[98]

This suggestion seemed to Sommerfeld to go "somewhat against the rightful credit due an author," but he bowed to the extent of offering a compromise proposal. Einstein had been mentioned five times, three times in connection with the special theory of relativity and twice with reference to the general theory of relativity. These first named citations could be struck, Sommerfeld wrote the Leipzig publisher; of the latter two, one "must unconditionally be left," and the other he would also "rather seen retained." Since the political watchdogs would "presumably not without your assistance" have inspected the manuscript, Sommerfeld left the matter up to the publisher. "You decide!"[99] Sommerfeld enclosed a carbon copy of this letter with his reply to Heisenberg. It would "satisfy you and your 'acquaintance,'" he added. He could not, however, keep from expressing his dismay that "in secret thorough pre-censorship is exercised."[100]

98 From Heisenberg, October 8, 1942. DMA, NL 89, 024, folder Nazizeit. Also in ASWB II.

99 To W. Becker, October 15, 1942. DMA, NL 89, 024, folder Nazizeit. Also in ASWB II.

100 To Heisenberg, October 14, 1942. München, Max-Planck-Institut für Physik, Heisenberg, alphabetical. Also in ASWB II.

The publisher followed the compromise suggestion. Einstein's name appears in the *Mechanics* only in connection with the general theory of relativity.[101] These concessions to the "zeitgeist" did not alter the appeal of the textbook, however. To meet demand, the publisher had to issue a second edition after just a year.[102] Following the war, too, demand for the book continued, so that in short order further editions were printed. This success mirrors the success Sommerfeld had achieved as a teacher. He wished, as he stressed in his Preface, "to provide [the reader] with a living picture of the rich material which, from the proper mathematical–physical viewpoint, the theory allows us to take in." In the process, it must have recalled to him that first impulse to coauthor a textbook on theoretical physics with Planck, for he contrasted his own lecture style with that of Planck. Unlike Planck, he had not sought an uninterrupted and systematic construction; rather, he had valued "a greater variety of material." He had wished "as quickly as possible to [press on] to the essential physical problems" and, in contrast to Planck, to set forth "a freer treatment of the mathematical apparatus."[103]

The memories stirred by writing up his lectures, and the success he achieved with it even in the middle of the war helped Sommerfeld transcend the bitterness caused him by the decline of his "nursery" and the compromises wrung from him by the "zeitgeist." Even from abroad, letters came in bespeaking undiminished admiration and recognition. "Your book on Mechanics is extremely successful here," a physicist from francophone Switzerland wrote him in July 1943; "indeed, my Lausanne students have already ordered the entire series."[104] In 1949, when Sommerfeld looked back at the publication of his lectures, he was still highly conscious of the mood in which, at the age of over 70, he had undertaken this mammoth task. "Without this work, I could hardly have endured the political turmoil of the war years."[105]

101 Sommerfeld, *Vorlesungen I*, 1943, pp. 15 and 203.

102 Sommerfeld, *Vorlesungen I*, 1944.

103 Sommerfeld, *Vorlesungen I*, 1943, p. VI.

104 From Stückelberg, July 21, 1943. DMA, NL 89, 013. ("Votre livre sur la mécanique trouve beaucoup de succès chez nous […] en effet mes étudiants de Lausanne ont déjà commandé toute la série.").

105 Autobiographical Sketch, ASGS IV, p. 679.

13 Carrying On

The end of the "obscenity" at Sommerfeld's former institute came in the summer of 1944 when the University was largely destroyed in bombing raids on Munich. Müller was evacuated to Garmisch-Partenkirchen. The city of Munich capitulated on April 30, 1945. Walther Wüst, the "Führer" Rector who had administered the University in the spirit of National Socialism for the past years, was arrested by the American Military Police and interned at Dachau. He was replaced by Albert Rehm (1871–1949), a known opponent of Nazism, who had occupied the post of Rector before 1933. On July 12, 1945, Müller was dismissed from his position. By the end of August, a total of 33 professors and 63 assistants had lost their positions, and in the following weeks and months, further dismissals ensued. Of a total of around 250 professors at the University, only 60 had not been Party members. Not until the spring of 1946 did the military government permit the various faculties of the University of Munich to resume instruction.[1]

Many Germans experienced the fall of the "Third Reich" as a catastrophe. Only for survivors of the concentration camps and opponents of the regime did the end of the war come as liberation. Even many who had remained distant from Nazism looked on the Allies more as conquering enemy than as liberators. The editor of the *Neue Physikalische Blätter* compared the relations between occupiers and Germans to a "game of cat and mouse."[2] The "game," known as "denazification," had been worked out by the Allies in July 1945 at the Potsdam Conference. Its goal was to root out National Socialism, with all its laws, organizations, and other manifestations, and to reeducate the German population accordingly. The path to this goal, however, differed considerably among the four occupation zones.

13.1 Denazification

In the American zone, which included Munich, the population was categorized according to the extent of Nazi activities into (1) Major Offenders; (2) Activists, Militants, and Profiteers, or Incriminated Persons; (3) Less Incriminated; (4) Followers, or Fellow Travelers; and (5) Exonerated, or Non-incriminated persons. Which of these five categories a person belonged to depended on a sometimes very prolonged bureaucratic process. It began with a questionnaire and ended with a

1 Huber, *Universität*, 1984; Müller, *Universitäten*, 1997; Boehm/Spörl, *Ludwig-Maximilians-Universität*, 1972, p. 369; Schreiber, *Walther Wüst*, 2008, p. 347; Litten, *Mechanik*, 2000, p. 159; Wiecki, *Denazification*, 2008, pp. 537 und 541.

2 Ernst Brüche zitiert in Hentschel, *Mentalität*, 2005, p. 32.

M. Eckert, *Arnold Sommerfeld: Science, Life and Turbulent Times 1868-1951*,
DOI 10.1007/978-1-4614-7461-6_13, © Springer Science+Business Media New York 2013

so-called denazification tribunal decision. Only then was it determined whether a person could continue in his profession at his accustomed post or would have to try to take up a new career. For the majority who were not consigned at the outset to the category of "Major Offenders" at one end of the spectrum of involvement or "Exonerated" at the other, this was the beginning of a period of waiting in limbo.

Even before details of the denazification procedure, codified in the "Law for Liberation from National Socialism and Militarism" of March 5, 1946,[3] became known, many people who had in one way or another been active Nazis began securing defense witnesses. "I assume that proceedings are underway against me towards removal from office on the grounds of Nazi activity," Tomaschek, for example, wrote to Sommerfeld in July 1945. "Since I believe that you, as the most important physicist and as an uncompromised witness, might be consulted decisively in this matter, may I be permitted to turn to you and forward a number of explanatory statements."[4] In 11 type-written pages, he pleaded for understanding for his activities during the "Third Reich." Reading this, Sommerfeld may well have been put in mind of Heisenberg's assessment in the quarrel over "German Physics" "that although Tomaschek was the only competent member of the opposition, in terms of character he was by far the nastiest."[5] Sommerfeld wrote Tomaschek that he had not yet been consulted on his situation and forwarded the letter to the "appropriate officials" at the Technical University.[6] Tomaschek was dismissed and saw himself compelled to make a fresh start in private industry in England. For Sommerfeld, Tomaschek's inquiry was the first indication that in the coming months he would be assuming a new role—as witness for denazification.

First, though, he had to undergo his own denazification. It began with a "registration form pursuant to the Law for Liberation from National Socialism and Militarism of March 5, 1946." The first question concerned membership in Nazi organizations; the registrant was obliged also to list Nazi awards and other benefits. To all these questions, Sommerfeld answered "no." Only on the question concerning monetary sums to the NSDAP "or any other Nazi organization" did he admit that contributed to the NS-Public Welfare and the Winter Aid Work. To the last question "in which category of the law do you count yourself?" he answered, "Exonerated."[7]

3 Law Nr. 104 for Liberation from National Socialism and Militarism http://www.verfassungen.de/de/bw/wuerttemberg-baden/wuertt-b-befreiungsgesetz46.htm (31 January 2013). On denazification in the American zone of occupation, see Tent, *Mission*, 1982.

4 From Tomaschek, July 25, 1945. DMA, NL 89, 013. Also in ASWB II.

5 From Heisenberg, January 5, 1941. DMA, HS 1977-28/A,136. Also in ASWB II.

6 To Tomaschek (draft), undated. DMA, NL 89, 013. Also in ASWB II. On Tomaschek's career at the TH Munich, see Wengenroth, *Aufruhr*, 1993.

7 Registration form, in DMA, NL 89, 008. The NS organizations listed were (in this order) the NSDAP, Allg.-SS, Waffen-SS, Gestapo, SD (Secret Service) of the SS, Secret Rural Police, SA, NSKK (NS.-Motor Corps), NSFK (NS.-Pilot Corps), NSF (NS.-Women's Organization), NSDSTB (NS.-Student League), NSDoB (NS.-Lecturers' League), Hitler Youth, and German Girls' League.

The occupation officials agreed with this assessment, and thus, Sommerfeld's denazification was completed. Since no doubts were raised about his responses, the tribunal waived the opening of a proceeding.[8]

Though at the age of 77, Sommerfeld himself held no official position at the University, he nonetheless concerned himself with the future of his former "nursery." First, there were the practical matters of reconstruction, and these began also with denazification. "Since at present the care for the interests of my former institute devolves upon me, I request that the dismissal of Chief Mechanic Selmayr be reversed," Sommerfeld wrote in August 1945 to the Rector. The surviving equipment in the "heavily bombed spaces" required expert restoration as only Selmayr, the "indispensable caretaker of the collection," could provide. Selmayr fell under the second rubric of denazification, "Activists, Militants, and Profiteers, or Incriminated Persons," since he had joined the NSDAP in 1932. Sommerfeld maintained, however, that he had done so only "for fear of losing his position at the institute on grounds of his earlier membership in the Majority Social Democratic Party when the expected overthrow came about." In fact, Selmayr had "always [been] an opponent of the Nazi Party, whose excesses he had become familiar with as a neighbor of Himmler in Haar." He had "demonstrated" his opposition also "in actions," thereby "exposing himself to grave dangers."[9]

Selmayr's denazification did not run as smoothly as Sommerfeld's, however. His membership in the NSDAP weighed heavily against him. Presumably in anticipation of a difficult exoneration, on May 2, 1945, Selmayr had already informed the American military government that he had compiled "an extensive collection of evidentiary material" concerning Müller's "Jew-baiting." In addition, he knew the hiding places of University Rector and SS Commander Wüst.[10] Moreover, as early as May 1945, Sommerfeld had drafted a declaration of exoneration for Selmayr, in which he explained his mechanic's joining the Party as follows: "In anticipation of Hitler's 1933 seizure of power, Selmayr joined the NSDAP in 1932 with the explicit motivation of being better able to attend to my interests and endeavors. Up to my departure in 1939, he helped me avoid difficulties based on his knowledge gained from within the Party." Selmayr had waged "a genuine campaign" against Müller. "Only his Party membership enabled Selmayr to carry this out."[11]

The denazification process also brought to light incidents from the past that illuminated daily life at the Sommerfeld institute from a political point of view. "My joining the Party had the purpose of shielding Counselor Sommerfeld and his institute from further difficulties in the impending rolling of heads because of my known social-democratic orientation," Selmayr wrote in his defense. He had wanted "to be

8 No tribunal file exists on Arnold Sommerfeld in the State Archive at Munich (StAM).

9 To the Rector, August 9, 1945. DMA, NL 89, 018.

10 "SS-Oberführer." Selmayr to Eisenhower Headquarters, May 2, 1945. DMA, NL 89, 020, folder 8,3.

11 Draft, May 1945. DMA, NL 89, 020, folder 8,3.

useful to the institute as a Party member" and to "oppose" the Nazi regime that was to be expected in 1932 "under cover of camouflage." At the time, there was a specific justification for this precaution. Sommerfeld had arranged for a new position at his institute for Fritz Kirchner (1896–1967), a lecturer at the Wien institute who had lost his academic backing at Wien's death in 1928. This had given rise to ambitions on Kirchner's part, according to Selmayr, "soon to become successor to Counselor Sommerfeld." Although Kirchner was an "excellent physicist," he had played the part of "a savage Nazi." "I constantly protested his political pronouncements," Selmayr explained, and Kirchner had tried to remove him from the institute as a political opponent. When Kirchner, "in the spring of 1933," learned of his Party membership, "his intention was foiled." His running battle against Müller had also been possible only as a Party member. "Except for my Party membership, this action would have been suicide." Selmayr could also cite credible witnesses for his anti-Nazi position. One recalled that Selmayr had refused an SA man a ride in his car once, saying "Out. SA marches."[12] The precision mechanic at the institute for physical chemistry, who had been active in Hans Conrad Leipelt's (1921–1945) resistance group at the neighboring chemical institute, recalled that the investigations of the SD, the Gestapo, and the Führer's Chancellery put Selmayr "into a tight squeeze because he had acquired the material for the aforementioned purpose [the fight against Müller] illegally." Selmayr had also supplied him with weapons and ammunition for the resistance group. Ultimately, the tribunal found the explanations adduced so persuasive that by a decree of March 24, 1948, Selmayr was classified as "Exonerated."[13]

Sommerfeld also wanted to speed the denazification process for Heinrich Welker, his last assistant. In August 1945, he asked the Rector to restore Welker to his former position so that, when the University reopened in the coming winter semester, he could "take over the requisite lectures and exercises in theoretical physics in consultation with me. . . With respect to his political positions, I can assure you, based on our intimate collaboration of many years, that he is a passionate opponent of the NS. For details, I refer you to the questionnaire he has filled out."[14] But Welker's questionnaire, too, showed entries that made an expeditious denazification problematic. To the question concerning membership in the NSDAP, he had responded "Yes (candidate)." In addition, he had been a member of the SA. Welker explained this by saying that as a student, he had joined the "Stahlhelm" and had then been transferred "by merger" to the "Active-SA." His candidacy for the NSDAP had also occurred automatically. He had broken off his connection to

12 An ironic allusion to the *Horst Wessel Song*, from 1930 to 1945, anthem of the Nazi Party, and from 1933 to 1945, the conational anthem of Germany (with *Deutschland über alles).* The second line is "SA marschiert mit ruhig festem Schritt," in English, "SA marches [or, SA is marching] . . .."

13 Spruchkammerakte Selmayr, Karton 1518, StAM. Wagner, *Hans Leipelt*, 2003.

14 To the Office of the Rector of the University of Munich, October 25, 1945. DMA, NL 89, 030, folder Hochschulangelegenheiten.

both organizations in 1939. "His actual relationship to the Party proceeded from the declaration by the University lecturer corps that his application for extension of his lectureship had no prospect of success, and from his dismissal as institute assistant on purely political grounds by the extreme party-liner Prof. Dr. W. Müller," Sommerfeld explained in his exoneration report on Welker's behalf. The mayor of the municipality Planegg, where Welker had last lived, certified that he "seems not to have been at all politically active" and that his "neighbors give assurances that he was anti-Nazi." The most decisive testimony of exoneration was provided by Miriam David, a member of the "White Rose" who had been arrested by the Gestapo in 1943 and had spent the years up to war's end in a concentration camp. She had met Welker in 1942 at the physical-chemical institute of the University, "where he had been taken on as a guest by Professor Clusius, after dismissal from the University of Munich on ideological grounds as a Sommerfeld student and anti-Nazi." Welker had been "not only an ideological opponent of Nazism, but had expressed his oppositional sentiments quite openly to both Nazis and anti-Nazis. I was often present when he discussed ways of fighting against and undermining the National Socialist system, and gave out news he had heard on foreign broadcasts." In April 1945, Welker had been a member of the resistance group "Bayern," which sabotaged actions of the Volkssturm and attempted to apprehend "dangerous Nazi partisans." The preliminary examination board categorized Welker as "Exonerated." However, the final tribunal decree of February 1947 categorized him as "fellow traveler," and imposed a "restitution fine" of 500 Marks.[15]

13.2 A Provisional Fresh Start

"Our list would be: 1.) Heisenberg, 2.) v. Weizsäcker, 3.) Hund," Sommerfeld wrote in February 1946 to Heisenberg concerning his plans for the restoration of theoretical physics at the University of Munich.[16] He was picking up the discussion of his succession from the 1930s, when Heisenberg had been his favorite candidate as well. The others appeared on the list in deference to the convention of putting forth three names. Carl Friedrich von Weizsäcker and Friedrich Hund were outstanding candidates for any academic chair in theoretical physics, to be sure, but compared to Heisenberg, it was clear they should come under consideration only if his appointment for some reason should fall through. Earlier, Heisenberg had related to Sommerfeld his experiences of internment at Farm Hall in England and described the peculiar situation making his acceptance of an appointment at Munich difficult. "The English and American politicians have arranged for the Kaiser Wilhelm Institutes for chemistry and physics to be moved to the English

15 Tribunal file on Welker, Box 1944. StAM.

16 To Heisenberg, February 17, 1946. München, Max-Planck-Institut für Physik. Also in ASWB II.

zone, though of course negotiations with the French, to be led not by us but by the politicians, still need to take place. So our fate is still quite uncertain. The English physicists are trying in every way to arrange reasonable work situations for us, but I will not end up in the American zone, at least not permanently."[17]

Thus, Sommerfeld could not count on a speedy appointment for Heisenberg and was forced to seek an interim solution. Were it up to him, Selmayr would have been set to intense reconstruction of his institute and Welker would have filled his academic chair until Heisenberg could be appointed, but the drawn-out denazification of Selmayr and Welker spoiled this plan. Welker was "rejected on political grounds (because of SA)," Sommerfeld wrote indignantly. "Gans is to assume the chair provisionally, and as such will certainly gladly serve however long we want," he informed Heisenberg about the emergency solution agreed upon at Munich.[18]

As a "non-Aryan," Richard Gans (1880–1954) had lost his professorship in theoretical physics at Königsberg in 1935 and had survived the "Third Reich" probably only thanks to lucky circumstances.[19] "I hereby request consideration for appointment to any teaching position in theoretical physics or related field that opens," he had written to the Bavarian Ministry of Culture after the end of the war, which he lived through in Upper Franconia.[20] The Ministry forwarded the application to the University of Munich. Sommerfeld had known Gans for many years and approved the application. He asked Gans "to step in," in case Welker were rejected on political grounds. "We are of course still hoping to be able to appoint Heisenberg permanently, or another prominent atomic physicist. My report that your intention is in any case to go to Argentina in the future was important to our provisional arrangement," Sommerfeld wrote concerning the background to the filling of the Munich academic chair.[21] Gans had taught at the University of La Plata from 1912 to 1925, and the field of physics in Argentina was greatly indebted to him.[22] In his written statement to the Rector, the Dean of the Faculty of Natural Sciences explicitly reemphasized that this was merely an interim solution, "since the faculty has other intentions regarding the final appointment to the academic chair in theoretical physics, about which it is my understanding both Prof. Gans and Counselor Sommerfeld have been informed."[23] On January 16, 1946, the Rector of the University approved the "provisional arrangement," so that, effective March 1, 1946, Gans could take up his duties.[24]

17 From Heisenberg, February 5, 1946. DMA, HS 1977-28/A,136. Also in ASWB II.

18 To Heisenberg, February 17, 1946. München, Max-Planck-Institut für Physik. Also in ASWB II.

19 Friendly physicists had arranged research assignments classified as "militarily critical" for him. For a comprehensive overview, see Swinne, *Richard Gans*, 1992.

20 Gans to the Bavarian Ministry of Culture, November 19, 1945. BayHStA, MK 69781.

21 To Gans, December 6, 1945. Quoted in Swinne, Richard Gans, 1992, pp. 133–134.

22 Reichenbach, *Richard Gans*, 2009.

23 Clusius to the Rector, January 4, 1946. UAM, OC-IX-070.

24 Bavarian Ministry of Culture to the Rector of the University of Munich, March 26, 1946. BayHStA, MK 69781.

Although it must have been clear to him that he was not the first choice for the Munich position, and that at the age of 66, he could no longer count on a long career, Gans still harbored the hope that his temporary appointment would, following his move to Munich, be converted to a permanent professorship. Gerlach encouraged him in this expectation,[25] but Sommerfeld did not let go the reins. Notwithstanding his age of nearly 78 years, he agreed to being officially installed again as Chairman of the institute, and arranged that Gans be authorized only to have access to the institute's funds.[26] Since the denazification of Selmayr and Welker dragged on, Gans soon felt exploited in his position as factotum.[27] At the end of the summer semester of 1946, he protested to the Dean of the faculty that it was an "intolerable situation" that not even one of the two open positions of assistant to which the institute was entitled had been filled, even though Welker's denazification was unproblematic.[28] After 9 months, he resigned the academic chair. For one thing, he wished to join his sons, who after imprisonment in Russia had just arranged their immigration to Argentina. For another, without assistants and a mechanic, he did not feel capable "of carrying out the pedagogical and research enterprise fruitfully, or of restoring the apparatuses and machine tools to working order," as he informed the Dean on December 24, 1946.[29] For his part, Welker soon thereafter moved with his family to Paris to establish a semiconductor laboratory at a French branch of the American electronics firm Westinghouse.[30]

Meanwhile, Sommerfeld was ruminating over who might succeed him at Munich should Heisenberg not be released by the authorities of the British Occupation Zone. Presumably, the same issues would stand in the way of an offer to Weizsäcker. "Up to now, in any case, our treatment in these matters has been precisely parallel," Weizsäcker wrote to Munich.[31] "If you put Weizsäcker and me on the list, you will see how the high and mighty lords react," Heisenberg suggested. "But there is not much prospect of their letting us come to Bavaria, and even if they do, one would have to have some sense of what the future holds to make a proper decision. So there is nothing for it but to wait and see what happens and how world history develops on both the large and small scales."[32]

"Heisenberg has written me a very nice letter, but the chances of getting him are minimal," Sommerfeld informed Bechert, whom he was also considering, but who, annoyed by the American denazification process, did not want to come to Munich,

25 Gerlach to Gans, November 12, 1946. Quoted in Swinne, *Richard Gans*, 1992, p. 139.
26 From the Bavarian Ministry of Culture, March 14, 1946; to the Rector, May 24, 1946. UAM, E-II-3187.
27 Gerlach to Gans, November 27, 1946. Cited in Swinne, *Richard Gans*, 1992, pp. 139–140.
28 Gans to the Dean, July 16, 1946. UAM, OC-VIII-3.
29 Gans to the Dean, December 24, 1946. UAM, E-II-01403.
30 Handel, *Halbleiterforschung*, 1998, pp. 122–125.
31 From Weizsäcker, November 16, 1946. DMA, NL 89, 014. Also in ASWB II.
32 From Heisenberg, February 7, 1947. DMA, HS 1977-28/A,136. Also in ASWB II.

and also declined an offer to become Minister of Culture in Hessen. "We have written to Hund as well," Sommerfeld reported further; "with his large family he is not likely to be able to leave the Russian zone. Jordan is too incriminated even to be taken into consideration. We are thinking also of Fues, who for his part would be ready a year from now."[33] Bethe also came under consideration. "Would you have the heart to return to Germany if Heisenberg is definitely unavailable to come to Munich?"[34] His reply reached Sommerfeld only months later. "If everything since 1933 could be undone, I would be very happy to accept this appointment," Bethe wrote back in May 1947. "Unfortunately, it is impossible to extinguish the last 14 years . . . For us who were driven from our positions in Germany, it is impossible to forget. The students of 1933 did not want to study theoretical physics with me (and it was a large group of students, perhaps even the majority), and even if the students of 1947 think differently, I cannot trust them. And what I hear about the reawakening of nationalistic sentiment among students at many universities, and among many other Germans, is not encouraging."[35]

While the search for a successor proceeded, at least the second assistant post at the institute was now filled with Paul August Mann.[36] Mann had been classified as "Exonerated." He had studied with Sommerfeld shortly before the war and thereafter as a physicist at Telefunken had been in frequent contact with Sommerfeld's son Ernst. "Your son Ernst recently wrote me that you had characterized me as a lucky dog because I had the opportunity to go to Switzerland," Mann wrote Sommerfeld in February 1947 concerning his plans. The military government had refused to approve his exit, however, so he asked Sommerfeld "whether there might not be some, if even a modest use to be made of me within your institute, at least until the exit regulations have eased."[37] His intended emigration never occurred, and Mann remained for many years as assistant at the institute.

Following the departure of Gans, there was, in the person of Ernst Lamla (1888–1986), once again a more senior physicist provisionally filling the leadership of the institute. Lamla had completed his doctorate in 1912 under Planck and had pursued a career as a high school teacher and superintendent. As a member of the SPD (Social-Democratic Party of Germany), he had been forced into retirement in 1933. During the 12 years of Nazi rule, he had struggled through as an "independent theoretical physicist" with miscellaneous research assignments. After the war, on the recommendation of the Social-Democrat and first Prime Minister of Bavaria Wilhelm Hoegner (1887–1980), he had come to Munich. "He would not be considered for a permanent position," it was noted at the Bavarian Ministry of Culture,

33 To Bechert, February 15, 1947. Bremen, private possession.
34 To Bethe, November 1, 1946. DMA, NL 89, 015. Also in ASWB II.
35 From Bethe, May 20, 1947. DMA, HS 1977-28/A,19. Also in ASWB II.
36 No birth and death dates found.
37 From Mann, Februar7 3, 1947. UAM, OC-VIII-3.

"but he is very well suited to give the lectures."[38] Effective from April 1, 1947, for the duration of the summer semester, Lamla was entrusted with the temporary appointment to the full professorship "in theoretical physics and applied mechanics," as the position at this time still was designated. Sommerfeld took this opportunity of petitioning for the revocation of the renaming of his professorship effected by Müller. His petition was granted without reservation by the Ministry of Culture: from May 1947, the addition "and applied mechanics" was dropped from the official designation of both professorship and institute.[39]

The matter of the permanent appointment to the academic chair remained open up to the summer of 1947. "The question of my succession remains problematic," Sommerfeld complained in March 1947 in a letter to his son; "no one wants to or is permitted to come to Munich."[40] The situation recalled that of the 1930s. "Efforts to get Werner Heisenberg as Sommerfeld's successor go back to the year 1935. They were thwarted by the authoritative decision of the Nazi League of Lecturers," wrote the Dean of the Faculty of Natural Sciences by way of introduction to his recommendation to the Ministry of Fritz Bopp (1909–1987) as ultimately the only candidate. There had also been "confidential inquiries made" of others who had been named as possible candidates in 1935. All had declined. "Heisenberg, von Weizsäcker, Hund, and Erwin Schrödinger, from Dublin, the discoverer of wave mechanics, have all replied similarly that they cannot accept, some because they were not free to make their own decisions, some because they had to await political and economic developments." Wentzel, Kronig, Bechert, and Bethe, "the last great student of Sommerfeld's," had also all declined. "Under these circumstances, the faculty believes it must waive appointment of a generally recognized, older scholar; it proposes the younger, very promising Dr. Fritz Bopp, born in 1909, for an associate professorship, naturally without thereby relinquishing the search for a permanent full professor."[41] Bopp was a student of Sommerfeld's student Erwin Fues and during the war had worked at the "Uranium Club" under the direction of Heisenberg. After the war, he was entrusted with the supervision of the remnants of the "Uranium Club" at Hechingen, which was situated in the French Occupation Zone. Subsequently, he taught theoretical physics at the University of Tübingen. It had been impossible for the Faculty, it was explained to the Ministry, to propose the customary list of three candidates, "since all the gentlemen under consideration have either declined or are not at liberty to carry out their decisions."[42]

38 Entry, January 9, 1947. BayHStA, MK 69781.

39 Ministry of Culture to the Office of the Rector of the University of Munich, May 28, 1947. BayHStA, MK 69781.

40 To Ernst, March 2, 1947.

41 Dean to the Ministry of Culture,. July 4, 1947. BayHStA, MK 69781.

42 Office of the Rector to the Ministry of Culture, July 7, 1947. UAM, E-II-00948.

Bopp had presumably come to Sommerfeld's attention through his student Gerhard Elwert (1912–1998), who in December 1946 had written rhapsodically about the "Hechinger seminar" and about Bopp's latest quantum theoretical papers.[43] In April 1947, Sommerfeld inquired of Bopp whether he would be prepared to come to Munich for the next winter semester "as associate professor, with the possible collaboration of Fues as honorary professor, or initially as provisional substitute." Sommerfeld's choice of words suggests that he did not yet regard Bopp as a final successor but rather as another provisional substitute. In Bopp's case, too, there were questions of Nazi incrimination. "Have you gone through the tribunal? Are you incriminated? Are you available to leave Hechingen at any time?"[44]

Bopp was "prepared in essence to come to Munich" but felt bound by a promise he had given Heisenberg, who had wanted to bring him to Göttingen as a collaborator. He could not yet show a tribunal decree, but Frédéric Joliot-Curie (1900–1958), who was serving in the French Occupation Zone as High Commissioner of Atomic Energy, had assured him verbally that he could "continue working." Bopp also assured Sommerfeld that he had not been a member of the Nazi Party. He had, though, belonged to a flying club in Breslau. As such, in 1937, he had been "automatically" transferred to the NSFK (National Socialist Flying Club) but had "from 1938 onward gradually left, which was confirmed *de jure* around the time of the outbreak of the war. I do not know how this situation would be judged in the State of Bavaria."[45] Sommerfeld guessed that membership in the NSFK would be regarded as innocuous, but advised Bopp for his part to do all he could to obtain a tribunal decision as soon as possible.[46] The Munich faculty likewise pressed the Ministry for a speedy decision. They cited "the crisis in the subject of theoretical physics in existence since 1940 and keenly felt by the student body, which had been heightened by the emigration of Prof. Gans and could only temporarily be eased somewhat by the personal intervention of Prof. Sommerfeld."[47] Although some back and forth in reservation and deliberation remained, so far as the question of Bopp's denazification and his transfer from the French to the American Occupation Zone was concerned, the appointment process ran a speedy course. On August 10, 1947, the Bavarian Ministry of Culture issued Bopp's appointment, and he accepted 1 week later.[48] On August 29, the tribunal's decision, by which Bopp was classified as "Exonerated," was added to his personnel file.[49] "His appointment has gone remarkably smoothly," Sommerfeld wrote to Heisenberg, greatly relieved. "That of Fues as honorary professor ought likewise to be accomplished quickly, although he

43 From Elwert, December 2, 1946. DMA, HS 1977-28/A,82.

44 To Bopp, April 25, 1947. DMA, NL 89, 006. Also in ASWB II.

45 From Bopp, May 10, 1947. DMA, NL 89, 006. Also in ASWB II.

46 To Bopp, May 15, 1947. DMA, NL 89, 006.

47 Dean to the Ministry of Culture, July 4, 1947. BayHStA, MK 69781.

48 Ministry of Culture to Bopp, August 10, 1947; Bopp to the Ministry of Culture. August 17, 1947. BayHStA, MK 69781.

49 UAM, E-II-00948.

remains at the head office in Stuttgart."[50] The teaching assignment and the honorary professorship for Fues was approved, "pending final decision of the tribunal," and will be issued retroactively following receipt of this document.[51]

Therewith, the provisional arrangement for the temporary appointments was completed. The winter semester of 1947/1948 was the start of a new era. Sommerfeld could hope that the tradition of theoretical physics at Munich would endure—if for now only in the form of an associate and an honorary professorship. Ewald at any rate thought Sommerfeld should be satisfied with this solution: "An associate professor with a future is worth more than a full professor with a past."[52]

13.3 "I'll Bet on the Anglo-Americans"

Even after the provisional resolution, the institute for theoretical physics at the University of Munich could no longer be seen as continuing the "golden age," when the pioneers of modern quantum mechanics were coming of age in the Sommerfeld "nursery." "Bopp does his work very well, and knows how to draw students up to the heights of abstract physics," Sommerfeld wrote Heisenberg in praise of his successor, but in the same breath he maintained "that nonetheless, we have not abandoned all hope for our original candidate."[53]

But after the war, not even Heisenberg could have restored the lofty reputation the institute had enjoyed in the 1920s. Germany had ceded the leading role in physics to the USA and Britain. This was clear already from the actual weapons introduced in World War II that had been developed by physicists in the radar, atom bomb, and other military projects of the Allies. In this regard, Sommerfeld did not have to rely on rumors. Ewald brought the "Smyth Report" to his attention, which described the Anglo-American atom bomb project.[54] Bethe had rejected the offer of the Sommerfeld succession not only because of his "negative memories of Germany" but also because of his positive experiences in the USA, as he wrote Sommerfeld. "I was permitted, as a quite recent immigrant, to work in the war-time laboratories, and in a prominent position at that. Now, after the war, Cornell has built a large new nuclear physics laboratory, essentially 'around me.' And 2 or 3 of the best American universities have made me tempting offers."[55] Pauli, who had spent the war years in Princeton and had himself not participated in military research, was able on his

50 To Heisenberg, September 24, 1947. Munich, Max-Planck-Institut für Physik.

51 Minister of Culture (Hundhammer) to the Rector, August 26, 1947, and January 15, 1948. UAM, E-II-01391.

52 From Ewald, November 28, 1947.

53 To Heisenberg, January 15, 1948. Munich, Max-Planck-Institut für Physik. Also in ASWB II.

54 From Ewald, April 1, 1947. DMA, 007. Also in ASWB II. Smyth, General Account, 1945.

55 From Bethe, May 20, 1947. DMA, HS 1977-28/A,19. Also in ASWB II.

return from the USA to report on advances achieved there with "modern radar technology," which gave new impetus to quantum electrodynamics.[56]

The discovery to which Pauli alluded in this letter went down in the history of modern physics as "the Lamb shift."[57] It describes a tiny energy shift in an electron of the hydrogen atom brought about through the interaction of the electron with its own electromagnetic field. During World War II, Willis Lamb (1913–2008), a student of Robert Oppenheimer (1904–1967), became an expert on the microwaves used in radar technology. At Columbia University in New York, after the war, together with Robert Retherford (1912–1981), he made his field of specialization the spectroscopic measurement of atoms in this previously inaccessible region of wavelengths. "The great wartime advances in microwave techniques in the neighborhood of three centimeters wavelength make possible the use of new physical tools for study of the n = 2 fine-structure states of the hydrogen atom," Lamb and Retherford wrote describing this transition from wartime to basic research.[58] They found that two of the excited energy levels in the hydrogen atom did not actually lie exactly where the Sommerfeld-Dirac fine-structure formula said they should. Before the war, with the instruments of optical spectroscopy available, no deviation from the theory had been observable. By irradiation of centimeter waves, it was possible to detect fundamentally finer energy differences than with light waves. When Lamb and Retherford published this deviation, it immediately became the subject of a physics conference in June 1947 on Shelter Island, a town at the eastern end of Long Island, near New York. There, the elite of American theoretical physics discussed the problems of quantum electrodynamics. Immediately following this conference, Bethe published the theoretical explanation of this effect.[59]

In the summer of 1948, Bethe returned to Europe for the first time since the war and paid a visit to his old teacher and his first venue of activity in physics in Munich. In a colloquium lecture, he reported to the Munich physicists on new microwave measurements at Columbia University by which the fine-structure constant could be determined with heretofore unattainable precision.[60] In this way, Sommerfeld learned firsthand the unimaginable currency the atomic theory he had established in 1916 had now gained in the USA. Two years later, when Lamb and Retherford sent him the manuscript of the comprehensive paper about their discovery, it was for Sommerfeld a moving testimonial to his life's work. "It was very thoughtful of you to send the 81-year-old great-grandfather of fine-structure your wonderful paper in advance of its publication," he wrote gratefully. "Shortly after your discovery, Bethe wrote me about it, and has lectured on it here."[61]

56 From Pauli, October 31, 1947. CERN (PLC). Also in ASWB II.

57 Schweber, *QED*, 1994, Chap. 5.

58 Lamb/Retherford, *Fine Structure*, 1947, p. 241, cited in Forman, *Swords*, 1995, p. 426

59 Schweber, *QED*, 1994, Chap. 5.6.

60 Josef Brandmüller and Eduard Rüchardt: Zum Problem der spektroskopischen Einheiten. Manuscript, December 4, 1948. DMA, NL 89, 042.

61 To Lamb and Retherford, May 30, 1950. Also in ASWB II, pp. 644–645.

Herzfeld also manifested his devotion to the old Munich "nursery" by taking a leave from Catholic University in Washington to serve alongside Bopp as a guest professor for the summer semester of 1948.[62] One year later, Edward G. Ramberg (1907–1995) came as a guest lecturer. As an American exchange student, Ramberg had come to Munich and taken his doctorate under Sommerfeld in 1932. Now he was bringing back to Munich in the form of lectures on electron optics what he had accomplished in his field as an industrial physicist at Radio Corporation of America. The Munich physicists thus learned in manifold ways about the achievements of American physics and learned that these were rooted to no small degree in the tradition of the Sommerfeld "nursery." With his translation into English of Volume 3 of Sommerfeld's lectures on electrodynamics, Ramberg also ensured that this tradition would live on in the USA long afterwards.[63]

As guest professor at Madison in 1922/1923, Sommerfeld had already sensed that the future of physics lay in the USA. On his visits to Pasadena in 1929 and Ann Arbor in 1931, this assessment of American physics was reinforced. Politically, too, America seemed to him to be the land of the future. A moderating intervention by the USA, he hoped, would protect Germany from French revanchism following World War I. The relaxed American lifestyle contributed to his affection for the USA. Shortly after the end of World War II, in an article for *Die Neue Zeitung*, a publication aimed by American editors at political reeducation in the American Occupation Zone, he made it clear that nothing had changed in his pro-American stance.[64] Neither denazification nor the broadly unpopular measures taken by the Americans for reconstruction of Germany diminished Sommerfeld's liking for the USA. "While the bourgeois Bavarian puts his money on Herr Semmler [sic.], I'll bet on the Anglo-Americans, from whom I have experienced much good," he wrote Heisenberg in January 1948.[65] He was alluding to the affair of the co-founder of the CSU (Christian Socialist Union) in Bavaria, Johannes Semler (1898–1973), a member of the business council of the Anglo-American bizone who had made himself ridiculous by his verbal gaffes against the American plans for the reconstruction of Germany[66] and had been replaced by Ludwig Erhard (1897–1977), who 1 year later became the first Economics Minister of the Federal Republic of Germany.

At this time, Sommerfeld was unaware that he was soon to "experience much good" from abroad once more. In November 1948, Jay William Buchta (1895–1966), President of the American Association of Physics Teachers, wrote that his organization wished to award him the Oersted Medal for the year 1949.[67] The Oersted Medal was the highest American award for achievement in the teaching of physics.

62 From Herzfeld, January 9, 1948. DMA, NL 89, 009.

63 Correspondence with Ramberg in DMA, NL 89, 012 and 043.

64 Sommerfeld, *Atomphysik*, 1945.

65 To Heisenberg, January 15, 1948. Munich, Max-Planck-Institut für Physik. Also in ASWB II.

66 *Der Spiegel*, 12/1951, p. 26.

67 From Buchta, November 26, 1948. DMA, NL 89, 005. Also in ASWB II.

It perhaps meant more to Sommerfeld than many other honors. That this news was conveyed to him by Buchta added yet an especially personal note to the award, for as a student, Buchta had attended Sommerfeld's lectures at the University of Wisconsin in the winter semester of 1922/1923 and had thereafter arranged for regular delivery of the *Astrophysical Journal* to the Munich institute.[68] In the days of raging inflation, the expenditure for this journal was just as valuable a gift as had been the CARE packages Sommerfeld received from American colleagues and friends after World War II.

13.4 Recognition for the Teacher

The award of the Medal took place on January 28, 1949, in McMillan Hall at Columbia University in New York. Initially, Sommerfeld had wanted to undertake the trip accompanied by his son, but this plan fell through. Instead, Edward Condon (1902–1974), who had studied at Göttingen and Munich as a Rockefeller Fellow in the 1920s, accepted the Medal on Sommerfeld's behalf. Condon gave a very pretty speech, one of the participants reported to him, and the huge McMillan Hall had been "jam-packed." The American publisher of his lectures also enthusiastically praised the "very spirited address" in which Condon paid homage to Sommerfeld's contributions.[69]

Sommerfeld was the first German ever awarded the Oersted Medal, and this just 4 years after the war. Eleven years passed before the award fell once more to a German (Robert Wichard Pohl). The American Association of Physics Teachers was known to be a "very staid, almost rather conventional club," Scherzer, who was familiar with its internal workings, wrote to Sommerfeld in advance of the event. "It is thus all the more amazing and gratifying that the award to a German is so seriously under consideration." The impetus had been given by Lloyd Preston Smith (1903–1988).[70] Smith had played a decisive role in bringing Bethe to Cornell in 1935. As first among the reasons Sommerfeld had been chosen, the chairman of the medal committee named *Atomic Structure and Spectral Lines.* This book more than any other had disseminated atomic theory. To make this theory accessible not only to a handful of specialists but to every physicist had required a messenger with a special pedagogical gift. "Sommerfeld was such a messenger. His *Ergänzungsband*, the second volume of *Atomic Structure and Spectral Lines*, was perhaps the most influential of the early scholarly interpretations of wave mechanics."[71]

68 From Buchta, January 13, 1923. DMA, NL 89, 019, folder 4,1; to Buchta, 14. December 14, 1948. DMA, NL 89, 043.

69 From Beth, February 6, 1949; from Jacoby, March 12, 1949. DMA, NL 89, 005.

70 From Scherzer, November 18, 1948. DMA, NL 89, 043.

71 Kirkpatrick, *Recipient*, 1949, pp. 313–314.

There followed a list of distinguished Sommerfeld students and a selection of excerpts from letters in which several of them characterized their teacher. For Houston, Sommerfeld was one of the most influential physics teachers of the years between the World Wars. Bechert regarded Sommerfeld as one of the most important living teachers of theoretical physics. Scherzer praised especially Sommerfeld's ability to attract and encourage young talent. For Laporte, more than anyone else Sommerfeld had influenced the current generation of physicists. Wentzel described the fascination Sommerfeld's instruction inspired by citing his ability to convey to his students the sense that science is a living thing, and that even a beginner can be a useful component of this organism. Never before had a physicist been honored for his service to teaching with such avidity. This was admittedly not so much an objective analysis of the pedagogical enterprise at Munich as the expression of views of Sommerfeld's students who could themselves look back on successful careers and thus felt indebted to their mentor. Nonetheless, it was an impressive demonstration of the broad influence of the Sommerfeld school, above all in the USA.

In return, Sommerfeld used the opportunity to honor his American students. The American Association of Physics Teachers had asked him for a report on his "teaching activity" for the *American Journal of Physics*, he informed Pauling. "In it, I have mentioned you as having attended my first lectures on wave mechanics, still in a nascent state at that time."[72] In the process, he had himself learned as much as Pauling and the other attendees, he confided to the readers of this American journal for physics teachers. He had also presented the electron theory of metals first in 1927 in a special lecture and thereafter published it with his American guests Carl Eckart and William V. Houston as coauthors. He mentioned also Isidor I. Rabi (1898–1988) and Edward U. Condon, who had attended his lectures during one semester. Particularly in the special lectures for advanced students, a good personal relationship had been an important prerequisite for successful pedagogy.[73]

American recognition of Sommerfeld as "teacher" took on added significance in another context, too. The need for physicists increased dramatically in the USA during the years of the Cold War. In the 5-year period after the war, the number of doctoral candidates increased tenfold to a rate of about 500 completed doctorates per year. The rate of production of American physicists during the Cold War showed the same relation as the notorious growth curves of stock prices preceding a crash—and was interpreted similarly.[74] "Project Paperclip" served to engage German scientists and engineers in the American Cold War programs also.[75] During the 1950s and 1960s, it became virtually the rule for physicists in the Federal Republic of Germany to spend a few years following their studies doing "post doc"

72 To Pauling, March 6, 1949. Corvallis, Oregon State University, Special Collections, Pauling Papers.

73 Sommerfeld, *Reminiscences*, 1949.

74 Kaiser, *Cold War*, 2002, p. 135, Fig. 2.

75 Gimbel, *Project Paperclip*, 1990.

research in the USA. In this context, Sommerfeld's award from the American Association of Physics Teachers appears to presage the increasing Americanization of physics in the Federal Republic, where as an ally of the USA in the Cold War, science, too, was patterned along American lines.[76]

13.5 The Eightieth Birthday

The news of the award of the Oersted Medal arrived almost as an eightieth birthday present for Sommerfeld. In the days before and after December 5, 1948, he learned that he was far from being "tossed onto the scrap heap." In his honor, the Munich physicists organized a self-produced theatrical and an exhibition interlarded with allusions to both the times and episodes from Sommerfeld's career at Munich.[77] In dozens of birthday letters, he learned how very much his students, colleagues, and friends throughout the world esteemed him. Millikan, who a few months earlier had celebrated his own eightieth birthday, recalled his time as an American exchange student at Göttingen when he had come to know Sommerfeld as a budding lecturer. "I can still see you," he wrote in his birthday greeting, "as I saw you in Göttingen in the spring of 1896, when you were Klein's assistant, always carrying your little portmanteau as you moved in and out of the class rooms in Göttingen in which I was visiting lectures by Voigt, Klein, and Nernst." Sommerfeld's visits to Pasadena in the 1920s also came alive again. "I wish you could be here again and meet in our discussions as you did in those memorable years. Your picture still hangs on the wall in that discussion room in the Bridge Laboratory, in which we all got so much out of your leadership."[78]

The astrophysicist Walter Grotrian recalled another incident from the distant past: "Do you know how you first made an impression on me? It was in the garden of my parents' house on Therseienstraße in Aachen; we children had been given a set of parallel bars by our parents on which we practiced our earliest gymnastic games. You happened to come one afternoon to visit my parents in the garden, and you showed us an elegant gymnastic exercise on the parallel bars that filled us youngsters with amazed admiration. Later, of course, it was much more your intellectual powers we admired and that became models for us." Sommerfeld's position vis à vis the Nazis was also memorable for him. "You were the first person who said to me in no uncertain terms that you believed the men in the leadership of the National Socialist regime were criminals. At that time, I was—this was on a visit to Munich during the war—deeply shaken by your view, and didn't really want to believe it."[79]

76 Metzler, *Internationale Wissenschaft*, 2000; Metzler, *Nationalismus*, 2002.

77 DMA, NL 89, 017, folder 2.5.

78 From Millikan, November 26, 1948. DMA, NL 89, 042. Also in ASWB II.

79 From Grotrian, November 20, 1948. DMA, NL 89, 042.

Wilhelm Magnus (1907–1990), who had done research with Sommerfeld during World War II on the propagation of electromagnetic waves, added another event from this time to the birthday bouquet of reminiscences. "I can still clearly see you as well as the stunned and confused looks on the faces of the exalted gentlemen from Telefunken and the Navy when in a wine restaurant—publicly, in the middle of the war—you raised your glass and calmly said, 'So then, Vive la France!'" He had not forgotten "what it meant at that time to utter these four words out loud: courage, independence, and a liberated vision beyond the scientific sphere."[80]

From another letter, Sommerfeld learned of a rather comic occurrence in connection with his 75th birthday at the Bavarian Ministry of Culture. As revealed to him by the university examiner who had discovered this event in the files of his predecessor, this birthday in the year 1943 had been something of a headache for the Nazi educational bureaucracy. Sommerfeld had not been sent a congratulatory letter, but some pains had been taken to characterize this not as mere negligence but rather as bureaucratically proper. "In the file the grounds adduced were, first of all, that it was better to observe only the even ten-year anniversaries, and second that there had been considerable political dispute about you [Sommerfeld]." The examiner at that time had added "consistent with National Socialist five-year planning" that "what is to be done at the 80th birthday can be deferred to a special review at the appropriate time."[81]

His 80th birthday was not just the occasion for nostalgia, however; it offered the opportunity of reviewing for Sommerfeld once more the broad influence of his work on current physics. The *Zeitschrift für Naturforschung* dedicated a festschrift to Sommerfeld that covered a wide spectrum of physical research topics. He had "educated virtually an entire generation of theoretical physicists who, dispersed widely across the globe now, think on this day of their mentor in Munich with affection and admiration," Heisenberg wrote in the Introduction to this festschrift. "In his pedagogy, he was not satisfied with presenting fundamental theoretical relations; rather, he showed students 'how it is done,' how one actually treats a physical problem mathematically through to its conclusion."[82]

What the festschrift authors from both sides of the Atlantic, in 34 essays, covering nearly 200 pages, presented as samples of their current research showed impressively where this or that problem had led over the course of time. With a paper on turbulence, Heisenberg described the arc from Sommerfeld's 1908 work on the stability of laminar flows (see Sect. 6.3) to statistical turbulence theory, which most recently had led to a promising line of research.[83] Using the latest knowledge from

80 From Magnus, December 4, 1948. DMA, NL 89, 042.

81 From Rheinfelder, December 2, 1948. DMA, NL 89, 042. The entry is to be found in Sommerfeld's personnel files dated November 23, 1943, under the rubric: Vormerkung betr. 75. Geburtstag von Arnold Sommerfeld (note regarding the 75th birthday of Arnold Sommerfeld). BayHStA, MK 35736.

82 Heisenberg, *Arnold Sommerfeld*, 1948.

83 Heisenberg, *Turbulenzproblem*, 1948; Davidson/Kaneda/Moffatt/Sreenivasan, *Voyage*, 2011.

Fig. 34 Even in advanced age, Sommerfeld took part in the development of theoretical physics. His eightieth birthday was for his students another opportunity to express their esteem and gratitude to him as teacher and scholar (Courtesy: Deutsches Museum, Munich, Archive)

research on turbulence, Weizsäcker analyzed the dynamics of cosmic gaseous masses.[84] With a paper on the continuous X-ray spectrum, Elwert elaborated Sommerfeld's "Pacific problem," which also achieved importance in astrophysics.[85] Welker wrote on superconductivity, a puzzle that had remained unsolved for decades.[86] But even the problems of classical physics, such as diffraction theory and the theory of plate vibrations, still offered material for new research.[87]

13.6 The Late Work

Even as an octogenarian, Sommerfeld authored scientific papers. Although he no longer opened up new problem areas, he showed that he followed current research in many areas with interest and endeavored to spin out the threads arising here and

84 Weizsäcker, *Rotation*, 1948.
85 Elwert, Absorptionskoeffizient, 1948.
86 Welker, *Modell*, 1948.
87 Meixner, Theorie der Beugung, 1948; Sauter, Schwingungstheorie, 1948; Niessen, Earth's Constants, 1948.

there from older work. In 1946, with a paper on quantum statistics, for example, he once more addressed questions which, with a view towards the puzzling behavior of superfluid helium, had long occupied him.[88] At very low temperatures, helium, like any gas, becomes fluid, but it behaves very unusually. It becomes a mixture of two fluids. One kind (He I) behaves like a normal fluid and at the border of its gaseous state takes on a vigorous bubbling. The other (He II) settles at the bottom of the container as a clear liquid and displays a behavior completely different from normal liquids. He II possesses almost no viscosity—which is what the term superfluidity designates—and is extremely heat conductive. Sommerfeld conjectured that this behavior could be explained theoretically by a modification of Bose-Einstein statistics introduced by Giovanni Gentile Jr. in 1940. He did not, however, go beyond rather vague indications. Although the "intermediary quantum statistics" à la Gentile and Sommerfeld was taken up and elaborated in various papers, it played ultimately only a marginal role in the theory of superfluidity.[89]

In another paper published in 1948, Sommerfeld corrected an error in his theory of "The Freely Vibrating Piston Diaphragm," published in 1943 in the *Annalen der Physik*.[90] The problem had presumably arisen in the context of "questions of underwater noise," which Fritz Renner, commissioned by the Navy's Experimental Telecommunications Command (Nachrichtenmittel-Versuchskommando), had worked up before being requisitioned by Sommerfeld as his assistant.[91] In this case, though, it was less the physics associated with the problem that made this an interesting subject for Sommerfeld, than the mathematical treatment. In studying more closely the pertinent technical literature on acoustics, he found that a procedure for solving the boundary value problem in question which to him seemed particularly useful appeared to be unknown in that domain. He corresponded about this also with Gustav Herglotz, a mathematician with a sense for physically based problems to whom he was very close from the standpoint of *modus operandi*, and promptly discovered a "big fat mistake" in his working out of it, about which he subsequently issued a correction.[92]

In 1950, the treatment of a different boundary value problem—from the realm of electrodynamics this time—became the subject of another publication in the *Annalen der Physik*.[93] He wrote it together with Edward Ramberg, who as translator of the volume on electrodynamics of Sommerfeld's lectures had presumably given the impetus to it and who characterized as an undeserved honor that Sommerfeld had referred to it in a letter as "our" work. Ramberg also prompted Sommerfeld (in collaboration with Bopp) to make an important matter in this context (calculation of magnetic forces) the subject of another publication.[94]

88 Sommerfeld, *Quantenstatistik*, 1942 and 1946.
89 Dingle, *Helium II*, 1952, p. 123.
90 Sommerfeld, *Kolbenmembran*, 1943; Sommerfeld, *Berichtigungen*, 1948.
91 To the Nachrichtenmittel-Versuchskommando, June 25, 1940. DMA, NL 89, 020, folder 7,2.
92 To Herglotz, October 2, 1943. SUB, Herglotz F 135.
93 Sommerfeld/Ramberg, *Drehmoment*, 1950.
94 From Ramberg, March 6, 1950. DMA, NL 89, 012. Sommerfeld/Bopp, *Problem*, 1950.

All these papers, however, dealt only with brief discussions or corrections of problems that had been treated earlier; they did not take up Sommerfeld's whole energy and attention. His principal focus was on completing the publication of his lectures. One year after the end of the war, two volumes had appeared, "a third, at the printer, a fourth in process," as Sommerfeld wrote Schrödinger in 1946. "This was the only way I could put away the tribulations of the war and the peace."[95] Volume 3, *Electrodynamics*, was published in 1948 by Verlag Dieterich, in Wiesbaden. "Up to now, from what I understand, bookstores in the West could not be supplied by you," Sommerfeld wrote in October 1948 to the Leipzig publisher of his first two volumes of lectures by way of justifying calling in the publisher in the West. "You will understand that I cannot undertake the great investment of labor I put into the publication of my lectures for just the limited circle of your distribution."[96]

This is how things came to a confusing duplication of East and West editions: Volume 1 (*Mechanics*) appeared in 1948 already in its fourth edition in the East and was reprinted in the West in 1949. Volume 2 (*Mechanics of Deformable Media*) and Volume 6 (*Partial Differential Equations of Physics*) appeared in the East in 1945 and in a second edition in 1947 and 1948 in the West. Volume 3 (*Electrodynamics*) appeared first in the West in 1948 and was issued unrevised in 1949 also as the East German edition. Volume 4 (*Optics*) and Volume 5 (*Thermodynamics and Statistics*) were issued in 1950 and 1952 as West German editions and only several years later as East German editions.[97] By 1947, various of the volumes were also in process of translation into English, Italian, and Russian. "We are proud that we can do our share to distribute your '*Lebenswerk*' throughout English-speaking countries," the editor of the Academic Press in New York wrote Sommerfeld in advance of the publication of the English translation of Volume 6, "and we are delighted to be able to tell you that the first reaction to our announcement of Partial Differential Equations In Physics met with a very enthusiastic reception."[98]

13.7 The Last Years

With three children long since grown to maturity and many grandchildren, the private life of Sommerfeld and his wife was also filled with stimulation and challenge. Her life long, Johanna Sommerfeld was a highly valued companion. Johanna as "Aunt Sommerfeld," like Sommerfeld himself, was among Ewald's closest correspondents, with whom she maintained regular contact even after his emigration from Germany.[99] In the Munich men's club "The Casual Ones," of which from 1948

95 To Schrödinger, May 31, 1946. DMA, NL 89, 015.
96 To Portig, 22. October 22, 1948. DMA, NL 89, 005.
97 See summary in ASGS IV, pp. 685–686.
98 From Jacoby, December 16, 1948. DMA, NL 89, 005.
99 From Ewald, November 16, 1949.

Sommerfeld's son was also a member, Sommerfeld's wife was paid the highest respect. When Aloys Wenzl (1887–1967) in his capacity as Rector of the University of Munich congratulated Sommerfeld and his wife on the occasion of their "golden wedding anniversary" on December 24, 1947, he did so "in the name of the faculty of our university, in the name of 'The Casual Ones,' the Association of the Friends of Munich, and last but not least in my own name."[100] Another "Casual Ones" member affirmed "that surely the name Sommerfeld has become that which it is in no small measure because of their fifty year marriage."[101]

With advancing age, Sommerfeld suffered from increasing hearing difficulty, which made his participation at conferences and public events more and more burdensome. As he wrote once to Bechert, his poor hearing spoiled especially "all pleasure in music. I hear the upper notes completely incorrectly—the human body does finally fail."[102] He suffered from "inner-ear hearing loss," according to the diagnosis of the Director of the University Clinic for Ear, Nose, and Throat Disease, where Sommerfeld had his hearing tested in April 1947. Unlike normal age-related hearing difficulty, his hearing loss affected "the entire acoustic range quite uniformly."[103]

Although events at which he was together with many people and had to depend on his hearing became ever more difficult, he did not want to withdraw completely into isolation. In July 1948, he traveled to Zürich to take part in an international physics conference, "under Herzfeld's wings," his wife wrote her sister. "Let's hope everything goes well."[104] Nor did he want to miss the events of "The Casual Ones." In a lecture on "Philosophy and Physics in Today's Worldview," which he had previously given on July 30, 1948, in an international vacation course at the University of Munich, he presented this "occasional product" to "The Casual Ones" as his "philosophical confession of faith."[105]

It was an open secret that he had not devoted himself very profoundly to the philosophical foundations of his field, as had Einstein and Planck, but he was actually more keenly interested in foundational questions than he sometimes made it appear. Although in agreement with Planck and Einstein on the matter of belief in unconditional natural law, he did not share their strict determinism. He spoke of a "causality relaxed by the Heisenberg uncertainty principle."[106] He was also skeptical of Einstein's fundamental conviction that ultimately everything, including elementary particles, had to be derived from a "physics of continuity." This conception had

100 From Wenzl, December 24, 1947.

101 From Anton Weiher, Christmas, 1947.

102 To Bechert, December 13, 1947. Bonn, Archive of the Ebert Foundation (Karl Bechert estate).

103 From Brünings, April 16, 1947. DMA, NL 89, 006.

104 To Helene Rhumbler, July 1, 1948.

105 To Landé, December, 1948. SBPK, Landé, 70 Sommerfeld; to Grimm, 7. October 7, 1948. DMA, HS 1978-12B/172; Sommerfeld, *Philosophie*, 1948; Zwanglose Gesellschaft, *Hundertfünfzig Jahre*, 1987, p. 157.

106 Sommerfeld, *Philosophie*, p. 100.

so far "not led to any palpable success," he wrote in an essay on the occasion of Einstein's seventieth birthday. "Most physicists today consider Einstein's goal unattainable, and have reconciled themselves to the dualism he himself first clearly set forth: wave-particle."[107] That he professed this majority view, and from it even derived deeper philosophical insights, he set forth clearly in his "confession of faith." He was "satisfied in many respects with this duality." The wave-particle dualism seemed to him "a useful contribution to one of the loftiest questions of general philosophy, the age-old problem of the relationship between matter and spirit, body and soul. Our entire life is determined by the dualism of physical and chemical processes, on the one hand, and mental processes on the other. Rather than 'dualism,' incidentally, we say with Bohr 'complementarity,' to indicate that only together do the two conceptions reflect the full nature of light and particle. The task of some Kant of the future would be to create a 'critique of pure reason' in which both conceptions have a place and mutually complete one another."[108]

When Sommerfeld read in the newspaper in August 1948 that Einstein was planning a world trip, he invited him to lecture on his latest work "to our Physical Society of Munich." Since he knew of Einstein's commitment against nuclear armament, he also wanted to arrange a lecture "at the Munich Association for Peace."[109] Einstein replied, however, that the reports in the newspapers had been "as usual" incorrect. "Now I am an old geezer, and no longer travel since I have come to know human beings sufficiently from all sides."[110] Sommerfeld found Einstein's formulation so amusing that he quoted it a year later in his article on the occasion of Einstein's seventieth birthday.[111] He himself did not feel like an "old geezer," although he was almost 10 years older than Einstein. As an octogenarian, Sommerfeld planned another trip to the USA, where he had been invited to attend the first International Mathematics Congress since the war.[112] The Congress took place in August 1950 in Cambridge, Massachusetts, and the opening lecture in the section on partial differential equations had been intended for him. In May 1950, his American friends still hoped he could visit them on this occasion.[113] Ultimately, he did not feel up to the rigors of the trip, so that John von Neumann (1903–1957) assumed his part at the Mathematics Congress.[114]

It may not have been just the anticipated rigors of the trip that caused Sommerfeld to cancel his trip to America but also the calendar proximity of the Cambridge Congress to the German Meeting of Physicists in October 1950 in Bad Nauheim.

107 Sommerfeld, *Albert Einstein*, 1949, pp. 145–146.
108 Sommerfeld, *Philosophie*, 1948, p. 100.
109 To Einstein, August 4, 1948. AEA, Einstein. Also in ASWB II.
110 From Einstein, September 5, 1948. AEA, Einstein. Also in ASWB II.
111 Sommerfeld, *Albert Einstein*, 1949, S. 145.
112 Draft reply to Jacoby, September 9, 1949. DMA, NL 89, 005.
113 From Ramberg, May 29, 1950. DMA, NL 89, 012.
114 From John von Neumann, September 18, 1950. DMA, NL 89, 011.

This meeting led to the creation of the Alliance of German Physical Societies, which since 1945 had pursued independent courses as regional societies in the various zones of occupation.[115] Sommerfeld wished not to miss this event. He also had a personal motivation for the journey to Bad Nauheim, for on this occasion the Max-Planck Medal for the year 1950 was to be awarded to Debye, and Sommerfeld was to deliver the laudatory address. "Given the close friendship that ties you and Prof. Debye, this is certainly the nicest solution," wrote Laue, who, from his situation in the British zone, had been working towards the reorganization of the German Physical Society.[116] Debye was not able to travel from the USA to this event, so Sommerfeld could not personally present the medal to his first assistant. "We keenly regret that we do not have the hero of the day among us today," he began his speech, and then projected onto the wall a photograph of the young Debye, so that Debye, "at least in effigy," could be present. "This is what he looked like when after graduation from high school in his home town of Maastricht he came to the Technical University of Aachen to study electrical engineering." The reminiscence of Debye became for the nearly 82-year-old Sommerfeld also a reminiscence of his own early career. At the conclusion of his speech, he presented the medal—made of bronze due to the materials shortage since the war—to an American physicist who was just then traveling through Germany and had volunteered to be the courier. "I have the honor of conveying the Planck Medal into the trusted hands of Prof. Mayer from Chicago. We regret that it has been converted, through a process of transmutation that is not nuclear-physically but politically well understood, from gold into bronze."[117]

Those attending the Physicists' Meeting in Bad Nauheim saw Sommerfeld for the last time in public. On March 28, 1951, walking near his home, he was struck by an automobile and taken to a Munich hospital with a broken lower leg and upper arm.

Aside from broken bones, no other injury was at first discernible. "He was completely calm, and there was no indication of brain damage or the like," Ernst Sommerfeld reported on his father's condition. "After about 10 days, however, a dark pall fell over his spirit, and everything was lost—power of thought, of memory, of discernment, and finally of speech." It was the doctors' view that he then lost consciousness, and only his "strong constitution" delayed the end. Sommerfeld died on April 26, 1951. "We believe he was spared the grief of parting from his beloved physics and from his loved ones."[118] Johanna Sommerfeld died 4 years later, on July 27, 1955. The couple is buried in the family plot in the Munich North Cemetery (Nordfriedhof).

115 Walcher, Physikalische Gesellschaften, 1995.
116 From Laue, August 26, 1950. DMA, NL 89, 010.
117 Sommerfeld, *Planck-Medaille*, 1950.
118 Ernst Sommerfeld to Bechert, April 27, 1951. Bonn, Archive of the Ebert Foundation (Karl Bechert estate).

14 Legacy

The posthumous recognition of scholars has a long history. The Paris Academy of Sciences traditionally honored its deceased members with eulogies, establishing thereby a ritual adopted subsequently by virtually every scientific society. Although both the hero worship of antiquity and the hagiography of the Middle Ages live on in this tradition, these eulogies, biographical memorials, commemorative speeches, and obituaries of academics are valuable sources for the history of science, for they reveal what the immediately following generation of an esteemed scientist regards as his most significant contributions. When the work of a scholar continues to be influential years after his death, hagiographical presentation makes way for another form of recognition. His textbooks are revised by his students and published in ever newer editions. Publication of his collected writings facilitates access to his work by new generations of scientists. Congresses at 10-year anniversaries serve to illuminate one or another pioneering accomplishment in light of developments that followed it. In few cases does interest in a scholar spread past this phase and reach circles outside his own sphere of influence. This does happen in the case of exceptional figures such as Einstein and Bohr, who achieved world renown even during their lifetimes, as well as of those whose influence spreads beyond their particular fields to other disciplines. Finally, there is yet another sort of scientific afterlife, when a law, a formula, or a natural phenomenon has been named for the scientist who postulated or discovered it.

Sommerfeld's continuing presence offers striking examples in each of these categories. From the obituaries of the year 1951 to contemporary concepts labeled with Sommerfeld's name, we see reflected quite various facets of the manner in which theoretical physicists deal with the history of their field.

14.1 Obituaries

Personal reminiscences dominated the first responses to the news of Sommerfeld's death. "I still remember so well how, when he stayed with us back then, Professor Sommerfeld began working on his book very early in the morning," wrote a physicist from the Philips research laboratory in Holland to Sommerfeld's widow. "As grateful as you must be to your late husband for having shared such a long life, we students are equally grateful to him for his textbooks and leadership and for the great pleasure reading his publications has always given."[1] A member of "The Casual

1 From Niessen, May 7, 1951. DMA, NL 89, 017, folder 2.6.

M. Eckert, *Arnold Sommerfeld: Science, Life and Turbulent Times 1868-1951*,
DOI 10.1007/978-1-4614-7461-6_14,

Ones" recalled in his condolence letter to Johanna that Sommerfeld had been a "true Casual One. . . He always had a large audience, and much that later found its way in print to publication was probably first uttered aloud to us." He had been "ever ready to lend a hand," and "quite a number of us have been the recipients of his good counsel, and many of his assistance, too. The loyalty of this man was a support, that bound the society closer together, all the more so as Arnold Sommerfeld also gladly gave himself up to high spirits and humor."[2]

Not everyone held Sommerfeld's memory in high esteem, however. "With his death, the terroristic tyranny of dissonant physics may well have lost the most powerful exponent it still possessed," Hugo Dingler wrote in a letter to the like-minded Bruno Thüring on the news of Sommerfeld's death.[3] Even 6 years after the fall of the "Third Reich," the war against modern theoretical physics had still not come to an end for these fanatics.

In the Sommerfeld obituaries, there was no hint of such enmity. Rather, they enumerated the lasting contributions he had made to his science, and they revealed in addition something of the individual relationship of the obituary writer to Sommerfeld. Pauli admired about Sommerfeld that in one person, he had "felicitously" embodied "the epitome of the scholar and the teacher."[4] For Heisenberg, Sommerfeld was the fatherly teacher who always showed sympathy for the problems and needs of his students and "considered the personal lives of the students with friendly interest, with the cheerful calm of the Munich professor who gladly eases tensions with a joking word, or readily overlooks inadequacies."[5]

The same tenor ran through the obituaries written by other Sommerfeld students. Each according to his personal experience adduced still other aspects. "Sommerfeld's success as a teacher was due to the clear and concrete expression of his ideas," Ewald explained to the readers of *Nature*. Even as a beginning student, one understood in Sommerfeld's lectures "that behind the domain of established theory lies a field of unsolved problems."[6] For Bechert, too, the secret of Sommerfeld's success lay "in his manner of teaching, of setting assignments, and in his willingness to support those working around him in their individuality." As longtime assistant to Sommerfeld, Bechert could also contribute a few items with respect to daily life in the Sommerfeld pedagogical enterprise. He cited Sommerfeld's concept of the large course that "Lectures should not be constructed and so smoothly delivered that the listener thinks he has understood everything. There must always be something left over which he needs to ponder." For advanced students and doctoral candidates the special lectures and the seminar had been most important. "Often, he gave the small lectures on a subject he wanted to get to know

2 From Anton Weiher, August 3, 1951. DMA, NL 89, 017, folder 2.7.
3 Dingler to Thüring, May 8, 1951. Aschaffenburg, Hofbibliothek, Dinglerestate.
4 Pauli, *Arnold Sommerfeld*, 1951.
5 Heisenberg, *Arnold Sommerfeld*, 1951.
6 Ewald, *Arnold Sommerfeld*, 1951.

himself, and not a few of his own papers grew out of such lectures." In the seminar, Sommerfeld had often interrupted the speakers and posed a question when something was not clear to him. "In discussions, he exhibited model generosity, and was open to any opinions opposing his; in scientific matters, only factual correctness interested him. In discussions, whether he had been right or wrong was not worth a second thought."[7]

In his obituary, Laue laid greatest stress on Sommerfeld's research but also honored Sommerfeld as teacher. As a student of Felix Klein, Sommerfeld had received superb preparation for his pedagogy. Klein's art of lecturing, "the art of representation in general," as well as "the art of knowing human beings and of knowing how to treat human beings" had been passed on "by the great teacher" to Sommerfeld. Laue also compared Sommerfeld to Planck and Einstein, whom he had known well for many years as Berlin colleagues. They had both concentrated "particularly on the fundamental principles of physics"; by contrast, "the model, or at least the concrete instance," had been Sommerfeld's focus. "In Planck, Einstein, and Sommerfeld we have representatives of two generally appearing, quite different scholar types."[8] Herewith, Laue was the first to formulate that dichotomy between orientation towards principle and orientation towards problem, in terms of which Sommerfeld's work later was often characterized (see Chap. 15).

Most obituaries filled only a few pages. With his ten printed pages, Laue had offered what was already a very detailed representation. Max Born, however, composed the most comprehensive tribute. As a member of the London Royal Society, Sommerfeld merited an entry in the *Obituary Notices* of this learned society, and Born, as no other scientist in Great Britain, was equipped to pay him this honor. It must have cost him some effort to meet the membership's high expectations, though, for the long tradition of obituary writing for the Royal Society set a lofty standard. It required a scrupulous listing of all scientific publications and a more comprehensive presentation of the scientific importance of the deceased than was usual in such obituaries.[9]

14.2 Leading Figure for the History of Physics

Ten years after Sommerfeld's death, the second phase of his legacy commenced. American physicists and historians of physics planned a trail-blazing project for the history of recent science: "Sources for History of Quantum Physics (SHQP)." The terrain was first to be explored with some "sample biographies." "The first, that of Arnold Sommerfeld, indicates what can be learned about a prominent physicist for whom numerous obituary notices have been written," the collaborators on the

7 Bechert, *Arnold Sommerfeld*, 1951.

8 Laue, *Sommerfelds Lebenswerk*, 1951, pp. 514 and 518.

9 Born, *Sommerfeld*, 1952.

project wrote in justifying the choice of Sommerfeld as the leading figure. The comprehensive obituary Born had published in *Obituary Notices* of the Royal Society furnished valuable grounds for the process going forward. The project's central focus was the collection of sources, and here, too, Sommerfeld proved a leading figure. Fritz Bopp put Sommerfeld's papers and other effects still in the possession of the institute for theoretical physics of the University of Munich at the disposal of the project. Ernst Sommerfeld contributed a portion of the correspondence he had found in his father's house. The papers thus conveyed were recorded on microfilm and, thereafter, returned to Germany. Together with materials from other estates, an extensive microfilm archive was thus assembled forming the basis of research into the history of modern quantum physics.[10]

With the SHQP Project, the modern history of physics was on the way to becoming a discipline within the history of science, committed to historical methodology and oriented towards primary sources. The works created during the 1960s within this project, under the leadership of Thomas Kuhn (1922–1996), in which Sommerfeld's part in the emergence of modern atomic and quantum physics was explored, belong to the milestones of the modern history of physics.[11] In the wake of the American collection of sources, research began in Germany into the history of quantum theory, and here too, Sommerfeld moved to the center of the research of historians of physics.[12]

The budding interest in Sommerfeld's contributions to quantum physics coincided with the wish of many physicists to observe Sommerfeld's 100th birthday in 1968 in a fitting manner. Bopp established a committee which prepared a four-volume edition of Sommerfeld's works commissioned by the Bavarian Academy of Sciences.[13] In September 1968, he organized a "double congress" at the University of Munich: an "Arnold Sommerfeld Centennial Memorial Meeting" and an "International Symposium on the Physics of the One- and Two-Electron Atoms."[14] Nor did Sommerfeld's textbooks lose their currency. "In deference to Prof. A. Sommerfeld's wish expressed shortly before his death, I have gladly assumed the task of working up Volume VI of his lectures for the present new edition," Fritz Sauer wrote in 1957 in the Introduction to the fourth edition of this volume. The fifth edition followed in 1961 and the sixth in 1965. In 1978 the Verlag Harri Deutsch took over the two volumes of *Atomic Structure and Spectral Lines* and the six volumes of Sommerfeld's *Lectures on Theoretical Physics*. Reprints of Volumes II and VI appeared as late as 1992.[15]

10 Kuhn/Heilbron/Forman/Allen, *Sources*, 1967. http://www.amphilsoc.org/guides/ahqp/index.htm (31January 2013).
11 Heilbron, *History*, 1964; Forman, *Environment*, 1967.
12 Hermann, *Diskussion*, 1967; Hermann, *Frühgeschichte*, 1969.
13 Sauter, ASGS, 1968.
14 Bopp/Kleinpoppen, *Physics*, 1969.
15 Sommerfeld, *Atombau*, 1992.

Given the extensive SHQP material, which includes numerous interviews with Sommerfeld's students, it was also possible to approach Sommerfeld biographically more intensively.[16] The contributions of Sommerfeld and his students to atomic theory now formed the subject of critical studies in the history of physics.[17] Sommerfeld's papers in the area of electron theory of metals also stirred great interest in the context of an international project begun in 1980 on the history of solid state physics. Like the SHQP Project, it took as its principal task the gathering and archiving of relevant sources.[18]

With these research initiatives, new source material came to light. Sommerfeld's estate proved so rich in material that it suggested an exhibition of selected documents to introduce broader circles to this scholarly life.[19] Sommerfeld and his "school" were also suited to the representation of the social context of theoretical physics in the first half of the twentieth century and the rise of this field to a science of the century.[20]

Herewith, a new phase of Sommerfeld's legacy opened up. His scientific correspondence proved a veritable treasure trove for the historical reconstruction of physical theories, giving impetus to extensive treatment.[21] Like the editions of the correspondence of Einstein, Bohr, and Pauli, the Sommerfeld correspondence also shows that modern theoretical physics is not the outcome merely of wrestling with ideas. It is, like other sciences, marked by social developments and historical trends, not discernible in the study of scientific publications, but which leave clear traces throughout the correspondence. The recent history of physics offers countless examples, also (not to say particularly) with respect to Sommerfeld.[22]

14.3 The Fine-Structure Constant

In the names wedding concepts in the natural sciences to their discoverers in near timeless eminence, we see a quite different form of legacy. The "Boltzmann constant," the "Planck quantum of action," and other constants and effects named for their discoverers immortalize their fame in the collective memory of physics. Apart from how Sommerfeld might otherwise be remembered, the "Sommerfeld fine-structure constant α" alone assures his name a lasting place in physics textbooks.

16 Benz, *Arnold Sommerfeld*, 1975; Forman/Hermann, *Sommerfeld*, 1975.
17 Heilbron, *Kossel-Sommerfeld Theory*, 1967; Nisio, *Formation*, 1973; Forman, *Doublet Riddle*, 1968; Forman, *Alfred Landé*, 1970; Cassidy, *Core Model*, 1979; Kragh, *Structure*, 1985.
18 Hoddeson/Baym/Eckert, *Development*, 1987; Eckert, *Propaganda*, 1987; Eckert, *Sommerfeld*, 1990.
19 Eckert/Pricha/Schubert/Torkar, *Geheimrat*, 1984.
20 Eckert, *Atomphysiker*, 1993.
21 Eckert/Märker, ASWB, 2000 and 2004.
22 Seth, *Crafting*, 2010.

Fig. 35: With this bust outside the "Arnold Sommerfeld Lecture Hall," the physics faculty of the Ludwig Maximilian University in Munich memorializes the fine-structure constant, whose introduction in 1916 is seen as Sommerfeld's greatest achievement (Courtesy: Deutsches Museum, Munich, Archive).

At first, α was simply an abbreviation for a quantity assembled from other natural constants, similar to the "Bohr radius" or the "Bohr magneton." By contrast to these quantities, which can be thought of as an elementary length or an elementary magnetic moment, α did not correspond to any elementary physical unit, since α is dimensionless, a number whose value lies very close to 1/137. In 1916, in his fine-structure theory, Sommerfeld had introduced this number as the relation of the "relativistic boundary moment" $p_0 = e^2 / c$ of the electron in the hydrogen atom to the first of n "quantum moments" $p_n = nh / 2\pi$. Sommerfeld had argued that $\alpha = p_0 / p_1 = 2\pi e^2 / hc$ would "play an important role in all succeeding formulas," he had argued.[23] In 1916, he had gone no further than to suggest that more fundamental physical questions might be tied to this "relational quantity." In *Atomic Structure and Spectral Lines*, α was given a somewhat clearer interpretation as the relation of the orbital speed of an electron "in the first Bohr orbit" of the hydrogen atom, to the speed of light.[24]

23 Sommerfeld, *Quantentheorie*, 1916, p. 51.

24 Sommerfeld, *Atombau*, 1919, p. 244.

Not until the development of quantum electrodynamics did the deeper meaning of the fine-structure constant emerge.[25] The correspondence between Heisenberg and Pauli in the 1930s bears witness to the intense—though unsuccessful—efforts in this area. A "true understanding of the numeric value of your constant," Heisenberg wrote in a letter to Sommerfeld, "lies far in the future still—I have come no further with it."[26] With the Lamb-shift experiment (Sect. 13.3), the fine-structure constant moved once again to the center of efforts at theoretical interpretation. The miniscule shift of the energy level was based "on the smallness of the so-called fine structure constant," wrote Pauli in his article on Sommerfeld's eightieth birthday. "The theoretical interpretation of your numerical value is one of the most important still unsolved problems of atomic physics."[27]

With quantum electrodynamics, theorizing about the elementary processes in physics moved in a new direction.[28] The electrodynamic interaction was thought to be a process in which light quanta were exchanged between electrically charged particles, where the fine-structure constant was recognized as a measure of the force of this interaction. For each of the fundamental natural forces—in addition to the electrical (or magnetic) force in electrodynamics, gravity, and the weak and strong nuclear forces—there is in quantum field theory a characteristic exchange particle and a coupling constant which express the force of the interaction. In the fine-structure constant, the magnitude of the electrical elementary charge presented the primary riddle. The universal nature of the elementary charge was mirrored in a mysterious way in the fine-structure constant and extended to the entire domain of electromagnetic interaction. "It is not only the coupling of the electrons with the light quanta that is determined by the fine structure constant, but the coupling of any arbitrary elementary particle with the electromagnetic radiation field," Heisenberg wrote in 1968 in his article on the occasion of the Sommerfeld Centennial. "So long as one did not understand that all elementary particles have charges that are integer multiples of the elementary charge, one could really not hope to derive the Sommerfeld constant. So an understanding, at least a qualitative understanding of the entire spectrum of elementary particles, was the precondition. I spoke with Sommerfeld about this in the years following the last war, too."[29] The fine-structure constant became the great riddle of the physics of elementary particles.

For the theoreticians, though, Sommerfeld's constant was not just a great puzzle so far as the ultimate bases of physics were concerned but also a stroke of luck that gave wings to their practical work. It also appeared as a coupling constant of quantum electrodynamics in the calculation of various interactions. By means of an elegant graphic technique (the Feynman diagram), the integrals derived—thanks

25 Kragh, *Magic Number*, 2003.
26 From Heisenberg, June 14, 1935. DMA, HS 1977-28/A,136. Also in ASWB II.
27 Pauli, *Beiträge*, 1948, p. 132.
28 Schweber, *QED*, Kapitel 2.
29 Heisenberg, *Ausstrahlung*, 1968, p. 536.

to the smallness of α—could be calculated by approximation. With respect to the agreement of theory and experiment, quantum electrodynamics belongs to the most exact theories in physics. It is no coincidence that Richard Feynman (1918–1988), awarded the Nobel Prize for his contributions in this area, like Heisenberg and Pauli, sang an encomium to the fine-structure constant. "It is one of the greatest mysteries of physics," he reasoned at the end of his book on quantum electrodynamics, "a magical number that exceeds the human grasp, as though written by the 'hand of God.'"[30]

The puzzle took a new turn at the end of the twentieth century when astrophysicists found in the spectra of quasars not the value measured on earth of 1/137.03599976 but 1/137.037. The difference may seem tiny, but in light of the otherwise exact correspondence of theory to experiment, it gives theoreticians pause. Light from the quasars was emitted billions of years ago. Can it be—some theoretical physicists ask themselves—that the natural constants are not at all constant but change their value over the course of time? As bizarre as this conjecture based on a discrepancy so infinitesimal might appear, the question of the immutability of the natural constants remains unanswered—and solidly on the test bench of new experimentation and theories.[31]

14.4 The "Sommerfeld Puzzle"

The fine-structure theory of 1916 pertained to an electron orbiting elliptically around an atomic nucleus. Twelve years later, the same fine-structure formula for the energy level of hydrogen-like atoms was derived from Dirac's theory of the electron, "with negligible alterations of terms," as Sommerfeld wrote in the *Wave Mechanical Supplement*[32]:

$$E = E_0 \left\{ 1 + \frac{(Z\alpha)^2}{\left(n - k + \sqrt{k^2 - (Z\alpha)^2}\right)^2} \right\}^{-\frac{1}{2}}$$

This formula in the *Handbook of Physics* from the year 1933 already incorporates those "negligible alterations."[33] For the difference of the quantum numbers $n - k$, the Sommerfeld fine-structure formula used the radial quantum number n_r and, in place of k, the azimuthal n_φ. E_0 is the stationary energy of the electron, Z the nuclear charge number, and α the fine-structure constant. How could such different

30 Feynman, *QED*, 1990, p. 148.
31 Fritzsch, *Konstanten*, 2003.
32 Sommerfeld, *Ergänzungsband*, 1929, p. 337.
33 Bethe, *Quantenmechanik*, 1933, p. 316.

theories yield the same fine-structure formula? In the Dirac electron theory, elliptical orbits were not taken into account. The very concept of orbit makes no sense at all in this theory. With the concept of "spin" a new degree of freedom made its appearance to which nothing corresponded in Sommerfeld's conception. Was it ultimately just a lucky coincidence that the two theories came to the same result?

For Sommerfeld, this was not a coincidental agreement. Spin was, after all, "according to the Dirac theory, a consequence of the relativistic wave equation. This yielded my fine structure formula exactly," he wrote retrospectively in 1942, outlining the basis of the agreement. But this was no explanation, only an indication of his conviction that the two very different theories had their common roots in relativity theory. "It is amazing that Sommerfeld's original 1916 formula for the energy levels can be derived from this new theory which takes account of electron spin," Pauli wrote in 1948 in his essay on the occasion of Sommerfeld's eightieth birthday.[34] Twenty years later, in his contribution for his teacher's centennial, Heisenberg also expressed amazement over it. "But as it were miraculously, Sommerfeld's formula, calculated for a spherically symmetrical electron on the basis of the old, inadequate quantum theory, has also proved itself as the exact solution of the quantum mechanical relativistic theory of a spinning electron. It would be a stimulating project to explore whether this is truly a miracle, or whether perhaps the group theoretical structure of the problem underlying the formulations of both Sommerfeld and Dirac itself leads already to this formula."[35]

Curiously enough, neither Sommerfeld nor Pauli nor Heisenberg ever took up this "stimulating project." Other prominent theoreticians, on the other hand, have time and again puzzled over it. "Sommerfeld's derivation of the fine-structure formula provides only fortuitously the result demanded by experiment," wrote Schrödinger in 1956 to the authors of a book on mathematical procedures in quantum theory. The analysis presented therein led to the conclusion that this was a case of a quite special kind of coincidence. In the Dirac theory, one arrives at the fine-structure formula by means of wave mechanics and spin; neither of these played a role in Sommerfeld's work. Had Sommerfeld employed only the relativistic wave mechanics (as Schrödinger at first had attempted), he would have arrived at a different formula. "Sommerfeld's explanation was successful," the authors concluded, "because the neglect of wave mechanics and the neglect of spin happen to cancel each other in the case of the hydrogen atom."[36]

This judgment, however, was refuted several years later by Lawrence C. Biedenharn (1922–1996), who had made a name for himself in particular with the application of group theoretical methods in physics. For him, the "Sommerfeld Puzzle" had nothing to do with a chance agreement, but rather—as Heisenberg had

34 Pauli, *Beiträge*, 1948, p. 131.
35 Heisenberg, *Ausstrahlung*, 1968, p. 534.
36 Yourgrau/Mandelstam, *Variational Principles*, 1968, S. 113–115.

surmised—with a deeper symmetry. The differential equations used to describe the fine structure in all their variety exhibit a mathematical structure that leads to the same result in both theories, so far as the meaning of the magnitudes appearing in them is concerned. "It is this symmetry which produces the most remarkable and detailed correspondence between the Sommerfeld procedure and the quantal solution, as discussed at length above in our resolution of the Sommerfeld puzzle."[37]

However, that had still not solved the puzzle of Sommerfeld's fine-structure formula. Or more precisely, other explanations emerged when the problem was approached "semiclassically." Semiclassical quantization permitted approximative treatment of problems that were unsolvable with normal quantum mechanics. With the emergence of chaos theory in the 1970s, it attracted particular interest as the boundary between quantum mechanics and chaotic systems in classical mechanics ("quantum chaos") began to be explored. Semiclassical quantization consists in generalization of the Bohr-Sommerfeld quantization conditions of the type $\oint pdq = nh$ by the addition of another quantum number ("Maslov-index" μ), so that the quantization condition takes the form $\oint pdq = \left(n + \frac{\mu}{4}\right)h$. In order to derive the fine-structure formula by the semiclassical method, yet another quantum number for the spin must be added to the quantization formula. It thus appears that these additions exactly cancel each other out.[38] It remains an open question, however, whether this result corresponds to Biedenharn's analysis or whether simply the identical substance is being described in two different mathematical languages.

In the "Sommerfeld Puzzle," an essential and perennially astonishing characteristic of theoretical physics is manifested. Previous results are occasionally confirmed by new theories, even though the physical understanding and its related mathematical procedures have fundamentally altered. Other papers by Sommerfeld experienced a late rebirth in this way. In nonlinear dynamics, for example, phenomena arising from feedback in the energy exchange between vibrating systems are designated as "Sommerfeld effect" or "Sommerfeld-Kononenko effect."[39] That Sommerfeld's name should attach to such phenomena goes back to his paper from the year 1902 in which he analyzed the vibrations caused by a motor driving an unbalanced weight (Sect. 5.3). The "rocking table" phenomenon showed manifestly that under given conditions, energy transmitted to the motor resulted not in higher revolutions but in stronger vibrations of the table. Relating this phenomenon to the real world Sommerfeld wrote, "This experiment corresponds roughly to the case in which a factory owner has a machine set on a poor foundation running at 30 horsepower. He achieves an effective level of just 1/3, however, because only 10 horsepower are doing useful work, while 20 horsepower are transferred to the

37 Biedenharn, *Sommerfeld Puzzle*, 1983, p. 32.

38 Keppeler, *Phase*, 2004.

39 Kononenko, *Vibrating Systems*, 1969; Krasnopolskaya /Shvets, *Chaos*, 1993.

foundational masonry."[40] Its practical significance assured that serious consideration was given to the theoretical analysis as well. Decades later, when the study of nonlinear systems underwent a meteoric rise, the "rocking table" achieved new fame. The physical processes it occasioned, along with their mathematical description, bore implications far beyond the case of the "rocking table" described by Sommerfeld—had it not, the phenomenon would hardly have been given his name and dubbed the "Sommerfeld effect."[41]

14.5 From the Pacific Problem to Dark Matter

Quite a different Sommerfeld effect provides for discussion among astrophysicists. Considerably more matter must be present in the universe than what makes up the stars and cosmic gas clouds and is visible to astronomers by electromagnetic radiation through telescopes and radio telescopes. The prevailing view among astrophysicists is that so-called weakly interacting massive particles (WIMPs) constitute this invisible "dark matter." According to the theory, already a nanosecond after the big bang, these weakly interacting massive particles should have ceased interacting with each other, except gravitationally, if they are far apart from each other. If they happen to collide, they should annihilate each other by the weak interaction. In the process, they ought to emit a γ-ray, observable in principle, thus serving indirectly as evidence of dark matter. The radiation predicted by the first model calculations is spread so thin, however, that proof of it is practically impossible. Later model calculations, however, showed that in certain regions of the galaxy the radiation engendered by the annihilation of WIMPs is far greater than expected. In a research report on the subject from the year 2009, we read that "The unambiguous detection of Galactic dark matter annihilation would unravel one of the most outstanding puzzles in particle physics and cosmology. Recent observations have motivated models in which the annihilation rate is boosted by the Sommerfeld effect, a non-perturbative enhancement arising from a long range attractive force."[42]

How does it happen that more than half a century after his death, Sommerfeld has the distinction of having discovered an effect that might lead to the proof of dark matter? In the "Sommerfeld enhancement," as this effect is also known, two processes act together: In the one, the intensity of the radiation is proportional to the particle stream of the colliding WIMPs; in the other, it is dependent on the cross section of its effect. The latter can be visualized as a disc laid crosswise to the motion of the colliding particles, whose extension gives the distance at which

40 Sommerfeld, *Beiträge*, 1902, p. 393.
41 Eckert, *Sommerfeld-Effekt*, 1996.
42 KuhlenMadau/Silk, *Dark Matter*, 2009.

the particles still act on each other. Since the WIMPs attract each other before their mutual annihilation, the cross section of the effect as well as the particle stream is boosted since they become too concentrated.[43] In general, the calculation of such particle interactions is possible only with the perturbation theoretical methods of quantum field theory developed after World War II. But in his Pacific problem, Sommerfeld had considered the nonrelativistic limit in which the methods of the Schrödinger wave mechanics suffice. In his paper "On the Diffraction and Braking of the Electrons" from the year 1931, he had carried out this procedure for the calculation of the X-ray bremsstrahlung.[44] The interaction of two mutually attracting particles of small kinetic energy treated therein is exactly such a "nonrelativistic quantum effect," as is also supposed to occur in the colliding of the WIMPs and their mutual annihilation in certain regions of the galaxy. Sommerfeld was dealing with the braking of an electron flying by an atomic nucleus that is entailed in the emission of X-ray bremsstrahlung. In the case of dark matter, one has to imagine that the WIMPs react with each other similarly when they slowly (nonrelativistically) collide. Their interaction occurs—in the language of Feynman diagrams—in the form of a ladder diagram, in which the rungs of the ladder express that the mutual attractive force of the WIMPs is mediated by exchange particles (vector bosons), which because of their slow (nonrelativistic) mutual motion are exchanged many times between the WIMPs, before they mutually annihilate one another, and radiate their energy in the form of γ-rays.[45]

When Sommerfeld put his work on the radiation of X-ray bremsstrahlung to paper in 1931, there was no talk yet of Feynman diagrams or exchange particles. Nonetheless, Sommerfeld already saw himself confronted with a long problem history. In 1909, he had described for the first time how at the braking of an electron at the anticathode of an X-ray tube electromagnetic radiation was emitted that was bundled, depending on the velocity of the impacting electron, more or less in the direction of radiation of the electron. In 1911, at the first Solvay Congress, he had undertaken the attempt to determine the braking duration in this process quantum theoretically—and soon had to concede that the problem could not be handled in that way (Sect. 6.5). He encountered the subject again in 1929 in discussions with Yoshikatsu Sugiura, who devoted himself at RIKEN to X-ray bremsstrahlung. On the crossing from Japan to California, he tackled the problem by means of wave mechanics, without completely coming to grips with it (Sect. 10.6). The Pacific problem became a perennial challenge for him. Even with his comprehensive treatment in 1931 he saw the problem as not yet solved. He passed it along as a challenge to several of his students and devoted a comprehensive presentation to the subject in 1939 in the new edition of the *Wave Mechanical Supplement* (Sect. 11.5). It is no

43 Iengo, *Sommerfeld Enhancement*, 2009.

44 Sommerfeld, *Beugung*, 1931.

45 Lattanzi/Silk, *WIMP Annihilation*, 2009.

coincidence that even in the translation of the "Pacific problem" into the quantum field theoretical language of Feynman diagrams, traces of the Sommerfeld tradition are still perceptible.[46] Concealed behind the "Sommerfeld effect" in modern astrophysics and particle physics, then, is a problem history that reaches back through an entire century.

14.6 The Nobel Prize Denied

Since 1900, the highest honor bestowing on scientists a durable afterlife far above his colleagues has been the Nobel Prize. It was denied Sommerfeld. What he himself thought about this, he expressed in a heartfelt statement in December 1928, when once more he saw his hopes dashed. "It is gradually becoming a public scandal that I have still not received the Prize," he wrote at that time.[47] There is no final clarity as to the conjecture expressed in this letter that he had been passed over due to "rivalry with Bohr." All that emerges from documents in the Nobel archive is that from 1917, he had been nominated for the prize virtually every year, and in 1924 was on the short list. Why even that year he did not receive the prize must be inferred from the report of his Swedish colleague Carl Wilhelm Oseen (1879–1944), who spoke for physics within the Nobel Committee. "Given Sommerfeld's strong—and conscious—disinclination to systematic thought, it is natural that his achievements are often ephemeral," it reads.[48]

With each award of the Nobel Prize, a choice among many prize-worthy research achievements is faced that almost always raises the question whether this or another person is not more deserving of the Prize. The nuclear physicist Valentine Telegdi (1922–2006), who had played a decisive part in the discovery of parity violation in the weak interaction and other significant nuclear-physical discoveries and, thereby, himself had nearly won a Nobel Prize, on several occasions took up the question why Sommerfeld had never received it. He subjected the report in which Oseen in 1924 had classified Sommerfeld's elaboration of the Bohr atomic theory as insufficient for a Nobel Prize to critical analysis. To the charge that Sommerfeld's theory had been insufficiently systematic and soon revised in many details, Telegdi countered that Bohr's trilogy had likewise presented "an altogether not very logical edifice" and yet had been deemed worthy of the Prize. Oseen had "disparaged"

46 Elwert/Haug, *Calculation*, 1969.

47 To Wieland, December 13, 1928.DMA, NL 57. Also in ASWB II.

48 Protokollvid Kungl. Vetenskapsakademiens Sammankomster för behandling af ärenden rörande Nobelstiftelsen år 1924, here p. 29. Nobel Archives, Royal Swedish Academy of Sciences, Stockholm.

Sommerfeld, he wrote, and thereby sought "to deflect definitively all suspicion that he had been influenced by Bohr."[49]

Oseen was friends with Bohr, but to infer a conspiracy against Sommerfeld therefrom is pure speculation. The judgments in the realm of physics expressed by Oseen in the Nobel Committee caused annoyance also in other cases, where the idea of Bohr's having exerted influence can hardly be posited.[50] Whatever the forces at work in the background, it was incomprehensible already for Sommerfeld's contemporaries that this recognition was denied him. Millikan had proposed Sommerfeld for the Prize in 1925 and 1930.[51] He thought that *Atomic Structure and Spectral Lines* alone justified this honor. "It is outstanding work which should have brought you the Nobel Prize long ago," he wrote Sommerfeld in 1948 on the occasion of his 80th birthday.[52]

When an examination of the Nobel files many years after Sommerfeld's death brought to light how often he had been nominated for the Prize, historians of science were also astonished. In a survey of the first 50 years of the awarding of the Nobel, Sommerfeld is recognized as the holder of an unhappy record: Of all candidates for the physics Prize, he had received the greatest number of nominations. "Arnold Sommerfeld must be the unluckiest man in physics," the survey's author notes, for with 81 nominations between 1917 and 1950, he had "the dubious honor of being the most-nominated physicist in the period 1901–1950, never to win a Nobel Prize."[53]

49 Telegdi: "Why did Arnold Sommerfeld never get the Nobel prize?" lecture on March 7, 2002 to the Physics Colloquium at CalTech, http://www.pma.caltech.edu/PhysColl/PhysColl01-02.html (31 January 2013). I am grateful to Valentine Telegdi for access to the text of a parallel colloquium lecture at the University of Munich. See also, Lippincott, *Conversation*, 2008, p. 106.

50 Friedman, *Politics*, 2001.

51 Crawford, *Nobel Population*, 2002.

52 From Millikan, November 26, 1948. DMA, NL 89, 042. Also in ASWB II.

53 Crawford, *Nobel Population*, 2001.

15 Epilogue

"Planck was the authority, Einstein the genius, and Sommerfeld the teacher." Thus, one writer at once trenchantly and succinctly summed up the roles of the most important exponents of theoretical physics in its "golden age."[1] Another writer stressed the orientation towards *problems* that typified the theoretical physics of Sommerfeld and his school, in contrast to the physics of a Planck, an Einstein, or a Bohr, whose physics was focused on *principles*.[2] And Sommerfeld's biography does offer many instances of this approach. "How this comes about remains utterly obscure. But the consequences of what is postulated have to be thought through."[3] Thus, in 1927 had Sommerfeld deferred all foundational questions regarding the Fermi-Dirac statistics so that, unencumbered by them, he could attack a string of unsolved problems in the electron theory of metals with this new statistics (Sect. 9.3). One might well preface numerous other works by Sommerfeld with the identical dictum.

Nonetheless, the essence of a long and many-faceted life in science cannot be condensed in such simple formulas. "Sommerfeld, the teacher," spotlights only one aspect of his personality; similarly, the rubric "problem-oriented research" must not be understood in an exclusionary sense. As in the case of the *h*-hypothesis (Sect. 6.5), Sommerfeld was capable of exhibiting a quite pronounced orientation towards principle. When a problem is closely bound up with the principles fundamental to its formulation, principle orientation and problem orientation are not mutually exclusive alternatives.

In any case, with his often lapidary formulations, Sommerfeld himself contributed to this characterization of him as representing a decided antithesis to his "principle-oriented" colleagues. "I can only contribute to the technical aspect of quantum theory; you must devise its philosophy," he once had written to Einstein enunciating his position.[4] One should not infer from this a lack of interest in the philosophical and epistemological issues arising from the relativity and quantum theories, however. When challenged with contentious questions in natural philosophy, he adopted a clear-enough stance. That he should have declared himself decidedly a "dogmatist on the point of natural laws"[5] becomes very understandable in the context of his earlier debate with exponents of the Vienna Circle, in which he had issued a clear rejection of the Mach-inspired positivism (Sect. 10.7). He had expressed himself similarly on May 1, 1933, in a lecture in Edinburgh at the invitation

1 Hermann, *Max Planck*, 1973, p. 56.
2 Seth, *Crafting*, 2010.
3 Sommerfeld, *Elektronentheorie der Metalle*, 1927, p. 825.
4 To Einstein, January 11, 1922. AEA, Einstein. Also in ASWB II.
5 To Moritz Schlick, October 17, 1932. DMA, NL 89, 025.

M. Eckert, *Arnold Sommerfeld: Science, Life and Turbulent Times 1868-1951*,
DOI 10.1007/978-1-4614-7461-6_15, © Springer Science+Business Media New York 2013

of the Royal Society.[6] His subject had been taken from a recently published book of speeches and lectures by Max Planck under the title "Paths to Physical Knowledge." Sommerfeld knew himself in agreement with Planck on many questions of natural philosophy. He referred also to Boltzmann, who, in the second volume of his *Electrodynamics*, had introduced Maxwell's equations with a familiar quotation from Goethe's *Faust*: "Was it a god who inscribed these symbols?" Belief in the harmony of the laws of nature was as deeply held a fundamental conviction for Sommerfeld as it was for Boltzmann, Planck, and Einstein. His philosophical statements may appear modest juxtaposed with those of Planck and Einstein, but they were nonetheless deeply felt (Sect. 13.7). Of all philosophical movements, he doubtless felt closest to the epistemology laid down by Kant, although he thought it in need of some revision in light of the general theory of relativity. "Certainly, it cannot remain in its original formulation," he wrote in 1948 about the Kantian conception that space and time are given "a priori." "Space and time acquire a physical structure a posteriori, stemming from the events playing out within them . . . A Kant of today would adjust his concepts to the doctrines of Einstein . . . Since Einstein, there is no longer any estrangement between physicists and philosophers. Physicists have become philosophers, while philosophers are careful not to conflict with physics."[7]

If in his scientific papers Sommerfeld broadly eschewed philosophical and ideological subjects, one should not judge this self-imposed reticence as a lack of interest in philosophy. But even the incorporation of philosophical convictions would not suffice to bring his activity as teacher and researcher to a common denominator. In the effort to compile a resume, music in particular must not be left out. His love of music formed a constant thread throughout his life, from his domestic private life to the convivial gatherings in the circle of his colleagues and, ultimately, to science itself. For him, atomic theory represented not just the challenge of solving problems and of thereby enhancing the careers in theoretical physics for a circle of his students. It was also (and perhaps primarily) "something aesthetic and harmonious that can be compared only to music." The number relations in the theory of spectral lines were "true quantum music," as he expressed it in 1924 in a lecture to the Prussian Academy of Sciences.[8]

Ultimately, all attempts to encapsulate Sommerfeld's life's work one way or another in a concise resume will fall short. One might rather recall a couplet from his favorite poet, Goethe:

"To find refreshment in the whole,
Seek the whole in the infinitesimal."[9]

6 Scott Lecture, May 1, 1933. Lecture text, Ms. in DMA, NL 89, 021, folder 9.9.

7 Sommerfeld, *Philosophie*, 1948, p. 98.

8 Sommerfeld, *Erforschung*, 1924, p. 875. Also in ASGS IV, p. 576.

9 "Willst du dich am Ganzen erquicken/So musst du das Ganze im Kleinsten erblicken." Goethe, Gedichte, 1827. *Werke*, 1981, p. 304 (I).

Abbreviations

AEA	Albert Einstein Archives, The Hebrew University of Jerusalem
AHQP	Archive for the History of Quantum Physics
AIP	American Institute of Physics, College Park, Maryland, USA
ASGS	Arnold Sommerfeld. Gesammelte Schriften. 4 Vols. Edited by F. Sauter and Bayerische Akademie der Wissenschaften. Braunschweig 1968
ASWB	Arnold Sommerfeld. Wissenschaftlicher Briefwechsel. Vol. I: 1892–1918; Vol. II: 1919–1951. Edited by Michael Eckert and Karl Märker. München, Berlin, Diepholz 2000 and 2004
BA	Bundesarchiv, Berlin
BA-MA	Bundesarchiv-Militärarchiv, Freiburg
BANL	Biblioteca dell'Accademia Nazionale dei Lincei e Corsiana, Rome
BayHStA	Bayerisches Hauptstaatsarchiv, München
BDM	Bund deutscher Mädchen
BSB	Bayerische Staatsbibliothek, München
CalTech	California Institute of Technology, Pasadena, California, USA
DMA	Deutsches Museum, Archiv, München
ESPC	Ecole superieure de physique et de chimie industrielles de la ville de Paris, Centre de ressources historiques, Paris
ETH	Eidgenössische Technische Hochschule, Zürich
Gestapo	Geheime Staatspolizei
GOAR	Göttingen, Archiv des Deutschen Zentrums für Luft- und Raumfahrt
GSA	Geheimes Staatsarchiv, Berlin
HJ	Hitlerjugend
KWKW	Kaiser-Wilhelm-Stiftung für Kriegstechnische Wissenschaft
MPGA	Max-Planck-Gesellschaft, Archiv, Berlin
NBA	Niels Bohr Archiv, Kopenhagen
NBCW	Niels Bohr Collected Works. 12 Vols. Amsterdam, New York, Oxford 1972–2006

M. Eckert, *Arnold Sommerfeld: Science, Life and Turbulent Times 1868-1951*,
DOI 10.1007/978-1-4614-7461-6, © Springer Science+Business Media New York 2013

NSDAP	Nationalsozialistische Deutsche Arbeiterpartei
NSFK	Nationalsozialistisches Fliegerkorps
NVK	Nachrichtenmittel-Versuchskommando
Pg	Parteigenosse (member of NSDAP)
RANH	Rijksarchief in Noord-Holland in Haarlem
Ransom	Harry Ransom Humanities Research Center, The University of Texas at Austin
RIKEN	Rikagaku Kenkyūjo (Physical-Chemical Institute) in Tokio
SBPK	Staatsbibliothek zu Berlin – Preußischer Kulturbesitz, Handschriftenabteilung
SD	Sicherheitsdienst des Reichsführers der SS
SHQP	Sources for the History of Quantum Physics
SPGK	Schriften der physikalisch-ökonomischen Gesellschaft zu Königsberg in Preußen
SS	Schutzstaffel der NSDAP
StAM	Staatsarchiv München
SUB	Staats- und Universitätsbibliothek, Niedersachsen, Göttingen
UAB	Universitätsarchiv der Humboldt-Universität Berlin
UAG	Universitätsarchiv, Göttingen
UAM	Universitätsarchiv der Ludwig-Maximilians-Universität, München
WIMPs	Weakly Interacting Massive Particles
WPWB	Wolfgang Pauli: Wissenschaftlicher Briefwechsel, 4 Vols. Edited by Karl von Meyenn. New York, Berlin 1979–2005

Bibliography

Abraham, Max: Dynamik des Electron In: Nachrichten von der Königl. Gesellschaft der Wissenschaften zu Göttingen. Mathematisch-physikalische Klasse (1902), 20–41.

Abraham, Max: Prinzipien der Dynamik des Elektron In: Annalen der Physik 10 (1903), 105–179.

Adam, Uwe Dietrich: Judenpolitik im Dritten Reich. Düsseldorf 2003.

Ahlheim, Hannah: „Deutsche, kauft nicht bei Juden! " Antisemitismus und politischer Boykott in Deutschland 1924 bis 1935. Göttingen 2011.

Aitken, Hugh G. J.: Syntony and Spark: The Origins of Radio. New York, NY, 1976.

Albrecht, Helmuth: „Max Planck: Mein Besuch bei Adolf Hitler" – Anmerkungen zum Wert einer historischen Quelle. In: Helmuth Albrecht (ed.): Naturwissenschaft und Technik in der Geschichte. Stuttgart 1993, 41–63.

Allis, William P./Morse, Philip M.: Theorie der Streuung langsamer Elektronen an Atomen. In: Zeitschrift für Physik 70 (1931), 567–582.

Anonym [L. B.]: George Hartley Bryan, 1864–1928. In: Obituary Notices of Fellows of the Royal Society 1:2 (1933), 139–142.

Appell, Paul: Sur l'équation et la théorie de la chaleur. In: Journal de Mathématiques 8 (1892), 187–216.

Ash, Mitchell G.: Wissenschaft – Krieg – Modernität: Einführende Bemerkungen. In: Berichte zur Wissenschaftsgeschichte 19 (1996), 69–75.

Assmus, Alexi: The Creation of Postdoctoral Fellowships and the Siting of American Scientific Research Student. In: Minerva 31 (1993), 151–183.

Bachmann, Wolf: Die Attribute der Bayerischen Akademie der Wissenschaften 1807–1827. Kallmünz 1966.

Baldus, Richard/Buchwald, Eberhard/Hase, Rudolf: Zur Geschichte der Richtwirkungs- und Peilversuche auf den Flugplätzen Döberitz und Lärz. In: Jahrbuch für drahtlose Telegraphie und Telephonie 15 (1920), 99–101.

Barkan, Diana Kormos: The Witches' Sabbath: The First International Solvay Congress in Physics. In: Science in Context 6 (1993), 59–82.

Bauer, Richard: Prinzregentenzeit. München und die Münchner in Fotografien. München 1988.

Bauer, Richard: Geschichte Münchens. München 2008.

Bechert, Karl: Arnold Sommerfeld. 5. Dezember 1868 – 26. April 1951. In: Experientia 7 (1951), 477–478.

Behrendt, Michael: Hans Nawiasky und die Münchner Studentenkrawalle von 1931. In: Kraus, Elisabeth (ed.): Die Universität München im Dritten Reich. Aufsätze. Teil 1. München 2006, 15–42.

Beller, Mara: Quantum Dialogue. The Making of a Revolution. Chicago, IL, London 1999.

M. Eckert, *Arnold Sommerfeld: Science, Life and Turbulent Times 1868-1951*,
DOI 10.1007/978-1-4614-7461-6,

Benz, Ulrich: Arnold Sommerfeld. Lehrer und Forscher an der Schwelle zum Atomzeitalter 1868–1951. Stuttgart 1975.

Berg, Matthias/Thiel, Jens/Walther, Peter (ed.): Mit Feder und Schwert. Militär und Wissenschaft – Wissenschaftler und Krieg. Stuttgart 2009.

Berghahn, Volker: Der Erste Weltkrieg. München 2003.

Bethe, Hans A.: Quantenmechanik der Ein- und Zwei-Elektronenprobleme. In: Handbuch der Physik 24:1 (1933), 273–560.

Bethe, Hans A.: Sommerfeld's Seminar. In: Physics in Perspective 2 (2000), 3–5.

Bethe, Hans A./Bacher, Robert F./Livingston, M. : Basic Bethe: Seminal Articles on Nuclear Physics, 1936–1937. New York, NY 1986.

Beyerchen, Alan D.: Wissenschaftler unter Hitler. Physiker im Dritten Reich. Berlin 1982.

Beyler, Richard/Eckert, Michael/Hoffmann, Dieter: Die Planck-Medaille. In: Hoffmann, Dieter/ Walker, Mark (ed.): Physiker zwischen Autonomie und Anpassung. Die Deutsche Physikalische Gesellschaft im Dritten Reich. Weinheim 2007, 217–235.

Biedenharn, Lawrence C.: The Sommerfeld Puzzle Revisited and Resolved. In: Foundations of Physics 13:1 (1983), 13–34.

Bieg-Brentzel, Rotraut: Die Tongji-Universität: zur Geschichte deutscher Kulturarbeit in Shanghai. Frankfurt am Main 1984.

Bittner, Lotte: Geschichte des Studienfachs Physik an der Wiener Universität in den letzten hundert Jahren. Dissertation an der Universität Wien. Wien 1949.

Böhm, Helmut: Von der Selbstverwaltung zum Führerprinzip. Die Universität München in den ersten Jahren des Dritten Reiches (1933–1936). Berlin 1995.

Boehm, Laetitia/Spörl, Johannes (ed.): Ludwig-Maximilians-Universität. Ingolstadt – Landshut – München, 1472–1972. Berlin 1972.

Bohlmann, Georg: Über Versicherungsmathematik. In: Felix Klein und Eduard Riecke: Über Angewandte Mathematik und Physik in ihrer Bedeutung für den Unterricht an den höheren Schulen. Nebst Erläuterung der bezüglichen Göttinger Universitätseinrichtungen. Leipzig 1900, 114–145.

Bohr, Niels: On the Series Spectrum of Hydrogen and the Structure of the Atom. In: Philosophical Magazine 29 (1915), 332–335.

Bohr, Niels: On the Quantum Theory of Radiation and the Structure of the Atom. In: Philosophical Magazine 30 (1915), 394–415.

Boltzmann, Ludwig: Über ein Medium, dessen mechanische Eigenschaften auf die von Maxwell für den Electromagnetismus aufgestellten Gleichungen führen. In: Annalen der Physik 48 (1893), 78–99.

Boltzmann, Ludwig: Vorlesungen über Maxwells Theorie der Elektricität und des Lichte 1. und 2. Teil. Graz, 1982.

Bonolis, Luisa: Giovanni Gentile Jr. a Milano. In: Atti del XXV Congresso Nazionale di Storia della Fisica e dell'Astronomia, Milano, 10–12 novembre 2005. Mailand 2008, C03.1–C03.6.

Bopp, Fritz/Kleinpoppen, Helmut (ed.): Physics of the One- and Two-Electron Atom Proceedings of the Arnold Sommerfeld Centennial Memorial Meeting and of the International Symposium on the Physics of the One- and Two-Electron Atoms, Munich, 10–14 September 1968. Amsterdam 1969.

Born, Max: Arnold Johannes Wilhelm Sommerfeld. 1868–1951. In: Obituary Notices of Fellows of the Royal Society 8 (1952), 274–296.

Born, Max/Heisenberg, Werner: Die Elektronenbahnen im angeregten Heliumatom. In: Zeitschrift für Physik 16 (1923), 229–243.

Bottazzini, Umberto: The Higher Calculus: A History of Real and Complex Analysis from Euler to Weierstrass. New York, NY 1986.

Brocke, Bernhard vom: Hochschul- und Wissenschaftspolitik in Preußen und im Deutschen Kaiserreich 1882–1907: Das "System Althoff". Stuttgart, 1980.

Broelmann, Jobst: Intuition und Wissenschaft in der Kreiseltechnik 1750–1930. München 2002.

Brush, Stephen G.: Nineteenth-Century Debates About the Inside of the Earth: Solid, Liquid or Gas? In: Annals of Science 36 (1979), 225–254.

Bryan, George H.: Allgemeine Grundlegung der Thermodynamik. In: Enzyklopädie der mathematischen Wissenschaften V:1 (1903), 71–160.

Burchfield, Joe D.: Darwin and the Dilemma of Geological Time. In: Isis 64 (1974), 301–321.

Burkhardt, Heinrich: Entwicklungen nach oscillierenden Functionen und Integration der Differentialgleichungen der mathematischen Physik. In: Jahresbericht der Deutschen Mathematiker-Vereinigung 10 (1908), 1–894 (Teil 1) und 895–1804 (Teil 2).

Busse, Detlev: Engagement oder Rückzug? Göttinger Naturwissenschaften im Ersten Weltkrieg. Schriften zur Göttinger Universitätsgeschichte. Band 1. Göttingen 2008. http://webdoc.sub.gwdg.de/univerlag/2008/SGU1_dbusse.pdf (15.11.2012).

Carson, Cathryn/Schweber, Silvan : Recent Biographical Studies in the Physical Science In: Isis 85 (1994), 284–292.

Cassidy, David C.: Heisenberg's First Core Model of the Atom: The Formation of a Professional Style. In: Historical Studies in the Physical Sciences 10 (1979), 187–224.

Cassidy, David C.: Uncertainty: The Life and Science of Werner Heisenberg. New York, NY 1992.

Cassidy, David C.: Beyond Uncertainty: Heisenberg, Quantum Physics, and the Bomb. New York, NY 2009.

Clausthal, Technische Universität (ed.): Die Königliche Bergakademie zu Clausthal. Clausthal 1883.

Clausthal, Technische Universität (ed.): Die Königliche Bergakademie zu Clausthal. Ihre Geschichte und ihre Neubauten. Festschrift zur Einweihung der Neubauten am 14. 15. und 16. Mai 1907. Leipzig 1907.

Coben, Stanley: The Scientific Establishment and the Transmission of Quantum Mechanics to the United States, 1919–32. In: The American Historical Review 76:2 (1971), 442–466.

Cochell, Gary G.: The Early History of the Cornell Mathematics Department: A Case Study in the Emergence of the American Mathematical Research Community. In: Historia Mathematica 25 (1998), 133–153.

Cochrane, Rexmond C.: Measures for Progres A History of the National Bureau of Standard Washington, D.C. 1966.

Conant, Jennet: Tuxedo Park. New York, NY u.a. 2002.

Cranz, Carl: Theoretische und experimentelle Untersuchungen über die Kreiselbewegungen der rotierenden Langgeschosse während ihres Fluge In: Zeitschrift für Mathematik und Physik 43 (1898), 133–162, 169–215.

Cranz, Carl: Äussere Ballistik. 5. Auflage. Berlin 1925.

Crawford, Elisabeth: Nobel Population 1901–50: Anatomy of a Scientific Elite. In: Physics World, November 2001. http://physicsworld.com/cws/article/print/3432 (9.1.2012).

Crawford, Elisabeth: The Nobel Population 1901–1950: A Census of the Nominators and Nominees for the Prizes in Physics and Chemistry. Tokio 2002.

Darrigol, Olivier: From c-Numbers to q-Numbers: The Classical Analogy in the History of Quantum Theory. Berkeley 1992.

Darrigol, Olivier: The Electrodynamic Origins of Relativity. In: Historical Studies in the Physical and Biological Sciences, 26:2 (1996), 214–312.

Darrigol, Olivier: Electrodynamics from Ampère to Einstein. Oxford 2000.

Darrigol, Olivier: The Historians' Disagreement over the Meaning of Planck's Quantum. In: Centaurus 43 (2001), 219–239.
Darrigol, Olivier: Worlds of flow. Oxford 2005.
Daston, Lorraine/Sibum, Otto (ed.): Scientific Personae and Their Historie In: Science in Context 16:1/2 (2003), 1–8.
Davidson, Peter A./Kaneda, Yukio/Moffatt, Keith/Sreenivasan, Katepalli R. (ed.): A Voyage Through Turbulence. Cambridge, MA 2011.
Debye, Peter: Das elektromagnetische Feld um einen Zylinder und die Theorie des Regenbogen In: Physikalische Zeitschrift 9 (1908), 775–778.
Debye, Peter: Der Lichtdruck auf Kugeln von beliebigem Material. In: Annalen der Physik 30 (1909), 57–136.
Debye, Peter: Quantenhypothese und Zeeman-Effekt. In: Nachrichten von der Königl. Gesellschaft der Wissenschaften zu Göttingen. Mathematisch-physikalische Klasse (1916), 142–153.
Debye, Peter: Quantenhypothese und Zeeman-Effekt. In: Physikalische Zeitschrift 17 (1916), 507–512.
Debye, Peter (ed.): Probleme der modernen Atomphysik. Arnold Sommerfeld zu seinem 60. Geburtstage gewidmet von seinen Schülern. Leipzig 1929.
Debye, Peter/Sommerfeld, Arnold: Theorie des lichtelektrischen Effektes vom Standpunkt des Wirkungsquantum In: Annalen der Physik 41 (1913), 873–930.
Dehlinger, Walter: Über spezifische Wärme zweiatomiger Kristalle. München 1915.
Dingle, R. B.: Theories of Helium II. In: Advances in Physics, 1:2 (1952), 111–168.
Dowson, Duncan: History of Tribology. London 1998.
Düwell, Kurt: Gründung und Entwicklung der Rheinisch-Westfälischen Technischen Hochschule Aachen bis zu ihrem Neuaufbau nach dem Zweiten Weltkrieg. Darstellung und Dokumente. In: Klinkenberg, Hans Martin (ed.): Rheinisch-Westfälische Technische Hochschule Aachen, 1870–1970. Stuttgart 1970, 19–77.
Dyck, Walther von (ed.): Katalog mathematischer und mathematisch-physikalischer Modelle, Apparate und Instrumente. München 1892. (Mit Nachtrag 1893.)
Eckart, Carl: Über die Elektronentheorie der Metalle auf Grund der Fermischen Statistik, insbesondere über den Volta-Effekt. In: Zeitschrift für Physik 47 (1928), 38–42.
Eckert, Michael: Propaganda in Science: Sommerfeld and the Spread of Electron Theory of Metal In: Historical Studies in the Physical Sciences 17:2 (1987), 191–233.
Eckert, Michael: Das ‚freie Elektronengas' – Vorquantenmechanische Theorien über die elektronischen Eigenschaften der Metalle. In: Wissenschaftliches Jahrbuch des Deutschen Museums (1989), 57–91.
Eckert, Michael: Sommerfeld und die Anfänge der Festkörperphysik. In: Wissenschaftliches Jahrbuch des Deutschen Museums (1990), 33–71.
Eckert, Michael: Die Atomphysiker. Eine Geschichte der theoretischen Physik am Beispiel der Sommerfeldschule. Braunschweig 1993.
Eckert, Michael: Der „Sommerfeld-Effekt": Theorie und Geschichte eines bemerkenswerten Resonanzphänomen In: European Journal of Physics 17 (1996), 285–289.
Eckert, Michael: Mathematik auf Abwegen: Ferdinand Lindemann und die Elektronentheorie. In: Centaurus 39 (1997), 121–140.
Eckert, Michael: Mathematics, Experiments, and Theoretical Physics: The Early Days of the Sommerfeld School. In: Physics in Perspective 1 (1999), 238–252.
Eckert, Michael: The Dawn of Fluid Dynamics. Weinheim 2006.
Eckert, Michael: Die Deutsche Physikalische Gesellschaft und die »Deutsche Physik«. In: Hoffmann, Dieter/Walker, Mark (ed.): Physiker zwischen Autonomie und Anpassung. Die Deutsche Physikalische Gesellschaft im Dritten Reich. Weinheim 2007, 139–172.

Eckert, Michael: Quantenmechanische Atommodelle zwischen musealer Didaktik und ideologischer Auseinandersetzung. In: Bigg, Charlotte/Hennig, Jochen (ed.): Atombilder. Ikonografie des Atoms in Wissenschaft und Öffentlichkeit des 20. Jahrhunderts. Göttingen 2009, 83–91.

Eckert, Michael: The Troublesome Birth of Hydrodynamic Stability Theory: Sommerfeld and the Turbulence Problem. In: European Physical Journal History 35:1 (2010), 29–51.

Eckert, Michael: Plancks Spätwerk zur Quantentheorie. In: Hoffmann, Dieter (ed.): Max Planck und die moderne Physik. Berlin, Heidelberg 2010, 119–134.

Eckert, Michael: Paul Peter Ewald (1888–1985) im nationalsozialistischen Deutschland: eine Studie über die Hintergründe einer Wissenschaftleremigration. In: Walker, Mark/Hoffmann, Dieter (ed.): Fremde Wissenschaftler im Dritten Reich. Die Debye-Affäre im Kontext. Göttingen 2011, 265–289.

Eckert, Michael: Disputed Discovery: The Beginnings of X-ray Diffraction in Crystals in 1912 and Its Repercussion In: Acta Crystallographica A68 (2012), 30–39. Auch in: Zeitschrift für Kristallographie 227 (2012), 27–35.

Eckert, Michael/Kaiser, Walter: An der Nahtstelle von Theorie und Praxi Arnold Sommerfeld und der Streit um die Wellenausbreitung in der drahtlosen Telegraphie. In: Schürmann, Astrid/Weiss, Burghard (ed.): Chemie – Kultur – Geschichte. Festschrift für Hans-Werner Schütt anlässlich seines 65. Geburtstage Berlin, Diepholz 2002, 203–212.

Eckert, Michael/Märker, Karl (ed.): Arnold Sommerfeld. Wissenschaftlicher Briefwechsel. Band 1: 1891–1918. München u.a. 2000. (Zitiert als ASWB I.)

Eckert, Michael/Märker, Karl (ed.): Arnold Sommerfeld. Wissenschaftlicher Briefwechsel. Band 2: 1919–1951. München u.a. 2004. (Zitiert als ASWB II.)

Eckert, Michael/Pricha, Willibald: Boltzmann, Sommerfeld und die Berufungen auf die Lehrstühle für theoretische Physik in Wien und München, 1890–1917. In: Mitteilungen der Österreichischen Gesellschaft für Geschichte der Naturwissenschaften 4 (1984), 101–119.

Eckert, Michael/Pricha, Willibald/Schubert, Helmut/Torkar, Gisela: Geheimrat Sommerfeld – Theoretischer Physiker: Eine Dokumentation aus seinem Nachlass. München 1984.

Eckert, Michael/Schubert, Helmut/Torkar, Gisela: The Roots of Solid State Physics Before Quantum Mechanic In: Hoddeson, Lillian/Braun, Ernest/Weart, Spencer/Teichmann, Jürgen (ed.): Out of the Crystal Maze. Chapters from the History of Solid State-Physics. New York, NY, Oxford 1992, 3–87.

Elwert, Gerhard: Verschärfte Berechnung von Intensität und Polarisation im kontinuierlichen Röntgenspektrum. In: Annalen der Physik 34 (1939), 178–208.

Elwert, Gerhard: Der Absorptionskoeffizient an der langwelligen Grenze des kontinuierlichen Röntgenspektrum In: Zeitschrift für Naturforschung 3a (1948), 477–481.

Elwert, Gerhard/Haug, Eberhard: Calculation of Bremsstrahlung Cross Sections with Sommerfeld-Maue Eigenfunction In: Physical Review 183:1 (1969), 90–105.

Enzyklopädie der mathematischen Wissenschaften. Band 5 (Physik). Leipzig 1903–1926. http://de.wikipedia.org/wiki/Enzyklopädie_der_mathematischen_Wissenschaften (15.11.2012).

Epstein, Paul : Kraftliniendiagramme für die Ausbreitung der Wellen in der drahtlosen Telegraphie bei Berücksichtigung der Bodenbeschaffenheit. In: Jahrbuch der drahtlosen Telegraphie 4 (1910), 176–187.

Epstein, Paul : Zur Theorie des Starkeffekts. In: Physikalische Zeitschrift 17 (1916), 148–150.

Eucken, Arnold (ed.): Die Theorie der Strahlung und der Quanten. Verhandlungen auf einer von E. Solvay einberufenen Zusammenkunft (30. Oktober bis 3. November 1911). Halle 1914.

Evans, Richard J.: Tod in Hamburg. Stadt, Gesellschaft und Politik in den Cholera-Jahren 1830–1910. Reinbek 1990.

Ewald, Paul P.: Bericht über die Tagung der British Association in Birmingham (10. bis 17. September 1913). In: Physikalische Zeitschrift 14 (1913), 1297–1307.

Ewald, Paul P.: Die Intensität der Interferenzflecke bei Zinkblende und das Gitter der Zinkblende. In: Annalen der Physik 44 (1914), 257–282.

Ewald, Paul P.: Arnold Sommerfeld. In: Nature 168 (1951), 364–366.

Ewald, Paul P. (ed.):. Fifty Years of X-Ray Diffraction. Utrecht 1962.

Ewald, Paul P.: Erinnerungen an die Anfänge des Münchener Physikalischen Kolloquium In: Physikalische Blätter 24 (1968), 538–542.

Ewald, Paul P.: The Myth of the Myths: Comments on P. Forman's Paper on ‚The Discovery of the Diffraction of x-Rays in Crystals'. In: Archive for History of Exact Science 6 (1969), 72–81.

Ewald, Paul P: Arnold Sommerfeld als Mensch, Lehrer und Freund. Rede, gehalten zur Feier der 100sten Wiederkehr seiner Geburt. In: Bopp, Fritz/Kleinpoppen, Hans (ed.): Physics of the One- and Two-Electron Atoms. Amsterdam 1969, 8–16.

Ewald, Paul P./Friedrich, Walter: Röntgenaufnahmen von kubischen Kristallen, insbesondere Pyrit. In: Annalen der Physik 44 (1914), 1183–1196.

Feiten, Willi: Der Nationalsozialistische Lehrerbund. Entwicklung und Organisation. Ein Beitrag zum Aufbau und zur Organisationsstruktur des nationalsozialistischen Herrschaftssystems. Weinheim 1981.

Felsch, Volkmar (ed.): Otto Blumenthals Tagebücher. Ein Aachener Mathematikprofessor erleidet die NS-Diktatur in Deutschland, den Niederlanden und Theresienstadt. Konstanz 2011.

Feynman, Richard P.: QED. Die seltsame Theorie des Lichts und der Materie. München, Zürich 1990 (am. Original 1985.)

Fischer, Joachim: Instrumente zur Mechanischen Integration. Ein Zwischenbericht. In: Schütt, Hans-Werner/Weiss, Burghard (ed.): Brückenschläge. 25 Jahre Lehrstuhl für Geschichte der exakten Wissenschaften und der Technik an der Technischen Universität Berlin 1969–1994. Berlin 1995, 111–156.

Fischer, Joachim: Instrumente zur Mechanischen Integration II. Ein (weiterer) Zwischenbericht. In: Schürmann, Astrid /Weiss, Burghard (ed.): Chemie – Kultur – Geschichte: Festschrift für Hans-Werner Schütt anlässlich seines 65. Geburtstages. Diepholz 2002, 143–155.

Flachowsky, Sören: Von der Notgemeinschaft zum Reichsforschungsrat. Wissenschaftspolitik im Kontext von Autarkie, Aufrüstung und Krieg. Stuttgart 2008.

Flitner, Andreas/Wittig, Joachim (ed.): Optik – Technik – Soziale Kultur. Siegfried Czapski, Weggefährte und Nachfolger Ernst Abbe Briefe, Schriften, Dokumente. Rudolstadt, Jena 2000.

Fölsing, Albrecht: Albert Einstein. Eine Biographie. Frankfurt am Main 1993.

Forman, Paul: The Environment and Practice of Atomic Physics in Weimar Germany: A Study in the History of Science. Dissertation an der University of California. Berkeley, CA 1967.

Forman, Paul: The Doublet Riddle and Atomic Physics Circa 1924. In: Isis, 59 (1968), 156–174.

Forman, Paul: The Discovery of the Diffraction of X-Rays by Crystals; A Critique of the Myths. In: Archive for the History of Exact Sciences 6 (1969), 38–71.

Forman, Paul: Alfred Landé and the Anomalous Zeeman Effect, 1919–1921. In: Historical Studies in Physical Sciences 2 (1970), 153–261.

Forman, Paul: Scientific Internationalism and the Weimar Physicists: The Ideology and Its Manipulation in Germany after World War I. In: Isis 64 (1973), 150–180.

Forman, Paul: The Financial Support and Political Alignment of Physicists in Weimar Germany. In: Minerva 12 (1974), 39–66.

Forman, Paul: Swords Into Ploughshares: Breaking New Ground with Radar Hardware and Technique in Physical Research After World War II. In: Reviews of Modern Physics 67 (1995), 397–455.

Forman, Paul: Die Naturforscherversammlung in Nauheim im September 1920. In: Hoffmann, Dieter/Walker, Mark (ed.): Physiker zwischen Autonomie und Anpassung. Die Deutsche Physikalische Gesellschaft im Dritten Reich. Weinheim 2007, 29–58.

Forman, Paul/Hermann, Armin: Sommerfeld, Arnold (Johannes Wilhelm). In: Dictionary of Scientific Biography 12 (1975), 525–532.
Föppl, August: Vorlesungen über Technische Mechanik, Vierter Band: Dynamik. Leipzig 1899.
Föppl, August: Das Pendeln parallel geschalteter Maschinen. In: Elektrotechnische Zeitschrift 23:4 (1902), 59–64.
Frank, Gelya: ‚Becoming the Other': Empathy and Biographical Interpretation. In: Biography 8:3 (1985), 189–210.
Frei, Günther (ed.): Der Briefwechsel David Hilbert – Felix Klein (1886–1918). Göttingen 1985.
Frei, Norbert: „Machtergreifung". Anmerkungen zu einem historischen Begriff. In: Vierteljahrshefte für Zeitgeschichte 31 (1983), 136–145.
Frewer, Magdalena: Das mathematische Lesezimmer der Universität Göttingen unter der Leitung von Felix Klein (1886–1922). Köln 1979.
Fröhlich, Herbert:. Zum Photoeffekt an Metallen. In: Annalen der Physik 7 (1930), 103–128.
Fröhlich, Herbert:. Elektronentheorie der Metalle. Berlin 1936.
Friedl, Johannes/Rutte, Heiner (ed.): Moritz Schlick: Die Wiener Zeit: Aufsätze, Beiträge, Rezensionen 1926–1936. Band 6 der Moritz Schlick Gesamtausgabe. Berlin 2007.
Friedman, Robert Marc: The Politics of Excellence: Behind the Nobel Prize in Science. New York, NY 2001.
Friedrich, Bretislav/Herschbach, Dudley: Stern and Gerlach at Frankfurt: Experimental Proof of Space Quantization. In: Trageser, Wolfgang (ed.): Stern-Stunden. Höhepunkte Frankfurter Physik. Frankfurt am Main 2005, 149–171.
Friedrich, Walter: Röntgenstrahlinterferenzen. In: Physikalische Zeitschrift 14 (1913), 1079–1087.
Friedrich, Walter/Knipping, Paul/Laue, Max: Interferenzerscheinungen bei Röntgenstrahlen. (Vorgelegt von A. Sommerfeld in der Sitzung am 8. Juni 1912). In: Sitzungsberichte der mathematematisch-physikalischen Klasse der K. B. Akademie der Wissenschaften zu München (1912), 303–322.
Fritzsch, Harald: Sind die fundamentalen Konstanten konstant? In: Physik Journal 2:4 (2003), 49–52.
Gause, Fritz: Die Geschichte der Stadt Königsberg in Preussen. 3 vols. Köln 1996.
Germania. Festschrift zu ihrem neunzigjährigen Stiftungsfeste. Königsberg 1933.
Gimbel, John: Project Paperclip: German Scientists, American Policy, and the Cold War. In: Diplomatic History 14 (1990), 343–366.
Goethe, Johann Wolfgang von: Werke. Band 1: Gedichte und Epen I. München 1981.
Goethe, Johann Wolfgang von: Torquato Tasso. Ein Schauspiel. Reclam Universalbibliothek. Ditzingen 2005.
Goldberg, Stanley: The Abraham Theory of the Electron: The Symbiosis of Experiment and Theory. In: Archive for History of Exact Sciences 7 (1970), 7–25.
Greenaway, Frank: Science International: A History of the International Council of Scientific Unions. Cambridge, MA 1996.
Grotrian, Walter: Die Entwirrung der komplizierten Spektren, insbesondere des Eisenspektrums. In: Die Naturwissenschaften 12 (1924), 945–955.
Grotrian, Walter: Graphische Darstellung der Spektren von Atomen und Ionen mit ein, zwei und drei Valenzelektronen. Berlin 1928.
Grundmann, Siegfried:. Einsteins Akte. Wissenschaft und Politik – Einsteins Berliner Zeit. Berlin 2004.
Handel, Kai: Anfänge der Halbleiterforschung und -entwicklung, dargestellt an den Biographien von vier deutschen Halbleiterpionieren. Dissertation RWTH Aachen. Aachen 1998.

Hankins, Thomas L.: In Defence of Biography: The Use of Biography in the History of Science. In: History of Science 17 (1979), 1–16.

Hashagen, Ulf: Walther von Dyck (1856–1934). Mathematik, Technik und Wissenschaftsorganisation an der TH München. Stuttgart 2003.

Hashagen, Ulf: Ein ausländischer Mathematiker im NS-Staat: Constantin Carathéodory als Professor an der Universität München. München 2010.

Haug, Eberhard/Nakel, Werner: The Elementary Process of Bremsstrahlung. Singapur 2004.

Heiber, Helmut: Universität unterm Hakenkreuz. Teil II: Die Kapitulation der Hohen Schulen. Band 1. München 1994.

Heilbron, John L.: A History of the Problem of Atomic Structure from the Discovery of the Electron to the Beginning of Quantum Mechanics. PhD thesis, University of California, Berkeley, CA 1964.

Heilbron, John L.: The Kossel-Sommerfeld Theory and the Ring Atom. In: Isis 58 (1967), 451–485.

Heilbron, John L.: H. G. J. Moseley: The Life and Letters of an English Physicist, 1887–1915. Berkeley, CA 1974.

Heilbron, John L.: The Dilemmas of an Upright Man. Max Planck as Spokesman for German Science. Berkeley, CA 1986.

Heisenberg, Werner: Zur Quantentheorie der Linienstruktur und der anomalen Zeemaneffekte. In: Zeitschrift für Physik 8 (1922), 273–297.

Heisenberg, Werner: Zur Quantentheorie der Multiplettstruktur und der anomalen Zeemaneffekte. In: Zeitschrift für Physik 32 (1925), 841–860.

Heisenberg, Werner: Über quantentheoretische Umdeutung kinematischer und mechanischer Beziehungen. In: Zeitschrift für Physik 33 (1925), 879–893.

Heisenberg, Werner: Die physikalischen Prinzipien der Quantentheorie. Leipzig, 1930.

Heisenberg, Werner: Arnold Sommerfeld zum 5. Dezember 1948. In: Zeitschrift für Naturforschung 3a (1948), 429.

Heisenberg, Werner: Bemerkungen zum Turbulenzproblem. In: Zeitschrift für Naturforschung 3a (1948), 434–437.

Heisenberg, Werner: Arnold Sommerfeld. In: Die Naturwissenschaften 38 (1951), 337–338.

Heisenberg, Werner: Ausstrahlung von Sommerfelds Werk in die Gegenwart. In: Physikalische Blätter 24 (1968), 530–537.

Heitler, Walter/London, Fritz: Wechselwirkung neutraler Atome und homöopolare Bindung nach der Quantenmechanik. In: Zeitschrift für Physik 44 (1927), 455–472.

Hendry, John: Bohr-Kramers-Slater: A Virtual Theory of Virtual Oscillators and Its Role in the History of Quantum Mechanics. In: Centaurus 25 (1981), 189–221.

Hensel, Susann: Die Auseinandersetzungen um die mathematische Ausbildung der Ingenieure an den Technischen Hochschulen in Deutschland Ende des 19. Jahrhunderts. In: Hensel, Susann/Ihmig, K. M./Otte, M. (ed.): Mathematik und Technik im 19. Jahrhundert in Deutschland. Soziale Auseinandersetzungen und philosophische Problematik. Göttingen 1989, 1–111.

Hentschel, Klaus: Die Mentalität deutscher Physiker in der frühen Nachkriegszeit. Heidelberg 2005.

Hentschel, Klaus/Hentschel, Ann (ed.): Physics and National Socialism: An Anthology of Primary Sources. Basel 1996.

Herglotz, Gustav: Zur Elektronentheorie. In: Nachrichten von der Königl. Gesellschaft der Wissenschaften zu Göttingen. Mathematisch-physikalische Klasse (1903), 357–382.

Hermann, Armin: Die frühe Diskussion zwischen Stark und Sommerfeld über die Quantenhypothese (1). In: Centaurus 12 (1967), 38–59.

Hermann, Armin: Frühgeschichte der Quantentheorie 1899–1913. Mosbach in Baden 1969.
Hermann, Armin: Max Planck in Selbstzeugnissen und Bilddokumenten. Reinbek 1973.
Hilz, Helmut: »Eine Bildungsanstalt für alle Stände unseres Volke« Die Bibliothek des Deutschen Museums in der Zeit des Nationalsozialismus. In: Vaupel, Elisabeth/Wolff, Stefan L. (ed.): Das Deutsche Museum in der Zeit des Nationalsozialismu Eine Bestandsaufnahme. Göttingen 2010, 244–286.
Hoddeson, Lillian/Baym, Gordon/Eckert, Michael: The Development of the Quantum Mechanical Electron Theory of Metals. In: Reviews of Modern Physics 59 (1987), 287–327.
Hoddeson, Lillian/Braun, Ernst/Teichmann, Jürgen/Weart, Spencer (ed.): Out of the Crystal Maze. Chapters from The History of Solid State Physics. Oxford, New York, NY 1992.
Hoerschelmann, Harald von: Über die Wirkungsweise des geknickten Marconischen Senders in der drahtlosen Telegraphie. In: Jahrbuch der drahtlosen Telegraphie und Telephonie 5 (1911), 14–34, 188–211.
Hoffmann, Dieter: Zur Etablierung der technischen Physik in Deutschland. In: Guntau, Martin/Laitko, Hubert (ed.): Der Ursprung der modernen Wissenschaften. Studien zur Entstehung wissenschaftlicher Disziplinen. Berlin 1987, 140–153.
Hoffmann, Dieter: Die Ramsauer-Ära und die Selbstmobilisierung der Deutschen Physikalischen Gesellschaft. In: Hoffmann, Dieter/Walker, Mark (ed.): Physiker zwischen Autonomie und Anpassung. Die Deutsche Physikalische Gesellschaft im Dritten Reich. Weinheim 2007, 173–215.
Hoffmann, Dieter: Max Planck: Die Entstehung der modernen Physik. München 2008.
Hoffmann, Dieter/Walker, Mark (ed.): Physiker zwischen Autonomie und Anpassung. Die Deutsche Physikalische Gesellschaft im Dritten Reich. Weinheim 2007.
Holton, Gerald: On the Hesitant Rise of Quantum Physics Research in the United States. In: Holton, Gerald: Thematic Origins of Scientific Thought: Kepler to Einstein. Cambridge, MA 1988, 147–187.
Hondros, Demetrios: Über elektromagnetische Drahtwellen. In: Annalen der Physik 30 (1909), 905–950.
Hondros, Demetrios/Debye, Peter: Elektromagnetische Wellen an dielektrischen Drähten. In: Annalen der Physik 32 (1910), 465–476.
Hong, Sungook: Wireless: from Marconi's Black-Box to the Audion. Cambridge, MA 2001.
Hopf, Ludwig: Turbulenz bei einem Flusse. In: Annalen der Physik 32 (1910), 777–808.
Houston, William V.: Elektrische Leitfähigkeit auf Grund der Wellenmechanik. In: Zeitschrift für Physik 48 (1928), 449–468.
Hoyer, Ulrich: Introduction. In: Hoyer, Ulrich (ed.): Niels Bohr Collected Works. Band 2. Amsterdam, New York, NY, Oxford 1981 (abbreviated as NBCW 2), 103–134.
Huber, Ursula: Die Universität München – Ein Bericht über den Fortbestand nach 1945. In: Prinz, Friedrich (ed.): Trümmerzeit in München: Kultur und Gesellschaft einer deutschen Großstadt im Aufbruch 1945–1949. München 1984, 156–160.
Hughes, Jeff: Radioactivity and Nuclear Physic In: Jo Nye, Mary (ed.): The Cambridge History of Science, Band 5: The Modern Physical and Mathematical Sciences. Cambridge, MA 2003, 350–374.
Iengo, Roberto: Sommerfeld Enhancement: General Results from Field Theory Diagrams. In: Journal of High Energy Physics 5 (2009), 24.
Ioffe, Abram F.: Begegnungen mit Physikern. Basel 1967.
Jackson, Derek/Launder, Brian: Osborne Reynolds and the Publication of His Papers on Turbulent Flow. In: Annual Review of Fluid Mechanics 39 (2007), 19–35.
Jacobs, Konrad (ed.): Felix Klein. Handschriftlicher Nachlass. Erlangen 1977.

Jammer, Max: The Conceptual Development of Quantum Mechanics. Los Angeles 1989 (Erstauflage 1966).

Janssen, Michel/Duncan, Tony: From Canonical Transformations to Transformation Theory, 1926–1927: The Road to Jordan's Neue Begründung. In: Studies in History and Philosophy of Modern Physics 40 (2009), 352–362.

Janssen, Michel/Mecklenburg, Matthew: From Classical to Relativistic Mechanics: Electromagnetic Models of the Electron. In: Hendricks, Vincent F./Jörgensen, Klaus F./ Lützen, Jesper/Pedersen, Stig A. (ed.): Interactions: Mathematics, Physics and Philosophy, 1860–1930. Dordrecht 2007, 65–134.

Jarausch, Konrad Hugo: Deutsche Studenten, 1800–1970. Frankfurt am Main 1984.

Jenkin, John: A Unique Partnership: William and Lawrence Bragg and the 1915 Nobel Prize in Physics. In: Minerva 39 (2001), 373–392.

Jo-Nye, Mary: Scientific Biography: History of Science by Another Means? In: Isis 97 (2006), 322–329.

Jungnickel, Christa/McCormmach, Russell: Intellectual Mastery of Nature: Theoretical Physics from Ohm to Einstein. Band 2: The Now Mighty Theoretical Physics 1870–1925. Chicago 1990.

Kaiser, David: Cold War Requisitions, Scientific Manpower, and the Production of American Physicists after World War II. In: Historical Studies in the Physical and Biological Sciences 33 (2002), 131–159.

Kaiserfeld, Thomas: When Theory Addresses Experiment. The Siegbahn-Sommerfeld Correspondence, 1917–1940. In: Lindqvist, Svante (ed.): Center on the Periphery. Historical Aspects of 20th-Century Swedish Physics. Canton, MA 1993, 306–324.

Kant, Horst: Albert Einstein, Max von Laue, Peter Debye und das Kaiser-Wilhelm-Institut für Physik in Berlin (1917–1939). In: Brocke, Bernhard vom/Laitko, Hubert (ed.): Das Harnack-Prinzip. Die Kaiser-Wilhelm-/Max-Planck-Gesellschaft und ihre Institute: Studien zur Geschichte. Berlin 1996, 227–243.

Keppeler, Stefan: Eine geometrische Phase rettete Sommerfelds Theorie der Feinstruktur. In: Physik Journal, 3:4 (2004), 45–49.

Kevles, Daniel J.: Into Hostile Political Camps: The Reorganization of International Science in World War I. In: Isis 62 (1971), 47–60.

Kevles, Daniel J.: The Physicist The History of a Scientific Community in Modern America. New York, NY 1979.

Kirkpatrick, Paul: Recipient of the 1948 Oersted Medal for Notable Contributions to the Teaching of Physic In: American Journal of Physics. 17 (1949), 312–314.

Klein, Felix: Über den mathematischen Unterricht an der Göttinger Universität im besonderen Hinblicke auf die Bedürfnisse der Lehramtskandidaten. In: Zeitschrift für mathematischen und naturwissenschaftlichen Unterricht 26 (1895), 383–388.

Klein, Felix: Ausgewählte Kapitel der Zahlentheorie. Band 1. Vorlesung, gehalten im Wintersemester 1895/96. Ausgearbeitet von A. Sommerfeld. Leipzig 1896.

Klein, Felix: Ausgewählte Kapitel der Zahlentheorie. Band 2. Vorlesung, gehalten im Sommersemester 1896. Ausgearbeitet von A. Sommerfeld und Ph. Furtwängler. Leipzig 1896.

Klein, Felix: Gesammelte Mathematische Abhandlungen. 3 vols. Berlin 1921–1923.

Klein, Felix: Riemannsche Flächen. Vorlesungen, gehalten in Göttingen 1891/92. Leipzig 1985. (Originalausgabe 1894; nachgedruckt 1906 und 1985 mit Kommentaren neu herausgegeben.) http://www.archive.org/details/rieflachvolesungooklierich (7.1. 2012).

Klein, Felix/Sommerfeld, Arnold: Über die Theorie des Kreisel Heft 1–4. Leipzig 1897–1910.

Klein, Felix: Ernst Ritter †. In: Jahresbericht der Deutschen Mathematiker-Vereinigung 4 (1895), 52–54.

Klein, Martin J.: Einstein and the Wave-Particle Duality. In: The Natural Philosopher 3 (1964), 3–49.

Klein, Martin J.: Paul Ehrenfest. Band 1: The Making of a Theoretical Physicist. Amsterdam 1970.

Kleinert, Andreas: Paul Weyland, der Berliner Einstein-Töter. In: Albrecht, Helmuth (ed.): Naturwissenschaft und Technik in der Geschichte. 25 Jahre Lehrstuhl für Geschichte der Naturwissenschaft und Technik am Historischen Institut der Universität Stuttgart. Stuttgart 1993, 198–232.

Kleinert, Andreas: Der Briefwechsel zwischen Philipp Lenard (1862–1947) und Johannes Stark (1874–1975). In: Jahrbuch 2000 der Deutschen Akademie der Naturforscher Leopoldina, R. 3, 46 (2001), 243–261.

Kleinert, Andreas: Die Axialität der Lichtemission und Atomstruktur. Johannes Starks Gegenentwurf zur Quantentheorie. In: Schürmann, Astrid/Weiss, Burghard (ed.): Chemie – Kultur – Geschichte. Festschrift für Hans-Werner Schütt anlässlich seines 65. Geburtstages. Berlin, Diepholz 2002, 213–222.

Kleinert, Andreas/Schönbeck, Charlotte: Lenard und Einstein. Ihr Briefwechsel und ihr Verhältnis vor der Nauheimer Diskussion von 1920. In: Gesnerus 35 (1973), 318–333.

Koch, Ernst-Eckhard: Das Konservatorenamt und die mathematisch-physikalische Sammlung der Bayerischen Akademie der Wissenschaften. Arbeitsbericht aus dem Institut für Geschichte der Naturwissenschaften der Universität München. München 1967.

Koch, Peter: Die Bedeutung Göttingens für die Entwicklung der Versicherungswissenschaft und -praxis. In: Versicherungswirtschaft. Beiträge zur Geschichte des deutschen Versicherungswesen Teil 2. Karlsruhe 2005, 33–38.

Kononenko, Viktor Olimpanovich: Vibrating Systems with a Limited Power Supply. London 1969.

Kossel, Walther: Bemerkung zur Absorption homogener Röntgenstrahlen. In: Verhandlungen der Deutschen Physikalischen Gesellschaft (1914), 898–909.

Kossel, Walther: Bemerkung zur Absorption homogener Röntgenstrahlen II. In: Verhandlungen der Deutschen Physikalischen Gesellschaft (1914), 953–963.

Kox, Anne J.: The Discovery of the Electron: II. The Zeeman Effect. In: European Journal of Physics 18 (1997), 139–144.

Kragh, Helge: Niels Bohr's Second Atomic Theory. In: Historical Studies in the Physical and Biological Sciences 10 (1979), 123–186.

Kragh, Helge: The Fine Structure of Hydrogen and the Gross Structure of the Physics Community, 1916–26. In: Historical Studies in the Physical Sciences 15 (1985), 67–125.

Kragh, Helge: Magic Number: A Partial History of the Fine-Structure Constant. In: Archive for History of Exact Sciences 57 (2003), 395–431.

Kramers, Hendrik A.: On the Theory of X-ray Absorption and of the Continuous X-ray Spectrum. In: Philosophical Magazine 46 (1923), 836–871.

Krasnopolskaya, Tatyana /Shvets, Alexander Yu.: Chaos in Vibrating Systems With a Limited Power-Supply. In: Chaos 3 (1993), 387–395.

Krauß, Oliver: Rüstung und Rüstungserprobung in der deutschen Marinegeschichte unter besonderer Berücksichtigung der Torpedoversuchsanstalt (TVA). Dissertation Universität Kiel. Kiel 2006.

Körber, Hans-Günther (ed.): Aus dem wissenschaftlichen Briefwechsel Wilhelm Ostwald 1. Teil. Berlin 1961.

Kuhlen, Michael/Madau, Piero/Silk, Joseph: Exploring Dark Matter with Milky Way Substructure. In: Science 325 (2009), 970–973.

Kuhn, Thomas : The Structure of Scientific Revolutions. Chicago 1962.

Kuhn, Thomas : Black-Body Theory and the Quantum Discontinuity, 1894–1912. Oxford 1978.

Kuhn, Thomas /Heilbron, John L./Forman, Paul/Allen, Lini: Sources for History of Quantum Physics. Philadelphia 1967. http://www.amphilsoc.org/guides/ahqp/ (15.11.2012).

Kulenkampff, Helmuth: Untersuchungen der kontinuierlichen Röntgenstrahlung dünner Aluminiumfolien. In: Annalen der Physik 87 (1928), 597–637.

Lamb, W. E. Jr./Retherford, R. C.: Fine Structure of the Hydrogen Atom by a Microwave Method. In: Physical Review 72 (1947), 241–243.

Landé, Alfred: Zur Methode der Eigenschwingungen der Quantentheorie. Dissertation an der Universität München. München 1914.

Landé, Alfred: Über den anomalen Zeemaneffekt. Teil I. In: Zeitschrift für Physik 5 (1921), 231–241.

Langevin, Paul/de Broglie, Maurice (ed.): La théorie du rayonnement et les quanta: Rapports et discussions de la réunion tenue à Bruxelles, du 30 octobre au 3 novembre 1911, sous les auspices de M. E. Solvay. Paris 1912.

Laporte, Otto: Multipletts im Spektrum des Vanadiums. In: Die Naturwissenschaften 11 (1923), 779–782.

Lattanzi, Massimiliano/Silk, Joseph: Can the WIMP Annihilation Boost Factor Be Boosted by the Sommerfeld Enhancement? In: Physical Review D 79 (2009). http://link.aporg/doi/10.1103/PhysRevD.79.083523 (15.11.2012).

Laue, Max von: Das Relativitätsprinzip. Braunschweig 1911.

Laue, Max von: Eine quantitative Prüfung der Theorie für die Interferenzerscheinungen bei Röntgenstrahlen. In: Annalen der Physik 41 (1913), 989–1002.

Laue, Max von: Über die Auffindung der Röntgenstrahlinterferenzen. Nobelvortrag. Stockholm, 3. Juni 1920. In: Max von Laue. Gesammelte Schriften und Vorträge. Band 3. Braunschweig 1961, 5–18. Enlish translation in http://www.nobelprize.org/nobel_prizes/physics/laureates/1914/present.html. (15.11.2012).

Laue, Max von: Röntgenstrahlinterferenzen. In: Physikalische Zeitschrift 14 (1913), 1075–1079.

Laue, Max von: Wellenoptik. In: Enzyklopädie der mathematischen Wissenschaften 5:3 (1915), 362–487.

Laue, Max von: Sommerfelds Lebenswerk. Nachruf, gehalten am 15. Juni 1951 vor der Physikalischen Gesellschaft zu Berlin. In: Die Naturwissenschaften 38 (1951), 513–518.

Laurent, Joseph: Die städtebauliche und bauliche Entwickelung der Bade- und Industriestadt Aachen von 1815–1915. Aachen 1920.

Lemmerich, Jost: Ein Angriff von Johannes Stark auf Werner Heisenberg über das Reichsministerium für Wissenschaft, Erziehung und Volksbildung. In: Kleimt, Christian/Wiemers, Gerald (ed.): Werner Heisenberg im Spiegel seiner Leipziger Schüler und Kollegen. Leipzig 2005, 213–221.

Lertes, P.: Die drahtlose Telegraphie und Telephonie. Dresden, Leipzig 1923.

Liebmann, Heinrich: Zur Erinnerung an Heinrich Burkhardt. In: Jahresbericht der Deutschen Mathematiker-Vereinigung 24 (1915), 185–195.

Lindemann, Ferdinand: Über Molekularphysik. Versuch einer einheitlichen dynamischen Behandlung der physikalischen und chemischen Kräfte. In: Schriften der physikalisch-ökonomischen Gesellschaft zu Königsberg in Pr. 29 (1888), 31–81.

Lindemann, Ferdinand: Über die Bewegung der Elektronen. Erster Teil: Die translatorische Bewegung. In: Abhandlungen der K. Bayer. Akademie der Wiss. II. Kl. (1907), 235–335.

Lindemann, Ferdinand: Über die Bewegung der Elektronen. Zweiter Teil: Stationäre Bewegung. In: Abhandlungen der K. Bayer. Akademie der Wiss. II. Kl. (1907), 339–375.

Lindemann, Ferdinand: Zur Elektronentheorie. In: Sitzungsberichte der mathematisch-physikalischen Klasse der K. B. Akademie der Wissenschaften zu München (1907), 177–209.

Lindemann, Ferdinand: Zur Elektronentheorie II. In: Sitzungsberichte der mathematisch-physikalischen Klasse der K. B. Akademie der Wissenschaften zu München (1907), 353–360.

Lindqvist, Svante (ed.): Center on the Periphery: Historical Aspects of 20th-Century Swedish Physics. Canton, MA 1993.

Lippincott, Sara: A Conversation with Valentine L.Telegdi – Part II. In: Physics in Perspective 10 (2008), 77–109.

Litten, Freddy: Die Trennung der Verwaltung der Wissenschaftlichen Sammlungen des Staates von der Bayerischen Akademie der Wissenschaften – Ein Beitrag zur Geschichte der Wissenschaftsorganisation in Bayern. In: Zeitschrift für Bayerische Landesgeschichte 55 (1992), 411–420.

Litten, Freddy: Astronomie in Bayern 1914–1945. Stuttgart 1992.

Litten, Freddy: Die Korn-Röntgen-Affäre. In: Kultur & Technik 17:4 (1993), 42–49.

Litten, Freddy: Die Carathéodory-Nachfolge in München 1938–1944. In: Centaurus 37 (1994), 154–172.

Litten, Freddy: Mechanik und Antisemitismus – Wilhelm Müller (1880–1968). München 2000.

Lorentz, Hendrik Antoon: Maxwells elektromagnetische Theorie. In: Enzyklopädie der mathematischen Wissenschaften, 5:2 (1904), 63–144.

Lorentz, Hendrik Antoon: Weiterbildung der Maxwellschen Theorie. Elektronentheorie. In: Enzyklopädie der mathematischen Wissenschaften, 5:2 (1904), 145–280.

Lorey, Wilhelm: Das Studium der Mathematik an den deutschen Universitäten seit Anfang des 19. Jahrhunderts. Leipzig 1916.

Ludwig, Karl-Heinz: Technik und Ingenieure im Dritten Reich. Düsseldorf 1974.

Maey, Eugen: Über die Beugung des Lichts an einem geraden, scharfen Schirmrande. In: Annalen der Physik 49 (1893), 69–104.

Maier, Helmut (ed.): Rüstungsforschung im Nationalsozialismu Organisation, Mobilisierung und Entgrenzung der Technikwissenschaften. Göttingen 2002.

Malley, Marjorie C.: Radioactivity: A History of a Mysterious Science. Oxford 2011.

Manegold, Karl-Heinz: Universität, Technische Hochschule und Industrie. Ein Beitrag zur Emanzipation der Technik im 19. Jahrhundert unter besonderer Berücksichtigung der Bestrebungen Felix Kleins. Berlin 1970.

Marsch, Ulrich: Notgemeinschaft der Deutschen Wissenschaft: Gründung und frühe Geschichte, 1920–1925. Frankfurt am Main 1994.

Massimi, Michela: Pauli's Exclusion Principle. Cambridge, MA 2005.

Maue, August Wilhelm: Das kontinuierliche und kontinuierlich-diskrete Röntgenspektrum nach der Theorie von Kramers und nach der Wellenmechanik. In: Annalen der Physik 13 (1932), 161–190.

McCormmach, Russell: H. A. Lorentz and the Electromagnetic View of Nature. In: Isis 61 (1970), 459–497.

Meggers, William F./Kiess, Carl C./Walters, Jr. Francis M.: The Displacement Law of Arc and Spark spectra. In: Journal of the Optical Society of America 9 (1924), 335–374.

Mehra, Jagdish: The Solvay Conferences on Physic Aspects of the Development of Physics Since 1911. Dordrecht 1975.

Mehra, Jagdish/Rechenberg, Helmut: The Historical Development of Quantum Theory. Band 1. New York, NY 1982.

Mehra, Jagdish/Rechenberg, Helmut: The Historical Development of Quantum Theory. Band 6. New York, NY 2001.

Meixner, Josef: Strenge Theorie der Beugung elektromagnetischer Wellen an der vollkommen leitenden Kreisscheibe. In: Zeitschrift für Naturforschung, 3a (1948), 506–518.

Mertens, Lothar: Bildungsprivileg und Militärdienst im Kaiserreich. In: Bildung und Erziehung 44 (1990), 217–228.

Merton, Robert K.: Auf den Schultern von Riesen. Ein Leitfaden durch das Labyrinth der Gelehrsamkeit. Frankfurt am Main 1980.

Metzler, Gabriele: Internationale Wissenschaft und nationale Kultur: Deutsche Physiker in der internationalen Community 1900–1960. Göttingen 2000.

Metzler, Gabriele: Nationalismus und Internationalismus in der Physik des 20. Jahrhundert Das deutsche Beispiel. In: Jessen, Ralph/Vogel, Jakob (ed.): Wissenschaft und Nation in der europäischen Geschichte. Frankfurt am Main 2002, 285–309.

Meyenn, Karl von: Paulis Weg zum Ausschließungsprinzip, Teil I. In: Physikalische Blätter 36 (1980), 293–298.

Meyenn, Karl von: Paulis Weg zum Ausschließungsprinzip, Teil II. In: Physikalische Blätter 37 (1981), 13–19.

Meyenn, Karl von (ed.): Wolfgang Pauli. Wissenschaftlicher Briefwechsel, Bd. 1–4. New York, NY, Berlin 1979–2005 (abbreviated as WPWB).

Meyenn, Karl von: Eine Entdeckung von ganz außerordentlicher Tragweite. Schrödingers Briefwechsel zur Wellenmechanik und zum Katzenparadoxon. Berlin 2011.

Minkowski, Hermann: Raum und Zeit. In: Jahresbericht der Deutschen Mathematiker-Vereinigung 18 (1909), 75–88.

Moore, Walter J.: Schrödinger, Life and Thought. Cambridge, MA 1989.

Müller, Georg: Carl Schnabel. Wissenschaftler und Musensohn. In: TUContact 7 (November 2000), 41–46. http://www.tu-clausthal.de/presse/tucontact/2000/November/tuc1/23.pdf (15.11.2012).

Müller, Wilhelm: Die Lage der theoretischen Physik an den Universitäten. In: Zeitschrift für die gesamte Naturwissenschaft 6 (1940), 281–298.

Müller, Winfried: Die Universitäten Erlangen, München und Würzburg nach 1945. Zur Hochschulpolitik in der amerikanischen Besatzungszone. In: Lanzinner, Maximilian (ed.): Landesgeschichte und Zeitgeschichte. Forschungsperspektiven zur Geschichte Bayerns nach 1945. Augsburg 1997, 53–87.

München, Landeshauptstadt (ed.): München – wie geplant. München 2004.

Nagel, Anne: „Er ist der Schrecken überhaupt der Hochschule": Der Nationalsozialistische Deutsche Dozentenbund in der Wissenschaftspolitik des Dritten Reich In: Scholtyseck, Joachim/Studt, Christoph (ed.): Universitäten und Studenten im Dritten Reich. Berlin 2008, 115–132.

Navarro, Luis/Pérez, Enric: Paul Ehrenfest: The Genesis of the Adiabatic Hypothesis, 1911–1914 . In: Archive for History of Exact Sciences 60 (2006), 209–267.

Needell, Allan A.: Irreversibility and the Failure of Classical Dynamics: Max Planck's Work on the Quantum Theory 1900–1915. Dissertation an der Yale University, New Haven. New Haven, CT 1980.

Niemann, Erich: Funkentelegraphie für Flugzeuge. Berlin 1921.

Niessen, Karel Frederik: The Earth's Constants From Combined Electric and Magnetic Measurements Partly in the Vicinity of the Emitter. In: Zeitschrift für Naturforschung, 3a (1948), 552–558.

Nisio, Sigeko: The Formation of the Sommerfeld Quantum Theory of 1916. In: Japanese Studies in the History of Science 12 (1973), 39–78.

Noether, Fritz: Über analytische Berechnung der Geschosspendelungen. In: Nachrichten von der Königl. Gesellschaft der Wissenschaften zu Göttingen. Mathematisch-physikalische Klasse (1919), 373–391.

Nollendorfs, Cora Lee: The First World War and the Survival of German Studies: With a Tribute to Alexander R. Hohlfeld. In: Benseler, D. (ed.): Teaching German in America: Prolegomena to a History. Madison, WI 1988, 176–195.

Nordheim, Lothar: Statistische und kinetische Theorie des metallischen Zustandes. In: Müller-Pouillets Lehrbuch der Physik, 4:4 (1934), 243–389.

Nordheim, Lothar: Quantentheorie des Magnetismus. In: Müller-Pouillets Lehrbuch der Physik, 4:4 (1934), 798–876.

Olenhusen, Albrecht Götz von: Die „nichtarischen" Studenten an den deutschen Hochschulen. In: Vierteljahrshefte für Zeitgeschichte 14 (1966), 175–206.

Olesko, Kathryn: Physics as a Calling. Discipline and Practice in the Königsberg Seminar for Physics. Ithaca, NY, New York, NY 1991.

Ostrowski, Alexander: Zur Entwicklung der numerischen Analysis. In: Jahresbericht der Deutschen Mathematiker-Vereinigung 68 (1966), 97–111.

Ozawa, Takeshi: Arnold Sommerfelds Aufenthalt in Japan. In: Historia Scientiarum 15:1 (2005), 44–65.

Parshall, Karen Hunger/Rowe, David E.: The Emergence of the American Mathematical Research Community 1876–1900: J. J. Sylvester, Felix Klein, and E. H. Moore. Providence, RI, London 1994.

Paschen, Friedrich: Über die durchdringenden Strahlen des Radiums. In: Annalen der Physik 14 (1904), 164–171.

Paschen, Friedrich: Über die Kathodenstrahlen des Radiums. In: Annalen der Physik 14 (1904), 389–405.

Paschen, Friedrich: Bohrs Heliumlinien. In: Annalen der Physik 50 (1916), 901–940.

Pauli, Wolfgang: Über das Wasserstoffspektrum vom Standpunkt der neuen Quantenmechanik. In: Zeitschrift für Physik 36 (1926), 336–363.

Pauli, Wolfgang: Über Gasentartung und Paramagnetismus. In: Zeitschrift für Physik 41 (1927), 81–102.

Pauli, Wolfgang: Sommerfelds Beiträge zur Quantentheorie. In: Die Naturwissenschaften 35 (1948), 129–132.

Pauli, Wolfgang: Arnold Sommerfeld. In: Zeitschrift für angewandte Mathematik und Physik 2 (1951), 301.

Pauli, Wolfgang: Wissenschaftlicher Briefwechsel, Bd. 1–4, herausgegeben von Karl von Meyenn. New York, NY, Berlin, 1979–2005 (abbreviated as WPWB).

Peierls, Rudolf: Elektronentheorie der Metalle. In: Ergebnisse der exakten Naturwissenschaften 11 (1932), 264–351.

Pérez, Enric: Ehrenfest's Adiabatic Theory and the Old Quantum Theory, 1916–1918. In: Archive for History of Exact Sciences 63 (2009), 81–125.

Planck, Max: Die Kaufmannschen Messungen der Ablenkbarkeit der β-Strahlen in ihrer Bedeutung für die Dynamik der Elektronen. In: Physikalische Zeitschrift 7 (1906), 753–761.

Pockels, Friedrich: Über die partielle Differentialgleichung und deren Auftreten in der mathematischen Physik. Leipzig 1891.

Pohl, Robert Wichard: Die Physik der Röntgenstrahlen. Vieweg, Brauschweig, 1912.

Poincaré, Henri: Sur la polarisation par diffraction. In: Acta Mathematica 16 (1892), 297–340.

Poincaré, Henri: Sur la polarisation par diffraction. In: Acta Mathematica 20 (1897), 313–355.

Popp, Emil: Zur Geschichte des Königsberger Studententums, 1900–1945. Beihefte zum Jahrbuch der Albertus-Universität Königsberg/Pr. XII. Würzburg 1955.

Prandtl, Ludwig: Kipperscheinungen: Ein Fall von instabilem elastischem Gleichgewicht. Dissertation an der der Universität München 1900. In: Ludwig Prandtl – Gesammelte Abhandlungen. Band 1. Berlin 1961, 10–74.

Preston, Diana: Lusitania: An Epic Tragedy. New York, NY 2002.

Prinz, Friedrich/Kraus, Marita (ed.): München – Musenstadt mit Hinterhöfen. Die Prinzregentenzeit 1886 bis 1912. München 1988.

Pyenson, Lewis/Skopp, Douglas: Educating Physicists in Germany Circa 1900. In: Social Studies of Science 7 (1977), 329–366.

Pyenson, Louis: Einstein's Early Scientific Collaboration. In: Historical Studies in the Physical Sciences 7 (1978), 83–124.

Pyenson, Louis: Physics in the Shadow of Mathematic The Göttingen Electron-Theory Seminar of 1905. In: Archive for History of Exact Science 21 (1979), 55–89.

Radkau, Joachim: Das Zeitalter der Nervosität. Deutschland zwischen Bismarck und Hitler. München 1998.

Rammer, Gerhard: Die Nazifizierung und Entnazifizierung der Physik an der Universität Göttingen. Dissertation an der Georg-August-Universität Göttingen. Göttingen 2004.

Rasch, Manfred: Wissenschaft und Militär: Die Kaiser Wilhelm Stiftung für Kriegstechnische Wissenschaft. In: Militärgeschichtliche Mitteilungen 44 (1991), 73–120.

Rechenberg, Helmut: Werner Heisenberg – Die Sprache der Atome: Leben und Wirken - Eine wissenschaftliche Biographie. Berlin 2010.

Reich, Karin: Die Rolle Arnold Sommerfelds bei der Diskussion um die Vektorrechnung, dargestellt anhand der Quellen im Nachlaß des Mathematikers Rudolf Mehmke. In: Dauben, J./ Folkerts, M./Knobloch, E./Wussing, H. (ed.): History of Mathematics: States of Art. Flores quadrivii – Studies in Honor of Christoph Scriba. San Diego, Boston u.a. 1995, 317–341.

Reichenbach, Maria Cecilia von: Richard Gans: The First Quantum Physicist in Latin America. In: Physics in Perspective 11 (2009), 302–317.

Reid, Constance: Hilbert. New York, NY 1996 (erste Auflage 1970).

Reiff, Richard: Elasticität und Electrizität. Freiburg 1893.

Reiff, Richard/Sommerfeld, Arnold: Standpunkt der Fernwirkung. Die Elementargesetze. In: Enzyklopädie der mathematischen Wissenschaften, V:2 (1904), 3–62.

Reinbothe, Roswitha: Deutsch als internationale Wissenschaftssprache und der Boykott nach dem Ersten Weltkrieg. Frankfurt am Main 2006.

Renn, Jürgen (ed.): Einstein's Annalen Paper The Complete Collection 1901–1922. Weinheim 2005.

Renner, Fritz: Zur Theorie des atomaren lichtelektrischen Effektes. In: Annalen der Physik 29 (1937), 11–24.

Richter, Steffen: Die Kämpfe innerhalb der Physik in Deutschland nach dem Ersten Weltkrieg. In: Sudhoffs Archiv 57 (1973), 195–207.

Ricking, Klaus: Der Geist bewegt die Materie. 125 Jahre Geschichte der RWTH Aachen. Aachen 1995.

Riesenberger, Dieter: Das Deutsche Rote Kreuz. Eine Geschichte 1864–1990. Paderborn 2002.

Robertson, Peter: The Early Years: The Niels Bohr Institute, 1921–1930. Kopenhagen 1979.

Rohmer, Gustav (ed.): Die Zwanglose Gesellschaft in München 1837–1937. München 1937.

Rowe, David E.:Felix Klein's "Erlanger Antrittsrede". A Transcription with English Translation and Commentary. In: Historia Mathematica 12 (1985), 123–141.

Rowe, David E.: Felix Klein, David Hilbert, and the Göttingen Mathematical Tradition. In: Osiris 5 (1989), 186–213.

Rowe, David E.: Felix Klein as Wissenschaftspolitiker. In: Umberto Bottazzini/Amy Dahan (ed.): Changing Images in Mathematics: From the French Revolution to the New Millennium, London 2001, 69–92.

Saldern, Adelheid von: Göttingen im Kaiserreich. In: Thadden, Rudolf von/Trittel, Günter J. (ed.): Göttingen. Geschichte einer Universitätsstadt. Band 3: Von der preußischen Mittelstadt zur südniedersächsischen Großstadt 1866–1989. Göttingen 1999, 5–62.

Sanchez-Ron, José Manuel: International Relations in Spanish Physics from 1900 to the Cold War. In: Historical Studies in Physical Sciences 33:1 (2002), 3–31.

Saunders, Frederick A.: Some Aspects of Modern Spectroscopy. In: Science 59 (1924), 47–53.

Sauter, Fritz: Bemerkungen zur Schwingungstheorie dünner elastischer Platten. In: Zeitschrift für Naturforschung 3a (1948), 548–552.

Sauter, Fritz (ed.): Arnold Sommerfeld. Gesammelte Schriften. 4 vols. Braunschweig 1968 (abbreviated as ASGS).

Scherzer, Otto: Über die Ausstrahlung bei der Bremsung von Protonen und schnellen Elektronen. In: Annalen der Physik 13 (1932), 137–160.

Scherzer, Otto: Physik im totalitären Staat. In: Flitner, Andreas (ed.): Deutsches Geistesleben und Nationalsozialismus. Tübingen 1965, 47–58.

Schilling, Friedrich: Über darstellende Geometrie. In: Felix Klein/Eduard Riecke: Über Angewandte Mathematik und Physik in ihrer Bedeutung für den Unterricht an den höheren Schulen. Nebst Erläuterung der bezüglichen Göttinger Universitätseinrichtungen. Leipzig 1900, 42–56.

Schirrmacher, Arne: Das leere Atom: Instrumente, Experimente und Vorstellungen zur Atomstruktur um 1903. In: Hashagen, Ulf/Blumtritt, Oskar/Trischler, Helmuth (ed.): Circa 1903. Artefakte in der Gründungszeit des Deutschen Museums. München 2003, 127–152.

Schirrmacher, Arne: Philipp Lenard: Erinnerungen eines Naturforscher Kritische annotierte Ausgabe des Originaltyposkriptes von 1931/1943. Berlin 2010.

Schirrmacher, Arne: Ein physikalisches Konzil. Wie die Solvay-Konferenz und das Solvay-Institut vor hundert Jahren nicht nur der Quantentheorie zum Durchbruch verhalfen. In: Physikjournal 11:1 (2012), 39–42.

Schlote, Karl-Heinz: Zu den Wechselbeziehungen zwischen Mathematik und Physik an der Universität Leipzig in der Zeit von 1830 bis 1904/05. Band 63:1, Abhandlungen der Sächsischen Akademie der Wissenschaften zu Leipzig, Mathematisch-naturwissenschaftliche Klasse. Leipzig, 2004.

Schmidt, Rainer F.: Rudolf Heß – Botengang eines Toren? Der Flug von Rudolf Heß nach Großbritannien vom 10. Mai 1941. Düsseldorf 1997.

Schmidt-Böcking, Horst/Reich, Karin: Otto Stern. Frankfurt am Main 2011.

Schmitz, Norbert: Adolf Kratzer (1893–1983). Münster 2011.

Schröder-Gudehus, Brigitte: Deutsche Wissenschaft und internationale Zusammenarbeit 1914–1928. Ein Beitrag zum Studium kultureller Beziehungen in politischen Krisenzeiten. Genf 1966.

Schröder-Gudehus, Brigitte: Internationale Wissenschaftsbeziehungen und auswärtige Kulturpolitik, 1919–1933. Vom Boykott und Gegen-Boykott zu ihrer Wiederaufnahme. In: Vierhaus, Rudolf/vom Brocke, Bernhard (ed.): Forschung im Spannungsfeld von Politik und Gesellschaft. Geschichte und Struktur der Kaiser-Wilhelm-/Max-Planck-Gesellschaft aus Anlaß ihres 75jährigen Bestehens. Stuttgart 1990, 858–885.

Schreiber, Maximilian: Walther Wüst. Dekan und Rektor der Universität München, 1935–1945. München 2008.

Schulz, Karl: Theodor Liebisch †. In: Centralblatt für Mineralogie, Geologie und Paläologie (1922), 417–434.

Schwarzschild, Karl: Zur Elektrodynamik I-III. In: Nachrichten von der Königl. Gesellschaft der Wissenschaften zu Göttingen. Mathematisch-physikalische Klasse 1903, 126–131, 132–141, 245–278.

Schwarzschild, Karl: Bemerkung zur Aufspaltung der Spektrallinien im elektrischen Feld. In: Verhandlungen der Deutschen Physikalischen Gesellschaft 16 (1914), 20–24.

Schwarzschild, Karl: Über die maximale Aufspaltung beim Zeemaneffekt. In: Verhandlungen der Deutschen Physikalischen Gesellschaft 16 (1914), 24–40.

Schwarzschild, Karl: Zur Quantenhypothese. In: Sitzungsberichte der Preußischen Akademie der Wissenschaften in Berlin (1916), 548–568.

Schweber, Samuel Silvan: The Empiricist Temper Regnant: Theoretical Physics in the United States 1920–1950. In: Historical Studies in the Physical and Biological Sciences 17:1 (1986), 55–98.

Schweber, Samuel Silvan: QED and the Men Who Made It: Dyson, Feynman, Schwinger, and Tomonaga. Princeton, NJ 1994.

Schweidler, Egon von: Zur experimentellen Entscheidung der Frage nach der Natur der γ-Strahlen. In: Physikalische Zeitschrift 11 (1910), 225–227, 614–619.

Schweizer Physikalische Gesellschaft (ed.): La théorie des électrons dans les métaux: Conférences internationales des sciences mathématiques Genève, 15–18 octobre 1934. In: Helvetica Physica Acta 7 (1934), Supplementum.

Seth, Suman: Quantum Theory and the Electromagnetic World-View. In: Historical Studies in the Physical Sciences 35:1 (2004), 67–93.

Seth, Suman: Crisis and the Construction of Modern Theoretical Physics. In: British Journal for the History of Science 40 (2007), 25–51.

Seth, Suman: Crafting the Quantum. Arnold Sommerfeld and the Practice of Theory, 1890–1926. Cambridge, MA 2010.

Siegmund-Schultze, Reinhard: Mathematicians Fleeing from Nazi Germany: Individual Fates and Global Impact. Princeton, NJ 2009.

Singh, Rajinder: Arnold Sommerfeld – The Supporter of Indian Physics in Germany. In: Current Science 81 (2001), 1489–1494.

Singh, Rajinder/Riess, Falk: Seventy Years Ago – the Discovery of the Raman Effect As Seen From German Physicists. In: Current Science 74 (1998), 1112–1115.

Smyth, Henry DeWolf: A General Account of the Development of Methods of Using Atomic Energy for Military Purposes Under the Auspices of the United States Government. London 1945.

Söderqvist, Thomas: Existential Projects and Existential Choice in Science: Science Biography as an Edifying Genre. In: Shortland, Michael/Yeo, Richard (ed.): Telling Lives in Science: Essays on Scientific Biography. Cambridge, MA 1996, 45–84.

Söderqvist, Thomas (ed.): The History and Poetics of Scientific Biography. Aldershot 2007.

Sommerfeld, Arnold: Über die Genossenschaft freiwilliger Krankenpfleger im Kriege. In: Burschenschaftliche Blätter 4 (1890), 220–223.

Sommerfeld, Arnold: Eine Maschine zur Entwickelung einer willkürlichen Function in Fourier'sche Reihen. In: Schriften der Physikalisch-ökonomischen Gesellschaft zu Königsberg i. Pr. 32 (1891), 28–33.

Sommerfeld, Arnold: Die Willkürlichen Functionen in der Mathematischen Physik. Doctoral dissertation, Universität Königsberg. Königsberg 1891. Also in ASGS I, 1–76.

Sommerfeld, Arnold: Mechanische Darstellung der electromagnetischen Erscheinungen in ruhenden Körpern. In: Annalen der Physik 46 (1892), 139–151.

Sommerfeld, Arnold: Zur mathematischen Theorie der Beugungserscheinungen. In: Nachrichten von der Königl. Gesellschaft der Wissenschaften zu Göttingen. Mathematisch-physikalische Klasse (1894), 338–342.

Sommerfeld, Arnold: Zur analytischen Theorie der Wärmeleitung. In: Mathematische Annalen 45 (1894), 263–277.

Sommerfeld, Arnold: Diffractionsprobleme in exacter Behandlung. In: Verhandlungen der Gesellschaft Deutscher Naturforscher und Ärzte 67 (1895), 34–35.

Sommerfeld, Arnold: Diffractionsprobleme in exacter Behandlung. In: Jahresbericht der Deutschen Mathematiker-Vereinigung 4 (1895), 172–174.

Sommerfeld, Arnold: Mathematische Theorie der Diffraction. In: Mathematische Annalen 47 (1896), 317–374.

Sommerfeld, Arnold: Über verzweigte Potentiale im Raume. In: Proceedings of the London Mathematical Society 28 (1897), 395–429.

Sommerfeld, Arnold: Geometrischer Beweis des Dupin'schen Theorems und seiner Umkehrung. In: Jahresbericht der Deutschen Mathematiker-Vereinigung 6 (1897), 123–128.

Sommerfeld, Arnold: Mathematische Annalen. Generalregister zu den Bänden 1–50. Leipzig 1898.

Sommerfeld, Arnold: Über das Problem der elektrodynamischen Drahtwellen. In: Jahresbericht der Deutschen Mathematiker-Vereinigung 7 (1898), 112–113.

Sommerfeld, Arnold: Über einige mathematische Aufgaben aus der Elektrodynamik. In: Verhandlungen der Gesellschaft Deutscher Naturforscher und Ärzte 70 (1898), 14.

Sommerfeld, Arnold: Über die numerische Auflösung transcendenter Gleichungen durch successive Approximationen. In: Nachrichten von der Königl. Gesellschaft der Wissenschaften zu Göttingen. Mathematisch-physikalische Klasse (1898), 360–369. (Vorgelegt von D. Hilbert in der Sitzung vom 10. Dezember 1898.)

Sommerfeld, Arnold: Über die Fortpflanzung elektrodynamischer Wellen längs eines Drahtes. In: Annalen der Physik 67 (1899), 233–290.

Sommerfeld, Arnold: Theoretisches über die Beugung der Röntgenstrahlen. (Vorläufige Mitteilung.). In: Physikalische Zeitschrift 1 (1900), 105–111.

Sommerfeld, Arnold: Neuere Untersuchungen zur Hydraulik. In: Verhandlungen der Gesellschaft Deutscher Naturforscher und Ärzte 72 (1900), 56.

Sommerfeld, Arnold: Beiträge zum dynamischen Ausbau der Festigkeitslehre. In: Zeitschrift des Vereines deutscher Ingenieure 46 (1902), 391–394.

Sommerfeld, Arnold: Zur Theorie der Eisenbahnbremsen. In: Denkschrift der Königlich Technischen Hochschule Aachen 1902, 58–71.

Sommerfeld, Arnold: Randwertaufgaben in der Theorie der partiellen Differentialgleichungen. In: Enzyklopädie der mathematischen Wissenschaften 2:7c (1904), 504–570 (abgeschlossen im April 1900).

Sommerfeld, Arnold: Das Pendeln parallel geschalteter Wechselstrommaschinen. In: Elektrotechnische Zeitschrift 25 (1904), 273–276, 291–295, 469.

Sommerfeld, Arnold: Zur hydrodynamischen Theorie der Schmiermittelreibung. In: Zeitschrift für Mathematik und Physik 50 (1904), 97–155.

Sommerfeld, Arnold: Bezeichnung und Benennung der elektromagnetischen Größen in der Enzyklopädie der mathematischen Wissenschaften V. In: Physikalische Zeitschrift 5 (1904), 467–470.

Sommerfeld, Arnold: Zur Elektronentheorie. I. Allgemeine Untersuchung des Feldes eines beliebig bewegten Elektrons. In: Nachrichten von der Königl. Gesellschaft der Wissenschaften zu Göttingen. Mathematisch-physikalische Klasse (1904), 99–130.

Sommerfeld, Arnold: Zur Elektronentheorie. II. Grundlagen für eine allgemeine Dynamik des Elektrons. In: Nachrichten von der Königl. Gesellschaft der Wissenschaften zu Göttingen. Mathematisch-physikalische Klasse (1904), 363–439.

Sommerfeld, Arnold: Vereenvoudigte Afleiding von Het Veldan, en de Krachten Werkende op een Elektren bij Willekeurige Beweging. In: Akad. Versl. Amsterdam 13 (1904), 431–452.

Sommerfeld, Arnold: Eine einfache Vorrichtung zur Veranschaulichung des Knickungsvorganges. In: Zeitschrift des Vereines deutscher Ingenieure 49 (1905), 1320–1323.

Sommerfeld, Arnold: Lissajous-Figuren und Resonanzwirkungen bei schwingenden Schraubenfedern; ihre Verwertung zur Bestimmung des Poissonschen Verhältnisses. In: Festschrift Adolph Wüllner gewidmet zum siebzigsten Geburtstage 13. Juni 1905 von der Königl. Technischen Hochschule zu Aachen, ihren früheren und jetzigen Mitgliedern. Leipzig 1905, 162–193.

Sommerfeld, Arnold: Zur Elektronentheorie. III. Ueber Lichtgeschwindigkeits- und Ueberlichtgeschwindigkeits-Elektronen. In: Nachrichten von der Königl. Gesellschaft der Wissenschaften zu Göttingen. Mathematisch-physikalische Klasse 1905, 201–235.

Sommerfeld, Arnold: Über die Mechanik der Elektronen. In: Verhandlungen des dritten Internationalen Mathematiker-Kongresses in Heidelberg vom 8. bis 13. August 1904. Leipzig 1905, 417–432.

Sommerfeld, Arnold: Die Knicksicherheit der Stege von Walzwerkprofilen. In: Zeitschrift des Vereines deutscher Ingenieure 50 (1906), 1104–1107.

Sommerfeld, Arnold: Bemerkungen zur Elektronentheorie. In: Jahresbericht der Deutschen Mathematiker-Vereinigung 15 (1906), 51–55.

Sommerfeld, Arnold: Nachtrag und Berichtigung zu der Abhandlung: Über die Knicksicherheit der Stege von Walzwerkprofilen. In: Zeitschrift für Mathematik und Physik 54 (1907), 318–324.

Sommerfeld, Arnold: Über die Bewegung der Elektronen. In: Sitzungsberichte der mathematisch-physikalischen Klasse der K. B. Akademie der Wissenschaften zu München (1907), 155–171

Sommerfeld, Arnold: Zur Diskussion über die Elektronentheorie. In: Sitzungsberichte der mathematisch-physikalischen Klasse der K. B. Akademie der Wissenschaften zu München (1907), 281.

Sommerfeld, Arnold: Ein Einwand gegen die Relativtheorie der Elektrodynamik und seine Beseitigung. In: Physikalische Zeitschrift 8 (1907), 841–842.

Sommerfeld, Arnold: Ein Beitrag zur hydrodynamischen Erklärung der turbulenten Flüssigkeitsbewegung. In: Atti del IV Congresso Internazionale dei Matematici (Roma, 6–11 Aprile 1908). Band 3 (1909), 116–124. http://mathunion.org/ICM/ICM1908.3/ICM1908.3.ocr.pdf (19.11.2012).

Sommerfeld, Arnold: Über die Ausbreitung der Wellen in der drahtlosen Telegraphie. In: Annalen der Physik, 28 (1909b), 665–736.

Sommerfeld, Arnold: Über die Ausbreitung der Wellen in der drahtlosen Telegraphie. In: Sitzungsberichte der mathematisch-physikalischen Klasse der K. B. Akademie der Wissenschaften zu München (1909), 1–19.

Sommerfeld, Arnold: Über die Verteilung der Intensität bei der Emission von Röntgenstrahlen. In: Physikalische Zeitschrift 10 (1909), 969–976.

Sommerfeld, Arnold: Zur Relativitätstheorie. I. Vierdimensionale Vektoralgebra. In: Annalen der Physik 32 (1910), 749–776.

Sommerfeld, Arnold: Zur Relativitätstheorie. II. Vierdimensionale Vektoranalysis. In: Annalen der Physik 33 (1910), 649–689.

Sommerfeld, Arnold: Ausbreitung der Wellen in der drahtlosen Telegraphie. Einfluß der Bodenbeschaffenheit auf gerichtete und ungerichtete Wellenzüge. In: Jahrbuch der drahtlosen Telegraphie 4 (1910), 157–176.

Sommerfeld, Arnold: Die Greensche Funktion der Schwingungsgleichung für ein beliebiges Gebiet. In: Physikalische Zeitschrift 11 (1910), 1057–1066.

Sommerfeld, Arnold: Über die Struktur der γ-Strahlen. In: Sitzungsberichte der mathematisch-physikalischen Klasse der K. B. Akademie der Wissenschaften zu München (1911), 1–60. (Vorgetragen in der Sitzung am 7. Januar 1911.)

Sommerfeld, Arnold: Das Plancksche Wirkungsquantum und seine allgemeine Bedeutung für die Molekularphysik. In: Verhandlungen der Gesellschaft Deutscher Naturforscher und Ärzte 83 (1911a), 31–50.
Sommerfeld, Arnold: Das Plancksche Wirkungsquantum und seine allgemeine Bedeutung für die Molekularphysik. In: Physikalische Zeitschrift 12 (1911b), 1057–1069.
Sommerfeld, Arnold: Sur l'application de la théorie de l'élément d'action aux phénomènes moléculaires non périodique In: Langevin, Paul/de Broglie, Maurice: La théorie du rayonnement et les quanta. Brüssel 1912, 313–392.
Sommerfeld, Arnold: Über die Beugung der Röntgenstrahlung. In: Annalen der Physik 38 (1912), 473–506.
Sommerfeld, Arnold: Unsere gegenwärtigen Anschauungen über Röntgenstrahlung. In: Die Naturwissenschaften 1 (1913), 705–713. (Vortrag bei der Versammlung des Vereins zur Förderung des Unterrichtes in der Mathematik und den Naturwissenschaften, München. Gehalten Pfingsten 1913.)
Sommerfeld, Arnold: Der Zeemaneffekt eines anisotrop gebundenen Elektrons und die Beobachtungen von Paschen-Back. In: Annalen der Physik 40 (1913), 748–774.
Sommerfeld, Arnold: Die Bedeutung des Wirkungsquantums für unperiodische Molekularprozesse in der Physik. In: Eucken, Die Theorie der Strahlung und der Quanten, 1914, 252–317.
Sommerfeld, Arnold: Probleme der freien Weglänge. In: Mathematische Vorlesungen an der Universität Göttingen 6 (1914), 123–166.
Sommerfeld, Arnold: Zur Voigtschen Theorie des Zeeman-Effekte In: Nachrichten von der Königlichen Gesellschaft der Wissenschaften zu Göttingen. Mathematisch-physikalische Klasse (1914), 207–229. (Vorgelegt von W. Voigt in der Sitzung vom 7. März 1914.)
Sommerfeld, Arnold: Zur Theorie der Balmerschen Serie. In: Sitzungsberichte der mathematisch-physikalischen Klasse der K. B. Akademie der Wissenschaften zu München (1915), 425–458.
Sommerfeld, Arnold: Die Feinstruktur der Wasserstoff- und der Wasserstoff-ähnlichen Linien. In: Sitzungsberichte der mathematisch-physikalischen Klasse der K. B. Akademie der Wissenschaften zu München (1915), 459–500.
Sommerfeld, Arnold: Zur Quantentheorie der Spektrallinien. In: Annalen der Physik 51 (1916), 1–94, 125–167.
Sommerfeld, Arnold: Zur Theorie des Zeemaneffektes der Wasserstofflinien, mit einem Anhang über den Starkeffekt. In: Physikalische Zeitschrift 17 (1916), 491–507.
Sommerfeld, Arnold: Besuch an der Universität Gent. Monatshefte für den naturwissenschaftlichen Unterricht aller Schulgattungen (1918), 57–61.
Sommerfeld, Arnold: Ein Besuch in Gent. In: Süddeutsche Monatshefte (1918), 44–46.
Sommerfeld, Arnold: Die Entwicklung der Physik in Deutschland seit H. Hertz. In: Deutsche Revue (1918), 122–132.
Sommerfeld, Arnold: Über die Feinstruktur der Kβ-Linie. In: Sitzungsberichte der mathematisch-physikalischen Klasse der K. B. Akademie der Wissenschaften zu München (1918), 367–372. (Vorgetragen am 1. Juni 1918.)
Sommerfeld, Arnold: Atombau und Spektrallinien. Braunschweig 1919.
Sommerfeld, Arnold: Ein Zahlenmysterium in der Theorie des Zeemaneffekte In: Die Naturwissenschaften 8 (1920), 61–64.
Sommerfeld, Arnold: Allgemeine spektroskopische Gesetze, insbesondere ein magnetooptischer Zerlegungssatz. In: Annalen der Physik 63 (1920), 221–263.
Sommerfeld, Arnold: Sur les photogrammes quaternaires et ternaires de la blende et le spectre du rayonnement de Röntgen. In: Institut International de Physique Solvay (ed.): La structure de

la matiere. Rapports et discussions du conseil de physique, tenu a Bruxelles du 27 au 31 octobre 1913. Paris 1921, 125–140.

Sommerfeld, Arnold: Atombau und Spektrallinien. 2. Auflage. Braunschweig 1921.

Sommerfeld, Arnold: Ursachen - Wirkungen! Die Lusitania-Medaille. In: Münchner Neueste Nachrichten 24. Juni 1921.

Sommerfeld, Arnold: Quantentheoretische Umdeutung der Voigt'schen Theorie des anomalen Zeeman-Effektes vom D-Linientypus. In: Zeitschrift für Physik 8 (1922), 257–297.

Sommerfeld, Arnold: Atombau und Spektrallinien. 3. Auflage. Braunschweig 1922.

Sommerfeld, Arnold: Über Linienstrukturen im Spektrum von Mangan. In: Verhandlungen der Deutschen Physikalischen Gesellschaft 3 (1922), 45. (Vortrag in Göttingen am 14. Juni 1922.)

Sommerfeld, Arnold: Über die Deutung verwickelter Spektren (Mangan, Chrom usw.) nach der Methode der inneren Quantenzahlen. In: Annalen der Physik 70 (1923), 32–62.

Sommerfeld, Arnold: Spektroskopische Magnetonenzahlen. In: Physikalische Zeitschrift 24 (1923), 360–364.

Sommerfeld, Arnold: Zur Theorie des Magnetons. In: Zeitschrift für Physik 19 (1923), 221–229.

Sommerfeld, Arnold: Atombau und Spektrallinien. 4. Auflage. Braunschweig 1924.

Sommerfeld, Arnold: Grundlagen der Quantentheorie und des Bohrschen Atommodells. In: Die Naturwissenschaften 12 (1924), 1047–1049.

Arnold Sommerfeld: Die Erforschung des Atoms. In: Strahlentherapie, Bd. 16, 1924, 873–882.

Sommerfeld, Arnold: Das Institut für theoretische Physik. In: Müller, Karl Alexander von (ed.): Die wissenschaftlichen Anstalten der Ludwig-Maximilians-Universität zu München. Chronik zur Jahrhundertfeier im Auftrag des akademischen Senats. München 1926, 290–292.

Sommerfeld, Arnold: Three Lectures on Atomic Physics. London 1926. (Übersetzung von Henry L. Brose.)

Sommerfeld, Arnold: Zur Elektronentheorie der Metalle. In: Die Naturwissenschaften 15 (1927), 825–832.

Sommerfeld, Arnold: Elektronentheorie der Metalle und des Voltaeffektes nach der Fermischen Statistik. In: Atti Congr. Intern. dei Fisici Como-Pavia-Roma, II (1927), 449–473.

Sommerfeld, Arnold: Warum ich Berlin abgelehnt habe? In: Süddeutsche Sonntagspost (1927), 3.

Sommerfeld, Arnold: Zur Elektronentheorie der Metalle auf Grund der Fermi'schen Statistik. 1. Allgemeines, Strömungs- und Austrittsvorgänge. In: Zeitschrift für Physik 47 (1928), 1–32.

Sommerfeld, Arnold: Zur Elektronentheorie der Metalle auf Grund der Fermi'schen Statistik. 2. Thermoelektrische, galvano-magnetische und thermomagnetische Vorgänge. In: Zeitschrift für Physik 47 (1928), 43–60.

Sommerfeld, Arnold: Zur Frage der Bedeutung der Atommodelle. In: Zeitschrift für Elektrochemie und angewandte physikalische Chemie 34 (1928), 426–430.

Sommerfeld, Arnold: Atombau und Spektrallinien. Wellenmechanischer Ergänzungsband. Braunschweig 1929.

Sommerfeld, Arnold: Indische Reiseeindrücke. In: Zeitwende 5 (1929), 289–298.

Sommerfeld, Arnold: Über die Entwicklung der Atomphysik in den letzten zwanzig Jahren. In: Tung-Chi Medizinische Monatsschrift (1929a), 75–88.

Sommerfeld, Arnold: Über die Entwicklung der Atomphysik in den letzten zwanzig Jahren. In: Japanisch-Deutscher Geistesaustausch 2 (1929b), 1–22.

Sommerfeld, Arnold: On the Production of X-Radiation, According to Wave Mechanics. In: Journal of the Franklin Institute 208 (1929a), 571–588.

Sommerfeld, Arnold: About the Production of the Continuous X-ray Spectrum. In: Proceedings of the National Academy of Sciences 15 (1929b), 393–400.

Sommerfeld, Arnold: Die Physik in Japan, Indien und Amerika. In: Verhandlungen der Deutschen Physikalischen Gesellschaft 10 (1929), 21–22.

Sommerfeld, Arnold: Bericht über Besuche bei den Akademien von Tokio und Washington. In: Sitzungsberichte der mathematisch-naturwissenschaftlichen Abteilung der Bayerischen Akademie der Wissenschaften zu München (1929), 11. (Sitzung am 6. Juli 1929.)

Sommerfeld, Arnold: Einige grundsätzliche Bemerkungen zur Wellenmechanik. In: Physikalische Zeitschrift 30 (1929), 866–871.

Sommerfeld, Arnold: Über die Elektronentheorie der Metalle und die Natur des Elektrons. In: Monatshefte für Mathematik und Physik 7 (1930), 183–198.

Sommerfeld, Arnold: Erwiderung auf die Angriffe von Herrn J. Stark. In: Annalen der Physik 7 (1930), 889–891.

Sommerfeld, Arnold: Über Anschaulichkeit in der modernen Physik. In: Unterrichtsblätter für Mathematik und Naturwissenschaften 36 (1930a), 161–167.

Sommerfeld, Arnold: Über Anschaulichkeit in der modernen Physik. In: Scientia 48 (1930b), 81–86.

Sommerfeld, Arnold: Über die Beugung und Bremsung der Elektronen. In: Annalen der Physik 11 (1931), 257–330.

Sommerfeld, Arnold: Sur Quelques Problèmes de la Méchanique Ondulatoire. In: Annales de l'Institut Poincaré (1931), 1–24.

Sommerfeld, Arnold: Zur Theorie des Ramsauer-Effektes. In: Zeitschrift für angewandte Chemie 44 (1931), 611.

Sommerfeld, Arnold: Magnetismus und Spektroskopie. In: Institut Internationale de Physique de Solvay (ed.): Le magnetisme. Rapports et discussions du sixième Conseil de Physique tenu à Bruxelles du 20 au 25 octobre 1930 sous les auspices de l'Institut International de Physique de Solvay. Paris 1932.

Sommerfeld, Arnold: Atombau und Spektrallinien. Band 2 (= 2. Auflage des Wellenmechanischen Ergänzungsbandes). Braunschweig 1939.

Sommerfeld, Arnold: Zwanzig Jahre spektroskopischer Theorie in München. In: Scientia (1942), 123–130.

Sommerfeld, Arnold: Die Quantenstatistik und das Problem des Heliums II. In: Berichte der Deutschen Chemischen Gesellschaft 75 (1942), 1988–1996.

Sommerfeld, Arnold: David Hilbert. Nachruf. In: Jahrbuch der Akademie der Wissenschaften in Göttingen, 1943, 87–92.

Sommerfeld, Arnold: Vorlesungen über theoretische Physik. Band 1–6. Leipzig 1943–1952.

Sommerfeld, Arnold: Die frei schwingende Kolbenmembran. In: Annalen der Physik 42 (1943), 389–420.

Sommerfeld, Arnold: Ludwig Boltzmann zum Gedächtni Zur hundertsten Wiederkehr seines Geburtstages (20. 2. 1944). In: Wiener Chemiker-Zeitung 47 (1944), 25–28.

Sommerfeld, Arnold: Vorlesungen über theoretische Physik. Band 1: Mechanik. 2. Auflage, Leipzig, 1944.

Sommerfeld, Arnold: Atomphysik in Amerika. In: Die Neue Zeitung, 18. November 1945.

Sommerfeld, Arnold: Die Quantenstatistik und das Problem des He II. In: Zeitschrift für Naturforschung 1 (1946), 120.

Sommerfeld, Arnold: Vorlesungen über theoretische Physik. Band 6: Partielle Differentialgleichungen der Physik. Leipzig, 1948.

Sommerfeld, Arnold: Philosophie und Physik seit 1900. In: Naturwissenschaftliche Rundschau 1 (1948), 97–100.

Sommerfeld, Arnold: Berichtigungen und Ergänzungen zu der Arbeit: Die frei schwingende Kolbenmembran. In: Annalen der Physik 2 (1948), 85–86.

Sommerfeld, Arnold: Vorlesungen über theoretische Physik. Band 3: Elektrodynamik. Leipzig 1949.

Sommerfeld, Arnold: Zum hundertsten Geburtstag von Felix Klein. In: Die Naturwissenschaften 36 (1949), 289–291.

Sommerfeld, Arnold: Some Reminiscences of My Teaching Career. In: American Journal of Physics 17 (1949), 315–316.

Sommerfeld, Arnold: Albert Einstein zum 70. Geburtstag. In: Deutsche Beiträge. Eine Zweimonatsschrift 3 (1949), 141–146.

Sommerfeld, Arnold: Aus den Lehrjahren von Walter Rogowski. In: Archiv für Elektrotechnik 40 (1950), 3.

Sommerfeld, Arnold: Überreichung der Planck-Medaille für Peter Debye durch A. Sommerfeld. In: Physikalische Blätter 6 (1950), 509–512.

Sommerfeld, Arnold: Vorlesungen über Theoretische Physik Band 6: Partielle Differentialgleichungen der Physik. Thun 1992.

Sommerfeld, Arnold:. Gesammelte Schriften. 4 vols. (edited by F. Sauter and the Bayerische Akademie der Wissenschaften). Braunschweig 1968 (abbreviated as ASGS).

Arnold Sommerfeld., Arnold: Wissenschaftlicher Briefwechsel. Band 1: 1892–1918; Band 2: 1919–1951. Herausgegeben von Michael Eckert und Karl Märker. München, Berlin, Diepholz 2000 und 2004 (abbreviated as ASWB I and ASWB II).

Sommerfeld, Arnold/Bethe, Hans: Elektronentheorie der Metalle. In: Handbuch der Physik, 24:2 (1933), 333–622.

Sommerfeld, Arnold/Bopp, Fritz: Zum Problem der Maxwellschen Spannungen. In: Annalen der Physik 8 (1950), 41–45.

Sommerfeld, Arnold/Frank, Nathaniel H.: Statistical Theory of Thermoelectric Galvano- and Thermomagnetic Phenomena in Metals. In: Reviews of Modern Physics 3 (1931), 1–42.

Sommerfeld, Arnold/Heisenberg, Werner: Die Intensität der Mehrfachlinien und ihrer Zeeman-Komponenten. In: Zeitschrift für Physik 11 (1922), 131–154.

Sommerfeld, Arnold/Kossel, Walther: Auswahlprinzip und Verschiebungssatz bei Serienspektren. In: Verhandlungen der Deutschen Physikalischen Gesellschaft 21 (1919), 240–259.

Sommerfeld, Arnold/Maue, August Wilhelm: Verfahren zur näherungsweisen Anpassung einer Lösung der Schrödinger- an die Dirac-Gleichung. In: Annalen der Physik 22 (1935), 629–642.

Sommerfeld, Arnold/Ramberg, Edward: Das Drehmoment eines permanenten Magneten im Felde eines permeablen Mediums. In: Annalen der Physik 8 (1950), 46–54.

Sommerfeld, Arnold/Renner, Fritz: Strahlungsenergie und Erdabsorption bei Dipolantennen. In: Annalen der Physik 59 (1942), 168–173.

Sommerfeld, Franz: Über die Familie der Quarze. In: Schriften der Physikalisch-ökonomischen Gesellschaft zu Königsberg i. Pr. 41 (1900), 6–9. (Erroneously attributed to Arnold Sommerfeld in abgedruckt in ASGS IV, 483–487).

Sopka, Katherine Russell: Quantum Physics in America: The Years Through 1935. Los Angeles, CA 1988.

Staley, Richard: Einstein's Generation: The Origins of the Relativity Revolution. Chicago, IL 2008.

Stark, Johannes: Die Axialität der Lichtemission und Atomstruktur. VII: Zur physikalischen Kritik eines Sommerfeldschen Theorems. In: Annalen der Physik 4 (1930), 710–724.

Stark, Johannes: Die Axialität der Lichtemission und Atomstruktur. IX. Die Axialitat der Valenzfelder des Kohlenstoff- und Stickstoffatoms. In: Annalen der Physik 6 (1930), 663–680.

Stark, Johannes: Über den Dogmatismus moderner Theorien in der Physik. In: Unterrichtsblätter für Mathematik und Naturwissenschaften 36 (1930), 305–309.

Stark, Johannes/Müller, Wilhelm: Jüdische und Deutsche Physik. Leipzig 1941.

Steen, Andreas: Deutsch-chinesische Beziehungen 1911–1927: vom Kolonialismus zur 'Gleichberechtigung': eine Quellensammlung. Berlin 2006.

Stieda, Ludwig: Gedächtnisrede auf den am 15. August 1889 verstorbenen Präsidenten der Gesellschaft, Geheimen Sanitätsrat Dr. Wilhelm Schiefferdecker. In: Schriften der physikalisch-ökonomischen Gesellschaft zu Königsberg in Pr. 30 (1889), 50–62.

Stieda, Ludwig: Zur Geschichte der physikalisch-ökonomischen Gesellschaft. Festrede, gehalten am 22. Februar 1890. In: Schriften der physikalisch-ökonomischen Gesellschaft zu Königsberg in Pr. 30 (1890), 38–82.

Stöltzner, Michael/Uebel, Thomas (ed.): Wiener Krei Texte zur wissenschaftlichen Weltauffassung von Rudolf Carnap, Otto Neurath, Moritz Schlick, Philipp Frank, Hans Hahn, Karl Menger, Edgar Zilsel und Gustav Bergmann. Hamburg 2006.

Stuewer, Roger H.: The Compton Effect. Turning Point in Physics. New York, NY 1975.

Sweetnam, George Kean: The Command of Light: Rowland's School of Physics and the Spectrum. Philadelphia 2000.

Swinne, Edgar: Richard Gans. Hochschullehrer in Deutschland und Argentinien. Berlin 1992.

Szabó, Anikó: Vertreibung, Rückkehr, Wiedergutmachung Göttinger Hochschullehrer im Schatten des Nationalsozialismus. Göttingen 2000.

Szöllösi-Janze, Margit: Fritz Haber. 1868–1934. Eine Biographie. München 1998.

Margit Szöllösi-Janze: Lebens-Geschichte – Wissenschafts-Geschichte. Vom Nutzen der Biographie für Geschichtswissenschaft und Wissenschaftsgeschichte. In: Berichte zur Wissenschaftsgeschichte 23 (2000), 17–35.

Tazzioli, Rossana: Green's Function in Some Contributions of 19th Century Mathematicians. In: Historia Mathematica 28 (2001), 232–252.

Tent, James F.: Mission on the Rhine. „Reeducation" and Denazification in American-Occupied Germany. Chicago 1982.

Thomson, William: Motion of a Viscous Liquid; Equilibrium or Motion of an Elastic Solid; Equilibrium or Motion of an Ideal Substance Called for Brevity Ether; Mechanical Representation of Magnetic Force. In: William Thomson: Mathematical and Physical Paper Band 3: Elasticity, Heat Electro-Magnetism. London 1890, 436–465.

Tobies, Renate: Mathematik als Bestandteil der Kultur - Zur Geschichte des Unternehmens "Encyklopädie der mathematischen Wissenschaften mit Einschluss ihrer Anwendungen". In: Mitteilungen der Österreichischen Gesellschaft für Wissenschaftsgeschichte 14 (1994), 1–90.

Tobies, Renate: Felix Klein und der Verein zur Förderung des mathematischen und naturwissenschaftlichen Unterrichts. In: Der Mathematikunterricht 46 (2000), 22–40.

Tobies, Renate: The Development of Göttingen into the Prussian Centre of Mathematics and the Exact Sciences. In: Nicolaas Rupke (ed.): Göttingen and the Development of the Natural Sciences. Göttingen 2002, 116–142.

Toepell, Michael: Mathematiker und Mathematik an der Universität München. 500 Jahre Lehre und Forschung. München 1996.

Torkar, Gisela: Sommerfeld's Meeting With Raman in Calcutta During a World Tour, 1928–29. In: Journal of Raman Spectroscopy 17 (1986), 13–15.

Trischler, Helmuth: Die neue Räumlichkeit des Krieges: Wissenschaft und Technik im Ersten Weltkrieg. In: Berichte zur Wissenschaftsgeschichte 19 (1996), 95–103.

Turner, Laura E.: The Mittag-Leffler Theorem: The Origin, Evolution, and Reception of a Mathematical Result, 1876–1884. Master's thesis, Simon Fraser University, Vancouver, 2007. http://people.math.sfu.ca/~tarchi/turnermsc2007.pdf (5.1.2012).

Uhlenbeck, George/Goudsmit, Samuel: Ersetzung der Hypothese vom unmechanischen Zwang durch eine Forderung bezüglich des inneren Verhaltens jedes einzelnen Elektrons. In: Die Naturwissenschaften 13 (1925), 953–954.

Ullrich, Volker: Die Revolution von 1918/19. München 2009.

Ungern-Sternberg, Jürgen von/Ungern-Sternberg, Wolfgang von: Der Aufruf „An die Kulturwelt!" Das Manifest der 93 und die Anfänge der Kriegspropaganda im Ersten Weltkrieg. Mit einer Dokumentation. Stuttgart 1996.

Unsöld, Albrecht: Beiträge zur Quantenmechanik der Atome. In: Annalen der Physik 82 (1927), 355–393.

Vogel-Prandtl, Johanna: Ludwig Prandtl: ein Lebensbild ; Erinnerungen, Dokumente. Göttingen 2005.

Voigt, Woldemar: Über die anomalen Zeemaneffekte. In: Annalen der Physik 40 (1913), 368–380.

Voigt, Woldemar: Weiteres zum Ausbau der Kopplungstheorie der Zeemaneffekte. In: Annalen der Physik 41 (1913), 403–440.

Voigt, Woldemar: Die anomalen Zeemaneffekte der Spektrallinien vom D-Typus. In: Annalen der Physik 42 (1913), 210–230.

Volkmann, Paul: Beiträge zur Wertschätzung der Königsberger Erdthermometer-Station 1872–1892. Schriften der physikalisch-ökonomischen Gesellschaft zu Königsberg in Pr. 34 (1893), 54–61.

Volkmann, Paul: Franz Neumann. Ein Beitrag zur Geschichte deutscher Wissenschaft. Leipzig 1896.

Vorländer, Herwart: Die NSV. Darstellung und Dokumentation einer nationalsozialistischen Organisation. Boppard am Rhein 1988.

Wagner, Hans-Ulrich: Hans Leipelt und Marie-Luise Jahn - Studentischer Widerstand in der Zeit des Nationalsozialismus am Chemischen Staatslaboratorium der Universität München. München 2003.

Walcher, Wilhelm: Physikalische Gesellschaften im Umbruch. Zusammenschlüsse der Physiker in den Nachkriegsjahren bis zur Wiedergründung der Deutschen Physikalischen Gesellschaft. In: Physikalische Blätter 51 (1995), F107-F133.

Walter, Scott: Minkowski, Mathematicians, and the Mathematical Theory of Relativity. In: Goenner, Hubert/Renn, Jürgen/Ritter, Jim/Sauer, Tilmann (ed.): The Expanding Worlds of General Relativity. Boston, MA, Basel, Berlin 1999, 45–86.

Walter, Scott: Minkowski's Modern World. In: Petkov, Vesselin (ed.): Minkowski Spacetime: A Hundred Years Later. Dordrecht, Heidelberg, Berlin, New York, NY 2010, 43–61.

Warburg, Emil: Bemerkungen zu der Aufspaltung der Spektrallinien im elektrischen Feld. Verhandlungen der Deutschen Physikalischen Gesellschaft 15 (1913), 1259–1266.

Emil Warburg/Laue, Max von/Sommerfeld, Arnold/Einstein, Albert (ed.): Zu Plancks sechzigstem Geburtstag. Ansprachen, gehalten am 26. April 1918 in der Deutschen Physikalischen Gesellschaft. Karlsruhe 1918.

Warwick, Andrew: Masters of Theory: Cambridge and the Rise of Mathematical Physic Chicago, IL 2003.

Watson, Alexander/Porter, Patrick: Bereaved and Aggrieved: Combat Motivation and the Ideology of Sacrifice in the First World War. In: Historical Research 83 (2008), 146–164.

Wazeck, Milena: Einsteins Gegner. Die öffentliche Kontroverse um die Relativitätstheorie in den 1920er Jahren. Frankfurt am Main 2009.

Weizsäcker, Carl Friedrich von: Die Rotation kosmischer Gasmassen. In: Zeitschrift für Naturforschung 3a (1948), 524–539.

Welker, Heinrich: Ein wellenmechanisches Modell des Supraleiters. In: Zeitschrift für Naturforschung 3a (1948), 461–469.

Wengenroth, Ulrich: Zwischen Aufruhr und Diktatur. Die Technische Hochschule 1918–1945. In: Wengenroth, Ulrich (ed.): Die Technische Universität München. Annäherungen an ihre Geschichte. München 1993, 215–260.

Wentzel, Gregor: Zur Quantentheorie des Röntgenbremsspektrums. In: Zeitschrift für Physik 27 (1924), 257–284.

Wheaton, Bruce R.: The Tiger and the Shark. Empirical Roots of Wave-Particle Dualism. Cambridge, MA 1983.

Widmalm, Sven: Science and Neutrality: The Nobel Prizes of 1919 and Scientific Internationalism in Sweden. In: Minerva 33 (1995), 339–360.

Wiecki, Stefan: The Denazification of Munich University, 1945–1948. In: Kraus, Elisabeth (ed.): Die Universität München im Dritten Reich. Aufsätze. Teil II. München 2008, 519–569.

Wien, Wilhelm: Über die Energie der Kathodenstrahlen im Verhältnis zur Energie der Röntgen- und Sekundärstahlen. In: Festschrift Adolph Wüllner gewidmet zum siebzigsten Geburtstage 13. Juni 1905 von der Königl. Technischen Hochschule zu Aachen, ihren früheren und jetzigen Mitgliedern. Leipzig 1905, 1–14.

Wirsching, Andreas (ed.): Das Jahr 1933. Die nationalsozialistische Machteroberung und die deutsche Gesellschaft. Göttingen 2009.

Wolff, Stefan L.: Physicists in the ‚Krieg der Geister': Wilhelm Wien's ‚Proclamation'. In: Historical Studies in the Physical Sciences 33:2 (2003), 337–368.

Wolff, Stefan L.: Die Ausgrenzung und Vertreibung der Physiker im Nationalsozialismus. In: Hoffmann, Dieter/Walker, Mark (ed.): Physiker zwischen Autonomie und Anpassung. Die Deutsche Physikalische Gesellschaft im Dritten Reich. Weinheim 2007, 91–138.

Wolff, Stefan L.: Die Konstituierung eines Netzwerkes reaktionärer Physiker in der Weimarer Republik. In: Berichte für Wissenschaftsgeschichte 31 (2008), 372–392.

Wulff, Georg: Über die Kristallröntgenogramme. In: Physikalische Zeitschrift 14 (1913), 217–220.

Yourgrau, Wolfgang/Mandelstam, Stanley: Variational Principles in Dynamics and Quantum Theory. Philadelphia, PA 1968.

Zehnder, Ludwig (ed.): W. C. Röntgen - Briefe an L. Zehnder. Zürich, Leipzig, Stuttgart 1935.

Zenda, Benjamin: Subtle Loyalty: German and German-Americans in Wisconsin During the First World War. Madison, WI 2010. http://digital.library.wisc.edu/1793/44582 (16.1.2012).

Zenneck, Jonathan: Über die Fortpflanzung ebener elektromagnetischer Wellen längs einer ebenen Leiterfläche und ihre Beziehung zur drahtlosen Telegraphie. In: Annalen der Physik 23 (1907), 846–866.

Zenneck, Jonathan: Leitfaden der drahtlosen Telegraphie. Stuttgart 1909.

Zinke, Otto/Brunswig, Heinrich: Hochfrequenztechnik 1. Berlin, Heidelberg 1999.

Zwanglose Gesellschaft (ed.): Hundertfünfzig Jahre Zwanglose Gesellschaft München, 1837–1987. München 1987.

[illegible]

Index

M. Eckert, *Arnold Sommerfeld: Science, Life and Turbulent Times 1868-1951*,
DOI 10.1007/978-1-4614-7461-6,

C

D

E

F

G

T

U

V